METHODS IN MOLECULAR BIOLOGY™

For further volumes:
http://www.springer.com/series/7651

Genomic Structural Variants

Methods and Protocols

Edited by

Lars Feuk

Department of Immunology, Genetics and Pathology, Rudbeck Laboratory, Uppsala University, Uppsala, Sweden

Editor
Lars Feuk, Ph.D.
Department of Immunology
Genetics and Pathology
Rudbeck Laboratory
Uppsala University
Uppsala, Sweden
lars.feuk@igp.uu.se

ISSN 1064-3745 e-ISSN 1940-6029
ISBN 978-1-61779-506-0 e-ISBN 978-1-61779-507-7
DOI 10.1007/978-1-61779-507-7
Springer New York Dordrecht Heidelberg London

Library of Congress Control Number: 2011944357

Printed on acid-free paper

Humana Press is part of Springer Science+Business Media (www.springer.com)

Preface

The completion of a consensus draft sequence for the human genome approximately 10 years ago was the starting point for more thorough investigations of individual genome variation. Initially, the focus of variation discovery was targeted toward microsatellites and single nucleotide polymorphisms (SNPs) for use as markers in linkage and association analysis. The development of array-based strategies made it possible to look at our genome in new ways and for new types of variation to be discovered and characterized. Application of comparative genomic hybridization (CGH) arrays used for detection of unbalanced rearrangements led to the discovery that copy number variation (CNV) is abundant in the human genome. Characterization of CNV and other forms of structural genetic variation has highlighted the complexity of human genetic variation, and also provided significant insight into the evolution and dynamic nature of the genome. Another important technical advance is the development of high-density SNP genotyping arrays. The SNP arrays have enabled the discovery of hundreds of genes associated with complex disease through whole genome association studies. The SNP arrays are also used for the identification of CNVs, and many studies are now designed to take both SNPs and CNVs into consideration when analyzing variation in patient and control cohorts.

Over the past decade, the introduction of array-based technologies has revolutionized genomics and genetic diagnostics. Array-based screening has replaced karyotyping as the primary analysis of patient samples in cytogenetic diagnostics, and this has led to a significant increase in the number of patients for whom a clinically relevant aberration can be detected. The number of patients referred for genetic screening is increasing rapidly, and diagnoses for referral have expanded from developmental delay and intellectual disability to a wider range of developmental disorders. Other diagnoses, such as epilepsy and congenital heart disease, may be added in the near future.

Now, we are on the brink of a new paradigm shift in genetics with the advent of massively parallel sequencing in research and diagnostics. In the next few years, we will witness the identification of causative genes for most of the monogenic disorders and achieve better insight into the true spectrum of variation contributing to complex disease.

This book provides an in-depth description of the developments in our understanding of structural genetic variation and its implications for human disease, from the introduction of microarrays up to current state-of-the-art sequencing strategies. The book covers the major technologies used for research and diagnostics, Web-based resources for variation data, and goes into depth regarding specific regions of the genome that differ in variation content. Specific patient groups where CNV has been shown to be of great importance are highlighted, and implications for both prenatal and standard diagnostics are described.

Uppsala, Sweden *Lars Feuk, Ph.D.*

Contents

Contributors

PAUL D. BRADY • *Centre for Human Genetics, K.U. Leuven, Leuven, Belgium*
CLAUDIA M.B. CARVALHO • *Department of Molecular and Human Genetics, Baylor College of Medicine, Houston, TX, USA*
SHANA CEULEMANS • *Applied Molecular Genomics Unit VIB Department of Molecular Genetics, Belgium; University of Antwerp (UA), Antwerpen, Belgium*
EVELYNE CHAIGNAT • *The Center for Integrative Genomics, University of Lausanne, Lausanne, Switzerland*
KEN CHEN • *The Genome Institute at Washington University School of Medicine, St. Louis, MO, USA*
DEANNA M. CHURCH • *NCBI, National Library of Medicine, National Institutes of Health, Bethesda, MD, USA*
JURGEN DEL-FAVERO • *Applied Molecular Genomics Unit VIB Department of Molecular Genetics, University of Antwerp (UA), Antwerpen, Belgium*
JAN DEPREST • *Foetal Medicine Unit, University Hospital Leuven, Leuven, Belgium*
KOENRAAD DEVRIENDT • *Centre for Human Genetics, K.U. Leuven, Leuven, Belgium*
LI DING • *The Genome Institute at Washington University School of Medicine, St. Louis, MO, USA*
JAN P. DUMANSKI • *Department of Immunology Genetics and Pathology, Uppsala University, Uppsala, Sweden*
HENDRIK F. VAN ESSEN • *University of Antwerp (UA), Antwerpen, Belgium*
LARS FEUK • *Department of Immunology, Genetics and Pathology, Rudbeck Laboratory, Uppsala University, Uppsala, Sweden*
JOSEPH A. GOGOS • *Department of Physiology and Cellular Biophysics, Columbia University Medical Center, New York, NY, USA; Department of Neuroscience, Columbia University Medical Center, New York, NY, USA*
IRA M. HALL • *Department of Biochemistry and Molecular Genetics, University of Virginia School of Medicine, Charlottesville, VA, USA; Center for Public Health Genomics, University of Virginia, Charlottesville, VA, USA*
LOUISE HAREWOOD • *The Center for Integrative Genomics, University of Lausanne, Lausanne, Switzerland*
EDWARD J. HOLLOX • *Department of Genetics, Adrian Building, University of Leicester, University Road, Leicester, UK*
MATTIAS JAKOBSSON • *Department of Evolutionary Biology, Evolutionary Biology Center, Uppsala University, Uppsala, Sweden*
ANNA C.V. JOHANSSON • *Department of Immunology, Genetics and Pathology, Rudbeck Laboratory, Uppsala University, Uppsala, Sweden*
MARIA KARAYIORGOU • *Department of Psychiatry, Columbia University Medical Center, New York, NY, USA; New York State Psychiatric Institute, New York, NY, USA*

DANIEL C. KOBOLDT • *The Genome Institute at Washington University School of Medicine, St. Louis, MO, USA*
DAVID E. LARSON • *The Genome Institute at Washington University School of Medicine, St. Louis, MO, USA*
REBECCA J. LEVY • *Department of Psychiatry, Columbia University Medical Center, New York, NY, USA; Doctoral Program in Neurobiology and Behavior, Columbia University, New York, NY, USA*
JAMES R. LUPSKI • *Department of Molecular and Human Genetics, Baylor College of Medicine, Houston, TX, USA; Department of Pediatrics, Baylor College of Medicine, Houston, TX, USA; Texas Children's Hospital, Houston, TX, USA*
CHRISTIAN R. MARSHALL • *The Centre for Applied Genomics, The Hospital for Sick Children, Toronto, ON, Canada; Program in Genetics and Genome Biology, The Hospital for Sick Children, Toronto, ON, Canada*
ROLPH PFUNDT • *Department of Human Genetics, Nijmegen Centre for Molecular Life Sciences, Radboud University Nijmegen Medical Centre, Nijmegen, The Netherlands*
ARKADIUSZ PIOTROWSKI • *Department of Biology and Pharmaceutical Botany, Medical University of Gdansk, Gdansk, Poland*
AARON R. QUINLAN • *Department of Biochemistry and Molecular Genetics, University of Virginia School of Medicine, Charlottesville, VA, USA*
JIANNIS RAGOUSSIS • *Wellcome Trust Centre for Human Genetics, Oxford University, Oxford, UK*
ALEXANDRE REYMOND • *The Center for Integrative Genomics, University of Lausanne, Lausanne, Switzerland*
M. KATHARINE RUDD • *Department of Human Genetics, Emory University School of Medicine, Atlanta, GA, USA*
STEPHEN W. SCHERER • *The Centre for Applied Genomics, The Hospital for Sick Children, Toronto, ON, Canada; Program in Genetics and Genome Biology, The Hospital for Sick Children, Toronto, ON, Canada; Department of Molecular Genetics, University of Toronto, Toronto, ON, Canada*
ALEXANDRA D. SIMMONS • *Biology Department, University of St Thomas, Houston, TX, USA*
PER SJÖDIN • *Department of Evolutionary Biology, EBC, Uppsala University, Uppsala, Sweden*
TAM P. SNEDDON • *NCBI, National Library of Medicine, National Institutes of Health, Bethesda, MD, USA*
PAWEŁ STANKIEWICZ • *Department of Molecular and Human Genetics, Baylor College of Medicine, Houston, TX, USA; Department of Medical Genetics, Institute of Mother and Child, Warsaw, Poland*
KARLIJN VAN DER VEN • *Applied Molecular Genomics Unit VIB Department of Molecular Genetics, Belgium; University of Antwerp (UA), Antwerpen, Belgium*
JORIS A. VELTMAN • *Department of Human Genetics, Nijmegen Centre for Molecular Life Sciences, Radboud University Nijmegen Medical Centre, Nijmegen, The Netherlands*
JORIS R. VERMEESCH • *Centre for Human Genetics, K.U. Leuven, Leuven, Belgium*
LISENKA E.L.M. VISSERS • *Department of Human Genetics, Nijmegen Centre for Molecular Life Sciences, Nijmegen Centre for Molecular Life Sciences, Radboud University Nijmegen Medical Centre, Nijmegen, The Netherlands*

Richard K. Wilson • *The Genome Institute at Washington University School of Medicine, St. Louis, MO, USA*
Laura Winchester • *Wellcome Trust Centre for Human Genetics, Oxford University, Oxford, UK*
Bin Xu • *Department of Psychiatry, Columbia University Medical Center, New York, NY, USA; Department of Physiology and Cellular Biophysics, Columbia University Medical Center, New York, NY, USA*
Bauke Ylstra • *VU Medical Center, Amsterdam, The Netherlands*

Chapter 1

What Have Studies of Genomic Disorders Taught Us About Our Genome?

Alexandra D. Simmons, Claudia M.B. Carvalho, and James R. Lupski

Abstract

The elucidation of genomic disorders began with molecular technologies that enabled detection of genomic changes which were (a) smaller than those resolved by traditional cytogenetics (less than 5 Mb) and (b) larger than what could be determined by conventional gel electrophoresis. Methods such as pulsed field gel electrophoresis (PFGE) and fluorescent in situ hybridization (FISH) could resolve such changes but were limited to locus-specific studies. The study of genomic disorders has rapidly advanced with the development of array-based techniques. These enabled examination of the entire human genome at a higher level of resolution, thus allowing elucidation of the basis of many new disorders, mechanisms that result in genomic changes that can result in copy number variation (CNV), and most importantly, a deeper understanding of the characteristics, features, and plasticity of our genome.

In this chapter, we focus on the structural and architectural features of the genome, which can potentially result in genomic instability, delineate how mechanisms, such as NAHR, NHEJ, and FoSTeS/MMBIR lead to disease-causing rearrangements, and briefly describe the relationship between the leading methods presently used in studying genomic disorders. We end with a discussion on our new understanding about our genome including: the contribution of new mutation CNV to disease, the abundance of mosaicism, the extent of subtelomeric rearrangements, the frequency of *de novo* rearrangements associated with sporadic birth defects, the occurrence of balanced and unbalanced translocations, the increasing discovery of insertional translocations, the exploration of complex rearrangements and exonic CNVs. In the postgenomic era, our understanding of the genome has advanced very rapidly as the level of technical resolution has become higher. This leads to a greater understanding of the effects of rearrangements present both in healthy subjects and individuals with clinically relevant phenotypes.

Key words: Copy number variation, Recombination, Gene dosage, Mosaicism, Insertional translocation, Comparative genomic hybridization

1. Introduction

In the last century, researchers viewed our genome as an imperturbable collection of base pairs in which small mistakes (i.e., single base pair changes) were the predominant cause of inherited disease. Today, we have a broader understanding of the plasticity of our genome.

Lars Feuk (ed.), *Genomic Structural Variants: Methods and Protocols*, Methods in Molecular Biology, vol. 838, DOI 10.1007/978-1-61779-507-7_1, © Springer Science+Business Media, LLC 2012

This will be made evident by the numerous examples of genomic alterations described in this book. Depending on the segments (and genes) involved one may encounter different results: if deviations from the normal diploid state convey a phenotype, these conditions are referred to as genomic disorders (1, 2). In other cases, segments that either do not contain genes or that contain genes or regulatory regions which are not dosage-sensitive, may be involved in rearrangements. These alterations have been associated with both small-scale changes reflected in individual genome variation (a few base pairs to a few Kb), and large-scale genomic changes that are obvious throughout evolution of primate and hominid genomes (3–7).

The first reports of diseases caused by mega base sized genomic rearrangements were published in the early 1990s (8). Thanks to technological and conceptual advancements, hundreds of distinct disorders, and thousands of copy number variant (CNV) (i.e., deviating from the normal diploid state; usually $n=2$) regions have now been described.

Genomic rearrangements include changes in the diploid genome that lead to duplication, deletion, insertion, inversion, or translocation of segments. Such changes can range from a few hundred base pairs to mega bases and may lead to neutral polymorphisms or disease phenotypes (9). It is also known that CNVs are ubiquitous, involving up to 12% of the human genome (10–20) and may arise during meiosis (21) or mitosis, as apparent from somatic CNV mosaicism studies (22–24).

The Human Genome Project generated the first reference haploid genome, which, together with the development of high-resolution genome analysis techniques, like array-based methods, and the refinement of those methods have enabled total genome-wide analyses to rapidly proceed. In less than a century, the substance of heredity was identified; its structure elucidated, the genetic code deciphered, the genome sequenced, and corresponding base pairs accurately described. Despite the overwhelming (and constantly growing) amounts of data, there are still many questions that remain unsolved.

Structural rearrangements of the human genome are of two general types: recurrent and nonrecurrent. Rearrangements with recurrent end-points show breakpoint clustering and junctions that are limited to the location of low copy repeats (LCRs) (described below) where homology is extensive. In this case, the LCRs both stimulate and mediate the rearrangement by acting as homologous recombination (HR) substrates. Nonrecurrent structural changes, on the other hand, have scattered breakpoints and the boundaries share limited or no nucleotide identity (i.e., homology). Here, the resulting rearrangement can be complex and tends to have one breakpoint that groups close to highly polymorphic LCR rich regions. In this case, it is possible that the presence of LCRs may stimulate the formation of secondary DNA structural conformations that can lead to genomic instability. Different LCR conformations can provide single-stranded regions that may result in collapsed DNA replication forks. Such events can generate one-ended, double-stranded DNA breaks or conformations that can result in two-ended double-stranded breaks and must be repaired by classical double-strand break repair (DSBR) (25).

The focus of this chapter is to illustrate *which mechanisms* lead to genomic rearrangements, the genomic disorders that result from such rearrangements and the *knowledge acquired* through such genomic studies.

2. Architectural Features of the Human Genome

A little over half our genome is made up of both *repeat sequences* and *highly repetitive elements.* The main difference between these two is primarily the frequency, with the former presenting very few copies (i.e., LCRs) that have arisen as segmental duplications of the original sequence and seem to make up independent sets (i.e., they originate from many different sequences each copied just a few times), and the latter appear to have arisen from duplication events of the same sequence (or same type of sequences) that has occurred several thousand times (see Table 1).

Table 1
Classification and approximate percentage of repeat sequences in the Human genome (International Human Genome Sequencing Consortium, Nature, 2001, 2004, Levy/Diploid Genome PloSB/2007)

Category	Type	Length of one unit	Copy number (per genome)	Approximate percentage of the human genome	
Transposon-derived repeats (Interspersed repeats)	LINEs	6 kb	850,000	21	45
	SINEs	100–400 bp	1,500,000	13	
	LTR retrotransposons	1.5–11 kb	450,000	8	
	DNA transposons	80 bp–3 kb	300,000	3	
Simple sequence repeats	Direct repeats of short *k*-mers (minisatellites, microsatellites)	1–500 bp	?	4.4	
Low copy repeats or segmental duplications	Blocks of DNA that have been copied from one region of the genome into another region	1–300 kb	?	5.3	
Blocks of tandemly repeated sequences	Centromeres, telomeres, short arms of acrocentric chromosomes and ribosomal gene clusters	?	?	8	

Sedimentation equilibrium centrifugation data provided the first line of evidence that demonstrated the presence of repeat DNA in an experimental fashion. Researchers found that for most eukaryotes DNA was divided into a main band (or peak) and any additional peaks were dubbed "satellite DNA." A few years later, reassociation kinetics assays proved that these peaks were made up of highly repetitive DNA (reviewed in (26)). Today, the term "satellite DNA" depicts tandemly repeated sequences.

Table 1 shows four categories of repeat DNA: Interspersed repeats, simple sequence repeats, low copy repeats and blocks of tandemly repeated sequences. *Interspersed repeats* are by far the most abundant, among them *LINE1* (LINE) and *Alu* (SINE) elements are prominent as these have 850,000 and 1,500,000 copies per haploid genome, and they comprise 18.9 and 10.6% of our genome, respectively. Among the *simple sequence repeats* (or microsatellites), dinucleotides are the most abundant and make up 0.5% of our genome, half of that being specifically AC repeats. These also include minisatellites (also called VNTRs, variable number tandem repeats) that are highly polymorphic. There is less information on *tandemly repeated sequences*, like centromeres, telom eres, pericentromeric and telomeric regions, as these have been purposefully underrepresented in sequencing projects mostly because of the difficulty to find the correct genomic location of clones containing such repetitive sequences (for more detailed info, see ref. 27).

LCRs, (28) or segmental duplications (SDs, (29)) are defined as DNA segments that occur more than twice in the haploid genome, have an extension larger than 1 kb (may extend to over 300 kb in size) and present more than 95% sequence identity between their paralogous copies (4, 30). LCRs are known to be one of the mediators of a large number of genomic rearrangements, however, all repetitive DNA can potentially be a substrate for rearrangements. The presence of these specific repetitive sequences and repeats as architectural features of our genome has been shown to convey genomic instability.

Some genomic loci have a complex genomic architecture, meaning that they bear complicated patterns of direct and inverted LCRs. These regions are associated with what are called CNV regions where the copy number can vary from the expected diploid genome number, usually $n=2$, one inherited from each parent (i.e., one copy of each segment per haploid genome) (25).

Genomic changes can occur at three distinct levels: (a) at the base pair level (SNPs, point mutations), (b) at the structural level (chromosomal rearrangements, genomic disorders) or (c) at the conformational level. Bacolla and Wells reviewed how specific DNA sequences can lead to conformational changes of DNA and the formation of non-B DNA structures. For example: inverted repeats, also called mirror repeats, form cruciforms; purine-pyrimidine tracts can form triplex DNA; alternating purine–pyrimidine tracts can form left handed Z-DNA; and four runs of closely spaced guanine seem to be able to form tetraplex DNA (31). A series of experiments using plasmids in which the LacZ-GFP system was incorporated as a reporter adjacent to a poly purine–pyrimidine tract (containing direct and inverted repeats) increased the observed frequency of mutational events. This was evident by the disruption of the GFP gene. Intra- and intermolecular recombination was observed. In most cases it lead to deletion, and in a particular instance, to an inversion. In all deletions examined, the breakpoints occurred at the predicted non-B DNA structures and left a microhomology *scar* (2–8 bp) at the breakpoint site. The authors suggest that non-B DNA conformations increase the frequency of appearance of gross rearrangements by increasing the number of double-strand breaks (DSBs): their data shows that a primary DSB in one site seems to trigger the appearance of a second DSB at a secondary site on the same or different DNA molecule to produce a recombination reaction between these two DSBs (32, 33).

Evidence for this is also provided in work done with yeast (34) and mammalian cells (35), where artificial induction of DSBs can lead to reciprocal translocation by the non-homologous end joining (NHEJ) pathway (explained below). Using chromosome conformation capture (3C) it has been shown that gross chromosomal rearrangements can be generated after the formation of DSBs in yeast chromosomes in which DSBs become sequestered by the telomerase machinery and the Mps3p nuclear envelope protein to the nuclear periphery, which limits the rate of HR. The results suggest that there might be a competition between alternative repair mechanisms and that the nuclear location might affect the outcome of the repair by modifying the interaction of the DSB with other chromatin fragments (36).

CNVs and structural change seem to cluster in regions were LCRs are present, in heterochromatic regions (centromeres and telomeres) (37–41), in replication origins and terminators (42), where scaffold attachment sequences are present (43, 44) and where LINE and SINE elements are prevalent (45).

In addition to the chromosomal architecture, the type of cell division (mitosis, meiosis), the stage of the cell cycle and the characteristics of the broken segment (two-ended vs. one-ended double-stranded DNA break) may lead to the recruitment of different factors that utilize specific types of repair mechanism, leading to genomic rearrangements.

3. Mechanisms Leading to Rearrangements Observed in Disease

There are a number of proposed mechanisms that lead to rearrangements of our genome that can generate CNV. These mechanisms are commonly classified as HR mechanisms and nonhomologous recombination mechanisms, the latter are further subdivided into replicative and nonreplicative mechanisms (reviewed in (25)). Our focus is on mechanisms that lead to rearrangements associated with disease, so we concentrate on *non-allelic homologous recombination* (NAHR), NHEJ, and *fork stalling and template switching* (FoSTeS)/*microhomology-mediated break-induced replication* (MMBIR).

3.1. NAHR

NAHR is a homologous recombination mechanism. HR is a type of DNA repair mechanism with other functions in dividing cells, like permitting ordered segregation of chromosomes and producing new combinations of linked alleles (meiosis). HR in mammalian and human cells (46, 47) requires around 200–300 bp of near identical sequence, a minimal efficient processing segment, and the participation of the Rad51 protein, which catalyzes the invasion of a 3′ end of ssDNA to a duplex sequence either on the sister chromatid or the homolog. HR is considered an accurate repair mechanism, but may lead to structural rearrangements because multiple paralogous LCR tracts may be present in contiguous regions, thus *confounding* the repair machinery. During repair, the utilization of paralogous LCRs as homology substrate sequences leads to NAHR (25).

The mechanism of NAHR as a cause of rearrangements associated with genomic disorders was first described in the early 1990s (48), but the term NAHR was not introduced until 2002 (30). Also know as ectopic homologous recombination, NAHR can cause a variety of rearrangements. This includes duplications, deletions and inversions, which may take place between LCRs on the same or different chromosomes, in direct or opposite orientation. Deletion or duplication can result when two LCRs are positioned in the genome in direct orientation allowing interchromatidal or interchromosomal exchanges. When the interacting LCRs cause intrachromatidal NAHR, only deletions are observed. If the LCRs involved in the rearrangement are on the same chromosome but in opposite orientation, NAHR will result in the inversion of the segment contained between LCR substrate copies. Translocations can result if the LCRs mediating the rearrangement are located on different chromosomes (1). Based on this information, we expect the frequency of deletions to be higher than the frequency of duplications, and single sperm PCR assays seem to support this hypothesis, at least for the loci examined (21, 49, 50).

Repetitive elements, such as SINEs, LINEs, and LTRs also seem to be able to mediate NAHR (11, 51, 52) even though the identity stretch between any two of these elements is usually shorter than what has been described for LCRs.

When NAHR is operative, the crossovers cluster in narrow hotspots (53). This has been observed in diseases like Neurofibromatosis Type 1 (NF1, MIM162200), Charcot-Marie-Tooth disease Type 1A (CMT1A, MIM118220) and Hereditary Neuropathy with Liability to Pressure Palsies (HNPP, MIM162500).

NAHR may occur during meiosis or mitosis, the first usually responsible for inherited disorders (like CMT1 and HNPP) and sporadic disorders (like PTLS or SMS), the latter responsible for somatic mosaicism, which may lead to cancerous tumor formation (53–55). There does not seem to be a distinction between the sequences involved in meiotic or mitotic NAHR (for examples, see ref. 54–56), but the actual recombination hotspots and the frequency of mitotic or meiotic recombination can differ (53). For example, a study on NF1 shows that mitotic and meiotic rearrangements seem to be responsible for different deletion sizes (i.e., uses different recombination substrates) and suggest different mechanisms may be operating in each case. NF1 can occur as a consequence of intragenic mutations or microdeletions involving the *NF1* gene. The most common deletion (type-1) spans 1.4 Mb on 17q11.2 and encompasses 14 genes, including *NF1*. Type-1 deletions predominantly involve the maternal chromosome by interchromosomal recombination during meiosis. Type-2 microdeletions spans 1.2 Mb, and hence the patients present a less severe clinical phenotype. Both type-1 and type-2 microdeletions seem to be mediated by NAHR. In addition, patients with atypical microdeletions have also been identified. Atypical NF1 deletions seem to arise as a consequence of a nonhomology-based mechanisms (like NHEJ) and take place on the paternal chromosome by an intrachromosomal mechanism during mitotic cell division in spermatogenesis (57).

The efficiency of NAHR may be affected by several factors including: (a) degree of sequence identity, (b) orientation of the LCRs, (c) distance between them, (d) their location (intra- or interchromosomal) provided that the recombination event occurs during mitosis or meiosis, and (e) on the sex of the individual (oogenesis or spermatogenesis) (53).

NAHR is distinguished from unequal crossing-over since the latter refers to the segregation of marker genotypes and the phenomena observed (recombinant chromosomes). NAHR is mechanistic; its products include inversions, and because of the architectural features involved, allow us to make specific outcome predictions depending on the HR substrate orientation.

When PTLS-associated uncommon recurrent duplications were investigated using aCGH and recombination hot spot analyses, it was found that the crossover occurred close to, within 400 nucleotides, a recently described homologous recombination hotspot motif. This *cis*-acting sequence appears to bind PRDM9, a protein with histone H3K4 trimethylase activity. The motif was defined during HapMap studies as an allelic homologous recombination (AHR) stimulating sequence associated with recombination "hotspots." However, it is found nearby both AHR and NAHR "hotspots" for crossovers (58).

3.2. NHEJ

NHEJ is a non-HR, nonreplicative repair mechanism. In mammalian and yeast cells, NHEJ is one of the primary repair mechanisms used to resolve DNA DSBs and seems to function in all phases of the cell cycle, but especially in G1 phase. DSB can occur because of the presence of reactive oxygen species, that may arise as a consequence of endogenous (by-products of cellular metabolism) or exogenous phenomena (X-rays, gamma-rays). Not all DSBs are stochastic events; some are actually programmed in cells as is the case for V(D)J recombination to generate antibody, T cell receptor diversity (mitotic) (59) and crossing-over (meiotic).

Mammals possess two mechanisms to repair DSBs: homologous repair (HR, described above) and NHEJ. NHEJ requires the binding of a Ku complex (specifically, Ku70 and Ku80, (25)), which recognizes the break, and the participation of numerous protein complexes. The most studied complexes are (a) DNA-dependent protein kinase (DNA-PK) apparently involved in the tethering of broken ends to facilitate rejoining and recruiting/activating proteins responsible for chromatin remodeling, DNA end-processing and ligation; (b) DNA Ligase IV-XRCC4 complex, which is present in all eukaryotes, stimulates DNA ligation; and (c) The participation of one or more end-processing enzymes (mainly exonucleases).

NHEJ is considered an error-prone repair mechanism since it does not rely on the presence of a homologous template. Unlike NAHR, NHEJ does not require homologous sequence substrates nor minimal efficient processing segment (MEPS), and as a consequence, small deletions (1–4 bp) may be apparent or a few nucleotides (2–34 bp) of free DNA (usually of mitochondrial or retrotransposon origin) can be added to the broken ends and remain in the junctional sequence (60–65).

While NHEJ does not depend on the presence of LCRs, the occurrence of repetitive DNA elements and sequences related to architectural modification of DNA seem to cause genome instability and susceptibility to DSBs which may be repaired by NHEJ. Examples of genomic disorders that can occur by NHEJ are some nonrecurrent rearrangements associated with Pelizaeus–Merzbacher disease (PMD, MIM312080, (66)) and Smith–Magenis syndrome (SMS, MIM182290, (51)).

3.3. FoSTeS/MMBIR

Studies of *E. coli* show that replication inhibition leads to the formation of DSBs (67). Normal human cells that are induced to replicative stress by using aphidicolin, a drug that is able to inhibit DNA polymerases associated with replication, show numerous, large copy number changes that have microhomology at the join points. Such occurrences could be explained by erroneous repair happening after the replication fork has stalled (68).

In many cases, genomic disorders seem to be caused by complex rearrangements in which duplicated and triplicated segments are interrupted with regions of no observable copy number change. Such is the case for several genes, like *PLP1*, *MECP2 APP*, *SNCA*, *RAI1*, *PMP22* (69–82), which show duplications and triplications that are causal of the disease and associated phenotypes. In 2007, Lee et al. (83) proposed a replication-based mechanism that could lead to such complex human genomic rearrangements: FoSTeS. The authors studied patients with PMD previously determined to have a duplication of the dosage-sensitive gene *PLP1* using a high-resolution oligonucleotide array (comparative genomic hybridization assay) and breakpoint sequence analysis. Many patients, in fact, had complex rearrangements, in which duplicated and triplicated segments were interspersed with normal-copy-number sequences. They sequenced breakpoints in 3 out of 17 patients and described extremely complex rearrangements, which could not be parsimoniously explained by either NAHR or NHEJ. FoSTeS/MMBIR is proposed to occur during mitosis. It also represents a type of non-HR mechanism since little or no homology is necessary.

According to the FoSTeS model, a nick (or ssDNA lesion) could lead to stalling of the replisome at the fork, and the lagging strand would disengage and switch to another active replication fork. Microhomology would be necessary for the priming of DNA replication on the switched template, and the direction of the newly inserted segment is dependent of the direction in which the replisome is advancing (5′ to 3′ or 3′ to 5′) (83).

The signature experimental observations described by Lee et al. (83) (i.e., complex rearrangements with microhomology at the breakpoint junctions) are similar to those described in *E. coli*, whereby starvation stress induces amplification of the *lac operon* to 20–100 copies that appear both in direct and inverted orientation (84–86). In this system, as shown by evidence evaluating the effects of the deletion or over-expression of a 3′ exonuclease gene (*xonA*), free 3′ DNA ends seem to be involved in the amplification of the *lac operon*. The fact that duplication is more prevalent in strains where *xonA* is deleted and similar experiments with a 5′ exonuclease show no relevant change in the amplification number, support the notion that 3′ ends are required for this mechanism (87).

Lac system assay in *E. coli* provided for an amplification model in which replication is restarted at sites of DSB repair of DNA, where template switching (to a different replication fork) was hypothesized to occur during replication restart at stalled replication forks (87). Also, double-strand cleavage of DNA close to the operon augments amplification rate (88).

Studies in yeast seem to support the hypothesis that repeats may arise in the genome as a consequence of repair during DNA replication (42). In this experiment, they used a topoisomerase I inhibitor to nick DNA. These nicked strands caused fork collapse, and ultimately increased the frequency of duplication formation. Interestingly, the authors also observed that fork stalling, as opposed to fork collapse, did not have a strong impact on the frequency of duplication (42). Experiments in which DNA was transfected into mammalian cells revealed microhomology and erratic sequence insertion at the junctions, which the authors explained by proposing a similar model also based on template switching (89).

MMBIR was recently proposed by Hastings et al. (84) to account for the microhomology observed in complex rearrangements, template switching, and the insertion of short sequences at the breakpoint junction that are templated from nearby genomic intervals. They suggest that a one-ended, double-stranded DNA molecules generated by a collapsed replication fork, from stalled transcription complexes, at excision repair tracts, or at secondary structures in DNA (i.e., cruciforms, hairpins), can be repaired by using available single-stranded DNA (also available in secondary DNA structures) if it shares very short homology with 3′ end of the broken strand that has escaped from the collapsed fork. At the molecular level, they propose that a stress-induced reduction of Rad51 leaves repair of the collapsed replication fork in the hands of Rad52 (and possibly other proteins), which requires only minimal homology and would use annealing of single DNA strands to prime DNA replication as the main mechanism available for such repair. In this model, MMBIR is substituting classical break-induced replication (BIR). The MMBIR model can explain complex rearrangements and can also explain the formation of simple duplications, deletions, inversions, translocations, and amplifications due to rolling circles depending on the location that the lagging strand switches to: a position behind the location where the fork collapsed, to a nonhomologous sequence, etc. The authors suggest numerous implications that support a replication repair model, starting with cancer formation and genomic disorders and moving all the way up to exon shuffling and evolution. MMBIR is a molecular mechanistic model based on studies in both human and model organisms, such as *E. coli* and yeast. The FoSTeS model was based upon the "phenomenology" observed in human genomic disorders; it does not provide the mechanistic detail as elucidated for MMBIR.

4. High-Resolution Genome Analyses Methods and Their Limitations

Initially, the only way to visualize our entire genome was by performing karyotypes and studying chromosomes. Some human disease phenotypes were found to be the consequence of whole chromosome aneuploidy or segmental aneuploidy, the presence of large deletions that spanned over 5 Mb in length. Fluorescence in situ hybridization (FISH) was the next approach used to identify and delimit duplications and deletions (90). In most cases, BAC clones (150 to 200 Kb) were used as probes, providing a resolution of deletions under 1 Mb in size (53). Fosmid clones (~40 kb) were also used (57).

Currently, chromosomal banding and FISH are still performed, but CNV, for both research and clinical purposes is mostly studied in diagnostic laboratories by array-based copy number analysis (ABCNA) due to its superior level of genome resolution. Two types of platforms are widely available to perform ABCNA: *array comparative genomic hybridization* (aCGH), in which the test sample is directly compared to a gender matched control sample (91), and the *noncomparative arrays* (NCA) that determine the relative copy number within a single genome, in a quantitative manner, without the use of a control sample in the same experiment.

An array consists of a collection of DNA fragments (BACs, PACs, cDNA, PCR products or synthetic oligonucleotides) that are attached to a glass or silica slide (92). The two leading companies that produce oligo CGH arrays are Agilent and NimbleGen (Roche). By labeling the patient and control DNA with different fluorescent dyes, one can evaluate copy number changes depending on the color and intensity of the signal (green vs. red) recorded for each interrogating probe on the array. Agilent allows for custom designed arrays that may span a segment of interest or the full genome. They offer single or multipack formats (1, 2, 4, or 8 arrays per slide) and up to 1 million interrogating 60-mer probes can be *printed* onto each piece of glass. It is possible to produce custom or predesigned arrays (http://www.chem.agilent.com). NimbleGen HD2 arrays also use long oligo probes (50–75 bases in length) with a maximum number of 2.1 million probes per array, making it the currently available highest density CGH. Their arrays can be ordered in multiplex formats (1, 3, or 12 arrays per slide). Custom-designed arrays can also be produced (http://www.nimblegen.com).

Advantages of using oligo CGH arrays are numerous, some of the most obvious being (a) dividing cells are not required and small amounts of DNA can be used (350 ng to 2 μg, depending on the format), (b) it detects unbalanced chromosomal abnormalities at a level that escapes banded karyotype resolution, (c) aCGH detects genomic mosaicism previously not detected by karyotype analysis, and (d) both LCRs and repetitive sequences may be excluded.

The main disadvantage of this technology is that it is a comparative methodology, in which the result is relative, instead of absolute. The final signal depends on the copy number status of the control DNA; for example, what looks like a gain on patient DNA, might actually be a loss in the control sample (93). Other disadvantages of aCGH include (a) the inability to provide positional and orientational information, which is reflected in the failure to detect balanced inversions and balanced translocations (93), and (b) the difficulty of identifying the total copy number of genes that have more than four copies. This last disadvantage represents a technical limitation of the method that can be explained as an overall decrease in dynamic range as the comparative relative copy number increases. This will certainly be one of the next challenges of copy number identification: dealing with disease states that result from copy numbers greater than 4 (and in most cases, even greater than 3). There are several genes that are present in multiple copies like immune function genes. Once this issue resolves the evaluation of some of complex traits like lupus and autoimmune disorders will become more feasible.

Recently, next-generation sequencing platforms (NGS) were used to develop a read-mapping algorithm (mrFAST) that allows assessment of CNV and accurately predicts the absolute copy number of multicopy genes and genomic regions (94). It is quite possible that the next level of genomic resolution might be whole genome sequencing and an even greater number of CNVs may be found: with the change in technology, the number of described CNVs seems to increase at least an order of magnitude (see Table 2).

NCAs are primarily produced by Affimetrix and Illumina. Affimetrix offers a hybrid genotyping array (Genome-Wide Human SNP Array 6.0) that places 906,600 SNPs and 946,000 nonpolymorphic probes on the same array. The probes are evenly spaced

Table 2
Genome resolution and copy number variation

Interrogating probes (pixels)	Total CNVs detected	Average number of CNVs in an individual	References
ROMA 85 K	221	11	(12)
BAC aCGH 3 K	255	12	(10)
BAC aCGH 25 K Affymetrix GeneChip Human Mapping 500 K	1,447	~24–70 depending on the platform	(13)
Oligonucleotide aCGH 42×10^3 K	11,700	~1,117–1,488	(138)

aCGH array-based comparative genomic hybridization, *BAC* bacterial artificial chromosome, *ROMA* representational oligonucleotide microarray analysis and target known genomic regions that exhibit CNV. The test sample is processed, labeled, and directly hybridized to the array, without the use of a control sample. Their software uses median absolute pairwise difference (MAPD) values to predict copy number.

Illumina offers Infinium HD BeadChips that provide from 300,000 to over a million markers. Illumina arrays are different from the other available commercial arrays in that the oligonucleotides are placed on 3-micron silica beads that self assemble in microwells on fiber optic bundles or planar silica slides (BeadArray Technology). The GenomeStudio Software allows for CNV data analysis by using the cnvPartition algorithm, which computes the output values of the log*R* ratio (LRR) and B allele frequency (BAF). Since LRR is the log ratio of observed probe intensity vs. expected intensity, deviations from zero are interpreted as a copy number change.

Both Affimetrix and Illumina offer options for developing custom arrays although these platforms are not as flexible as the aCGH format for custom arrays. The advantages/disadvantages of these platforms are very similar to those described for oligo CGH arrays, except that they are not a comparative assay. Advantages over aCGH include the ability to: identify consanguinity relationships, infer loss (LOH) or absence (AOH) of heterozygosity and differentiate alleles and parental origin using the SNP information.

Being able to analyze our genome at this new level of resolution has permitted a more thorough study of normal and deleterious copy number changes and rearrangements. It has also allowed us to delve into the mechanisms resulting in rearrangements and even to propose new mechanisms that may lead to genomic rearrangements. Such was the case for FoSTeS (53, 83) and MMBIR (84).

5. What Have We Learned from High-Resolution Analyses of the Human Genome?

The increase in the level of resolution at which we can examine the human genome has been proportional to the gain in knowledge relative to the structure, fluidity, and mechanisms leading to changes in our genome, both normal and related to disease. Some of the most unanticipated but interesting findings are described below (see Fig. 1).

5.1. Mosaicism

Chromosomal mosaicism is defined as having more than one cell line with distinct karyotypes in different cells of one individual. When karyotypes and chromosomal banding are used for this type of screening, the identification of mosaicism is limited to the cell type used and percent required for detection. Conventional chromosome studies use stimulated peripheral blood cultures that

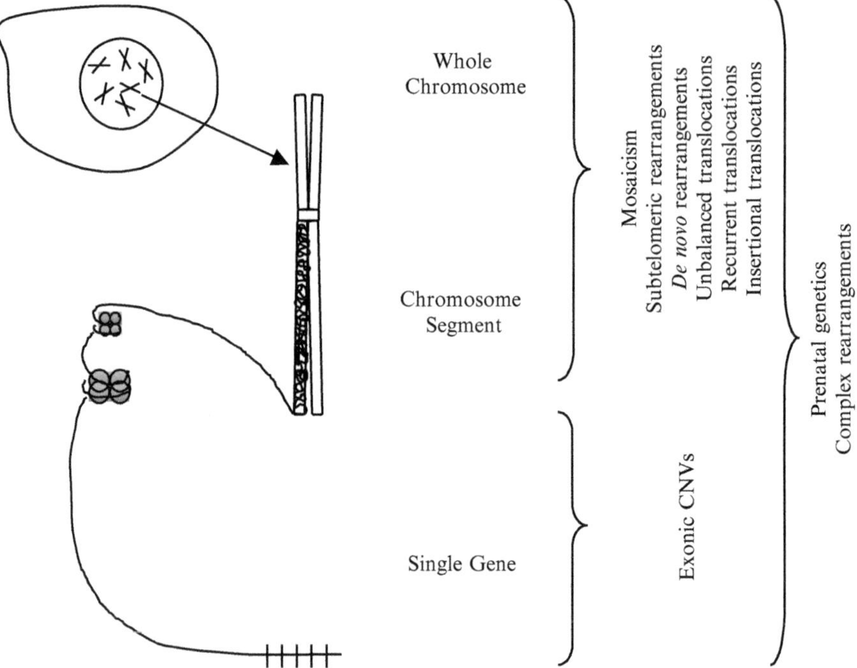

Fig. 1. What studies of genomic disorders have taught us about our genome: Findings according to the size of segments involved in the rearrangements.

depend on the use of artificially added mitogens, a selection step that limits the use of conventional chromosome analysis in mosaicism studies.

Targeted BAC clone arrays were used to screen 2,585 clinical samples and reported finding 12 (0.46%) that presented chromosomal mosaicism (95). Of those, 10 (0.39%) patients were previously reported to have a normal blood chromosome analysis. Also using array CGH, (96) reported 18 cases where mosaicism was detected, 14 of which (8% of all abnormal cases evaluated) were previously unknown. Lu et al. reported chromosomal mosaicism in 1.9% of neonates with birth defects (97).

In a more recent study, mosaicism sensitivity was measured by using artificially derived whole chromosome and segmental aneuploidies. Surprisingly, the authors found that oligonucleotide aCGH can detect low-level mosaicism in both cases, about 10% mosaicism for whole chromosome and 20–30% mosaicism of segmental aneuploidies (98). The higher mosaicism detection rates seem to be due to the fact that (a) DNA is extracted from all white blood cell lineages to perform microarray analysis, (b) cell "population" analysis of arrays versus the single cell events studied by karyotypes, and (c) the greater ability to detect subtle copy number changes with arrays; making oligonucleotide aCGH a powerful tool for mosaicism detection.

Cumulative evidence supports the notion that many of us might be mosaics due to somatic rearrangements occurring in different tissues. Studies of monozygotic twins with either concordant or discordant clinical phenotypes showed the presence of unlike CNVs within twin-pairs in both groups (22). In the past, we have assumed that normal cells are genetically identical and that most CNV must be generated during meiosis. One study analyzed CNVs in 34 tissue samples (brain, skin, heart, kidney, smooth muscle, liver, etc.) from three subjects and found CNVs that ranged in size from 82 to 176 kb, which affected a single organ, or one or more tissues of the same subject (99). These data indicate that humans might regularly present somatic mosaicism and suggests the involvement of somatic CNVs in tissue-specific disorders, such as cancer. In a study on mouse embryonic stem cells (23), extensive and recurrent CNVs generated in the clonal isolates derived from common parental lines were observed. They proposed that novel CNVs likely arose during mitosis and that all somatic tissues in individuals may show mosaicism in the form of variants of the zygotic genome. A study on chromosomal inversions in human DNA derived from blood (24) reported recurrent genomic inversions were found at a relatively high frequency in blood cells using long-range PCR assays. The rearrangements were more abundant in adults than in newborns, concluding that cell populations should be considered as mosaics with regard to their genomic structure and that mosaicism increases with age. The case for antibody diversity occurring as a consequence of programmed rearrangements (59) is an example of how mitotic rearrangements in specific cell types may represent developmental biology pathways or systems that utilizes genomic rearrangements to implement a specific cell-type programming repertoire.

5.2. Subtelomeric Rearrangements

Developmental delay, mental retardation, dysmorphic features, and congenital anomalies may be caused by subtelomeric rearrangements. Subtelomeric regions are gene-rich and because of the presence of repeat sequences are susceptible to genomic rearrangements (100, 101). Traditional methods to assess subtelomeric rearrangements include karyotype, chromosome banding and subtelomeric FISH. The use of aCGH or chromosomal microarray analysis (CMA) with extended subtelomeric coverage has enhanced the detection of rearrangements in these genomic regions.

Targeted aCGH with extended coverage at subtelomeric regions, up to 10 Mb of the 41 subtelomeric regions, was first introduced at Baylor College of Medicine in 2004 as a clinical test coupled with FISH verification (102). In one study, CMA was applied to assess 5,380 clinical patients and identified 499 (9.3%) cases with subtelomeric imbalances, of which 236 (4.4%) were pathogenic. This represents close to half of all genomic abnormalities detected using those specific arrays (37), and was significantly higher than that reported by other groups using other technologies with less genome resolving capability. Approximately 2.5% pathogenic subtelomeric rearrangements were detected by subtelomeric FISH (103, 104).

A high proportion of cytogenetic abnormalities are due to rearrangements of subtelomeric regions and aCGH is a sensitive and robust platform that can be used to detect them (37, 105–108).

Subtelomeric rearrangements of 9q34.3 were never identified prior to subtelomere FISH as they were too small to be seen by conventional microscopy. Analysis of 43 breakpoints within the 9q34 region of patients that presented subtelomeric deletions using array-CGH showed that short repetitive elements, such as SINE, LINE, LTRs, and STRs, were present at or near the breakpoints. Such elements are susceptible both to DSBs, because of the formation of secondary structures, and also to accumulation of single-strand breaks in the replication fork. The presence of these short repetitive elements in subtelomeric regions has been proposed to aid in the stabilization of terminal deletions. In one patient, P6, the presence of an interrupted deletion/inverted duplication structure is proposed to be the scar of a breakage-fusion-bridge cycle mechanisms. However, other complex structures can be more parsimoniously explained by a FoSTeS/MMBIR mechanism than by multiple DSBs healed by NHEJ (38).

5.3. De Novo Genomic Rearrangements and Sporadic Birth Defects

Conventional diagnostic genetic analyses usually perform single locus (gene) testing by DNA sequencing with the expectation to find a point mutation that will be causal for a specific phenotype. It has been described that copy number changes can be a common cause of genetic disorders, where *de novo* locus-specific mutation rates for genomic rearrangements are between 10^{-6} and 10^{-4}, 100- to 10,000-fold greater than that of point mutations (110).

According to the 2008 report generated by the National Center for Health Statistics (http://www.cdc.gov/nchs/hus.htm), the leading cause of death in the neonatal period (less than 28 days of life) are disorders related to short gestation, low birth weight and congenital malformations. The increase in genome resolution has provided clinical geneticists a finer tool, enabling them to identify a causal CNV in patients in which the presence of a chromosomal abnormality is suspected. When using traditional cytogenetic analysis, structural rearrangements have been detected in 0.5% of newborns (110). When a physician suspects a chromosome syndrome, the detection rate of a chromosome abnormality is about 21% (111–113).

In a study published in 2008 (97), a total of 638 neonates with different birth defects were evaluated using CMA. This is one of the largest published studies that used aCGH to estimate the frequencies of genomic imbalances testing neonates with birth defects. The results are astonishing. The reported pathological CNV detection rate for subject samples referred with an indication of "suspected chromosomal abnormality" was 66.7%. When compared to other studies published during the previous 15 years (111–114), this number is about three times greater (66.7 vs. 21.6%). It is clear that these rearrangements could not have been identified nor defined by GTG-banded karyotype and illustrates the importance in resolution of the genome for diagnostics.

5.4. Prenatal Genetics

Balanced de novo chromosomal rearrangements have been observed in about 6% of prenatal evaluations (115) and are commonly found in both patients with phenotypic abnormalities and in seemingly normal individuals. Of these, ostensibly 1 out of every 5 is a de novo event (110); *de novo* apparently balanced translocations have a greater chance of being associated with an abnormal clinical outcome (115).

In 2006, an analysis of 30 uncultured, previously characterized prenatal samples using two different BAC and PAC arrays to detect prenatal chromosome abnormalities was performed (105). Two types of arrays were used: a "large" array covered the entire genome at a 1-Mb resolution, and a "small" array, specifically targeted regions of clinical interest. The use of the small array allowed the correct diagnosis of 29/30 samples (the exception was a case of triploidy). The authors concluded that aCGH could potentially replace conventional cytogenetics for the majority of prenatal diagnosis, but warned that using large arrays could generate difficulties in interpretation until more was learned about genomic CNV. Aggregate data has shown in recent years that all rearrangements do not lead to disease, and many appear to represent polymorphic variants in the population (see Database of Genomic Variants, http://projects.tcag.ca/variation/). Similar findings were reported by (106).

The practicability of using targeted array CGH for genomic imbalance assessment of current pregnancies was evaluated (116). Ninety-eight samples of amniotic fluid, chorionic villi, and cultured cell were analyzed, and they all showed complete concordance between karyotype and array results. The authors demonstrated the feasibility of using aCGH for prenatal diagnosis and suggested that the use of arrays could increase the detection of abnormalities relative to risk. They also reported a shorter turnaround time (6–16 days). Chromosome banding and microscopic karyotyping is a labor intensive technique that is not amenable to automation and, because of the need to culture cells, has an average turnaround time of 2 weeks.

Restricting the analysis of a prenatal sample to only kayrotype or G-banding may limit the information available for counseling and informed decision making on behalf of the parents. This was nicely shown using G-banded chromosome analysis (107) to detect a *de novo* cytogenetically balanced translocation between chromosome 2 and chromosome 9; t(2;9)(q11.2;q34.3). Array CGH was then used to uncover a submicroscopic 2.7 Mb deletion of a subtelomeric region of 9q34.3, demonstrating that the rearrangement was unbalanced. The results were confirmed using FISH. Effective analysis of dosage-sensitive genomic regions is of high importance to prenatal care.

In a larger study, 300 prenatal samples were analyzed. The most common indications were advanced maternal age and abnormal ultrasound findings. Copy number changes were detected in approximately 58 (19%) of the samples, with only 15 (5%) bearing pathological significance (108). They concluded that aCGH has improved diagnostics for prenatal chromosomal tests.

5.5. Apparently Balanced, Unbalanced Translocations

The use of DNA microarrays has shown that cytogenetically visible, apparently "balanced translocations" can actually be unbalanced and present complex rearrangements, like deletions, inversions, and insertions at or near one or both breakpoints. One study showed that six out of ten patients with abnormal phenotypes but what was thought to be balanced translocations presented previously unrecognized imbalances on the chromosomes involved in the translocation (117). Another study compared phenotypically normal and abnormal individuals that were *balanced* translocation carriers and found that in all affected individuals, there were genomic imbalances at the breakpoints or elsewhere (118).

5.6. Recurrent Translocations

NAHR is mostly thought of as a mechanism that leads to interstitial microdeletions, microduplications, and inversions, but translocations might also be stimulated by genomic architectural features found on different chromosomes. Such seems to be the case for two of the recurrent constitutional translocations t(11;22)(q23;q11), in which palindromic AT rich repeats (PATTR) are responsible for the rearrangement (119–124) and t(4;8)(p16;p23), in which olfactory receptor-gene cluster LCRs mediate the translocation via NAHR (125, 126).

A more recent publication studied three patients with an unbalanced translocation der(4)t(4;11)(p16.2;p15.4). In all three cases, they found that the breakpoints occurred within LCRs that extended over 200 kb in length and had over 94% DNA sequence identity. Further analysis of two other previously reported patients with the same translocation documented that in all (five out of five) recurrent t(4;11) translocations, the rearrangement had occurred by NAHR, mediated by interchromosomal LCRs. Subsequently, using computational methods, they analyzed the genome-wide presence of interchromosomal LCRs greater than 10 kb in length, and a "recurrent translocation map" was constructed. The map predicted the existence of ~400 interchromosomal LCRs (20 kb in length) potentially able to act as HR substrates and mediate recurrent translocations by NAHR. Upon the generation of the "recurrent translocation map," the authors reevaluated the public and clinical laboratory databases and found six other examples of recurrent translocations which reported breakpoints predicted by the map (127). These results suggest that human genomic architecture plays a role in recurrent translocations.

5.7. Insertional Translocations

When interpreting aCGH results, one must always remember that a comparative analysis for copy number is performed, and that arrays do not provide information regarding position or orientation of the change. The term "gain" is preferred over "duplication" because in some cases the genomic interval showing apparent gain (when compared to control) is not inserted adjacent to the original sequence. In some instances, segments that become duplicated might be inserted into another genomic location, a phenomenon known as insertional translocation. These events are considered rare as they require that at least three breaks occur (as opposed to one or two breaks required for deletions, duplications, or terminal translocations). In this situation, the use of microarrays is insufficient to confirm the location of the duplicated sequence, and FISH is required to identify the location of the additional material (128).

The previously estimated frequency of insertional translocation was about 1:80,000 karyotypes (129, 130). Using aCGH together with FISH analyses, a total of 40/18,000 cases were identified in which insertional translocation had occurred (a frequency of about 1:500) 160 times greater than previously established in the literature using techniques of limited resolution (128).

5.8. Complex Rearrangements

Genomic rearrangements can be complex. Examples include triplications within duplications, noncontiguous duplications, insertions at the breakpoint junction of deletions and duplications, etc. Complex rearrangements are now being found in numerous loci across the genome (reviewed by Zhang et al. (131)).

PMD, an X-linked dysmyelinating disorder of the central nervous system is caused by point mutations or genomic rearrangements involving the *PLP1* gene. Previously, it was thought that the PMD associated with *PLP1* duplication occurred mainly through coupled homologous and nonhomologous recombination mechanisms (66, 132). In 2007, a study (83) used high resolution arrays and breakpoint sequencing to study the rearrangements of 17 patients with *PLP1* duplications; remarkably, 65% of the samples presented complexities, such as interspersed stretches of DNA of normal copy number amid the duplications, triplications within duplications, insertions at the breakpoint junctions, etc.

Deletions in 17p13.3 cause Miller–Dieker syndrome. Individuals with submicroscopic duplication of the same region present an increased risk for macrosomia, mild developmental delay, and pervasive developmental disorder with a characteristic facial dysmorphology. In 2009, a report described seven patients with microduplications, all of which were nonrecurrent, and three (about 42%) presented complex rearrangements (133).

Duplication of the *MECP2* gene on Xq28 is one of the most common rearrangements identified in males that present developmental delay and the most common subtelomeric duplication (37). Xq28 shows intricate genomic architecture and presents multiple direct and inverted LCRs, causing both polymorphic structural variation and disease in the population (71, 134, 135). Using a 4-Mb tiling-path oligonucleotide array, an analysis of 30 patients with duplications, including the *MECP2* gene (134), reported finding nonrecurrent rearrangements that ranged in size from 250 kb to 2.6 Mb; complex rearrangements were found in 27% of patients (six subjects showed triplications within the duplications and two samples with stretches of nonduplicated segments embedded in the duplicated region). Complex rearrangements in these regions were not described previously as karyotype analysis and/or BAC arrays did not provide the necessary level of human genome resolution. The use of tiling oligonucleotide arrays has enabled reevaluation of many patient samples and suggests that complexity is prevalent in numerous nonrecurrent rearrangements that cause genomic disorders (134).

Using high-density aCGH, two sample sets with rearrangements involving 17p11.2 or 17p12 were analyzed. The first group consisted of 14 nonrecurrent PTLS-associated duplications (17p11.2) that varied in size from 3.5 to 19.6 Mb. Interestingly, complex rearrangements were found in over half of those patients (57%). The second group was made of seven samples with previously identified rearrangements involving the CMT1A/HNPP region (17p12) bearing multiplex ligation-dependent probe amplification patterns inconsistent with rearrangements mediated by NAHR. One sample was shown to have a complex rearrangement and the other six showed exonic deletions. In both cohorts, the data shows that complex rearrangements may be generated by FoSTeS/MMBIR, that these can range in size from a few base pairs to about 20 Mb, and that rearrangements mediated by FoSTeS can originate mitotically (136).

When PTLS-associated nonrecurrent rearrangements were investigated, over 50% were shown to be complex rearrangements. Using aCGH and breakpoint sequence analysis a study analyzing 21 individuals with nonrecurrent *PMP22* CNVs revealed that various mechanisms (NHEJ, *Alu–Alu* mediated recombination FoSTeS/MMBIR) could be responsible for generating the nonrecurrent 17p12 rearrangements associated with neuropathy. Among the 21 patients, three (~14%) presented deletions that involved one or more exons of the *PMP22* gene. Seven other patients presented partial *PMP22* deletions, showing that partial *PMP22* deletion can also result in loss-of-function mutations and haploinsufficiency of the PMP22 protein, causing neuropathy (82).

5.9. Exonic CNVs

Rearrangements have often been considered as deletions, duplications, inversions, or translocations of large blocks of DNA, but it was recently shown (136) that this might not always be the case since rearrangements can involve single exons. In some examples presented in Subheading 5.8, authors report finding small rearrangements that included one or more exons: Zhang et al. (136) reported 86% of samples with rearrangements in 17p12 had exonic CNVs. When analyzing *PMP22* CNVs, 34% showed rearrangements involving one or a few exons (82).

Using 180 K exon-targeted arrays to examine 2,550 samples, Cheung et al. reported 15 cases (0.59%) of intragenic rearrangements involving exons of different genes (*FMA58A*, *PTEN*, *CREBBP*, *DLG3*, among others). The phenotypes of the patients were consistent with the syndrome described for the affected gene. Such findings would have been missed using non-exon-targeted aCGH and other molecular diagnostic methods, reaffirming the importance of using high-resolution techniques to evaluate patients with unexplained mental retardation and congenital anomalies (137, 138).

6. Final Remarks

Throughout this chapter, the methods used to study genomic rearrangements related to human diseases were described (128) (chromosome banding, FISH, microarrays), and their ability to resolve changes in the human genome of increasingly smaller sizes, including exons of only a few hundred base pairs. The use of high resolution human genome analysis in clinical diagnostics, and its limitations (e.g., positional, orientational information) have been enumerated. Interestingly, there is a direct relationship between the technological advancements and the ability to resolve genome changes and the discovery and description of the mechanisms (NAHR, NHEJ, and FoSTeS/MMBIR) that may be causal to such genomic changes, and hence the understanding of our genome. We now know that NAHR typically leads to recurrent rearrangements of the same size in different patients, which is expected considering that this mechanism reflects the architecture of the region and the requirement for HR substrates. On the other hand, nonrecurrent rearrangements can be generated by NHEJ or FoSTeS/MMBIR, the latter responsible for producing complex alterations (e.g., triplications within duplications), which were seldom reported before the use of high-resolution CGH. It has also been suggested that FoSTeS/MMBIR may have a role in the molecular evolutionary process underlying events, such as exon shuffling, gene fusion/fission, and exon accretion, and thus represent an important mechanism for evolving our genome.

References

1. Lupski, J. R. (1998) Genomic disorders: structural features of the genome can lead to DNA rearrangements and human disease traits, *Trends Genet 14*, 417–422.
2. Lupski, J. R. (2009) Genomic disorders ten years on, *Genome Med 1*, 42.
3. Nahon, J. L. (2003) Birth of 'human-specific' genes during primate evolution, *Genetica 118*, 193–208.
4. Bailey, J. A., and Eichler, E. E. (2006) Primate segmental duplications: crucibles of evolution, diversity and disease, *Nat Rev Genet 7*, 552–564.
5. Stankiewicz, P., Shaw, C. J., Withers, M., Inoue, K., and Lupski, J. R. (2004) Serial segmental duplications during primate evolution result in complex human genome architecture, *Genome Res 14*, 2209–2220.
6. Fortna, A., Kim, Y., MacLaren, E., *et al.* (2004) Lineage-specific gene duplication and loss in human and great ape evolution, *PLoS Biol 2*, E207.
7. Dumas, L., Kim, Y. H., Karimpour-Fard, A., *et al.* (2007) Gene copy number variation spanning 60 million years of human and primate evolution, *Genome Res 17*, 1266–1277.
8. Lupski, J. R., de Oca-Luna, R. M., Slaugenhaupt, S., *et al.* (1991) DNA duplication associated with Charcot-Marie-Tooth disease type 1A, *Cell 66*, 219–232.
9. Lupski, J. R., and Stankiewicz, P. (2005) Genomic disorders: molecular mechanisms for rearrangements and conveyed phenotypes, *PLoS Genet 1*, e49.
10. Iafrate, A. J., Feuk, L., Rivera, M. N., *et al.* (2004) Detection of large-scale variation in the human genome, *Nat Genet 36*, 949–951.
11. Korbel, J. O., Urban, A. E., Affourtit, J. P., *et al.* (2007) Paired-end mapping reveals extensive structural variation in the human genome, *Science 318*, 420–426.
12. Sebat, J., Lakshmi, B., Troge, J., *et al.* (2004) Large-scale copy number polymorphism in the human genome, *Science 305*, 525–528.
13. Redon, R., Ishikawa, S., Fitch, K. R., *et al.* (2006) Global variation in copy number in the human genome, *Nature 444*, 444–454.
14. Wong, K. K., deLeeuw, R. J., Dosanjh, N. S., *et al.* (2007) A comprehensive analysis of common copy-number variations in the human genome, *Am J Hum Genet 80*, 91–104.
15. Khaja, R., Zhang, J., MacDonald, J. R., *et al.* (2006) Genome assembly comparison identifies structural variants in the human genome, *Nat Genet 38*, 1413–1418.
16. Newman, T. L., Tuzun, E., Morrison, V. A., *et al.* (2005) A genome-wide survey of structural variation between human and chimpanzee, *Genome Res 15*, 1344–1356.
17. Fiegler, H., Redon, R., Andrews, D., *et al.* (2006) Accurate and reliable high-throughput detection of copy number variation in the human genome, *Genome Res 16*, 1566–1574.
18. Komura, D., Shen, F., Ishikawa, S., *et al.* (2006) Genome-wide detection of human copy number variations using high-density DNA oligonucleotide arrays, *Genome Res 16*, 1575–1584.
19. Lupski, J. R. (2007) Structural variation in the human genome, *N Engl J Med 356*, 1169–1171.
20. Tuzun, E., Sharp, A. J., Bailey, J. A., *et al.* (2005) Fine-scale structural variation of the human genome, *Nat Genet 37*, 727–732.
21. Turner, D. J., Miretti, M., Rajan, D., *et al.* (2008) Germline rates of *de novo* meiotic deletions and duplications causing several genomic disorders, *Nat Genet 40*, 90–95.
22. Bruder, C. E., Piotrowski, A., Gijsbers, A. A., *et al.* (2008) Phenotypically concordant and discordant monozygotic twins display different DNA copy-number-variation profiles, *Am J Hum Genet 82*, 763–771.
23. Liang, Q., Conte, N., Skarnes, W. C., and Bradley, A. (2008) Extensive genomic copy number variation in embryonic stem cells, *Proc Natl Acad Sci USA 105*, 17453–17456.
24. Flores, M., Morales, L., Gonzaga-Jauregui, C., *et al.* (2007) Recurrent DNA inversion rearrangements in the human genome, *Proc Natl Acad Sci USA 104*, 6099–6106.
25. Hastings, P. J., Lupski, J. R., Rosenberg, S. M., and Ira, G. (2009) Mechanisms of change in gene copy number, *Nat Rev Genet 10*, 551–564.
26. Britten, R. J., and Kohne, D. E. (1968) Repeated sequences in DNA. Hundreds of thousands of copies of DNA sequences have been incorporated into the genomes of higher organisms, *Science 161*, 529–540.
27. Lander, E. S., Linton, L. M., Birren, B., *et al.* (2001) Initial sequencing and analysis of the human genome, *Nature 409*, 860–921.
28. Edelmann, L., Pandita, R. K., and Morrow, B. E. (1999) Low-copy repeats mediate the common 3-Mb deletion in patients with velo-cardio-facial syndrome, *Am J Hum Genet 64*, 1076–1086.
29. Bailey, J. A., Yavor, A. M., Massa, H. F., Trask, B. J., and Eichler, E. E. (2001) Segmental

duplications: organization and impact within the current human genome project assembly, *Genome Res 11*, 1005–1017.

30. Stankiewicz, P., and Lupski, J. R. (2002) Genome architecture, rearrangements and genomic disorders, *Trends Genet 18*, 74–82.
31. Bacolla, A., and Wells, R. D. (2004) Non-B DNA conformations, genomic rearrangements, and human disease, *J Biol Chem 279*, 47411–47414.
32. Bacolla, A., Jaworski, A., Larson, J. E., *et al.* (2004) Breakpoints of gross deletions coincide with non-B DNA conformations, *Proc Natl Acad Sci USA 101*, 14162–14167.
33. Bacolla, A., Wojciechowska, M., Kosmider, B., Larson, J. E., and Wells, R. D. (2006) The involvement of non-B DNA structures in gross chromosomal rearrangements, *DNA Repair (Amst) 5*, 1161–1170.
34. Yu, X., and Gabriel, A. (2004) Reciprocal translocations in Saccharomyces cerevisiae formed by nonhomologous end joining, *Genetics 166*, 741–751.
35. Elliott, B., Richardson, C., and Jasin, M. (2005) Chromosomal translocation mechanisms at intronic *Alu* elements in mammalian cells, *Mol Cell 17*, 885–894.
36. Oza, P., Jaspersen, S. L., Miele, A., Dekker, J., and Peterson, C. L. (2009) Mechanisms that regulate localization of a DNA double-strand break to the nuclear periphery, *Genes Dev 23*, 912–927.
37. Shao, L., Shaw, C. A., Lu, X. Y., *et al.* (2008) Identification of chromosome abnormalities in subtelomeric regions by microarray analysis: a study of 5,380 cases, *Am J Med Genet A 146A*, 2242–2251.
38. Yatsenko, S. A., Brundage, E. K., Roney, E. K., Cheung, S. W., Chinault, A. C., and Lupski, J. R. (2009) Molecular mechanisms for subtelomeric rearrangements associated with the 9q34.3 microdeletion syndrome, *Hum Mol Genet 18*, 1924–1936.
39. Zhang, L., Lu, H. H., Chung, W. Y., Yang, J., and Li, W. H. (2005) Patterns of segmental duplication in the human genome, *Mol Biol Evol 22*, 135–141.
40. She, X., Horvath, J. E., Jiang, Z., *et al.* (2004) The structure and evolution of centromeric transition regions within the human genome, *Nature 430*, 857–864.
41. Nguyen, D. Q., Webber, C., and Ponting, C. P. (2006) Bias of selection on human copy-number variants, *PLoS Genet 2*, e20.
42. Payen, C., Koszul, R., Dujon, B., and Fischer, G. (2008) Segmental duplications arise from Pol32-dependent repair of broken forks through two alternative replication-based mechanisms, *PLoS Genet 4*, e1000175.
43. Nobile, C., Toffolatti, L., Rizzi, F., *et al.* (2002) Analysis of 22 deletion breakpoints in dystrophin intron 49, *Hum Genet 110*, 418–421.
44. Visser, R., Shimokawa, O., Harada, N., *et al.* (2005) Identification of a 3.0-kb major recombination hotspot in patients with Sotos syndrome who carry a common 1.9-Mb microdeletion, *Am J Hum Genet 76*, 52–67.
45. de Smith, A. J., Walters, R. G., Coin, L. J., *et al.* (2008) Small deletion variants have stable breakpoints commonly associated with alu elements, *PLoS One 3*, e3104.
46. Liskay, R. M., Letsou, A., and Stachelek, J. L. (1987) Homology requirement for efficient gene conversion between duplicated chromosomal sequences in mammalian cells, *Genetics 115*, 161–167.
47. Reiter, L. T., Hastings, P. J., Nelis, E., De Jonghe, P., Van Broeckhoven, C., and Lupski, J. R. (1998) Human meiotic recombination products revealed by sequencing a hotspot for homologous strand exchange in multiple HNPP deletion patients, *Am J Hum Genet 62*, 1023–1033.
48. Pentao, L., Wise, C. A., Chinault, A. C., Patel, P. I., and Lupski, J. R. (1992) Charcot-Marie-Tooth type 1A duplication appears to arise from recombination at repeat sequences flanking the 1.5 Mb monomer unit, *Nat Genet 2*, 292–300.
49. Lam, K. W., and Jeffreys, A. J. (2006) Processes of copy-number change in human DNA: the dynamics of {alpha}-globin gene deletion, *Proc Natl Acad Sci USA 103*, 8921–8927.
50. Lam, K. W., and Jeffreys, A. J. (2007) Processes of *de novo* duplication of human alpha-globin genes, *Proc Natl Acad Sci USA 104*, 10950–10955.
51. Shaw, C. J., and Lupski, J. R. (2005) Non-recurrent 17p11.2 deletions are generated by homologous and non-homologous mechanisms, *Hum Genet 116*, 1–7.
52. Kidd, J. M., Cooper, G. M., Donahue, W. F., *et al.* (2008) Mapping and sequencing of structural variation from eight human genomes, *Nature 453*, 56–64.
53. Gu, W., Zhang, F., and Lupski, J. R. (2008) Mechanisms for human genomic rearrangements, *Pathogenetics 1*, 4.
54. Barbouti, A., Stankiewicz, P., Nusbaum, C., *et al.* (2004) The breakpoint region of the most common isochromosome, i(17q), in human neoplasia is characterized by a complex

genomic architecture with large, palindromic, low-copy repeats, *Am J Hum Genet 74*, 1–10.

55. Carvalho, C. M., and Lupski, J. R. (2008) Copy number variation at the breakpoint region of isochromosome 17q, *Genome Res 18*, 1724–1732.
56. Mendrzyk, F., Korshunov, A., Toedt, G., *et al.* (2006) Isochromosome breakpoints on 17p in medulloblastoma are flanked by different classes of DNA sequence repeats, *Genes Chromosomes Cancer 45*, 401–410.
57. Kehrer-Sawatzki, H., Schmid, E., Funsterer, C., Kluwe, L., and Mautner, V. F. (2008) Absence of cutaneous neurofibromas in an NF1 patient with an atypical deletion partially overlapping the common 1.4 Mb microdeleted region, *Am J Med Genet A 146A*, 691–699.
58. Zhang, F., Potocki, L., Sampson, J. B., *et al.* (2010) Identification of uncommon recurrent Potocki-Lupski syndrome-associated duplications and the distribution of rearrangement types and mechanisms in PTLS, *Am J Hum Genet 86*, 462–470.
59. Krangel, M. S., Hernandez-Munain, C., Lauzurica, P., McMurry, M., Roberts, J. L., and Zhong, X. P. (1998) Developmental regulation of V(D)J recombination at the TCR alpha/delta locus, *Immunol Rev 165*, 131–147.
60. Burma, S., Chen, B. P., and Chen, D. J. (2006) Role of non-homologous end joining (NHEJ) in maintaining genomic integrity, *DNA Repair (Amst) 5*, 1042–1048.
61. Weterings, E., and van Gent, D. C. (2004) The mechanism of non-homologous end-joining: a synopsis of synapsis, *DNA Repair (Amst) 3*, 1425–1435.
62. Lieber, M. R. (2008) The mechanism of human nonhomologous DNA end joining, *J Biol Chem 283*, 1–5.
63. Daley, J. M., Palmbos, P. L., Wu, D., and Wilson, T. E. (2005) Nonhomologous end joining in yeast, *Annu Rev Genet 39*, 431–451.
64. Haviv-Chesner, A., Kobayashi, Y., Gabriel, A., and Kupiec, M. (2007) Capture of linear fragments at a double-strand break in yeast, *Nucleic Acids Res 35*, 5192–5202.
65. Yu, X., and Gabriel, A. (2003) Ku-dependent and Ku-independent end-joining pathways lead to chromosomal rearrangements during double-strand break repair in *Saccharomyces cerevisiae*, *Genetics 163*, 843–856.
66. Lee, J. A., Inoue, K., Cheung, S. W., Shaw, C. A., Stankiewicz, P., and Lupski, J. R. (2006) Role of genomic architecture in *PLP1* duplication causing Pelizaeus-Merzbacher disease, *Hum Mol Genet 15*, 2250–2265.
67. Michel, B., Ehrlich, S. D., and Uzest, M. (1997) DNA double-strand breaks caused by replication arrest, *EMBO J 16*, 430–438.
68. Arlt, M. F., Mulle, J. G., Schaibley, V. M., *et al.* (2009) Replication stress induces genome-wide copy number changes in human cells that resemble polymorphic and pathogenic variants, *Am J Hum Genet 84*, 339–350.
69. Cabrejo, L., Guyant-Marechal, L., Laquerriere, A., *et al.* (2006) Phenotype associated with APP duplication in five families, *Brain 129*, 2966–2976.
70. Chartier-Harlin, M. C., Kachergus, J., Roumier, C., *et al.* (2004) Alpha-synuclein locus duplication as a cause of familial Parkinson's disease, *Lancet 364*, 1167–1169.
71. del Gaudio, D., Fang, P., Scaglia, F., *et al.* (2006) Increased *MECP2* gene copy number as the result of genomic duplication in neurodevelopmentally delayed males, *Genet Med 8*, 784–792.
72. Farrer, M., Kachergus, J., Forno, L., *et al.* (2004) Comparison of kindreds with parkinsonism and alpha-synuclein genomic multiplications, *Ann Neurol 55*, 174–179.
73. Ibanez, P., Bonnet, A. M., Debarges, B., *et al.* (2004) Causal relation between alpha-synuclein gene duplication and familial Parkinson's disease, *Lancet 364*, 1169–1171.
74. Meins, M., Lehmann, J., Gerresheim, F., *et al.* (2005) Submicroscopic duplication in Xq28 causes increased expression of the MECP2 gene in a boy with severe mental retardation and features of Rett syndrome, *J Med Genet 42*, e12.
75. Nishioka, K., Hayashi, S., Farrer, M. J., *et al.* (2006) Clinical heterogeneity of alpha-synuclein gene duplication in Parkinson's disease, *Ann Neurol 59*, 298–309.
76. Polymeropoulos, M. H., Lavedan, C., Leroy, E., *et al.* (1997) Mutation in the alpha-synuclein gene identified in families with Parkinson's disease, *Science 276*, 2045–2047.
77. Potocki, L., Bi, W., Treadwell-Deering, D., *et al.* (2007) Characterization of Potocki-Lupski syndrome (dup(17)(p11.2p11.2)) and delineation of a dosage-sensitive critical interval that can convey an autism phenotype, *Am J Hum Genet 80*, 633–649.
78. Rovelet-Lecrux, A., Hannequin, D., Raux, G., *et al.* (2006) APP locus duplication causes autosomal dominant early-onset Alzheimer disease with cerebral amyloid angiopathy, *Nat Genet 38*, 24–26.
79. Singleton, A. B., Farrer, M., Johnson, J., *et al.* (2003) Alpha-Synuclein locus triplication causes Parkinson's disease, *Science 302*, 841.

80. Van Esch, H., Bauters, M., Ignatius, J., *et al.* (2005) Duplication of the *MECP2* region is a frequent cause of severe mental retardation and progressive neurological symptoms in males, *Am J Hum Genet 77*, 442–453.
81. Vissers, L. E., Stankiewicz, P., Yatsenko, S. A., *et al.* (2007) Complex chromosome 17p rearrangements associated with low-copy repeats in two patients with congenital anomalies, *Hum Genet 121*, 697–709.
82. Zhang, F., Seeman, P., Pengfei, L., *et al.* (2010) Mechanisms for nonrecurrent genomic rearrangements associated with CMT1A or HNPP: rare CNVs as a cause for missing heritability, *Am J Hum Genet 86*, 892–903.
83. Lee, J. A., Carvalho, C. M., and Lupski, J. R. (2007) A DNA replication mechanism for generating nonrecurrent rearrangements associated with genomic disorders, *Cell 131*, 1235–1247.
84. Hastings, P. J., Ira, G., and Lupski, J. R. (2009) A microhomology-mediated break-induced replication model for the origin of human copy number variation, *PLoS Genet 5*, e1000327.
85. Hastings, P. J., Bull, H. J., Klump, J. R., and Rosenberg, S. M. (2000) Adaptive amplification: an inducible chromosomal instability mechanism, *Cell 103*, 723–731.
86. Hastings, P. J., Slack, A., Petrosino, J. F., and Rosenberg, S. M. (2004) Adaptive amplification and point mutation are independent mechanisms: evidence for various stress-inducible mutation mechanisms, *PLoS Biol 2*, e399.
87. Slack, A., Thornton, P. C., Magner, D. B., Rosenberg, S. M., and Hastings, P. J. (2006) On the mechanism of gene amplification induced under stress in *Escherichia coli*, *PLoS Genet 2*, e48.
88. Ponder, R. G., Fonville, N. C., and Rosenberg, S. M. (2005) A switch from high-fidelity to error-prone DNA double-strand break repair underlies stress-induced mutation, *Mol Cell 19*, 791–804.
89. Merrihew, R. V., Marburger, K., Pennington, S. L., Roth, D. B., and Wilson, J. H. (1996) High-frequency illegitimate integration of transfected DNA at preintegrated target sites in a mammalian genome, *Mol Cell Biol 16*, 10–18.
90. Borg, I., Delhanty, J. D., and Baraitser, M. (1995) Detection of hemizygosity at the elastin locus by FISH analysis as a diagnostic test in both classical and atypical cases of Williams syndrome, *J Med Genet 32*, 692–696.
91. Yatsenko, S. A., Shaw, C. A., Ou, Z., *et al.* (2009) Microarray-based comparative genomic hybridization using sex-matched reference DNA provides greater sensitivity for detection of sex chromosome imbalances than array-comparative genomic hybridization with sex-mismatched reference DNA, *J Mol Diagn 11*, 226–237.
92. Beaudet, A. L., and Belmont, J. W. (2008) Array-based DNA diagnostics: let the revolution begin, *Annu Rev Med 59*, 113–129.
93. Stankiewicz, P., and Beaudet, A. L. (2007) Use of array CGH in the evaluation of dysmorphology, malformations, developmental delay, and idiopathic mental retardation, *Curr Opin Genet Dev 17*, 182–192.
94. Alkan, C., Kidd, J. M., Marques-Bonet, T., *et al.* (2009) Personalized copy number and segmental duplication maps using next-generation sequencing, *Nat Genet 41*, 1061–1067.
95. Cheung, S. W., Shaw, C. A., Scott, D. A., *et al.* (2007) Microarray-based CGH detects chromosomal mosaicism not revealed by conventional cytogenetics, *Am J Med Genet A 143A*, 1679–1686.
96. Ballif, B. C., Rorem, E. A., Sundin, K., *et al.* (2006) Detection of low-level mosaicism by array CGH in routine diagnostic specimens, *Am J Med Genet A 140*, 2757–2767.
97. Lu, X. Y., Phung, M. T., Shaw, C. A., *et al.* (2008) Genomic imbalances in neonates with birth defects: high detection rates by using chromosomal microarray analysis, *Pediatrics 122*, 1310–1318.
98. Scott, S. A., Cohen, N., Brandt, T., Toruner, G., Desnick, R. J., and Edelmann, L. (2010) Detection of low-level mosaicism and placental mosaicism by oligonucleotide array comparative genomic hybridization, *Genet Med 12*, 85–92.
99. Piotrowski, A., Bruder, C. E., Andersson, R., *et al.* (2008) Somatic mosaicism for copy number variation in differentiated human tissues, *Hum Mutat 29*, 1118–1124.
100. Linardopoulou, E. V., Williams, E. M., Fan, Y., Friedman, C., Young, J. M., and Trask, B. J. (2005) Human subtelomeres are hot spots of interchromosomal recombination and segmental duplication, *Nature 437*, 94–100.
101. Riethman, H., Ambrosini, A., and Paul, S. (2005) Human subtelomere structure and variation, *Chromosome Res 13*, 505–515.
102. Cheung, S. W., Shaw, C. A., Yu, W., *et al.* (2005) Development and validation of a CGH microarray for clinical cytogenetic diagnosis, *Genet Med 7*, 422–432.
103. Ravnan, J. B., Tepperberg, J. H., Papenhausen, P., *et al.* (2006) Subtelomere FISH analysis of 11 688 cases: an evaluation of the frequency and pattern of subtelomere rearrangements in

individuals with developmental disabilities, *J Med Genet 43*, 478–489.

104. Ballif, B. C., Sulpizio, S. G., Lloyd, R. M., *et al.* (2007) The clinical utility of enhanced subtelomeric coverage in array CGH, *Am J Med Genet A 143A*, 1850–1857.
105. Rickman, L., Fiegler, H., Shaw-Smith, C., *et al.* (2006) Prenatal detection of unbalanced chromosomal rearrangements by array CGH, *J Med Genet 43*, 353–361.
106. Bi, W., Breman, A. M., Venable, S. F., *et al.* (2008) Rapid prenatal diagnosis using uncultured amniocytes and oligonucleotide array CGH, *Prenat Diagn 28*, 943–949.
107. Simovich, M. J., Yatsenko, S. A., Kang, S. H., *et al.* (2007) Prenatal diagnosis of a 9q34.3 microdeletion by array-CGH in a fetus with an apparently balanced translocation, *Prenat Diagn 27*, 1112–1117.
108. Van den Veyver, I. B., Patel, A., Shaw, C. A., *et al.* (2009) Clinical use of array comparative genomic hybridization (aCGH) for prenatal diagnosis in 300 cases, *Prenat Diagn 29*, 29–39.
109. Lupski, J. R. (2007) Genomic rearrangements and sporadic disease, *Nat Genet 39*, S43-47.
110. Jacobs, P. A., Browne, C., Gregson, N., Joyce, C., and White, H. (1992) Estimates of the frequency of chromosome abnormalities detectable in unselected newborns using moderate levels of banding, *J Med Genet 29*, 103–108.
111. Shah, V. C., Murthy, D. S., and Murthy, S. K. (1990) Cytogenetic studies in a population suspected to have chromosomal abnormalities, *Indian J Pediatr 57*, 235–243.
112. Kenue, R. K., Raj, A. K., Harris, P. F., and el-Bualy, M. S. (1995) Cytogenetic analysis of children suspected of chromosomal abnormalities, *J Trop Pediatr 41*, 77–80.
113. Goud, M. T., Al-Harassi, S. M., Al-Khalili, S. A., Al-Salmani, K. K., Al-Busaidy, S. M., and Rajab, A. (2005) Incidence of chromosome abnormalities in the Sultanate of Oman, *Saudi Med J 26*, 1951–1957.
114. Kim, S. S., Jung, S. C., Kim, H. J., Moon, H. R., and Lee, J. S. (1999) Chromosome abnormalities in a referred population for suspected chromosomal aberrations: a report of 4117 cases, *J Korean Med Sci 14*, 373–376.
115. Warburton, D. (1991) *De novo* balanced chromosome rearrangements and extra marker chromosomes identified at prenatal diagnosis: clinical significance and distribution of breakpoints, *Am J Hum Genet 49*, 995–1013.
116. Sahoo, T., Cheung, S. W., Ward, P., *et al.* (2006) Prenatal diagnosis of chromosomal abnormalities using array-based comparative genomic hybridization, *Genet Med 8*, 719–727.
117. Gribble, S. M., Prigmore, E., Burford, D. C., *et al.* (2005) The complex nature of constitutional *de novo* apparently balanced translocations in patients presenting with abnormal phenotypes, *J Med Genet 42*, 8–16.
118. Baptista, J., Mercer, C., Prigmore, E., *et al.* (2008) Breakpoint mapping and array CGH in translocations: comparison of a phenotypically normal and an abnormal cohort, *Am J Hum Genet 82*, 927–936.
119. Zackai, E. H., and Emanuel, B. S. (1980) Site-specific reciprocal translocation, t(11;22) (q23;q11), in several unrelated families with 3:1 meiotic disjunction, *Am J Med Genet 7*, 507–521.
120. Shaikh, T. H., Budarf, M. L., Celle, L., Zackai, E. H., and Emanuel, B. S. (1999) Clustered 11q23 and 22q11 breakpoints and 3:1 meiotic malsegregation in multiple unrelated t(11;22) families, *Am J Hum Genet 65*, 1595–1607.
121. Kurahashi, H., and Emanuel, B. S. (2001) Long AT-rich palindromes and the constitutional t(11;22) breakpoint, *Hum Mol Genet 10*, 2605–2617.
122. Edelmann, L., Spiteri, E., Koren, K., *et al.* (2001) AT-rich palindromes mediate the constitutional t(11;22) translocation, *Am J Hum Genet 68*, 1–13.
123. Ashley, T., Gaeth, A. P., Inagaki, H., *et al.* (2006) Meiotic recombination and spatial proximity in the etiology of the recurrent t(11;22), *Am J Hum Genet 79*, 524–538.
124. Kato, T., Inagaki, H., Yamada, K., *et al.* (2006) Genetic variation affects *de novo* translocation frequency, *Science 311*, 971.
125. Giglio, S., Calvari, V., Gregato, G., *et al.* (2002) Heterozygous submicroscopic inversions involving olfactory receptor-gene clusters mediate the recurrent t(4;8)(p16;p23) translocation, *Am J Hum Genet 71*, 276–285.
126. Maas, N. M., Van Vooren, S., Hannes, F., *et al.* (2007) The t(4;8) is mediated by homologous recombination between olfactory receptor gene clusters, but other 4p16 translocations occur at random, *Genet Couns 18*, 357–365.
127. Ou, Z., Stankiewicz, P., Xia, Z., *et al.* (2011) Observation and prediction of recurrent human translocations mediated by NAHR between nonhomologous chromosomes. *Genome Res 21*, 33–46.
128. Kang, S. H., Shaw, C., Ou, Z., *et al.* (2010) Insertional translocation detected using FISH confirmation of array-comparative genomic hybridization (aCGH) results, *Am J Med Genet A 152A*, 1111–1126.

129. Fryns, J. P., Kleczkowska, A., and Kenis, H. (1984) *De novo* complex chromosomal rearrangement (CCR) in a severely mentally retarded boy, *Ann Genet 27*, 62–64.
130. Van Hemel, J. O., and Eussen, H. J. (2000) Interchromosomal insertions. Identification of five cases and a review, *Hum Genet 107*, 415–432.
131. Zhang, F., Carvalho, C. M., and Lupski, J. R. (2009) Complex human chromosomal and genomic rearrangements, *Trends Genet 25*, 298–307.
132. Woodward, K. J., Cundall, M., Sperle, K., *et al.* (2005) Heterogeneous duplications in patients with Pelizaeus-Merzbacher disease suggest a mechanism of coupled homologous and nonhomologous recombination, *Am J Hum Genet 77*, 966–987.
133. Bi, W., Sapir, T., Shchelochkov, O. A., *et al.* (2009) Increased *LIS1* expression affects human and mouse brain development, *Nat Genet 41*, 168–177.
134. Carvalho, C. M., Zhang, F., Liu, P., *et al.* (2009) Complex rearrangements in patients with duplications of *MECP2* can occur by fork stalling and template switching, *Hum Mol Genet 18*, 2188–2203.
135. Small, K., Iber, J., and Warren, S. T. (1997) Emerin deletion reveals a common X-chromosome inversion mediated by inverted repeats, *Nat Genet 16*, 96–99.
136. Zhang, F., Khajavi, M., Connolly, A. M., Towne, C. F., Batish, S. D., and Lupski, J. R. (2009) The DNA replication FoSTeS/MMBIR mechanism can generate genomic, genic and exonic complex rearrangements in humans, *Nat Genet 41*, 849–853.
137. Boone, P. M., Bacino, C. A., Shaw, C. A., *et al.* Detection of clinically relevant exonic copy-number changes by array CGH, *Hum Mutat 31*, 1326–1342.
138. Conrad, D. F., Pinto, D., Redon, R., *et al.* (2009) Origins and functional impact of copy number variation in the human genome, *Nature 464*, 704–712.

Chapter 2

Microdeletion and Microduplication Syndromes

Lisenka E.L.M. Vissers and Paweł Stankiewicz

Abstract

During the past decade, widespread use of microarray-based technologies, including oligonucleotide array comparative genomic hybridization (aCGH) and single nucleotide polymorphism (SNP) genotyping arrays have dramatically changed our perspective on genome-wide structural variation. Submicroscopic genomic rearrangements or copy-number variation (CNV) have proven to be an important factor responsible for primate evolution, phenotypic differences between individuals and populations, and susceptibility to many diseases. The number of diseases caused by chromosomal microdeletions and microduplications, also referred to as genomic disorders, has been increasing at a rapid pace. Microdeletions and microduplications are found in patients with a wide variety of phenotypes, including Mendelian diseases as well as common complex traits, such as developmental delay/intellectual disability, autism, schizophrenia, obesity, and epilepsy. This chapter provides an overview of common microdeletion and microduplication syndromes and their clinical phenotypes, and discusses the genomic structures and molecular mechanisms of formation. In addition, an explanation for how these genomic rearrangements convey abnormal phenotypes is provided.

Key words: Copy-number variation, Recurrent and nonrecurrent genomic rearrangements, Contiguous gene deletion/duplication syndrome, Higher-order genomic structure, NAHR, NHEJ, MMBIR, DNA microhomology, Genomic disorders

1. Introduction

Genetic variation of the human genome ranges from single nucleotide differences to microscopically visible gross chromosomal aberrations. Genomic rearrangements of intermediate size, i.e., between ~30 kb and 5 Mb, were initially underestimated due to technological limitations not allowing the systematic detection of these rearrangements on a genome-wide scale. A few common and well-known disease-causing rearrangements in this size-range, referred to as chromosomal microdeletions and microduplications or genomic disorders (1, 2), were assayed for using fluorescence in situ hybridization (FISH) targeting the region of interest. In the

Lars Feuk (ed.), *Genomic Structural Variants: Methods and Protocols*, Methods in Molecular Biology, vol. 838, DOI 10.1007/978-1-61779-507-7_2, past 15 years, technological advances, such as array-based comparative genomic hybridization (aCGH) (3, 4), have enabled genome-wide screening and the detection of this intermediate size range rearrangements which resulted in a previously unrecognized large scale of structural genomic variation known as copy-number variation (CNV) (5–7). CNVs have been shown to be involved in human evolution, genetic diversity between individuals, and a rapidly growing number of genomic disorders (8–11, 223).

Although CNVs are found throughout the genome, some regions are more prone to rearrangements than others. This phenomenon is typically due to local genomic architectural features, such as low-copy repeats (LCRs) (12), or segmental duplications (13, 14). LCRs can mediate recurrent common-sized rearrangements, such as deletions, duplications (Table 1), and inversions, through chromosome or chromatid misalignment followed by nonallelic homologous recombination (NAHR) (12). For overlapping nonrecurrent CNVs of different size (Table 2), local genomic architecture may also stimulate their formation, but potentially through a novel replication-based mechanism referred to as microhomology-mediated break-induced replication (MMBIR) and fork stalling and template switching (FoSTeS) (15–17).

The clinical consequences of CNVs are determined mainly by the size of the genomic rearrangement, the total number and status of genes within the CNV, and the mode of inheritance. The vast majority of disease-causing CNVs harbor a dosage-sensitive/haploinsufficient gene that conveys an abnormal phenotype by a decreased or increased amount of encoded protein. In contrast to monogenic disorders harboring a single such gene, contiguous gene deletion/duplication syndromes (18) result from CNVs containing two or more dosage sensitive genes within the rearranged interval and usually have a more complex phenotype.

Some phenotypic traits resulting from microdeletions or microduplications can show non-Mendelian inheritance, such as genomic imprinting (e.g., in Prader-Willi and Angelman syndromes) or mosaicism (e.g., in neurofibromatosis type 1). Of note, microduplications (with an additional allele) can be considered as representing a triallelic mode of inheritance (19).

This chapter focuses on the clinical aspects of most common microdeletion and microduplication syndromes and the role of genomic architecture in generating CNV. In addition, it also provides an outline for the identification of novel microdeletion syndromes as well as the identification of dosage-sensitive/haploinsufficient genes.

Table 1
Selected recurrent microdeletion and microduplication syndromes

Syndrome	OMIM	Locus	CNV	Dosage-sensitive gene(s)	Phenotype	References
Angelman syndrome	105830	15q11–q12	mat del, pat UPD15	*UBE3A*	Severe DD/ID, profound speech impairment, ataxic gait, excitable personality, and inappropriately happy effect, microcephaly, hypopigmentation and seizures	(38)
AZFa	415000	Yq11.2	Deletion	*USP9Y, DB9*	Male infertility, characterized by azoospermia/severe oligospermia	(116)
Charcot-Marie-Tooth disease, type 1A (CMT1A)/ Hereditary neuropathy with liability to pressure palsies (HNPP)	118220/162500	17p12	Duplication/ Deletion	*PMP22*	Decreased motor nerve conduction velocity (less than 38 m/s), distal limb muscle weakness/atrophy, pes cavus, hammer toes, foot deformities, depressed reflexes// Recurrent focal pressure nerve palsies, such as carpal tunnel syndrome and/or peroneal palsy with foot drop, prolongation of median nerve conduction latency	(117, 118)
DGS/VCFS// 22q11.2 duplication syndrome	188400/192430/ 608363	22q11.2	Deletion/ Duplication	Contiguous gene deletion syndrome, incl. *TBX1* and *COMT*	DD/ID, outflow tract heart defects, immune deficiency, transient/severe neonatal hypocalcemia, velopharyngeal insufficiency, absence of thymus gland//Not a clinically recognizable syndrome, diagnosis follows molecular testing. DD/ID, delayed psychomotor development, growth retardation, and muscular hypotonia	(119–121)
Neurofibromatosis, type 1 (NF1)	613113	17q11.2	Deletion	*NF1*	Cafe-au-lait spots, Lisch nodules in the eye, fibromatous tumors of the skin, generalized elevated susceptibility to the development of benign and malignant tumors	(122)

(continued)

Table 1 (continued)

Syndrome	OMIM	Locus	CNV	Dosage-sensitive gene(s)	Phenotype	References
Prader-Willi syndrome	176270	15q11.2–q12	pat del, mat UPD15	Contiguous gene deletion syndrome incl. *SNRPN*	Moderate ID, postnatal hypotonia and feeding difficulties; failure to thrive in infancy; obesity due to insatiable appetite and food seeking/hoarding, hypogonadotrophic hypogonadism, sexual activity is uncommon and fertility is rare	(123, 124)
Smith-Magenis syndrome/Potocki-Lupski syndrome	182290	17p11.2	Deletion/Duplication	*RAI1*	DD, learning disability, behavioral disturbance (sleep disturbance and self-injurious behavior), infantile hypotonia, short stature, obesity with small hands and feet, facial characteristics, including flat, square, rather heavy face./Infantile hypotonia, failure to thrive, ID, autistic features, sleep apnea, and structural cardiovascular anomalies	(125, 126)
Sotos syndrome/5q35 duplication syndrome	606681	5q35	Deletion/Duplication	*NSD1*	Macrocrania, dolichocephaly, prominent forehead, apparent hypertelorism, high-arched palate, increased birth length and weight, excessive growth in childhood, disproportionately large hands and feet, hypotonia, DD and speech delay/short stature, microcephaly and speech delay	(41, 127)
Steroid suphatase deficiency/X-linked ichthyosis	308100	Xp22.32	Deletion	*STS*	Generalized scaling beginning soon after birth, corneal opacity, increased incidence of cryptorchidism	(128)

Williams-Beuren syndrome/ 7q11.23 duplication syndrome	194050/609757	7q11.23	Deletion/ Duplication	Contiguous gene deletion syndrome, incl. *ELN*	DD/ID, behavioral characteristics (anxiety, ADHD), overfriendliness, congenital heart disease (SVAS, supravalvular pulmonary stenosis), facial characteristics, including periorbital fullness, bulbous nasal tip, long philtrum, wide mouth, full lips, full cheeks and small widely spaced teeth//Mild growth retardation, subtle facial dysmorphism, including high and broad nose, short philtrum, severe expressive language delay (especially syntax and phonology)	(129, 130)
7q11.23 distal deletion syndrome	613729	7q11.23	Deletion	Contiguous gene deletion syndrome, incl. *HIP1* and *YWHAG*	ID, epilepsy neurobehavioral problems	(223)
1q21.1 deletion/ duplication syndromes	612474/612475	1q21.1	Deletion/ duplication	Contiguous gene deletion syndrome, incl. *GJA5*	Highly variable phenotype from asymptomatic to severe DD. Autism, schizophrenia, congenital heart abnormality, ligamentous laxity or joint hypermobility, hypotonia, seizures, cataracts. Deletions associated with reduced brain size; duplications associated with increased brain size	(36, 58, 59)
1q21.1 deletion with thrombocytopenia absent radius (TAR)	274000	1q21	Deletion	Contiguous gene deletion syndrome, incl. *HFE2*	Hypomegakaryocytic thrombocytopenia and bilateral radial aplasia with the presence of both thumbs	(131)
3q29 deletion/ duplication syndromes	609425/611936	3q29	Deletion/ Duplication	*PAK2, DLG1*	Variable. May include mild/moderate ID with mild dysmorphic facial features, including long and narrow face, short philtrum and high nasal bridge, autism, ataxia, chest-wall deformity, long, tapering fingers, microcephaly/Variable	(132, 133)

(continued)

Table 1 (continued)

Syndrome	OMIM	Locus	CNV	Dosage-sensitive gene(s)	Phenotype	References
8p23.1 deletion/ duplication syndromes	600576	8p23.1	Deletion/ Duplication	*GATA4*, *SOX7*, *CLDN23*	Congenital heart disease, congenital diaphragmatic hernia, DD and a characteristic behavior profile with hyperactivity and impulsiveness.//Variable, incl prominent forehead and arched eyebrows, adrenal insufficiency, tetralogy of Fallot, partial 2/3 syndactyly of the toes and cleft palate	(134, 135, 166)
10q11.21q11.23 deletion/ duplication syndromes	–	10q11.21q11.23	Deletion/ duplication	Contiguous gene deletion/ duplication syndrome, incl. *CHAT* and *SLC18A3*	DD/ID, hypotonia, sleep apnea, chronic constipation, gastroesophageal and vesicoureteral refluxes, epilepsy, ataxia, dysphagia, nystagmus, and ptosis	(224)
10q22-q23 deletion syndrome	–	10q22–q23	Deletion	Contiguous gene deletion syndrome incl. *NRG3* and *GRID1*	Cognitive impairment, autism, hyperactivity, and possibly psychiatric disease	(136)
15q11.2 BP1/BP2 microdeletion	–	15q11.2	Deletion	*NIPA1*	DD/ID, epilepsy	(227)
15q12 duplication/ triplication syndrome	608636	15q11.2–q13	Duplication	Contiguous gene duplication syndrome	ID, seizures, behavioral difficulties, including impaired social interactions and impaired use of nonverbal communication, impaired language development	(137, 138)
15q13.3 deletion	118511	15q13.3	Deletion/ Duplication	*CHRNA7*	DD with mild/moderated learning disability, seizures, autism, schizophrenia	(61–64)

15q24 deletion/ duplication syndromes	–	15q24	Deletion/ Duplication	Contiguous gene deletion/ duplication syndrome, incl. *CPLX3, SEMA7A*	Mild ID, growth retardation, microcephaly, digital abnormalities, hypospadias, and loose connective tissue. High anterior hair line, broad medial eyebrows, hypertelorism, downslanted palpebral fissures, broad nasal base, long smooth philtrum, and full lower lip/mild DD, hypotonia, tapering fingers, characteristic facial features, and prominent ears	(139, 140)
15q25.2 deletion syndrome	–	15q25.2	Deletion	Contiguous gene deletion syndrome	Congenital diaphragmatic hernia, cognitive deficits, Diamond-Blackfan anemia	(141)
16p11.2 deletion syndrome	–	16p11.2	Deletion	Contiguous gene deletion syndrome, incl. *SH2B1*	Obesity	(56, 142)
16p11.2 deletion/ duplication syndromes	611913	16p11.2	Deletion/ Duplication	Contiguous gene deletion/ duplication syndrome, incl. *TBX6*	Susceptibility locus for autism (deletion)/ schizophrenia (duplication), global DD, behavioral problems, dysmorphism, epilepsy, and abnormal head size. Deletions associated with increased head size; duplications associated with reduced head size	(66–68)
16p11.2p12.2 deletion/ duplication syndrome	–	16p11.2–p12.2	Deletion/ Duplication	Contiguous gene deletion syndrome, incl. *CDR2, EEF2K*	DD/ID, dysmorphic facial features, including flat facies, downslanting palpebral fissures, low-set ears, and eye anomalies	(143)
16p13.11 deletion/ duplication syndromes	–	16p13.11	Deletion/ Duplication	Contiguous gene deletion syndrome, incl. *MYH11*	Risk factor for autism, ID, idiopathic-generalized epilepsies//Variable	(57, 144–146)
17q12 deletion/ duplication syndromes schizophrenia.// Cognitive	137920	17q12	Deletion/ Duplication	*TCF2*	Renal disease, diabetes, seizures, high risk of autism and schizophrenia.//Cognitive impairment and behavioral abnormalities, increased risk of epilepsy	(147, 148)

(continued)

Table 1 (continued)

Syndrome	OMIM	Locus	CNV	Dosage-sensitive gene(s)	Phenotype	References
17q21.31 deletion/duplication	610443	17q21.31	Deletion/Duplication	Contiguous gene deletion/duplication syndrome, incl. *MAPT*	Moderate DD/learning disability, neonatal hypotonia, poor feeding in infancy (often requiring naso-gastric feeding for a period), oromotor dyspraxia, epilepsy, heart defects (ASD, VSD), kidney/urological anomalies. Silvery depigmentation of strands of hair, with age there is an apparent coarsening of facial features, friendly/amiable behavior	(36, 149–151)
17q23.1q23.2 deletion/duplication syndromes	–	17q23.1q23.2	Deletion	Contiguous gene syndrome incl. *TBX2* and *TBX4*	Heart defects and limb abnormalities	(152)
22q11 distal deletion syndrome	611867	22q11	Deletion	Contiguous gene deletion syndrome, incl. *MAPK1*	Prematurity, pre- and postnatal growth delay, DD, characteristic facial appearance with arched eyebrows, deep-set eyes, a smooth philtrum, a thin upper lip, hypoplastic alae nasi, and a small, pointed chin	(153)
22q13.3 deletion syndrome/Phelan-McDermid syndrome	606230	22q13.3	Deletion	*SHANK3*	Severe expressive language delay, behavioral disturbance, including hyperactivity and aggressive outbursts, hypotonia, ID/DD	(154)
Xp11.22-p11.23 deletion syndrome	–	Xp11.2	Deletion	Contiguous gene deletion syndrome	ID, speech delay, and EEG anomalies	(155)

Table 2
Selected nonrecurrent microdeletions and microduplication syndromes of variable size

Syndrome	OMIM	Locus	CNV	Dosage-sensitive gene(s)	Phenotype	References
1q41-q42 microdeletion syndrome	607502	1q41	Deletion	Contiguous gene deletion syndrome, incl. *DISP1*	DD/ID, midline defects, including cleft palate, diaphragmatic hernia, seizures	(156)
1q43-q44 microdeletion syndrome	611223	1q44	Deletion	Contiguous gene deletion syndrome, incl. *AKT3* *ZNF238*, *CEP170*	Corpus callosum agenesis, microcephaly, cerebellar vermis hypoplasia	(157, 228, 229)
2q23.1 deletion syndrome	–	2q23.1	Deletion	Contiguous gene deletion/duplication syndrome, incl. *MBD5*	ID with pronounced speech delay and additional abnormalities, including short stature, seizures, microcephaly and coarse facies	(158)
2q31.1 deletion/duplication syndromes	613681	2q31.1	Deletion/Duplication	Contiguous gene deletion syndrome, incl. *HOXD13*, *MTX2*, *EVX2*, *DLX1*, and *DLX2*	Limb anomalies ranging from monodactylous/duplication ectrodactyly, brachydactyly and syndactyly to camptodactyly, heart defects, ocular anomalies.//Mesomelic dysplasia	(225, 230)
5q14.3 deletion syndrome	600662	5q14.3	Deletion	*MEF2C*	Severe ID, seizures, hypotonia	(159)

(continued)

Table 2 (continued)

Syndrome	OMIM	Locus	CNV	Dosage-sensitive gene(s)	Phenotype	References
8q21.11 deletion syndrome	–	8q21.11	Deletion/ duplication	Contiguous gene deletion syndrome, incl. *ZFHX4*	ID, round face with full cheeks, a high forehead, ptosis, cornea opacities, an underdeveloped alae, a short philtrum, a cupid's bow of the upper lip, down-turned corners of the mouth, micrognathia, low-set and prominent ears, and mild finger and toe anomalies hearing loss, and unusual behavior	(226)
9q34 deletion syndrome	610253	9q34.3	Deletion	*EHMT1*	Severe ID, hypotonia, brachy(micro) cephaly, epileptic seizures, flat face with hypertelorism, synophrys, anteverted nares, everted lower lip, carp mouth with macroglossia, and heart defects	(160)
14q12 deletion syndrome	164874	14q13	Deletion	*FOXG1*	Rett-like	(161)
22q13.3 deletion syndrome (Phelan-McDermid syndrome)	606232	22q13.3	Deletion	*SHANK3/PROSAP2*	Neonatal hypotonia, global DD, normal to accelerated growth, absent to severely delayed speech, autistic behavior and minor dysmorphic features	(162)

Adrenal hypoplasia, congenital, with hypogonadotropic hypogonadism/ dosage-sensitive sex reversal	300200/ 300018	Xp21.3–p21.2	Deletion/ Duplication	*NR0B1*	Failure to thrive, hypogonadotropic hypogonadism, cryptorchidism, hyperpigmentation, primary adreno-cortical failure, adrenal insufficiency, hypoplastic adrenal glands, absence, delayed pubertal development and puberty//Gonadal dysgenesia, male-to-female sex reversal	(163, 164)
Alagille syndrome	118450	20p12	Deletion	*JAG1*	Cholestasis, cardiac disease, skeletal abnormalities, ocular abnormalities, and a characteristic facial phenotype	(165)
Alveolar capillary dysplasia with misalignment of pulmonary veins (ACD/MPV)	265380	16q24	Deletion	*FOXF1*	Pulmonary hypertension, right-to-left shunt via the foramen ovale or ductus arteriosus, malposition of pulmonary vein branches, increased muscularization of arterioles, maldevelopment of pulmonary lobules and deficient capillarization of airspace walls	(47)
Alzheimer disease	104300	21q21	Duplication	*APP*	Early onset progressive dementia	(71)
Aniridia, Wilms tumor, WAGR	607108	11p13	Deletion	*PAX6, WT1*	Aniridia, decreased vision, cataract, glaucoma, Peter's anomaly, corneal clouding	(165)
Basal cell nevus syndrome/ Gorlin-Goltz syndrome	109400	9q22.3	Deletion	*PTCH1*	Odontogenic keratocysts of jaws, skeletal anomalies, nevoid basal cell carcinoma	(167)

(continued)

Table 2 (continued)

Syndrome	OMIM	Locus	CNV	Dosage-sensitive gene(s)	Phenotype	References
Branchio-oto-renal syndrome	113650	8q13.3	Deletion	*EYA1*	Sensorineural, conductive, or mixed hearing loss, structural defects of the outer, middle, and inner ear, branchial fistulas or cysts, renal abnormalities ranging from mild hypoplasia to complete absence	(168)
Buschke-Ollendorff syndrome	166700	12q14	Deletion, Duplication	*LEMD3*	Osteopoikilosis, spotty skin lesions, connective tissue nevi	(169)
Campomelic dysplasia	608160	17q24.3–q25.1	Deletion, Duplication	*SOX9*	Prenatal bowing of the long bones, short legs, 11 pairs of ribs, hypoplastic scapulae, congenital heart defect, sex reversal, Pierre-Robin sequence, laryngotracheomalacia	(170)
Cat-eye syndrome	115470	22q11.2	Triplication (mosaic)	Contiguous gene duplication/triplication syndrome	Coloboma of the iris, anal atresia with fistula, downslanting palpebral fissures, preauricular tags and/or pits, frequent occurrence of heart and renal malformations, near-normal intellectual development	(171)
CHARGE syndrome	214800	8q12.1	Deletion	*CHD7*	Coloboma, choanal atresia, heart malformation, ID, genital and/or urinary abnormalities and ear abnormalities, including deafness	(43)

Cleidocranial dysplasia	600211	6p21	Deletion	*RUNX2*	Persistently open skull sutures with bulging calvaria, hypoplasia, or aplasia of the clavicles, wide pubic symphysis, short middle phalanx of the fifth fingers, dental anomalies, and vertebral malformation	(172)
Congenital diaphragmatic hernia (CDH)	142340	15q23.1	Deletion	*CHD2*, *NR2F2*	Congenital diaphragmatic hernia	(173)
Cornelia de Lange syndrome	608667	5p13.1	Deletion	*NIPBL*	Characteristic facies (low anterior hairline, synophrys, anteverted nares, maxillary prognathism, long philtrum, “carp” mouth), prenatal and postnatal growth retardation, ID and upper limb anomalies	(44, 45)
Cowden disease	158350	10q23.31	Deletion	*PTEN*	Progressive macrocephaly, “birdlike” facies, hearing loss, multiple hamartomas most commonly found on the skin and mucous membranes	(174)
Cri-du-Chat syndrome	123450	5p15.2	Deletion	Contiguous gene deletion syndrome, incl. *TERT*	High-pitched monotonous cry, microcephaly, round face, hypertelorism, epicanthic folds, micrognathia, impaired growth, severe DD and learning disability	(175)
Currarino syndrome	176450	7q36.3	Deletion	*MNX1*	Partial sacral agenesis with intact first sacral vertebra (“sickle-shaped sacrum”), a presacral mass, and anorectal malformation	(176)
Duane-Radial Ray syndrome	607323	20q13.2	Deletion	*SALL4*	Upper limb anomalies, ocular anomalies, and, in some cases, renal anomalies, sensorineural deafness and gastrointestinal anomalies, such as imperforate anus	(177)

(continued)

Table 2 (continued)

Syndrome	OMIM	Locus	CNV	Dosage-sensitive gene(s)	Phenotype	References
Duchenne muscular dystrophy	300377	Xp21.2	Deletion, Duplication	*DMD*	Progressive proximal muscular dystrophy with characteristic pseudohypertrophy of the calves, myopathy	(178)
Early infantile epileptic encephalopathy-2/X-linked dominant infantile spasm syndrome-2	300672	Xp22	Deletion	*CDKL5*	Rett-like syndrome, early onset seizures	(179)
Feingold syndrome	164280	2p24.1	Deletion	*MYCN*	Esophageal atresia, duodenal atresia, microcephaly, learning disability, syndactyly, and cardiac	(46)
Focal dermal hypoplasia/Goltz syndrome	305600	Xp11.23	Deletion	*PORCN*	Atrophy and linear pigmentation of the skin, herniation of fat through the dermal defects, multiple papillomas of the mucous membranes or skin, digital anomalies (syndactyly, polydactyly, camptodactyly), lip papillomas, hypoplastic teeth, ocular anomalies (coloboma of iris and choroid, strabismus, microphthalmia), ID	(180, 181)
Greig cephalopolysyndactyly syndrome	175700	7p13	Deletion/ Duplication	*GLI3*	Craniosynostosis, (poly)syndactyly, macrocephaly, scaphocephaly, high forehead, frontal bossing, hypertelorism, downslanting palpebral fissures, broad nasal root, normal intelligence	(182)

Holt-Oram syndrome	601620	12q24.1	Deletion	*TBX5*	Thumb anomalies (absent, triphalangeal, nonopposable or finger-like) and atrial septal defect	(183)
Hypoparathyroidism, sensorineural deafness, renal disease (HDR)/ Barakat syndrome	146255	10p14	Deletion	*GATA3*	Hypoparathyroidism, sensorineural deafness, and renal dysplasia	(184)
Jacobsen syndrome	147791	11q23	Deletion	Contiguous gene deletion syndrome, incl. *BSX* and *NRGN*	IUGR, ID, micro/macrocephaly, micrognathia, low set ears, diverse ocular manifestations (hypertelorism/ telecanthus, palpebral fissures), depressed nasal bridge, short nose, VSD, missing ribs, pectus excavatum, pyloris stenosis, genitourinary anomalies (hypospadia, hypoplastic female external genitalia), joint contractures, clinodactyly, brachydactyly, hypotonia	(185)
Kallmann syndrome 1	308700	Xp22.31	Deletion	*KAL1*	Congenital, isolated, idiopathic hypogonadotropic hypogonadism and anosmia	(186)
Langer-Giedion syndrome	190351	8q24.12	Deletion	Contiguous gene deletion syndrome, incl. *TRPS1* and *EXT1*	Short stature, long flat philtrum, protruding ears, pear-shaped nose, thin upper lip, short hands/feet, cone-shaped epiphyses, severe brachydactyly, short metacarpals, short phalanges, sparse hair, multiple cartilaginous exostoses, laterally sparse eyebrow, ID.	(187)
Leri-Weill dyschondrosteosis	127300	Xpter-p22.32	Deletion	*SHOX*	Short stature, mesomelic limb shortening, characteristic "Madelung" deformity of the forearms (bowing of the radius and restriction of pronation/supination of the forearm)	(188, 189)

(continued)

Table 2 (continued)

Syndrome	OMIM	Locus	CNV	Dosage-sensitive gene(s)	Phenotype	References
Lissencephaly, X-linked	300067	Xq22.3–q23	Deletion	*DCX*	Classic lissencephaly with ID (males)/ subcortical band heterotopias (females); ID, delayed motor development, ataxia, dysarthria, limb spasticity, pachygeria/agyria, corpus callosum agenesis	(190, 191)
Lubs syndrome	300260	Xq28	Duplication, Triplication	*MECP2*	Recurrent infections, infantile hypotonia giving way to spasticity in childhood, severe ID, and lack of speech acquisition, gastroesophageal reflux, swallowing difficulties, facial hypotonia and excessive drooling, and inability or limited ability to walk	(192)
Miller-Dieker-syndrome/ 17p13.3 duplication syndrome	247200/ 613215	17p13.3	Deletion/ Duplication	*PAFAH1B1 (LIS1), YWAHE*	Lissencephaly, microcephaly, wrinkled skin over the glabella and frontal suture, prominent occiput, narrow forehead, downslanting fissures, small nose and chin, cardiac malformations, hypoplastic male external genitalia, growth retardation, and DD/ID with seizures and EEG abnormalities// Mild brain structural abnormalities, moderate to severe DD and failure to thrive, macrosomia, and pervasive developmental disorder, and facial dysmorphologies	(25, 193)

Mowat-Wilson syndrome	235730	2q22.3	Deletion	*ZEB2*	ID, delayed motor development, epilepsy, and a wide spectrum of clinically heterogeneous features suggestive of neurocristopathies at the cephalic, cardiac, and vagal levels	(194)
Nail-patella syndrome	161200	9q34.1	Deletion	*LMX1B*	Nail dysplasia, hypoplastic patellae, iliac horns, abnormality of the elbows interfering with pronation and supination, nephropathy	(195)
Neurofibromatosis, type 2	607379	22q12.2	Deletion	*NF2*	Tumors of the eighth cranial nerve (usually bilateral), meningiomas of the brain, and schwannomas of the dorsal roots of the spinal cord	(196)
Norrie disease	310600	Xp11.3	Deletion	*NDP*	Very early childhood blindness due to degenerative and proliferative changes of the neuroretina, Progressive ID often with psychotic features, sensorineural deafness, growth failure and seizures	(197)
Opitz GBBB syndrome	300000	Xp22.2	Deletion	*MID1*	Midline malformation syndrome characterized by hypertelorism, hypospadias, cleft lip/palate, laryngotracheoesophageal abnormalities, imperforate anus, DD, and cardiac defects	(198)
Ornithine transcarbamylase deficiency	311250	Xp11.4	Deletion	*OTC*	Hyperammonemia, encephalopathy, and respiratory alkalosis	(199)

(continued)

Table 2 (continued)

Syndrome	OMIM	Locus	CNV	Dosage-sensitive gene(s)	Phenotype	References
Pallister-Killian syndrome	601803	12p	Triplication (mosaic)	Contiguous gene triplication syndrome	Profound ID, seizures, streaks of hypo- or hyper-pigmentation, transient temporofrontal balding or sparseness with abundant hair over the top of the head, facial anomalies, including prominent forehead with sparse anterior scalp hair, flat occiput, hypertelorism, short nose with anteverted nostrils, flat nasal bridge, and short neck	(200)
Parkinson disease	168601	4q22.1	Duplication, Triplication	*SNCA*	Resting tremor, muscular rigidity, bradykinesia, postural abnormality/instability, dysautonomia, dystonic cramps, and dementia	(201)
Pelizaeus-Merzbacher syndrome	312080	Xq22.2	Duplication, Deletion	*PLP1*	Hypomyelinative leukodystrophy, nystagmus, spastic quadriplegia, ataxia, and DD	(202)
Pitt-Hopkins syndrome	610954	18q21.1	Deletion	*TCF4*	Severe ID, hypotonia, postnatal growth retardation, microcephaly, abnormal breathing, large-beaked nose, cup-shaped ears with broad helices, a wide mouth, cupid's-bow upper lip, wide and shallow palate, and broad or clubbed fingertips	(203)
Potocki-Shaffer syndrome	601224	11p11.2	Deletion	Contiguous gene deletion syndrome incl. *ALX4, EXT2*	Foramina parietalia permagna, multiple exostoses, and in some cases, craniofacial dysostosis and ID.	(204)

Retinoblastoma	180200	13q14.1–q14.2	Deletion	*RB1*	Retinoblastoma, retinal calcification, osteogenic sarcoma, pinealoma, leukemia, lymphoma, Ewing sarcoma, prominent eyebrows, broad nasal bridge, bulbous tip of the nose, large mouth with thin upper lip, and long philtrum	(205)
Rett syndrome	312750	Xq28	Deletion	*MECP2*	Arrested development between 6 and 18 months of age, regression of acquired skills, loss of speech, stereotypical movements (classically of the hands), microcephaly, seizures, and ID	(40, 206)
Rubinstein-Taybi syndrome	180849	16p13.3	Deletion	*CREBBP*	ID, broad thumbs and toes, congital/juvenile glaucoma, cardiac anomalies and facial abnormalities	(207)
Saethre-Chotzen syndrome	601622	7p21	Deletion	*TWIST1*	Asymmetry of the face, craniosynostosis, small ears with long and prominent ear crus, syndactyly (often 3rd-4th toes and/or 2nd-3rd finger), DD	(208)
Split-hand/foot malformation 1 (SHFM1)	183600	7q21.3	Deletion	*SHFM1*	Limb malformation involving the central rays of the autopod, syndactyly, median clefts of the hands and feet, aplasia and/or hypoplasia of the phalanges, metacarpals, and metatarsals, ID, ectodermal and craniofacial findings, and orofacial clefting	(209)
Split-hand/foot malformation 3 (SHFM3)	246560	10q24.1	Duplication	*FBXW4*	Limb malformation involving the central rays of the autopod, syndactyly, median clefts of the hands and feet, aplasia and/or hypoplasia of the phalanges, metacarpals, and metatarsals, ectodermal and craniofacial findings and orofacial clefting	(210)

(continued)

Table 2 (continued)

Syndrome	OMIM	Locus	CNV	Dosage-sensitive gene(s)	Phenotype	References
Townes-Brocks syndrome	107480	16q12.1	Deletion	*SALL1*	Imperforate anus, triphalangeal thumbs, fusion of metatarsals, absent bones, supernumerary thumbs, mild sensorineural deafness, and lop ears	(211)
Severe infantile polycystic kidney disease with tuberous sclerosis I	600273	16p13.3–p13.12	Deletion	*PKD1*, *TSC2*	Tuberous sclerosis, renal cysts, liver cysts, and intracranial aneurysm	(212)
Tuberous sclerosis-1	605284	9q34	Deletion	*TSC1*	Hamartomata in multiple organ systems with epilepsy, learning difficulties, behavioral problems, and skin lesions	(213)
Ulnar-mammary syndrome	181450	12q24.1	Deletion	*TBX3*	Ulnar finger and fibular toe ray defects; delayed growth and onset of puberty, obesity, hypogenitalism and diminished sexual activity; and hypoplasia of nipples and apocrine glands with subsequent diminished ability to perspire	(214)
van der Woude syndrome	119300	1q32–q41	Deletion	*IRF6*	Pits and/or sinuses of the lower lip, and cleft lip and/or cleft palate, large gray matter volumes of the anterior cerebrum	(215)
Wolf-Hirschhorn syndrome	194190	4p16.3	Deletion	Contiguous gene deletion syndrome	Low birth-weight and postnatal failure to thrive, microcephaly, DD, hypotonia, seizures, facial appearance with sagging everted lower eyelids, a “Greek-helmet” profile, a short nose and very short philtrum, iris colobomas	(216)

X-linked lissencephaly with ambiguous genitalia	300215	Xp21.3	Deletion	*ARX*	Lissencephaly with a posterior-to-anterior gradient and only moderate increase in thickness of the cortex, absent corpus callosum, neonatal-onset epilepsy, hypothalamic dysfunction, including deficient temperature regulation, and ambiguous genitalia in genotypic males	(217)
X-linked lymphoproliferative (XLP)	308240	Xq25	Deletion	*SH2D1A*	Extreme sensitivity to infection with Epstein-Barr virus, which results in a complex phenotype manifested by severe or fatal mononucleosis, acquired hypogammaglobulinema, and malignant lymphoma, aplastic anemia, red cell aplasia, and lymphomatoid granulomatosis.	(218)
X-linked mental retardation 21	300143	Xp21.3	Deletion	*IL1RAPL1*	Mild facial dysmorphism, hypotonic midface, upslanter palpebral fissures, hypertelorism, short nose, thickened alae nasi and columella, open mouth, tented upper lip, crowded dentition, ID, autistic features	(219)
X-linked mental retardation with isolated growth hormone deficiency	300123	Xq27.1	Deletion/ Duplication	*SOX3*	ID, panhypopituitarism	(220)
X-linked mental retardation with microcephaly and disproportionate pontine and cerebellar hypoplasia	300749	Xp11.4	Deletion	*CASK*	Postnatal microcephaly, severe ID, and disproportionate pontine and cerebellar hypoplasia	(221)

2. Materials

2.1. Copy-Number Variation

Recently, high-resolution genome-wide analyses of the human genome sequence have revealed extensive submicroscopic structural variation in addition to previously reported single nucleotide variation (SNVs), or single nucleotide polymorphisms (SNPs). This structural variation involves CNVs, such as deletions, duplications, triplications, insertions, and unbalanced cryptic translocations, as well as balanced genomic inversions. Recent comparative genome-wide studies have shown that the cumulative number of base-pairs distinguishing two individual genomes by CNVs exceeds those encompassing SNPs. In more detail, any two individual genomes differ by over 3.5 million of SNPs (0.1%) and, to a larger extent, by at least ~1,000 CNVs, ranging in size from ~500 bp to 1 Mb. Recent analyses have revealed over 38,000 CNVs (greater than 100 bp) that occupy more than 29% of the reference human genome (20). Despite the availability of novel technologies enabling whole-genome analyses, the total number, position, size, gene content, and population distribution of CNVs remain largely unknown, often limited by the lack of breakpoint resolution (genomic coordinates) and the multiple assay methods, so these estimates are still expected to be conservative.

A continuously updated summary of CNVs can be found in the Toronto Database of Genomic Variants (http://projects.tcag.ca/variation), and clinically relevant CNVs are categorized in the DatabasE of Chromosomal Imbalance and Phenotype in Humans using Ensembl Resources (DECIPHER; https://decipher.sanger.ac.uk/information) (21) as well as in the European Cytogeneticists Association Register of Unbalanced Chromosome Aberrations (ECARUCA; http://www.ecaruca.net).

2.2. Genomic Architecture and Molecular Mechanisms of Formation

2.2.1. Low-Copy Repeats

Computational analyses of the DNA sequence generated in the Human Genome Project, (HGP; http://www.ornl.gov/sci/techresources/Human_Genome/home.shtml) have revealed that LCRs, defined as DNA sequence identity greater than 90% and larger than 1 kb (encompassing both repetitive and unique sequences) occupy approximately 4–5% of the haploid human genome with most of them having arisen during primate speciation (13).

2.2.2. Nonallelic Homologous Recombination

LCRs longer than 10 kb and of >95–97% sequence identity can introduce local genomic instability by predisposing to misalignment of chromosomes or chromatids and subsequently mediating NAHR, also known as unequal crossing-over (12). NAHR between directly oriented LCRs results in unbalanced deletions or reciprocal duplications of the genomic segment located between them, whereas NAHR between the LCRs in opposite orientation leads to a balanced inversion of the intervening genomic segment (1). This molecular mechanism has been shown to be responsible for the vast majority of recurrent rearrangements occurring during meiosis; however, LCRs can also stimulate nonrecurrent genomic rearrangements during meiosis as well as replication-based mitotic (somatic) events.

2.2.3. Nonhomologous End Joining

Until recently, little has been known about the molecular mechanisms of the formation of nonrecurrent genomic rearrangements. In addition to recombination-based mechanisms for single-stranded DNA repair, nonhomologous end joining (NHEJ) can be used for repair of DNA double-strand breaks (DSBs). In NHEJ, DSBs are detected, both broken DNA ends are bridged, modified, and finally ligated (22). In contrast to NAHR, the NHEJ mechanism does not require the presence of specific genomic architecture to repair the DSB, but can be stimulated by it. The products of repair often contain additional random nucleotides at the DNA end junction, the so-called molecular scar.

2.2.4. Microhomology-Mediated Replication Error Mechanisms

Due to the absence of overt genomic architecture in close proximity of the breakpoints, NHEJ had been expected to play a major role in the formation of nonrecurrent genomic rearrangements. However, the novel DNA replication error-based mechanisms MMBIR and FoSTeS have also been shown to be responsible for the formation of nonrecurrent CNVs (15, 16). MMBIR/FoSTeS DNA repair mechanisms have been readily demonstrated mostly for complex structure genomic rearrangements, e.g., deletions and/or duplications interrupted by either normal copy-number or triplicated genomic segments. In these models, the DNA replication fork stalls or collapses, the lagging strand disengages from the original template and anneals to another replication fork in physical proximity, utilizing microhomology at the 3′ end, "priming" or reinitiating DNA synthesis (23).

MMBIR/FoSTeS have been proposed in the Pelizaeus-Merzbacher diseases region on chromosome Xq22 (16), *MECP2* duplication in Xq28 (24), deletions and duplications in 17p13.3 (25, 26), deletions and duplications in 17p11.2p12 (23), and deletions and duplications in 9q34 (27). In addition, Vissers et al. (28) demonstrated that MMBIR/FoSTeS mechanisms may be involved in the formation of rare pathology associated CNV that occurred as *de novo* genomic rearrangements. Detailed computational analyses of the genomic architecture surrounding the breakpoint of these rearrangements resulted in the identification of several genomic features, such as the potential to adopt non-B DNA conformations and enrichment of specific sequence motifs and repetitive elements, potentially predisposing to CNV formation (28).

2.3. Prevalence and Population Differences of Genomic Disorders

Genomic disorders are common diseases ~1 per 100–1,000 newborns and are often sporadic, due to de novo microdeletion or microduplication CNVs (29). Locus-specific mutation rates for CNVs range between 10^{-4} and 10^{-5} and thus are 1,000 to 10,000-fold more frequent than point mutations at a specific locus (30).

Except for the X chromosome, pathogenic duplications are less prevalent than deletions in clinically ascertained samples. This phenomenon is likely due to several factors, including ascertainment bias (duplications have been more challenging to detect), milder conferred phenotypes by duplications than the reciprocal deletions, and the lack of reciprocal duplication products in intra-chromatid NAHR events (12, 31).

In general, genomic disorders and CNVs at a specific locus occur with similar frequencies in different populations. However, significant differences in prevalence have been encountered for specific syndromes, such as the 17q21.31 microdeletion syndrome and Sotos syndrome (32, 33). In these cases, population-specific structural variants have been identified in the patients' parents. For instance, 17q21.31 microdeletion syndrome is expected to have a higher prevalence in the Caucasian population than the Asian population, as the deletion is associated with an 900 kb inversion present in ~20% of the European population compared to ~1% of the Asian population (34). Similarly, microdeletions leading to Sotos syndrome occur at a higher frequency in the Japanese population (32).

2.4. Defining Novel Genomic Disorders

2.4.1. "Genomotype-First" Approach

The worldwide use of high-resolution platforms on large patient cohorts has been instrumental for the identification of novel microdeletion and microduplication syndromes (Tables 1 and 2). Here, the classical paradigm of a "phenotype-first" approach, referring to the identification of a disease-gene in a preselected patient-cohort, has been shifted to a "genotype-first," or reverse genomics approach, wherein the overlapping CNVs identified in clinically heterogeneous patient cohort are analyzed first. Subsequently, more careful examination of the patients' phenotypes may reveal phenotypic overlap not expected for such a heterogeneous disease, hence allowing the definition of a new syndrome.

2.4.2. Recurrent Genomic Disorders Predicted and Identified Based on Genomic Structure

An alternative approach that has led to the successful identification of novel microdeletion and microduplication syndromes was based on the knowledge of higher-order human genome architecture and the "rules" of the NAHR mechanism. Sharp et al. (35) designed and developed a microarray with greater than 2000 BAC clones specific for 130 genomic regions that were computationally predicted to be unstable due to the flanking directly oriented LCRs. Using this approach, five novel recurrent pathogenic imbalances on chromosomes 1q21.1, 15q13.3, 15q24, 17q12, and 17q21.31 have been identified (36). All of these five microdeletions, and the subsequently identified reciprocal microduplications, have been described in a larger number of patients with different complex traits, including developmental delay/intellectual disability (DD/ID), autism, epilepsy, MODY, and schizophrenia.

2.5. Identification of Dosage-Sensitive/Haploinsufficient Genes

In some disorders, the phenotypic spectrum observed can be attributed to the malfunctioning (deletion, duplication, disruption, or point mutations) of a single dosage-sensitive gene. The lack of protein encoded by one allele of a gene with the reduction of the amount of the protein to approximately 50%, leading to an abnormal clinical phenotype is termed *haploinsufficiency*, a state in which a decreased level of the protein is insufficient to sustain the normal function of the protein. In reverse, duplications of the entire gene or gain-of-function mutations create proteins that exhibit an increase in constitutive amount or activity, respectively, even in the absence of a physiological activator, or that create insensitivity to negative regulators.

In contrast to the single gene disorders, it has been shown that several conditions, especially those including DD/ID and congenital/developmental abnormalities, result from submicroscopic chromosome rearrangements encompassing several genes, with at least two of them being dosage-sensitive but typically functionally unrelated to each other. In such contiguous gene deletion or duplication syndromes (e.g., Langer-Giedion syndrome or Potocki-Shaffer syndrome), the phenotype of the disorder results from an inappropriate dosage of only the dosage-sensitive genes located within the rearranged genomic region (18).

As CNVs often contain many genes, the identification of a CNV in a patient may point toward the identification of a novel dosage-sensitive gene related to disease. Detailed clinical evaluation of affected patients, development of an animal model(s) recapitulating the human phenotype, and/or identification of point mutation(s) in the candidate gene(s) are necessary to determine whether the phenotype can be attributed to a single gene within the CNV or whether multiple genes are likely contributors. Examples of successful candidate gene identification in microdeletion and microduplication syndromes include *CREBBP* in Rubinstein-Taybi syndrome (37), *UBE3A* in Angelman syndrome (38), *JAG1* in Alagille syndrome (39), *MECP2* in Rett syndrome (40), *NSD1* in Sotos syndrome (41), *RAI1* in Smith-Magenis syndrome (42), *CHD7* in CHARGE syndrome (43), *NIPBL* in Cornelia de Lange syndrome (44, 45), *MYCN* in Feingold syndrome (46), and *FOXF1* in alveolar capillary dysplasia (47).

2.6. Common Complex Traits and Rare De Novo CNVs

Although many common diseases—such as schizophrenia, DD/ID, and autism spectrum disorder—occur at high frequencies in the general population and show an overall high heritability, the genetic contribution in these diseases has been explained only partially. The application of genome-wide CNV detection methods has enabled the identification of rare pathogenic *de novo* CNVs in many of these disorders (48–57). Among these, several genomic regions have been identified in more than one individual, including microdeletions and microduplications in 1q21.1 in patients with DD/ID, autism, schizophrenia, and micro- and macrocephaly, respectively (53, 58–60), microdeletions in 15q13.3 involving the *CHRNA7* gene in patients with a range of neurobehavioral and neurodevelopmental abnormalities, including DD/ID, speech delay, epilepsy (~1%), schizophrenia, bipolar disorder, anxiety, and mood disorder, and antisocial behaviors (53, 61–64) and microdeletions and microduplications in 16p11.2 in patients with autism (~1%) and schizophrenia (65–68). Interestingly, autism is associated more frequently with microdeletions in 16p11.2 and macrocephaly, whereas schizophrenia is found in patients with microduplications in 16p11.2 and microcephaly, a correlation opposite or reciprocal to that in patients with microdeletions and microduplications in 1q21.1. This phenomenon has been defined as a diametric model of autism-spectrum and psychotic-spectrum behavioral phenotypes in reciprocal "sister" genomic disorders (69). In this model, diametric CNVs can generate contrasting phenotypes associated with autistic- and psychotic-spectrum conditions that may represent evolution of the social brain (70).

Other common traits for which the pathogenic or susceptibility CNVs have been reported include Alzheimer disease (duplications of the amyloid precursor *APP* gene) (71), Parkinson disease (duplications and triplications of the *SNCA* gene) (72–74), pancreatitis (75), Crohn disease (76, 77), glomerulonephritis (78), psoriasis (79), and systemic lupus erythematosus (80).

2.7. Clinical Consequences

2.7.1. Mechanisms of Conveying Clinical Phenotypes

The most common mechanism of conveying a phenotype by microdeletion or microduplication is altering the copy-number of a dosage-sensitive gene(s) located within the rearranged region (43, 47, 81). However, in some cases, CNVs convey phenotypes by other mechanisms (82). First, the breakpoints of a CNV can disrupt the coding sequence of a gene (83), or can generate a novel fusion gene (84), the latter typically encountered with somatic rearrangement events associated with cancers. Second, CNVs can delete or duplicate gene regulatory elements (e.g., tissue-specific enhancers) changing gene function but leaving the protein coding sequence intact. This position effect has been described mainly in balanced constitutional translocations (86, 87) and microdeletions (87, 88), with breakpoints mapping as far as 1 Mb upstream or downstream from the protein-coding region of the causative gene (89). Recently, microduplications of regulatory elements have been reported also to cause specific phenotypes. Klopocki et al. (90) identified a 590 kb duplication involving the regulatory element termed ZRS in intron 5 of the *LMBR1* gene (85) located ~1 Mb

5′ from the intact *SHH* gene associated with triphalangeal thumb and polysyndactyly. Dathe et al. (91) described an ~5.5 kb duplication in a noncoding sequence located ~110 kb downstream of the *BMP2* gene in chromosome 20p12.2 in two families with brachydactyly type A2. Kurth et al. (92) reported an ~2 Mb duplication of regulatory sequences in a gene desert upstream of *SOX9* in four unrelated families with features of Cooks syndrome.

A recessive mutation or functional polymorphisms can be unmasked on the other allele as a consequence of a deletion CNV (93). Finally, deletion or duplication CNVs can lead to abnormal phenotypes by disrupting a gene that is not dosage-sensitive, e.g., intragenic rearrangements in an autosomal dominant gene or X-linked dominant gene.

3. Methods

3.1. High-Resolution Genome-Wide Copy-Number Screening

3.1.1. Array-Based Platforms

Microarray-based genome profiling technologies have dramatically changed the nature of human genome analysis by combining the targeted high-resolution approach of FISH technology and the whole-genome approach of the karyotyping technology. Initially, genomic microarrays were developed in academia and contained mostly genomic fragments from bacterial artificial chromosomes (BACs), with a genome-wide spacing of one clone per 1 Mb or later with a tiling resolution of one clone per 100 kb. In the past few years, private enterprises have implemented array production and now offer microarrays for genome-wide copy-number profiling containing more than two million oligonucleotides, targeting unique sequences, SNPs or a combination thereof. Nowadays, the commercial companies provide various array options, such as full custom design arrays for any portion of the genome, exon-targeted arrays, and specific arrays for diagnostic use targeting only regions with known clinical significance. Oligonucleotide microarrays targeting SNPs enable the identification of regions with absence-of-heterozygosity (due to uniparental disomy or parental consanguinity), that may harbor pathogenic recessive alleles, or enable determination of parentage.

3.1.2. Validation Methods

Verification of genomic variants detected in the initial screening and the discrimination between “benign” CNVs and potential disease-causing CNVs have become increasingly important and at the same time challenging. The commonly used validation techniques include, among others, metaphase and interphase FISH and different PCR-based methods, such as multiplex ligation-dependent probe amplification (MLPA), genomic quantitative PCR, or long-range PCR.

In some cases, alternative higher-resolution (custom design) arrays are used to validate CNVs. The choice of validation technique depends on the costs involved, the turn-around time, technological limitations, the need to test additional patients for the same CNV, and the size of the rearrangement. For example, duplications smaller than 100–200 kb are difficult to identify by FISH because of the lower resolution limit in the interphase nuclei (94, 95), thus PCR-based technologies, such as MLPA, qPCR, or long-range PCR, are used as an alternative assay.

4. Discussion

4.1. Alternative Genomic Elements Contributing to Disease

4.1.1. Nongenic Conserved Sequences

In addition to the functional genes, 99% of all currently reported CNVs overlap conserved noncoding sequences (CNSs), stretches of DNA that are not genic, i.e., they do not produce transcripts (coding or noncoding) with functional properties (7). There are approximately 327,000 CNSs occupying 1–2% of the human genome. The majority of CNSs are intergenic (96, 97). Evolutionary analyses strongly support a conserved functional role for these CNSs (98, 99) and mutations in CNSs are known to cause disease, such as thalassemia, preaxial polydactyly, and X-linked deafness type 3 (85, 100–104). As CNSs may act as dosage-sensitive elements, CNV involving CNSs can contribute to phenotypic variation and/or disease. Categorizing CNSs located in CNV loci according to their functional characteristics will be a valuable step in furthering our understanding the role of CNSs, and their relation to CNV in disease.

4.1.2. MicroRNAs

An increasing number of studies have been performed to unravel the role of microRNAs (miRNAs) in disease. miRNAs are highly conserved, nonprotein-coding RNAs that function to regulate gene expression in homeostatic processes, such as development, cell proliferation, and cell death. Recently, dysregulation of miRNAs was identified as a cause of cancer initiation and progression (105). As such, it might be expected that any form of dysregulation of miRNAs can also be an important role in developmental disorders.

4.2. Next-Generation Sequencing

Most recently, a number of different next-generation sequencing (NGS) technologies, using chemistries other than the traditional Sanger dideoxy chain termination, have been developed and successfully implemented, allowing sequencing of the entire human genome at significantly less expense. The main differences between the methods are read length, number of reads per run, and the costs involved (106). All NGS methods are in principle capable of detecting both single base mutations (shotgun) and structural variation both read number as a surrogate measure of copy-number and mate-pairs aligned to a reference genome to surmise structural variation.

With shotgun sequencing, the genome is shredded (*shotgunning*) into smaller fragments of DNA which can be massively sequenced in parallel. Next, the sequenced fragments are assembled into contigs based on the overlap in the sequence reads (de novo assembly) or, alternatively, are individually mapped back to a reference genome. In the latter situation, single base pair changes compared to a reference genome can be identified and as such may the lead to disease-gene identification. This application of NGS has recently resulted in the first individual whole genome sequences (107, 108). The analyses of the first two available human genomes have demonstrated how challenging it still is to draw clinically or biologically relevant conclusions from individual sequences. More genomes need to be sequenced to learn how genotypes correlate with phenotypes. The "1000 human genomes" project is now in progress, with a goal to create a reference standard for the analysis of human genomic variations that are expected to contribute to studies of disease (http://www.1000genomes.org/). Nonetheless, the immediate applications and relevance of NGS techniques in the clinical genetics field have been demonstrated already, by the identification of several novel genes for X-linked MR using a resequencing strategy of the entire X chromosome (109). In parallel, projects are ongoing developing methods for targeted sequencing of all protein-coding regions of the genome ("exome" sequencing) to increase the chance of finding highly penetrant disease variants. A proof-of-principle study, using the autosomal dominant Freeman-Sheldon syndrome, has already shown that genes for Mendelian disorders can be identified using this approach (110). The first whole-genome sequence that enabled a personal genomics approach to identify the specific cause of neuropathy in a family segregating recessive Charcot-Marie-Tooth disease locus has also recently been reported (111).

The most specific NGS application to identify CNV is done using paired-end mapping or mate-pair library sequencing, as it directly provides detailed positional information. Moreover, this application not only identifies the unbalanced variants but also allows for the identification of balanced rearrangements, including (cryptic) translocations and inversions. For mate-pair runs, genomic DNA is randomly sheared and size selected. After several processing steps, shotgun reads are obtained by sequencing both ends of size-selected DNA library, whose approximate size is known. This information constrains the placement of the reads within the reference genome. Deviations from this expected size distribution may point to deletions, duplications, insertions, or inversions. The first two genomes, an African and putative European genome, analyzed using this paired-end approach showed >1,000 structural variants, both shared with and divergent to the reference genome, of which many potentially affect gene function (112). A more recent analysis of eight human genomes indicated ~1,700 intermediate-sized structural variants, of which 50% were detected in more than one individual, and involved regions not previously described as variants (113). Moreover, 525 new insertions, not represented in the reference genome were identified (114). This latter statement is further extended by the recent estimation that the human pan-genome is missing 19–40 Mb of genomic sequence, containing coding and conserved CNSs (114).

5. Conclusions

In the past few years, genome-wide analyses using microarray-based technologies have enabled the identification of several novel microdeletion and microduplication syndromes and, in some cases, the pathogenic dosage-sensitive genes. However, despite the continuously growing number of analyses, the list of dosage-sensitive genes is far from complete, and many genes still await recognition. Furthermore, these technologies revealed the previously underappreciated and underestimated extent of individual genomic variation in humans, with CNVs being responsible for this variation to a larger extent than SNPs.

Without any doubt, the implementation of NGS technologies and medical resequencing strategies will continue to change our understanding of structural variation. Moreover, these technologies will ultimately extend our ability to unravel the "genomic load" of CNVs (115) and the function of the noncoding portion of the human genome to decipher the full "genomic code" in health and disease.

Acknowledgments

This work was supported by grants from the Netherlands Organization for Health Research and Development (ZonMW 916.86.016 to L.E.L.M.V.) and from the Polish Ministry of Science and Higher Education (R13-0005-04/2008 to P.S.). We appreciate the critical reviews of Drs. J.R. Lupski and F. Probst. We thank Z. Xia for editorial assistance. We apologize to colleagues and the authors of relevant papers who could not be cited owing to space limitations.

References

1. Lupski, J.R. (1998) Genomic disorders: structural features of the genome can lead to DNA rearrangements and human disease traits. *Trends Genet* **14**, 417–22.
2. Lupski, J.R. (2009) Genomic disorders ten years on. *Genome Med* **24**, 1–42.
3. Pinkel, D., Segraves, R., Sudar, D., Clark, S., Poole, I., Kowbel, D., Collins, C., Kuo, W.-L., Chen, C., Zhai, Y., Dairkee, S.H., Ljung, B.M., Gray, J.W., Albertson, D.G. (1998) High resolution analysis of DNA copy-number variation using comparative genomic hybridization to microarrays. *Nat Genet* **20**, 207–11.
4. Solinas-Toldo, S., Lampel, S., Stilgenbauer, S., Nickolenko, J., Benner, A., Dohner, H., Cremer, T., Lichter, P. (1997) Matrix-based comparative genomic hybridization: biochips to screen for genomic imbalances. *Genes Chromosomes Cancer* **20**, 399–407.
5. Iafrate, A.J., Feuk, L., Rivera, M.N., Listewnik, M.L., Donahoe, P.K., Qi, Y., Scherer, S.W., Lee, C. (2004) Detection of large-scale variation in the human genome. *Nat Genet* **36**, 949–51.
6. Sebat, J., Lakshmi, B., Troge, J., Alexander, J., Young, J., Lundin, P., Månér, S., Massa, H., Walker, M., Chi, M., Navin, N., Lucito, R., Healy, J., Hicks, J., Ye, K., Reiner, A., Gilliam, T.C., Trask, B., Patterson, N., Zetterberg, A., Wigler, M. (2004) Large-scale copy-number polymorphism in the human genome. *Science* **305**, 525–8.
7. Redon, R., Ishikawa, S., Fitch, K.R., Feuk, L., Perry, G.H., Andrews, T.D., Fiegler, H., Shapero, M.H., Carson, A.R., Chen, W., Cho, E.K., Dallaire, S., Freeman, J.L., González, J.R., Gratacòs, M., Huang, J., Kalaitzopoulos, D., Komura, D., MacDonald, J.R., Marshall, C.R., Mei, R., Montgomery, L., Nishimura, K., Okamura, K., Shen, F., Somerville, M.J., Tchinda, J., Valsesia, A., Woodwark, C., Yang, F., Zhang, J., Zerjal, T., Zhang, J., Armengol, L., Conrad, D.F., Estivill, X., Tyler-Smith, C., Carter, N.P., Aburatani, H., Lee, C., Jones, K.W., Scherer, S.W., Hurles, M.E. (2006) Global variation in copy-number in the human genome. *Nature* **444**, 444–54.
8. Stankiewicz, P., Beaudet, A.L. (2007) Use of arrayCGHin the evaluation of dysmorphology, malformations, developmental delay, and idiopathic mental retardation. *Curr Opin Genet Dev* **17**, 182–92.
9. Slavotinek, A.M. (2008) Novel microdeletion syndromes detected by chromosome microarrays. *Hum Genet* **124**, 1–17.
10. Stankiewicz, P., Lupski, J.R. (2010) Structural variation in the human genome and its role in disease. *Annu Rev Med* **61**, 437–55.
11. Carvalho, C.M.B., Zhang, F., Lupski JR. (2010) *Evolution in Health and Medicine Sackler Colloquium*: Genomic disorders: A window into human gene and genome evolution. *PNAS* **107**, 1765–71.
12. Stankiewicz, P., Lupski, J.R. (2002) Genome architecture, rearrangements and genomic disorders. *Trends Genet* **18**, 74–82.
13. Bailey, J.A., Yavor, A.M., Massa, H.F., Trask, B.J., Eichler, E.E. (2001) Segmental duplications: organization and impact within the current Human Genome Project assembly. *Genome Res* **11**, 1005–17.
14. Eichler, E.E. (2001) Recent duplication, domain accretion and the dynamic mutation of the human genome. *Trends Genet* **17**, 661–9.
15. Slack, A., Thornton, P.C., Magner, D.B., Rosenberg, S.M., Hastings, P.J. (2006) On the mechanism of gene amplification induced under stress in Escherichia coli. *PLoS Genet* **2**, e48.
16. Lee, J.A., Carvalho, C.M., Lupski, J.R. (2007) A DNA replication mechanism for generating nonrecurrent rearrangements associated with genomic disorders. *Cell* **131**, 1235–47.
17. Hastings, P.J., Ira, G., Lupski, J.R. (2009) A microhomology-mediated break-induced replication model for the origin of human copy number variation. *PLoS Genet* **5**, e1000327.
18. Schmickel, R.D. (1986) Contiguous gene syndromes: a component of recognizable syndromes. *J Pediatr* **109**, 231–41.
19. Katsanis, N., Ansley, S.J., Badano, J.L., Eichers, E.R., Lewis, R.A., Hoskins, B.E., Scambler, P.J., Davidson, W.S., Beales, P.L., Lupski, J.R. (2001) Triallelic inheritance in Bardet-Biedl syndrome, a Mendelian recessive disorder. *Science* **293**, 2256–9.
20. Zhang F, Gu W, Hurles ME, Lupski JR. (2009) Copy number variation in human health, disease, and evolution. Annu Rev Genomics Hum Genet 10, 451–81.
21. Firth, H.V., Richards, S.M., Bevan, A.P., Clayton, S., Corpas, M., Rajan, D., Van Vooren, S., Moreau, Y., Pettett, R.M., Carter, N.P. (2009) DECIPHER: Database of Chromosomal Imbalance and Phenotype in Humans Using Ensembl Resources. *Am J Hum Genet* **84**, 524–33.
22. Pfeiffer, P., Goedecke, W., Obe, G. (2000) Mechanisms of DNA double-strand break repair and their potential to induce chromosomal aberrations. *Mutagenesis* **15**, 289–302.

23. Zhang, F., Khajavi, M., Connolly, A.M., Towne, C.F., Batish, S.D., Lupski, J.R. (2009) The DNA replication FoSTeS/MMBIR mechanism can generate human genomic, genic, and exonic complex rearrangements. *Nat Genet* **41**, 849–53.

24. Carvalho, C.M., Zhang, F., Liu, P., Patel, A., Sahoo, T., Bacino, C.A., Shaw, C., Peacock, S., Pursley, A., Tavyev, Y.J., Ramocki, M.B., Nawara, M., Obersztyn, E., Vianna-Morgante, A.M., Stankiewicz, P., Zoghbi, H.Y., Cheung, S.W., Lupski, J.R. (2009) Complex rearrangements in patients with duplications of MECP2 can occur by fork stalling and template switching. *Hum Mol Genet* **18**, 2188–203.

25. Bi, W., Sapir, T., Shchelochkov, O.A., Zhang, F., Withers, M.A., Hunter, J.V., Levy, T., Shinder, V., Peiffer, D.A., Gunderson, K.L., Nezarati, M.M., Shotts, V.A., Amato, S.S., Savage, S.K., Harris, D.J., Day-Salvatore, D.L., Horner, M., Lu, X.Y., Sahoo, T., Yanagawa, Y., Beaudet, A.L., Cheung, S.W., Martinez, S., Lupski, J.R., Reiner, O. (2008) LIS1 increased expression affects human and mouse brain development. *Nat Genet* **41**, 168–77.

26. Nagamani, S.C.S., Zhang, F., Shchelochkov, O.A., Bi, W., Ou, Z., Scaglia, F., Probst, F.J., Shinawi, M., Eng, C., Hunter, J.V., Sparagana, S., Lagoe, E., Fong, C.T., Pearson, M., Doco-Fenzy, M., Landais, E., Mozelle, M., Chinault, A.C., Patel, A., Bacino, C.A., Sahoo, T., Kang, S.H., Cheung, S.W., Lupski, J.R., Stankiewicz, P. (2009) Microdeletions including *YWHAE* in the Miller-Dieker syndrome region on chromosome 17p13.3 result in facial dysmorphisms, growth restriction, and cognitive impairment. *J Med Genet* **46**, 825–33.

27. Yatsenko, S.A., Brundage, E.K., Roney, E.K., Cheung, S.W., Chinault, A.C., Lupski, J.R. (2009) Molecular mechanisms for subtelomeric rearrangements associated with the 9q34.3 microdeletion syndrome. *Hum Mol Genet* **18**, 1924–36.

28. Vissers, L.E.L.M., Bhatt, S.S., Janssen, I.M., Xia, Z., Lalani, S.R., Pfundt, R., Derwinska, K., de Vries, B.B.A., Gilissen, C., Hoischen, A., Nesteruk, M., Wisniowiecka-Kowalnik, B., Smyk, M., Brunner, H.G., Cheung, S.W., van Kessel, A.D, Veltman, J.A., Stankiewicz, P. (2009) Rare pathogenic microdeletions and tandem duplications are microhomology-mediated and stimulated by local genomic architecture. *Hum Molec Genet* **18**, 3579–93.

29. Shaffer, L.G., and Lupski J.R. (2000) Molecular mechanisms for constitutional chromosomal rearrangements in humans. *Annu Rev Genet* **34**, 297–329.

30. Lupski, J.R. (2007) Genomic rearrangements and sporadic disease. *Nat Genet* **39**, S43–7.

31. Turner, D.J., Miretti, M., Rajan, D., Fiegler, H., Carter, N.P., Blayney, M.L., Beck, S., Hurles, M.E. (2008) Germline rates of de novo meiotic deletions and duplications causing several genomic disorders. *Nat Genet* **40**, 90–5.

32. Kurotaki, N., Harada, N., Shimokawa, O., Miyake, N., Kawame, H., Uetake, K., Makita, Y., Kondoh, T., Ogata, T., Hasegawa, T., Nagai, T., Ozaki, T., Touyama, M., Shenhav, R., Ohashi, H., Medne, L., Shiihara, T., Ohtsu, S., Kato, Z., Okamoto, N., Nishimoto, J., Lev, D., Miyoshi, Y., Ishikiriyama, S., Sonoda, T., Sakazume, S., Fukushima, Y., Kurosawa, K., Cheng, J.F., Yoshiura, K., Ohta, T., Kishino, T., Niikawa, N., Matsumoto, N. (2003) Fifty microdeletions among 112 cases of Sotos syndrome: low copy repeats possibly mediate the common deletion. *Hum Mutat* **22**, 378–87.

33. Koolen, D.A., Sharp, A.J., Hurst, J.A., Firth, H.V., Knight, S.J., Goldenberg, A., Saugier-Veber, P., Pfundt, R., Vissers, L.E., Destree, A., Grisart, B., Rooms, L., Van der, A.N., Field, M., Hackett, A., Bell, K., Nowaczyk, M.J., Mancini, G.M., Poddighe, P.J., Schwartz, C.E., Rossi, E., De, G.M., ntonacci-Fulton, L.L., McLellan, M.D., Garrett, J.M., Wiechert, M.A., Miner, T.L., Crosby, S., Ciccone, R., Willatt, L., Rauch, A., Zenker, M., Aradhya, S., Manning, M.A., Strom, T.M., Wagenstaller, J., Krepischi-Santos, A.C., Vianna-Morgante, A.M., Rosenberg, C., Price, S.M., Stewart, H., Shaw-Smith, C., Brunner, H.G., Wilkie, A.O., Veltman, J.A., Zuffardi, O., Eichler, E.E., de Vries, B.B. (2008) Clinical and molecular delineation of the 17q21.31 microdeletion syndrome. *J Med Genet* **45**, 710–20.

34. Stefansson, H., Helgason, A., Thorleifsson, G., Steinthorsdottir, V., Masson, G., Barnard, J., Baker, A., Jonasdottir, A., Ingason, A., Gudnadottir, V.G., Desnica, N., Hicks, A., Gylfason, A., Gudbjartsson, D.F., Jonsdottir, G.M., Sainz, J., Agnarsson, K., Birgisdottir, B., Ghosh, S., Olafsdottir, A., Cazier, J.B., Kristjansson, K., Frigge, M.L., Thorgeirsson, T.E., Gulcher, J.R., Kong, A., Stefansson, K. (2005) A common inversion under selection in Europeans. *Nat Genet* **37**, 129–37.

35. Sharp, A.J., Locke, D.P., McGrath, S.D., Cheng, Z., Bailey, J.A., Vallente, R.U., Pertz, L.M., Clark, R.A., Schwartz, S., Segraves, R., Oseroff, V.V., Albertson, D.G., Pinkel, D., Eichler, E.E. (2005) Segmental duplications

and copy-number variation in the human genome. *Am J Hum Genet* 77, 78–88.

36. Sharp, A.J., Hansen, S., Selzer, R.R., Cheng, Z., Regan, R., Hurst, J.A., Stewart, H., Price, S.M., Blair, E., Hennekam, R.C., Fitzpatrick, C.A., Segraves, R., Richmond, T.A., Guiver, C., Albertson, D.G., Pinkel, D., Eis, P.S., Schwartz, S., Knight, S.J., Eichler, E.E. (2006) Discovery of previously unidentified genomic disorders from the duplication architecture of the human genome. *Nat Genet* **38**, 1038–42.
37. Petrij, F., Giles, R. H., Dauwerse, H. G., Saris, J. J., Hennekam, R. C. M., Masuno, M., Tommerup, N., van Ommen, G. J. B., Goodman, R. H., Peters, D. J. M., Breuning, M. H. (1995) Rubinstein-Taybi syndrome caused by mutations in the transcriptional co-activator CBP. *Nature* **376**, 348–51.
38. Matsuura, T., Sutcliffe, J.S., Fang, P., Galjaard, R.-J., Jiang, Y., Benton, C.S., Rommens, J.M., Beaudet, A.L. (1997) De novo truncating mutations in E6-AP ubiquitin-protein ligase gene (*UBE3A*) in Angelman syndrome. *Nat Genet* **15**, 74–7.
39. Li, L., Krantz, I.D., Deng, Y., Genin, A., Banta, A.B., Collins, C.C., Qi, M., Trask, B.J., Kuo, W.L., Cochran, J., Costa, T., Pierpont, M.E., Rand, E.B., Piccoli, D.A., Hood, L., Spinner, N.B. (1997) Alagille syndrome is caused by mutations in human *Jagged1*, which encodes a ligand for Notch1. *Nat Genet* **16**, 243–51.
40. Amir, R.E., Van den Veyver, I.B., Wan, M., Tran, C.Q., Francke, U., Zoghbi, H.Y. (1999) Rett syndrome is caused by mutations in X-linked *MECP2*, encoding methyl-CpG-binding protein 2. *Nat Genet* **23**, 185–8.
41. Kurotaki, N., Imaizumi, K., Harada, N., Masuno, M., Kondoh, T., Nagai, T., Ohashi, H., Naritomi, K., Tsukahara, M., Makita, Y., Sugimoto, T., Sonoda, T., Hasegawa, T., Chinen, Y., Tomita Ha, H.A., Kinoshita, A., Mizuguchi, T., Yoshiura Ki, K., Ohta, T., Kishino, T., Fukushima, Y., Niikawa, N., Matsumoto, N. (2002) Haploinsufficiency of *NSD1* causes Sotos syndrome. *Nat Genet* **30**, 365–6.
42. Slager, R.E., Newton, T.L., Vlangos, C.N., Finucane, B., Elsea, S.H. (2003) Mutations in RAI1 associated with Smith-Magenis syndrome. *Nat Genet* **33**, 466–8.
43. Vissers, L.E.L.M., van Ravenswaaij, C.M., Admiraal, R., Hurst, J.A., de Vries, B.B., Janssen, I.M., van der Vliet, W.A., Huys, E.H., de Jong, P.J., Hamel, B.C., Schoenmakers, E.F., Brunner, H.G., Veltman, J.A., van Kessel, A.G. (2004) Mutations in a new member of the chromodomain gene family cause CHARGE syndrome. *Nat Genet* **36**, 955–7.
44. Krantz, I. D., McCallum, J., DeScipio, C., Kaur, M., Gillis, L. A., Yaeger, D., Jukofsky, L., Wasserman, N., Bottani, A., Morris, C. A., Nowaczyk, M. J. M., Toriello, H., Bamshad, M.J., Carey, J.C., Rappaport, E., Kawauchi, S., Lander, A.D., Calof, A.L., Li, H.H., Devoto, M., Jackson, L.G. (2004) Cornelia de Lange syndrome is caused by mutations in *NIPBL*, the human homolog of Drosophila melanogaster *Nipped-B*. *Nat Genet* **36**, 631–5.
45. Tonkin, E. T., Wang, T.-J., Lisgo, S., Bamshad, M. J., Strachan, T. (2004) NIPBL, encoding a homolog of fungal Scc2-type sister chromatid cohesion proteins and fly *Nipped-B*, is mutated in Cornelia de Lange syndrome. *Nat Genet* **36**, 636–41.
46. van Bokhoven, H., Celli, J., van Reeuwijk, J., Rinne, T., Glaudemans, B., van Beusekom, E., Rieu, P., Newbury-Ecob, R.A., Chiang, C., Brunner, H.G. (2005) MYCN haploinsufficiency is associated with reduced brain size and intestinal atresias in Feingold syndrome. *Nat Genet* **37**, 465–7.
47. Stankiewicz, P., Sen, P., Bhatt, S.S., Storer, M., Xia, Z., Bejjani, B.A., Ou, Z., Wiszniewska, J., Driscoll, D.J., Maisenbacher, M.K., Bolivar, J., Bauer, M., Zackai, E.H., McDonald-McGinn, D., Nowaczyk, M.M., Murray, M., Hustead, V., Mascotti, K., Schultz, R., Hallam, L., McRae, D., Nicholson, A.G., Newbury, R., Durham-O'Donnell, J., Knight, G., Kini, U., Shaikh, T.H., Martin, V., Tyreman, M., Simonic, I., Willatt, L., Paterson, J., Mehta, S., Rajan, D., Fitzgerald, T., Gribble, S., Prigmore, E., Patel, A., Shaffer, L.G., Carter, N.P., Cheung, S.W., Langston, C, Shaw-Smith, C. (2009) Genomic and genic deletions of the FOX gene cluster on 16q24.1 and inactivating mutations of *FOXF1* cause alveolar capillary dysplasia and other malformations. *Am J Hum Genet* **84**, 780–91.
48. de Vries, B.B., Pfundt, R., Leisink, M., Koolen, D.A., Vissers, L.E., Janssen, I.M., Reijmersdal, S., Nillesen, W.M., Huys, E.H., Leeuw, N., Smeets, D., Sistermans, E.A., Feuth, T., van Ravenswaaij-Arts, C.M., van Kessel, A.G., Schoenmakers, E.F., Brunner, H.G., Veltman, J.A. (2005) Diagnostic genome profiling in mental retardation. *Am J Hum Genet* 77, 606–16.
49. Schoumans, J., Ruivenkamp, C., Holmberg, E., Kyllerman, M., Anderlid, B.M., Nordenskjold, M. (2005) Detection of chromosomal imbalances in children with

idiopathic mental retardation by array based comparative genomic hybridisation (array-CGH). *J Med Genet* **42**, 699–705.

50. Autism Genome Project Consortium (AGPC)., Szatmari, P., Paterson, A.D., Zwaigenbaum, L., Roberts, W., Brian, J., Liu, X.Q., Vincent, J.B., Skaug, J.L., Thompson, A.P., Senman, L., Feuk, L., Qian, C., Bryson, S.E., Jones, M.B., Marshall, C.R., Scherer, S.W., Vieland, V.J., Bartlett, C., Mangin, L.V., Goedken, R., Segre, A., Pericak-Vance, M.A., Cuccaro, M.L., Gilbert, J.R., Wright, H.H., Abramson, R.K., Betancur, C., Bourgeron, T., Gillberg, C., Leboyer, M., Buxbaum, J.D., Davis, K.L., Hollander, E., Silverman, J.M., Hallmayer, J., Lotspeich, L., Sutcliffe, J.S., Haines, J.L., Folstein, S.E., Piven, J., Wassink, T.H., Sheffield, V., Geschwind, D.H., Bucan, M., Brown, W.T., Cantor, R.M., Constantino, J.N., Gilliam, T.C., Herbert, M., Lajonchere, C., Ledbetter, D.H., Lese-Martin, C., Miller, J., Nelson, S., Samango-Sprouse, C.A., Spence, S., State, M., Tanzi, R.E., Coon, H., Dawson, G., Devlin, B., Estes, A., Flodman, P., Klei, L., McMahon, W.M., Minshew, N., Munson, J., Korvatska, E., Rodier, P.M., Schellenberg, G.D., Smith, M., Spence, M.A., Stodgell, C., Tepper, P.G., Wijsman, E.M., Yu, C.E., Rogé, B., Mantoulan, C., Wittemeyer, K., Poustka, A., Felder, B., Klauck, S.M., Schuster, C., Poustka, F., Bölte, S., Feineis-Matthews, S., Herbrecht, E., Schmötzer, G., Tsiantis, J., Papanikolaou, K., Maestrini, E., Bacchelli, E., Blasi, F., Carone, S., Toma, C., Van Engeland, H., de Jonge, M., Kemner, C., Koop, F., Langemeijer, M., Hijmans, C., Staal, W.G., Baird, G., Bolton, P.F., Rutter, M.L., Weisblatt, E., Green, J., Aldred, C., Wilkinson, J.A., Pickles, A., Le Couteur, A., Berney, T., McConachie, H., Bailey, A.J., Francis, K., Honeyman, G., Hutchinson, A., Parr, J.R., Wallace, S., Monaco, A.P., Barnby, G., Kobayashi, K., Lamb, J.A., Sousa, I., Sykes, N., Cook, E.H., Guter, S.J., Leventhal, B.L., Salt, J., Lord, C., Corsello, C., Hus, V., Weeks, D.E., Volkmar, F., Tauber, M., Fombonne, E., Shih, A., Meyer, K.J. (2007) Mapping autism risk loci using genetic linkage and chromosomal rearrangements. *Nat Genet* **39**, 19–28.
51. Sebat, J., Lakshmi, B., Malhotra, D., Troge, J., Lese-Martin, C., Walsh, T., Yamrom, B., Yoon, S., Krasnitz, A., Kendall, J., Leotta, A., Pai, D., Zhang, R., Lee, Y.H., Hicks, J., Spence, S.J., Lee, A.T., Puura, K., Lehtimaki, T., Ledbetter, D., Gregersen, P.K., Bregman, J., Sutcliffe, J.S., Jobanputra, V., Chung, W., Warburton, D., King, M.C., Skuse, D., Geschwind, D.H., Gilliam, T.C., Ye, K., Wigler, M. (2007) Strong association of de novo copy-number mutations with autism. *Science* **316**, 445–9.
52. Marshall, C.R., Noor, A., Vincent, J.B., Lionel, A.C., Feuk, L., Skaug, J., Shago, M., Moessner, R., Pinto, D., Ren, Y., Thiruvahindrapduram, B., Fiebig, A., Schreiber, S., Friedman, J., Ketelaars, C.E., Vos, Y.J., Ficicioglu, C., Kirkpatrick, S., Nicolson, R., Sloman, L., Summers, A., Gibbons, C.A., Teebi, A., Chitayat, D., Weksberg, R., Thompson, A., Vardy, C., Crosbie, V., Luscombe, S., Baatjes, R., Zwaigenbaum, L., Roberts, W., Fernandez, B., Szatmari, P., Scherer, S.W. (2008) Structural variation of chromosomes in autism spectrum disorder. *Am J Hum Genet* **82**, 477–88.
53. International Schizophrenia Consortium. (2008) Rare chromosomal deletions and duplications increase risk of schizophrenia. *Nature* **455**, 237–41.
54. Lu, X.Y., Phung, M.T., Shaw, C.A., Pham, K., Neil, S.E., Patel, A., Sahoo, T., Bacino, C.A., Stankiewicz, P., Kang, S.H., Lalani, S., Chinault, A.C., Lupski, J.R., Cheung, S.W., Beaudet, A.L. (2008) Genomic imbalances in neonates with birth defects: high detection rates by using chromosomal microarray analysis. *Pediatrics* **122**, 1310–8.
55. McMullan, D.J., Bonin, M., Hehir-Kwa, J.Y., de Vries, B.B., Dufke, A., Rattenberry, E., Steehouwer, M., Moruz, L., Pfundt, R., de, L.N., Riess, A., tug-Teber, O., Enders, H., Singer, S., Grasshoff, U., Walter, M., Walker, J.M., Lamb, C.V., Davison, E.V., Brueton, L., Riess, O., Veltman, J.A. (2009) Molecular karyotyping of patients with unexplained mental retardation by SNP arrays: A multicenter study. *Hum Mutat* **30**, 1082–92.
56. Bochukova, E.G., Huang, N., Keogh, J., Henning, E., Purmann, C., Blaszczyk, K., Saeed, S., Hamilton-Shield, J., Clayton-Smith, J., O'Rahilly, S., Hurles, M.E., Farooqi, I.S. (2010) Large, rare chromosomal deletions associated with severe early-onset obesity. *Nature*, **463**, 666–70
57. de Kovel, C.G., Trucks, H., Helbig, I., Mefford, H.C., Baker, C., Leu, C., Kluck, C., Muhle, H., von Spiczak, S., Ostertag, P., Obermeier, T., Kleefuss-Lie, A.A., Hallmann, K., Steffens, M., Gaus, V., Klein, K.M., Hamer, H.M., Rosenow, F., Brilstra, E.H., Trenité, D.K., Swinkels, M.E., Weber, Y.G., Unterberger, I., Zimprich, F., Urak, L., Feucht, M., Fuchs, K., Møller, R.S., Hjalgrim, H., De Jonghe, P., Suls, A., Rückert, I.M., Wichmann, H.E., Franke, A., Schreiber, S., Nürnberg, P., Elger, C.E., Lerche, H.,

Stephani, U., Koeleman, B.P., Lindhout, D., Eichler, E.E., Sander, T. (2010) Recurrent microdeletions at 15q11.2 and 16p13.11 predispose to idiopathic generalized epilepsies. *Brain* **133**, 23–32.

58. Brunetti-Pierri, N., Berg, J.S., Scaglia, F., Belmont, J., Bacino, C.A., Sahoo, T., Lalani, S.R., Graham, B., Lee, B., Shinawi, M., Shen, J., Kang, S.H., Pursley, A., Lotze, T., Kennedy, G., Lansky-Shafer, S., Weaver, C., Roeder, E.R., Grebe, T.A., Arnold, G.L., Hutchison, T., Reimschisel, T., Amato, S., Geragthy, M.T., Innis, J.W., Obersztyn, E., Nowakowska, B., Rosengren, S.S., Bader, P.I., Grange, D.K., Naqvi, S., Garnica, A.D., Bernes, S.M., Fong, C.T., Summers, A., Walters, W.D., Lupski, J.R., Stankiewicz, P., Cheung, S.W., Patel, A. (2008) Recurrent reciprocal 1q21.1 deletions and duplications associated with microcephaly or macrocephaly and developmental and behavioral abnormalities. *Nat Genet* **40**, 1466–71.

59. Mefford, H.C., Sharp, A.J., Baker, C., Itsara, A., Jiang, Z., Buysse, K., Huang, S., Maloney, V.K., Crolla, J.A., Baralle, D., Collins, A., Mercer, C., Norga, K., de Ravel, T., Devriendt, K., Bongers, E.M., de Leeuw, N., Reardon, W., Gimelli, S., Bena, F., Hennekam, R.C., Male, A., Gaunt, L., Clayton-Smith, J., Simonic, I., Park, S.M., Mehta, S.G., Nik-Zainal, S., Woods, C.G., Firth, H.V., Parkin, G., Fichera, M., Reitano, S., Lo Giudice, M., Li, K.E., Casuga, I., Broomer, A., Conrad, B., Schwerzmann, M., Räber, L., Gallati, S., Striano, P., Coppola, A., Tolmie, J.L., Tobias, E.S., Lilley, C., Armengol, L., Spysschaert, Y., Verloo, P., De Coene, A., Goossens, L., Mortier, G., Speleman, F., van Binsbergen, E., Nelen, M.R., Hochstenbach, R., Poot, M., Gallagher, L., Gill, M., McClellan, J., King, M.C., Regan, R., Skinner, C., Stevenson, R.E., Antonarakis, S.E., Chen, C., Estivill, X., Menten, B., Gimelli, G., Gribble, S., Schwartz, S., Sutcliffe, J.S., Walsh, T., Knight, S.J., Sebat, J., Romano, C., Schwartz, C.E., Veltman, J.A., de Vries, B.B., Vermeesch, J.R., Barber, J.C., Willatt, L., Tassabehji, M., Eichler, E.E. (2008) Recurrent rearrangements of chromosome 1q21.1 and variable pediatric phenotypes. *N Engl J Med* **359**, 1685–99.

60. Stefansson, H., Rujescu, D., Cichon, S., Pietiläinen, O.P., Ingason, A., Steinberg, S., Fossdal, R., Sigurdsson, E., Sigmundsson, T., Buizer-Voskamp, J.E., Hansen, T., Jakobsen, K.D., Muglia, P., Francks, C., Matthews, P.M., Gylfason, A., Halldorsson, B.V., Gudbjartsson, D., Thorgeirsson, T.E., Sigurdsson, A., Jonasdottir, A., Jonasdottir, A., Bjornsson, A., Mattiasdottir, S., Blondal, T., Haraldsson, M., Magnusdottir, B.B., Giegling, I., Möller, H.J., Hartmann, A., Shianna, K.V., Ge, D., Need, A.C., Crombie, C., Fraser, G., Walker, N., Lonnqvist J., Suvisaari, J., Tuulio-Henriksson, A., Paunio, T., Toulopoulou, T., Bramon, E., Di Forti, M., Murray, R., Ruggeri, M., Vassos, E., Tosato, S., Walshe, M., Li, T., Vasilescu, C., Mühleisen, T.W., Wang, A.G., Ullum, H., Djurovic, S., Melle, I., Olesen, J., Kiemeney, L.A., Franke, B., GROUP., Sabatti, C., Freimer, N.B., Gulcher, J.R., Thorsteinsdottir, U., Kong, A., Andreassen, O.A., Ophoff, R.A., Georgi, A., Rietschel, M., Werge, T., Petursson, H., Goldstein, D.B., Nöthen, M.M., Peltonen, L., Collier, D.A., St Clair, D., Stefansson, K. (2008) Large recurrent microdeletions associated with schizophrenia. *Nature* **455**, 232– 6.

61. Sharp, A.J., Mefford, H.C., Li, K., Baker, C., Skinner, C., Stevenson, R.E., Schroer, R.J., Novara, F., De Gregori, M., Ciccone, R., Broomer, A., Casuga, I., Wang, Y., Xiao, C., Barbacioru, C., Gimelli, G., Bernardina, B.D., Torniero, C., Giorda, R., Regan, R., Murday, V., Mansour, S., Fichera, M., Castiglia, L., Failla, P., Ventura, M., Jiang, Z., Cooper, G.M., Knight, S.J., Romano, C., Zuffardi, O., Chen, C., Schwartz, C.E., Eichler, E.E. (2008) A recurrent 15q13.3 microdeletion syndrome associated with mental retardation and seizures. *Nat Genet* **40**, 322–8.

62. Ben-Shachar, S., Lanpher, B., German, J.R., Qasaymeh, M., Potocki, L., Nagamani, S.C., Franco, L.M., Malphrus, A., Bottenfield, G.W,, Spence, J.E., Amato, S., Rousseau, J.A., Moghaddam, B., Skinner, C., Skinner, S.A., Bernes, S., Armstrong, N., Shinawi, M., Stankiewicz, P., Patel, A., Cheung, S.W., Lupski, J.R., Beaudet, A.L., Sahoo, T. (2009) Microdeletion 15q13.3: a locus with incomplete penetrance for autism, mental retardation, and psychiatric disorders. *J Med Genet* **46**, 382–8.

63. van Bon, B.W., Mefford, H.C., Menten, B., Koolen, D.A., Sharp, A.J., Nillesen, W.M., Innis, J.W., de Ravel, T.J., Mercer, C.L., Fichera, M., Stewart, H., Connell, L.E., Ounap, K., Lachlan, K., Castle, B., Van der Aa, N., van Ravenswaaij, C., Nobrega, M.A., Serra-Juhé, C., Simonic, I., de Leeuw, N., Pfundt, R., Bongers, E.M., Baker, C., Finnemore, P., Huang, S., Maloney, V.K., Crolla, J.A., van Kalmthout, M., Elia, M., Vandeweyer, G., Fryns, J.P., Janssens, S., Foulds, N., Reitano, S., Smith, K., Parkel, S., Loeys, B., Woods, C.G., Oostra, A., Speleman, F., Pereira, A.C., Kurg, A., Willatt, L., Knight, S.J., Vermeesch, J.R., Romano, C., Barber, J.C., Mortier, G., Pérez-Jurado, L.A., Kooy,

F., Brunner, H.G., Eichler, E.E., Kleefstra, T., de Vries, B.B. (2009) Further delineation of the 15q13 microdeletion and duplication syndromes: a clinical spectrum varying from non-pathogenic to a severe outcome. *J Med Genet* **46**, 511–23.

64. Miller, D.T., Shen, Y., Weiss, L.A., Korn, J., Anselm, I., Bridgemohan, C., Cox, G.F., Dickinson, H., Gentile, J., Harris, D.J., Hegde, V., Hundley, R., Khwaja, O., Kothare, S., Luedke, C., Nasir, R., Poduri, A., Prasad, K., Raffalli, P., Reinhard, A., Smith, S.E., Sobeih, M.M., Soul, J.S., Stoler, J., Takeoka, M., Tan, W.H., Thakuria, J., Wolff, R., Yusupov, R., Gusella, J.F., Daly, M.J., Wu, B.L. (2009) Microdeletion/duplication at 15q13.2q13.3 among individuals with features of autism and other neuropsychiatric disorders. *J Med Genet* **46**, 242–8.

65. Kumar, R.A., KaraMohamed, S., Sudi, J., Conrad, D.F., Brune, C., Badner, J.A., Gilliam, T.C., Nowak, N.J., Cook, E.H. Jr., Dobyns, W.B., Christian, S.L. (2008) Recurrent 16p11.2 microdeletions in autism. *Hum Mol Genet* **17**, 628–38.

66. Weiss, L.A., Shen, Y., Korn, J.M., Arking, D.E., Miller, D.T., Fossdal, R., Saemundsen, E., Stefansson, H., Ferreira, M.A., Green, T., Platt, O.S., Ruderfer, D.M., Walsh, C.A., Altshuler, D., Chakravarti, A., Tanzi, R.E., Stefansson, K., Santangelo, S.L., Gusella, J.F., Sklar, P., Wu, B.L., Daly, M.J. (2008) Autism Consortium. Association between microdeletion and microduplication at 16p11.2 and autism. *N Engl J Med* **358**, 667–75.

67. McCarthy, S.E., Makarov, V., Kirov, G., Addington, A.M., McClellan, J., Yoon, S., Perkins, D.O., Dickel, D.E., Kusenda, M., Krastoshevsky, O., Krause, V., Kumar, R.A., Grozeva, D., Malhotra, D., Walsh, T., Zackai, E.H., Kaplan, P., Ganesh, J., Krantz, I.D., Spinner, N.B., Roccanova, P., Bhandari, A., Pavon, K., Lakshmi, B., Leotta, A., Kendall, J., Lee, Y.H., Vacic, V., Gary, S., Iakoucheva, L.M., Crow, T.J., Christian, S.L., Lieberman, J.A., Stroup, T.S., Lehtimäki, T., Puura, K., Haldeman-Englert, C., Pearl, J., Goodell, M., Willour, V.L., Derosse, P., Steele, J., Kassem, L., Wolff, J., Chitkara, N., McMahon, F.J., Malhotra, A.K., Potash, J.B., Schulze, T.G., Nöthen, M.M., Cichon, S., Rietschel, M., Leibenluft, E., Kustanovich, V., Lajonchere, C.M., Sutcliffe, J.S., Skuse, D., Gill, M., Gallagher, L., Mendell, N.R., Wellcome Trust Case Control Consortium., Craddock, N., Owen, M.J., O'Donovan, M.C., Shaikh, T.H., Susser, E., Delisi, L.E., Sullivan, P.F., Deutsch, C.K., Rapoport, J., Levy, D.L., King, M.C., Sebat, J. (2009) Microduplications of 16p11.2 are associated with schizophrenia. *Nat Genet* **41**, 1223–7.

68. Shinawi, M., Liu, P., Kang, S.H., Shen, J., Belmont, J.W., Scott, D.A., Probst, F.J., Craigen, W.J., Graham, B., Pursley, A., Clark, G., Lee, J., Proud, M., Stocco, A., Rodriguez, D., Kozel, B., Sparagana, S., Roeder, E., McGrew, S., Kurczynski, T., Allison, L., Amato, S., Savage, S., Patel, A., Stankiewicz, P., Beaudet, A., Cheung, S.W., Lupski, J.R. (2010) Recurrent reciprocal 16p11.2 rearrangements associated with global developmental delay, behavioral problems, dysmorphism, epilepsy, and abnormal head size. *J Med Genet* **47**, 332–41.

69. Crespi, B., Summers, K., Dorus, S. (2009) Genomic sister-disorders of neurodevelopment: an evolutionary approach. *Evolutionary Applications* **2**, 81–100.

70. Crespi, B., Stead, P., Elliot, M. (2010) Evolution in health and medicine sackler colloquium: comparative genomics of autism and schizophrenia. *Proc Natl Acad Sci USA* **107**, Suppl: 1736–41.

71. Rovelet-Lecrux, A., Hannequin, D., Raux, G., Le Meur, N., Laquerrière, A., Vital, A., Dumanchin, C., Feuillette, S., Brice, A., Vercelletto, M., Dubas, F., Frebourg, T., Campion, D. (2006) *APP* locus duplication causes autosomal dominant early-onset Alzheimer disease with cerebral amyloid angiopathy. *Nat Genet* **38**, 24–6.

72. Singleton, A.B., Farrer, M., Johnson, J., Singleton, A., Hague, S., Kachergus, J., Hulihan, M., Peuralinna, T., Dutra, A., Nussbaum, R., Lincoln, S., Crawley, A., Hanson, M., Maraganore, D., Adler, C., Cookson, M.R., Muenter, M., Baptista, M., Miller, D., Blancato, J., Hardy, J., Gwinn-Hardy, K. (2003) Alpha-synuclein locus triplication causes Parkinson's disease. *Science* **302**, 841.

73. Chartier-Harlin, M.C., Kachergus, J., Roumier, C., Mouroux, V., Douay, X., Lincoln, S., Levecque, C., Larvor, L., Andrieux, J., Hulihan, M., Waucquier, N., Defebvre, L., Amouyel, P., Farrer, M., Destée, A. (2004) Alpha-synuclein locus duplication as a cause of familial Parkinson's disease. *Lancet* **364**, 1167– 9.

74. Ibáñez, P., Bonnet, A.M., Debarges, B., Lohmann, E., Tison, F., Pollak, P., Agid, Y., Dürr, A., Brice, A. (2004) Causal relation between alpha-synuclein gene duplication and familial Parkinson's disease. *Lancet* **364**, 1169–71.

75. Le Maréchal, C., Masson, E., Chen, J.M., Morel, F., Ruszniewski, P., Levy, P., Férec, C. (2006) Hereditary pancreatitis caused by

triplication of the trypsinogen locus. *Nat Genet* **38**, 1372–4.

76. Fellermann, K., Stange, D.E., Schaeffeler, E., Schmalzl, H., Wehkamp, J., Bevins, C.L., Reinisch, W., Teml, A., Schwab, M., Lichter, P., Radlwimmer, B., Stange, E.F. (2006) A chromosome 8 gene-cluster polymorphism with low human beta-defensin 2 gene copy-number predisposes to Crohn disease of the colon. *Am J Hum Genet* **79**, 439–48.

77. McCarroll, S.A., Huett, A., Kuballa, P., Chilewski, S.D., Landry, A., Goyette, P., Zody, M.C., Hall, J.L., Brant, S.R., Cho, J.H., Duerr, R.H., Silverberg, M.S., Taylor, K.D., Rioux, J.D., Altshuler, D., Daly, M.J., Xavier, R.J. (2008) Deletion polymorphism upstream of *IRGM* associated with altered *IRGM* expression and Crohn's disease. *Nat Genet* **40**, 1107–12.

78. Aitman, T.J., Dong, R., Vyse, T.J., Norsworthy, P.J., Johnson, M.D., Smith, J., Mangion, J., Roberton-Lowe, C., Marshall, A.J., Petretto, E., Hodges, M.D., Bhangal, G., Patel, S.G., Sheehan-Rooney, K., Duda, M., Cook, P.R., Evans, D.J., Domin, J., Flint, J., Boyle, J.J., Pusey, C.D., Cook, H.T. (2006) Copy-number polymorphism in *Fcgr3* predisposes to glomerulonephritis in rats and humans. *Nature* **439**, 851–5.

79. Hollox, E.J., Huffmeier, U., Zeeuwen, P.L., Palla, R., Lascorz, J., Rodijk-Olthuis, D., van de Kerkhof, P.C., Traupe, H., de Jongh, G., den Heijer, M., Reis, A., Armour, J.A., Schalkwijk, J. (2008) Psoriasis is associated with increased beta-defensin genomic copy-number. *Nat Genet* **40**, 23–5.

80. Willcocks, L.C., Lyons, P.A., Clatworthy, M.R., Robinson, J.I., Yang, W., Newland, S.A., Plagnol, V., McGovern, N.N., Condliffe, A.M., Chilvers, E.R., Adu, D., Jolly, E.C., Watts, R., Lau, Y.L., Morgan, A.W., Nash, G., Smith, K.G. (2008) Copy-number of *FCGR3B*, which is associated with systemic lupus erythematosus, correlates with protein expression and immune complex uptake. *J Exp Med* **205**, 1573–82.

81. Lupski, J.R., Chance, P.F., Garcia, C.A. (1993) Inherited primary peripheral neuropathies. Molecular genetics and clinical implications of CMT1A and HNPP. *JAMA* **270**, 2326–30.

82. Lupski, J.R., Stankiewicz, P. (2005) Genomic disorder mechanisms elucidated by breakpoint analysis of 17p rearrangements. *PLoS Genet* **1**, e49.

83. Rujescu, D., Ingason, A., Cichon, S., Pietiläinen, O.P., Barnes, M.R., Toulopoulou, T., Picchioni, M., Vassos, E., Ettinger, U., Bramon, E., Murray, R., Ruggeri, M., Tosato, S., Bonetto, C., Steinberg, S., Sigurdsson, E., Sigmundsson, T., Petursson, H., Gylfason, A., Olason, P.I., Hardarsson, G., Jonsdottir, G.A., Gustafsson, O., Fossdal, R., Giegling, I., Möller, H.J., Hartmann, A.M., Hoffmann, P., Crombie, C., Fraser, G., Walker, N., Lonnqvist, J., Suvisaari, J., Tuulio-Henriksson, A., Djurovic, S., Melle, I., Andreassen, O.A., Hansen, T., Werge, T., Kiemeney, L.A., Franke, B., Veltman, J., Buizer-Voskamp, J.E., GROUP Investigators., Sabatti, C., Ophoff, R.A., Rietschel, M., Nöthen, M.M., Stefansson, K., Peltonen, L., St Clair, D., Stefansson, H., Collier, D.A. (2009) Disruption of the *neurexin 1* gene is associated with schizophrenia. *Hum Mol Genet* **18**, 988–96.

84. Lifton, R.P., Dluhy, R.G., Powers, M., Rich, G.M., Cook, S., Ulick, S., Lalouel, J.-M. (1992) A chimaeric 11-beta-hydroxylase/aldosterone synthase gene causes glucocorticoid-remediable aldosteronism and human hypertension. *Nature* **355**, 262–5.

85. Lettice, L.A., Horikoshi, T., Heaney, S.J.H., van Baren, M.J., van der Linde, H.C., Breedveld, G.J., Joosse, M., Akarsu, N., Oostra, B.A., Endo, N., Shibata, M., Suzuki, M., Takahashi, E., Shinka, T., Nakahori, Y., Ayusawa, D., Nakabayashi, K., Scherer, S.W., Heutink, P., Hill, R.E., Noji, S. (2002) Disruption of a long-range cis-acting regulator for Shh causes preaxial polydactyly. *Proc Natl Acad Sci USA* **99**, 7548–53.

86. Velagaleti, G.V., Bien-Willner, G.A., Northup, J.K., Lockhart, L.H., Hawkins, J.C., Jalal, S.M., Withers, M., Lupski, J.R., Stankiewicz, P. (2005) Position effects due to chromosome breakpoints that map approximately 900 Kb upstream and approximately 1.3 Mb downstream of *SOX9* in two patients with campomelic dysplasia. *Am J Hum Genet* **76**, 652–62.

87. Pop, R., Conz, C., Lindenberg, K.S., Blesson, S., Schmalenberger, B., Briault, S., Pfeifer, D., Scherer, G. (2004) Screening of the 1 Mb *SOX9* 5' control region by arrayCGH identifies a large deletion in a case of campomelic dysplasia with XY sex reversal. *J Med Genet* **41**, e47.

88. Beysen, D., Raes, J., Leroy, B.P., Lucassen, A., Yates, J.R., Clayton-Smith, J., Ilyina, H., Brooks, S.S., Christin-Maitre, S., Fellous, M., Fryns, J.P., Kim, J.R., Lapunzina, P., Lemyre, E., Meire, F., Messiaen, L.M., Oley, C., Splitt, M., Thomson, J., Van de Peer, Y., Veitia, R.A., De Paepe, A., De Baere, E. (2005) Deletions involving long-range conserved nongenic sequences upstream and downstream of *FOXL2* as a novel disease-causing mechanism in blepharophimosis syndrome. *Am J Hum Genet* **77**, 205–18.

89. Kleinjan, D.A., van Heyningen, V. (2005) Long-Range Control of Gene Expression. Emerging Mechanisms and Disruption in Disease. *Am J Hum Genet* 76, 8–32.

90. Klopocki, E., Ott, C. E., Benatar, N., Ullmann, R., Mundlos, S., Lehmann, L. (2008) A microduplication of the long range SHH limb regulator (ZRS) is associated with triphalangeal thumb-polysyndactyly syndrome. *J Med Genet* **45**, 370–5.

91. Dathe, K., Kjaer, K.W., Brehm, A., Meinecke, P., Nürnberg, P., Neto, J.C., Brunoni, D., Tommerup, N., Ott, C.E., Klopocki, E., Seemann, P., Mundlos, S. (2009) Duplications involving a conserved regulatory element downstream of *BMP2* are associated with brachydactyly type A2. *Am J Hum Genet* **84**, 483–92.

92. Kurth, I., Klopocki, E., Stricker, S., van Oosterwijk, J., Vanek, S., Altmann, J., Santos, H.G., van Harssel, J.J., de Ravel, T., Wilkie, A.O., Gal, A., Mundlos, S. (2009) Duplications of noncoding elements 5' of *SOX9* are associated with brachydactyly-anonychia. *Nat Genet* **41**, 862–3.

93. Kurotaki, N,, Shen, J.J., Touyama, M., Kondoh, T., Visser, R., Ozaki, T., Nishimoto, J., Shiihara, T., Uetake, K., Makita, Y., Harada, N., Raskin, S., Brown, C.W., Höglund, P., Okamoto, N., Lupski, J.R. (2005) Phenotypic consequences of genetic variation at hemizygous alleles: Sotos syndrome is a contiguous gene syndrome incorporating coagulation factor twelve (*FXII*) deficiency. *Genet Med* 7, 479–83.

94. Trask, B., Pinkel, D., Van den Engh, G.J. (1989) The proximity of DNA sequences in interphase cell nuclei is correlated to genomic distance and permits ordering of cosmids spanning 250 kilobase pairs. Genomics 5, 710–7.

95. Van den Engh, G., Van den Sachs, R., Trask, B.J. (1992) Estimating genomic distance from DNA sequence location in cell nuclei by a random walk model. *Science* **257**, 1410–2.

96. Dermitzakis, E.T., Reymond, A., Scamuffa, N., Ucla, C., Kirkness, E., Rossier, C., Antonarakis, S.E. (2003) Evolutionary discrimination of mammalian conserved non-genic sequences (CNGs). *Science* **302**, 1033–5.

97. Dermitzakis, E.T., Reymond, A., Antonarakis, S.E. (2005) Conserved non-genic sequences - an unexpected feature of mammalian genomes. *Nat Rev Genet* **6**, 151–7.

98. Tanabe, H. Muller, S., Neusser, M., von Hase, J., Calcagno, E., Cremer, M., Solovei, I., Cremer, C., Cremer, T. (2002) Evolutionary conservation of chromosome territory arrangements in cell nuclei from higher primates. *Proc Natl Acad Sci USA* **99**: 4424–9.

99. Glazko, G.V., Koonin, E.V., Rogozin, I.B., Shabalina, S.A. (2003). A significant fraction of conserved noncoding DNA in human and mouse consists of predicted matrix attachment regions. *Trends Genet* **19**, 119–24.

100. Kioussis, D., Vanin, E., deLange, T., Flavell, R.A., Grosveld, F.G. (1983) β-*globin* gene inactivation by DNA traslocation un γ-β-thalassaemia. *Nature* **306**, 662–6.

101. Driscoll, M.C., Dobkin, C.S., Alter, B.P. (1989). γ-δ-β-Thalassaemia due to a de novo mutation deleting the 5' β-*globin* gene activation-region hypersensitive sites. *Proc Natl Acad Sci USA* **86**, 7470–4.

102. de Kok, YJ., van der Maarel, S.M., Bitner-Glindzicz, M., Huber, I., Monaco, A.P., Malcolm, S., Pembrey, M.E., Ropers, H.H., Cremers, F.P. (1995). Association between X-linked mixed deafness and mutations in the POU domain gene *POU3F4*. *Science* **267**, 685–8.

103. de Kok, Y.J., Vossenaar, E.R., Cremers, C.W., Dahl, N., Laporte, J., Hu, L.J., Lacombe, D., Fischel-Ghodsian, N., Friedman, R.A., Parnes, L.S., Thorpe, P., Bitner-Glindzicz, M., Pander, H.J., Heilbronner, H., Graveline, J., den Dunnen, J.T., Brunner, H.G., Ropers, H.H., Cremers, F.P. (1996). Identification of a hot spot for microdeletions in patients with X-linked deafness type 3 (DFN3) 900 kb proximal to the DFN3 gene *POU3F4*. *Hum Mol Genet* **5**, 1229–35.

104. Lettice, L.A., Heaney, S.J., Purdie, L.A., Li, L., de Beer, P., Oostra, B.A., Goode, D., Elgar, G., Hill, R.E., de Graaff, E. (2003) A long-range Shh enhancer regulates expression in the developing limb and fin and is associated with preaxial polydactyly. *Hum Mol Genet* **12**, 1725–35.

105. Garzon, R., Calin, G.A., Croce, C.M. (2009) MicroRNAs in Cancer. *Annu Rev Med* **60**, 167–79.

106. Mardis, E.R. (2008) Next-generation DNA sequencing methods. *Annu Rev Genomics Hum Genet* **9**, 387–402.

107. Levy, S., Sutton, G., Ng, P.C., Feuk, L., Halpern, A.L., Walenz, B.P., Axelrod, N., Huang, J., Kirkness, E.F., Denisov, G., Lin, Y., MacDonald, J.R., Pang, A.W., Shago, M., Stockwell, T.B., Tsiamouri, A., Bafna, V., Bansal, V., Kravitz, S.A., Busam, D.A., Beeson, K.Y., McIntosh, T.C., Remington, K.A., Abril, J.F., Gill, J., Borman, J., Rogers, Y.H., Frazier, M.E., Scherer, S.W., Strausberg, R.L., Venter, J.C. (2007) The diploid genome sequence of an individual human. *PLoS Biol* **5**, e254.

108. Wheeler, D.A., Srinivasan, M., Egholm, M., Shen, Y., Chen, L., McGuire, A., He, W., Chen, Y.J., Makhijani, V., Roth, G.T., Gomes, X., Tartaro, K., Niazi, F., Turcotte, C.L., Irzyk, G.P., Lupski, J.R., Chinault, C., Song, X.Z., Liu, Y., Yuan, Y., Nazareth, L., Qin, X., Muzny, D.M., Margulies, M., Weinstock, G.M., Gibbs. R.A., Rothberg, J.M. (2008) The complete genome of an individual by massively parallel DNA sequencing. *Nature* **452**, 872–6.

109. Tarpey, P.S., Smith, R., Pleasance, E., Whibley, A., Edkins, S., Hardy, C., O'Meara, S., Latimer, C., Dicks, E., Menzies, A., Stephens, P., Blow, M., Greenman, C., Xue, Y., Tyler-Smith, C., Thompson, D., Gray, K., Andrews, J., Barthorpe, S., Buck, G., Cole, J., Dunmore, R., Jones, D., Maddison, M., Mironenko, T., Turner, R., Turrell, K., Varian, J., West, S., Widaa, S., Wray, P., Teague, J., Butler, A., Jenkinson, A., Jia, M., Richardson, D., Shepherd, R., Wooster, R., Tejada, M.I., Martinez, F., Carvill, G., Goliath, R., de Brouwer, A.P., van Bokhoven, H., Van Esch, H., Chelly, J., Raynaud, M., Ropers, H.H., Abidi, F.E., Srivastava, A.K., Cox, J., Luo, Y., Mallya, U., Moon, J., Parnau, J., Mohammed, S., Tolmie, J.L., Shoubridge, C., Corbett, M., Gardner, A., Haan, E., Rujirabanjerd, S., Shaw, M., Vandeleur, L., Fullston, T., Easton, D.F., Boyle, J., Partington, M., Hackett, A., Field, M., Skinner, C., Stevenson, R.E., Bobrow, M., Turner, G., Schwartz, C.E., Gecz, J., Raymond, F.L., Futreal, P.A., Stratton, M.R. (2009) A systematic, large-scale resequencing screen of X-chromosome coding exons in mental retardation. *Nat Genet* **41**, 535–43.

110. Ng, S.B., Turner, E.H., Robertson, P.D., Flygare, S.D., Bigham, A.W., Lee, C., Shaffer, T., Wong, M., Bhattacharjee, A., Eichler, E.E., Bamshad, M., Nickerson, D.A., Shendure, J. (2009) Targeted capture and massively parallel sequencing of 12 human exomes. *Nature* **461**, 272–6.

111. Lupski, J.R., Reid, J.G., Gonzaga-Jauregui, C., Deiros, D.R., Chen, D.C.Y., Nazareth, L., Bainbridge, M., Dinh, H., Jing, C., Wheeler, D.A., McGuire, A.L., Zhang, F., Stankiewicz, P., Halperin, J.J., Yang, C., Gehman, C., Guo, D., Irikat, R.K., Tom, W., Fantin, N.J., Muzny, D.M., Gibbs, R.A. (2010) Whole-genome sequencing in a patient with Charcot-Marie Tooth neuropathy. *N Engl J Med* **362**, 1181–91.

112. Korbel, J.O., Urban, A.E., Affourtit, J.P., Godwin, B., Grubert, F., Simons, J.F., Kim, P.M., Palejev, D., Carriero, N.J., Du, L., Taillon, B.E., Chen, Z., Tanzer, A., Saunders, A.C., Chi, J., Yang, F., Carter, N.P., Hurles, M.E., Weissman, S.M., Harkins, T.T., Gerstein, M.B., Egholm, M., Snyder, M. (2007) Paired-end mapping reveals extensive structural variation in the human genome. *Science* 318, 420–6.

113. Kidd, J.M., Cooper, G.M., Donahue, W.F., Hayden, H.S., Sampas, N., Graves, T., Hansen, N., Teague, B., Alkan, C., Antonacci, F., Haugen, E., Zerr, T., Yamada, N.A., Tsang, P., Newman, T.L., Tüzün, E., Cheng, Z., Ebling, H.M., Tusneem, N., David, R., Gillett, W., Phelps, K.A., Weaver, M., Saranga, D., Brand, A., Tao, W., Gustafson, E., McKernan, K., Chen, L., Malig, M., Smith, J.D., Korn, J.M., McCarroll, S.A., Altshuler, D.A., Peiffer, D.A., Dorschner, M., Stamatoyannopoulos, J., Schwartz, D., Nickerson, D.A., Mullikin, J.C., Wilson, R.K., Bruhn, L., Olson, M.V., Kaul, R., Smith, D.R., Eichler, E.E. (2008) Mapping and sequencing of structural variation from eight human genomes. *Nature* **453**, 56–64.

114. Li, R., Li, Y., Zheng, H., Luo, R., Zhu, H., Li, Q., Qian, W., Ren, Y., Tian, G., Li, J., Zhou, G., Zhu, X., Wu, H., Qin, J., Jin, X., Li, D., Cao, H., Hu, X., Blanche, H., Cann, H., Zhang, X., Li, S., Bolund, L., Kristiansen, K., Yang, H., Wang, J., Wang, J. (2010) Building the sequence map of the human pan-genome. *Nat Biotechnol* **28**, 57–63.

115. Lupski, J.R. (2007) Structural variation in the human genome. *N Engl J Med* **356**, 1169–7.

116. Hurles, M.E., Willey, D., Matthews, L., Hussain, S.S. (2004) Origins of chromosomal rearrangement hotspots in the human genome: evidence from the AZFa deletion hotspots. *Genome Biol* **5**, R55.

117. Lupski, J.R., Montes de Oca-Luna, R., Slaugenhaupt, S., Pentao, L., Guzzetta, V., Trask, B. J., Saucedo-Cardenas, O., Barker, D. F., Killian, J. M., Garcia, C. A., Chakravarti, A., Patel, P. I. (1991) DNA duplication associated with Charcot-Marie-Tooth disease type 1A. Cell **66**, 219–32.

118. Chance, P.F., Alderson, M.K., Leppig, K.A., Lensch, M.W., Matsunami, N., Smith, B., Swanson, P.D., Odelberg, S.J., Disteche, C.M., Bird, T.D. (1993) DNA deletion associated with hereditary neuropathy with liability to pressure palsies. *Cell* **72**, 143–51.

119. McDermid, H.E., Morrow, B.E. (2002) Genomic disorders on 22q11. *Am J Hum Genet* **70**, 1077–88.

120. Yagi, H., Furutani, Y., Hamada, H., Sasaki, T., Asakawa, S., Minoshima, S., Ichida, F., Joo, K., Kimura, M., Imamura, S., Kamatani, N., Momma, K., Takao, A., Nakazawa, M.,

Shimizu, N., Matsuoka, R. (2003) Role of TBX1 in human del22q11.2 syndrome. *Lancet* **362**, 1366–73.

121. Ensenauer, R.E., Adeyinka, A., Flynn, H.C., Michels, V.V., Lindor, N.M., Dawson, D.B., Thorland, E.C., Lorentz, C.P., Goldstein, J.L., McDonald, M.T., Smith, W.E., Simon-Fayard, E., Alexander, A.A., Kulharya, A.S., Ketterling, R.P., Clark, R.D., Jalal, S.M. (2003) Microduplication 22q11.2, an emerging syndrome: clinical, cytogenetic, and molecular analysis of thirteen patients. *Am J Hum Genet* **73**, 1027–40.
122. Riva, P., Corrado, L., Natacci, F., Castorina, P., Wu, B.-L., Schneider, G. H., Clementi, M., Tenconi, R., Korf, B. R., Larizza, L. (2000) NF1 microdeletion syndrome: refined FISH characterization of sporadic and familial deletions with locus-specific probes. *Am J Hum Genet* **66**, 100–9.
123. Chai, J.H., Locke, D.P., Greally, J.M., Knoll, J.H., Ohta, T., Dunai, J., Yavor, A., Eichler, E.E. (2003) Nicholls RD Identification of four highly conserved genes between breakpoint hotspots BP1 and BP2 of the Prader-Willi/Angelman syndromes deletion region that have undergone evolutionary transposition mediated by flanking duplicons. *Am J Hum Genet* **73**, 898–925.
124. Sahoo T, del Gaudio D, German JR, Shinawi M, Peters SU, Person RE, Garnica A, Cheung SW, Beaudet AL. (2008) Prader-Willi phenotype caused by paternal deficiency for the HBII-85 C/D box small nucleolar RNA cluster. *Nat Genet* **40**, 719–21.
125. Potocki, L., Shaw, C.J., Stankiewicz, P., Lupski, J.R. (2003) Variability in clinical phenotype despite common chromosomal deletion in Smith-Magenis syndrome (del(17)(p11.2p11.2)). *Genet Med* **5**, 430–4.
126. Potocki, L., Bi, W., Treadwell-Deering, D., Carvalho, C.M., Eifert, A., Friedman, E.M., Glaze, D., Krull, K., Lee, J.A., Lewis, R.A., Mendoza-Londono, R., Robbins-Furman, P., Shaw, C., Shi, X., Weissenberger, G., Withers, M., Yatsenko, S.A., Zackai, E.H., Stankiewicz, P., Lupski, J.R. (2007) Characterization of Potocki-Lupski syndrome (dup(17)(p11.2p11.2)) and delineation of a dosage-sensitive critical interval that can convey an autism phenotype. *Am J Hum Genet* **80**, 633–49.
127. Franco, L.M., de Ravel, T., Graham, B.H., Frenkel, S.M., Van Driessche, J., Stankiewicz, P., Lupski, J.R., Vermeesch, J.R., Cheung, S.W. (2010) A syndrome of short stature, microcephaly and speech delay is associated with duplications reciprocal to the common Sotos syndrome deletion. Eur J Hum Genet 18, 258–61.
128. Hernández-Martín, A., González-Sarmiento, R., De Unamuno, P. (1999) X-linked ichthyosis: an update. *Br J Dermatol* **141**, 617–27.
129. Tassabehji, M. (2003) Williams-Beuren syndrome: a challenge for genotype-phenotype correlations. *Hum Mol Genet* **2**, R229–37.
130. Somerville, M.J., Mervis, C.B., Young, E.J., Seo, E.J., del Campo, M., Bamforth, S., Peregrine, E., Loo, W., Lilley, M., Pérez-Jurado, L.A., Morris, C.A., Scherer, S.W., Osborne, L.R. (2005) Severe expressive-language delay related to duplication of the Williams-Beuren locus. *N Engl J Med* **353**, 1694–701.
131. Klopocki, E., Schulze, H., Strauss, G., Ott, C.E., Hall, J., Trotier, F., Fleischhauer, S., Greenhalgh, L., Newbury-Ecob, R.A., Neumann, L.M., Habenicht, R., König, R., Seemanova, E., Megarbane, A., Ropers, H.H., Ullmann, R., Horn, D., Mundlos, S. (2007) Complex inheritance pattern resembling autosomal recessive inheritance involving a microdeletion in thrombocytopenia-absent radius syndrome. *Am J Hum Genet* **80**, 232–40.
132. Willatt, L., Cox, J., Barber, J., Cabanas, E.D., Collins, A., Donnai, D., Fitzpatrick, D.R., Maher, E., Martin, H., Parnau, J., Pindar, L., Ramsay, J., Shaw-Smith, C., Sistermans, E.A., Tettenborn, M., Trump, D., de Vries, B.B., Walker, K., Raymond, F.L. (2005) 3q29 microdeletion syndrome: clinical and molecular characterization of a new syndrome. *Am J Hum Genet* 77, 154–60.
133. Ballif, B.C., Theisen, A., Coppinger, J., Gowans, G.C., Hersh, J.H., Madan-Khetarpal, S., Schmidt, K.R., Tervo, R., Escobar, L.F., Friedrich, C.A., McDonald, M., Campbell, L., Ming, J.E., Zackai, E.H., Bejjani, B.A., Shaffer, L.G. (2008) Expanding the clinical phenotype of the 3q29 microdeletion syndrome and characterization of the reciprocal microduplication. *Mol Cytogenet* **1**, 8.
134. Devriendt, K., Matthijs, G., Van Dael, R., Gewillig, M., Eyskens, B., Hjalgrim, H., Dolmer, B., McGaughran, J., Bröndum-Nielsen, K., Marynen, P., Fryns, J.P., Vermeesch, J.R. (1999) Delineation of the critical deletion region for congenital heart defects, on chromosome 8p23.1. *Am J Hum Genet* **64**, 1119–26.
135. Barber, J.C., Maloney, V.K., Huang, S., Bunyan, D.J., Cresswell, L., Kinning, E., Benson, A., Cheetham, T., Wyllie, J., Lynch, S.A., Zwolinski, S., Prescott, L., Crow, Y., Morgan, R., Hobson, E. (2008) 8p23.1 dupli-

cation syndrome: A novel genomic condition with unexpected complexity revealed by array CGH. *Eur J Hum Genet* **16**, 18–27.

136. Balciuniene, J., Feng, N., Iyadurai, K., Hirsch, B., Charnas, L., Bill, B.R., Easterday, M.C., Staaf, J., Oseth, L., Czapansky-Beilman, D., Avramopoulos, D., Thomas, G.H., Borg, A., Valle, D., Schimmenti, L.A., Selleck, S.B. (2007) Recurrent 10q22-q23 deletions: a genomic disorder on 10q associated with cognitive and behavioral abnormalities. *Am J Hum Genet* **80**, 938–47.
137. Cook, E.H. Jr., Lindgren, V., Leventhal, B.L., Courchesne, R., Lincoln, A., Shulman, C., Lord, C., Courchesne, E. (1997) Autism or atypical autism in maternally but not paternally derived proximal 15q duplication. *Am J Hum Genet* **60**, 928–34.
138. Itsara, A., Cooper, G. M., Baker, C., Girirajan, S., Li, J., Absher, D., Krauss, R.M., Myers, R.M., Ridker, P.M., Chasman, D. I., Mefford, H., Ying, P., Nickerson, D.A.; Eichler, E.E. (2009) Population analysis of large copy number variants and hotspots of human genetic disease. *Am J Hum Genet* **84**, 148–61.
139. Sharp, A.J., Selzer, R.R., Veltman, J.A., Gimelli, S., Gimelli, G., Striano, P., Coppola, A., Regan, R., Price, S.M., Knoers, N.V., Eis, P.S., Brunner, H.G., Hennekam, R.C., Knight, S.J., de Vries, B.B., Zuffardi, O., Eichler, E.E. (2007) Characterization of a recurrent 15q24 microdeletion syndrome. *Hum Mol Genet* **16**, 67–72.
140. El-Hattab, A.W., Smolarek, T.A., Walker, M.E., Schorry, E.K., Immken, L.L., Patel, G., Abbott, M.A., Lanpher, B.C., Ou, Z., Kang, S.H., Patel, A., Scaglia, F., Lupski, J.R., Cheung, S.W., Stankiewicz, P. (2009) Redefined genomic architecture in 15q24 directed by patient deletion/duplication breakpoint mapping. *Hum Genet* **126**, 589–602.
141. Wat, M.J., Enciso, V.B., Wiszniewski, W., Resnick, T., Bader, P., Roeder, E.R., Freedenberg, D., Brown, C., Stankiewicz, P., Cheung, S.W., Scott, D.A. (2010) Recurrent microdeletions of 15q25.2 are associated with increased risk of congenital diaphragmatic hernia, cognitive deficits, and possibly Diamond-Blackfan anemia. *J Med Genet* **47**, 777–81.
142. Walters, R.G., Jacquemont, S., Valsesia, A., de Smith, A.J., Martinet, D., Andersson, J., Falchi, M., Chen, F., Andrieux, J., Lobbens, S., Delobel, B., Stutzmann, F., El-Sayed Moustafa, J.S., Chèvre, J.C., Lecoeur, C., Vatin, V., Bouquillon, S., Buxton, J.L., Boute, O., Holder-Espinasse, M., Cuisset, J.M., Lemaitre, M.P., Ambresin, A.E., Brioschi, A., Gaillard, M., Giusti, V., Fellmann, F., Ferrarini, A., Hadjikhani, N., Campion, D., Guilmatre, A., Goldenberg, A., Calmels, N., Mandel, J.L., Le Caignec, C., David, A., Isidor, B., Cordier, M.P., Dupuis-Girod, S., Labalme, A., Sanlaville, D., Béri-Dexheimer, M., Jonveaux, P., Leheup, B., Ounap, K., Bochukova, E.G., Henning, E., Keogh, J., Ellis, R.J., Macdermot, K.D., van Haelst, M.M., Vincent-Delorme, C., Plessis, G., Touraine, R., Philippe, A., Malan, V., Mathieu-Dramard, M., Chiesa, J., Blaumeiser, B., Kooy, R.F., Caiazzo, R., Pigeyre, M., Balkau, B., Sladek, R., Bergmann, S., Mooser, V., Waterworth, D., Reymond, A., Vollenweider, P., Waeber, G., Kurg, A., Palta, P., Esko, T., Metspalu, A., Nelis, M., Elliott, P., Hartikainen, A.L., McCarthy, M.I., Peltonen, L., Carlsson, L., Jacobson, P., Sjöström, L., Huang, N., Hurles, M.E., O'Rahilly, S., Farooqi, I.S., Männik, K., Jarvelin, M.R., Pattou, F., Meyre, D., Walley, A.J., Coin, L.J., Blakemore, A.I., Froguel, P., Beckmann, J.S. (2010) A new highly penetrant form of obesity due to deletions on chromosome 16p11.2. *Nature* **463**, 671–5.
143. Ballif, B.C., Hornor, S.A., Jenkins, E., Madan-Khetarpal, S., Surti, U., Jackson, K.E., Asamoah, A., Brock, P.L., Gowans, G.C., Conway, R.L., Graham, J.M., Medne, L., Zackai, E.H., Shaikh, T.H., Geoghegan, J., Selzer, R.R., Eis, P.S., Bejjani, B.A., Shaffer, L.G. (2007) Discovery of a previously unrecognized microdeletion syndrome of 16p11.2-p12.2. *Nat Genet* **39**, 1071–3.
144. Ullmann, R., Turner, G., Kirchhoff, M., Chen, W., Tonge, B., Rosenberg, C., Field, M., Vianna-Morgante, A.M., Christie, L., Krepischi-Santos, A.C., Banna, L., Brereton, A.V., Hill, A., Bisgaard, A.M., Müller, I., Hultschig, C., Erdogan, F., Wieczorek, G., Ropers, H.H. (2007) Array CGH identifies reciprocal 16p13.1 duplications and deletions that predispose to autism and/or mental retardation. *Hum Mutat* **28**, 674–82.
145. Hannes, F.D., Sharp, A.J., Mefford, H.C., de Ravel, T., Ruivenkamp, C.A., Breuning, M.H., Fryns, J.P., Devriendt, K., Van Buggenhout, G., Vogels, A., Stewart, H.H., Hennekam, R.C., Cooper, G.M., Regan, R., Knight, S.J., Eichler, E.E., Vermeesch, J.R. (2008) Recurrent reciprocal deletions and duplications of 16p13.11: The deletion is a risk factor for MR/MCA while the duplication may be a rare benign variant. *J Med Genet* **46**, 223–32.
146. Mefford, H.C., Cooper, G.M., Zerr, T., Smith, J.D., Baker, C., Shafer, N., Thorland, E.C., Skinner, C., Schwartz, C.E., Nickerson, D.A., Eichler, E.E. (2009) A method for

rapid, targeted CNV genotyping identifies rare variants associated with neurocognitive disease. *Genome Res* **19**, 1579–85.

147. Mefford, H.C., Clauin, S., Sharp, A.J., Moller, R.S., Ullmann, R., Kapur, R., Pinkel, D., Cooper, G.M., Ventura, M., Ropers, H.H., Tommerup, N., Eichler, E.E., Bellanne-Chantelot, C. (2007) Recurrent reciprocal genomic rearrangements of 17q12 are associated with renal disease, diabetes, and epilepsy. *Am J Hum Genet* **81**, 1057–69.
148. Sreenath Nagamani, S.C., Erez, A., Shen, J., Li, C., Roeder, E., Cox, S., Karaviti, L., Pearson, M., Kang, S.-H.L., Sahoo, T., Lalani, S.R., Stankiewicz, P., Sutton, V.R., Cheung, SW. (2010) Clinical spectrum associated with recurrent genomic rearrangements in chromosome 17q12. *Eur J Hum Genet* **18**, 278–84.
149. Koolen DA, Vissers LE, Pfundt R, de Leeuw N, Knight SJ, Regan R, Kooy RF, Reyniers E, Romano C, Fichera M, Schinzel A, Baumer A, Anderlid BM, Schoumans J, Knoers NV, van Kessel AG, Sistermans EA, Veltman JA, Brunner HG, de Vries BB. (2006) A new chromosome 17q21.31 microdeletion syndrome associated with a common inversion polymorphism. *Nat Genet* **38**, 999–1001.
150. Shaw-Smith, C., Pittman, A.M., Willatt, L., Martin, H., Rickman, L., Gribble, S., Curley, R., Cumming, S., Dunn, C., Kalaitzopoulos, D., Porter, K., Prigmore, E., Krepischi-Santos, A.C., Varela, M.C., Koiffmann, C.P., Lees, A.J., Rosenberg, C., Firth, H.V., de Silva, R., Carter, N.P. (2006) Microdeletion encompassing *MAPT* at chromosome 17q21.3 is associated with developmental delay and learning disability. *Nat Genet* **38**, 1032–7.
151. Grisart, B., Willatt, L., Destrée, A., Fryns, J.P., Rack, K., de Ravel, T., Rosenfeld, J., Vermeesch, J.R., Verellen-Dumoulin, C., Sandford, R. (2009) 17q21.31 microduplication patients are characterised by behavioural problems and poor social interaction. *J Med Genet* **46**, 524–30.
152. Ballif, B.C., Theisen, A., Rosenfeld, J.A., Traylor, R., Gastier-Foster, J., Thrush, L.D., Astbury, C., Bartholomew, D., McBride, K., Pyatt, R., Shane, K., Smith, W.E., Banks, V., Gallentine, W.B., Brock, P., Rudd, M.K., Adam, M.P., Keene, J.A., Phillips III, J.A., Pfotenhauer, J.P., Gowans, G.C., Stankiewicz, P., Bejjani, B.A., Shaffer, L.G. (2010) Identification of a recurrent microdeletion on 17q23.1q23.2 flanked by segmental duplications associated with heart defects and limb abnormalities. *Am J Hum Genet* **86**, 454–61.
153. Ben-Shachar, S., Ou, Z., Shaw, C.A., Belmont, J.W., Patel, M.S., Hummel, M., Amato, S., Tartaglia, N., Berg, J., Sutton, V.R., Lalani, S.R., Chinault, A.C., Cheung, S.W., Lupski, J.R., Patel, A. (2008) 22q11.2 distal deletion: a recurrent genomic disorder distinct from DiGeorge syndrome and velocardiofacial syndrome. *Am J Hum Genet* **82**, 214–21.
154. Phelan, M.C. (2008) Deletion 22q13.3 syndrome. *Orphanet J Rare Dis* **3**, 14.
155. Giorda, R., Bonaglia, M.C., Beri, S., Fichera, M., Novara, F., Magini, P., Urquhart, J., Sharkey, F.H., Zucca, C., Grasso, R., Marelli, S., Castiglia, L., Di Benedetto, D., Musumeci, S.A., Vitello, G.A., Failla, P., Reitano, S., Avola, E., Bisulli, F., Tinuper, P., Mastrangelo, M., Fiocchi, I., Spaccini, L., Torniero, C., Fontana, E., Lynch, S.A., Clayton-Smith, J., Black, G., Jonveaux, P., Leheup, B., Seri, M., Romano, C., dalla Bernardina, B., Zuffardi, O. (2009) Complex segmental duplications mediate a recurrent dup(X)(p11.22-p11.23) associated with mental retardation, speech delay, and EEG anomalies in males and females. *Am J Hum Genet* **85**, 394–400.
156. Shaffer, L.G., Theisen, A., Bejjani, B.A., Ballif, B.C., Aylsworth, A.S., Lim, C., McDonald, M., Ellison, J.W., Kostiner, D., Saitta, S., Shaikh, T. (2007) The discovery of microdeletion syndromes in the post-genomic era: review of the methodology and characterization of a new 1q41q42 microdeletion syndrome. *Genet Med* **9**, 607–16.
157. Boland, E., Clayton-Smith, J., Woo, V.G., McKee, S., Manson, F.D., Medne, L., Zackai, E., Swanson, E.A., Fitzpatrick, D., Millen, K.J., Sherr, E.H., Dobyns, W.B., Black, G.C. (2007) Mapping of deletion and translocation breakpoints in 1q44 implicates the serine/threonine kinase *AKT3* in postnatal microcephaly and agenesis of the corpus callosum. *Am J Hum Genet* **81**, 292–303.
158. van Bon BW, Koolen DA, Brueton L, McMullan D, Lichtenbelt KD, Adès LC, Peters G, Gibson K, Novara F, Pramparo T, Dalla Bernardina B, Zoccante L, Balottin U, Piazza F, Pecile V, Gasparini P, Guerci V, Kets M, Pfundt R, de Brouwer AP, Veltman JA, de Leeuw N, Wilson M, Antony J, Reitano S, Luciano D, Fichera M, Romano C, Brunner HG, Zuffardi O, de Vries BB. (2010) The 2q23.1 microdeletion syndrome: clinical and behavioural phenotype. *Eur J Hum Genet* **18**, 163–70.
159. Le Meur, N., Holder-Espinasse, M., Jaillard, S., Goldenberg, A., Joriot, S., Amati-Bonneau, P., Guichet, A., Barth, M., Charollais, A., Journel, H., Auvin, S., Boucher, C., Kerckaert, J.-P., David, V., Manouvrier-Hanu, S., Saugier-Veber, P.,

Frébourg, T., Dubourg, C., Andrieux, J., Bonneau, D. (2009) *MEF2C* haploinsufficiency caused either by microdeletion of the 5q14.3 region or mutation is responsible for severe mental retardation with stereotypic movements, epilepsy and/or cerebral malformations. *J Med Genet* **47**, 22–9.

160. Kleefstra, T., Brunner, H. G., Amiel, J., Oudakker, A. R., Nillesen, W. M., Magee, A., Genevieve, D., Cormier-Daire, V., van Esch, H., Fryns, J. P., Hamel, B. C. J., Sistermans, E. A., de Vries, B. B. A., van Bokhoven, H. (2006) Loss-of-function mutations in Euchromatin histone methyl transferase 1 (*EHMT1*) cause the 9q34 subtelomeric deletion syndrome. *Am J Hum Genet* **79**, 370–7.

161. Ariani, F., Hayek, G., Rondinella, D., Artuso, R., Mencarelli, M.A., Spanhol-Rosseto, A., Pollazzon, M., Buoni, S., Spiga, O., Ricciardi, S., Meloni, I., Longo, I., Mari, F., Broccoli, V., Zappella, M., Renieri, A. (2008) *FOXG1* is responsible for the congenital variant of Rett syndrome. *Am J Hum Genet* **83**, 89–93.

162. Durand, C. M., Betancur, C., Boeckers, T. M., Bockmann, J., Chaste, P., Fauchereau, F., Nygren, G., Rastam, M., Gillberg, I. C., Anckarsater, H., Sponheim, E., Goubran-Botros, H., Delorme, R., Chabane, N., Mouren-Simeoni, M.C., de Mas, P., Bieth, E., Rogé, B., Héron, D., Burglen, L., Gillberg, C., Leboyer, M., Bourgeron, T. (2007) Mutations in the gene encoding the synaptic scaffolding protein *SHANK3* are associated with autism spectrum disorders. *Nat Genet* **39**, 25–7.

163. Wang, J., Killinger, D. W., Hegele, R. A. (1999) A microdeletion within DAX-1 in X-linked adrenal hypoplasia congenita and hypogonadotrophic hypogonadism. *J Invest Med* **47**, 232–5.

164. Bardoni, B., Zanaria, E., Guioli, S., Floridia, G., Worley, K.C., Tonini, G., Ferrante, E., Chiumello, G., McCabe, E.R., Fraccaro, M., Zuffardi, O.,Camerino, G. (1994) A dosage sensitive locus at chromosome Xp21 is involved in male to female sex reversal. *Nat Genet* **7**, 497–501.

165. Li, L., Krantz, I. D., Deng, Y., Genin, A., Banta, A. B., Collins, C. C., Qi, M., Trask, B. J., Kuo, W. L., Cochran, J., Costa, T., Pierpont, M.E., Rand, E.B., Piccoli, D. A., Hood, L., Spinner, N. B. (1997) Alagille syndrome is caused by mutations in human Jagged1, which encodes a ligand for Notch1. *Nat Genet* **16**, 243–51.

166. Garg, V., Kathiriya, I. S., Barnes, R., Schluterman, M. K., King, I. N., Butler, C. A., Rothrock, C. R., Eapen, R. S., Hirayama-Yamada, K., Joo, K., Matsuoka, R., Cohen, J. C., Srivastava, D. (2003) *GATA4* mutations cause human congenital heart defects and reveal an interaction with TBX5. *Nature* **424**, 443–7.

167. Shimkets, R., Gailani, M. R., Siu, V. M., Yang-Feng, T., Pressman, C. L., Levanat, S., Goldstein, A., Dean, M., Bale, A. E. (1996) Molecular analysis of chromosome 9q deletions in two Gorlin syndrome patients. *Am J Hum Genet* **59**, 417–22.

168. Chang, E.H., Menezes, M., Meyer, N.C., Cucci, R.A., Vervoort, V.S., Schwartz, C.E., Smith, R.J. (2004) Branchio-oto-renal syndrome: the mutation spectrum in *EYA1* and its phenotypic consequences. *Hum Mutat* **23**, 582–9.

169. Hellemans, J., Preobrazhenska, O., Willaert, A., Debeer, P., Verdonk, P. C. M., Costa, T., Janssens, K., Menten, B., Van Roy, N., Vermeulen, S. J. T., Savarirayan, R., Van Hul, W., Vanhoenacker, F., Huylebroeck, D., De Paepe, A., Naeyaert, J.M., Vandesompele, J., Speleman, F., Verschueren, K., Coucke, P.J., Mortier, G.R. (2004) Loss-of-function mutations in *LEMD3* result in osteopoikilosis, Buschke-Ollendorff syndrome and melorheostosis. *Nat Genet* **36**, 1213–8.

170. Wagner, T., Wirth, J., Meyer, J., Zabel, B., Held, M., Zimmer, J., Pasantes, J., Dagna Bricarelli, F., Keutel, J., Hustert, E., Wolf, U., Tommerup, N., Schempp, W., Scherer, G. (1994) Autosomal sex reversal and campomelic dysplasia are caused by mutations in and around the SRY-related gene *SOX9*. *Cell* **79**, 1111–20.

171. McDermid, H.E., Duncan, A.M.V., Brasch, K.R., Holden, J.J.A., Magenis, E., Sheehy, R., Burn, J., Kardon, N., Noel, B., Schinzel, A., Teshima, I., White, B.N. (1986) Characterization of the supernumerary chromosome in cat eye syndrome. *Science* **232**, 646–8.

172. Mundlos, S., Otto, F., Mundlos, C., Mulliken, J.B., Aylsworth, A.S., Albright, S., Lindhout, D., Cole, W.G., Henn, W., Knoll, J.H.M., Owen, M.J., Mertelsmann, R., Zabel, B.U., Olsen, B.R. (1997) Mutations involving the transcription factor CBFA1 cause cleidocranial dysplasia. *Cell* **89**, 773–9.

173. Klaassens, M., van Dooren, M., Eussen, H.J., Douben, H., den Dekker, A.T., Lee, C., Donahoe, P.K., Galjaard, R.J., Goemaere, N., de Krijger, R.R., Wouters, C., Wauters, J., Oostra, B.A., Tibboel, D., de Klein, A. (2005) Congenital diaphragmatic hernia and chromosome 15q26: determination of a candidate region by use of fluorescent in situ hybridization and array-based comparative genomic hybridization. *Am J Hum Genet* **76**, 877–82.

174. Nelen, M.R., van Staveren, W.C.G., Peeters, E.A.J., Ben Hassel, M., Gorlin, R.J., Hamm, H., Lindboe, C.F., Fryns, J.-P., Sijmons, R.H., Woods, D.G., Mariman, E.C.M., Padberg, G.W., Kremer, H. (1997) Germline mutations in the PTEN/MMAC1 gene in patients with Cowden disease. *Hum Mol Genet* **6**, 1383–7.

175. Mainardi, P.C., Perfumo, C., Cali, A., Coucourde, G., Pastore, G., Cavani, S., Zara, F., Overhauser, J., Pierluigi, M., Bricarelli, F.D. (2001) Clinical and molecular characterisation of 80 patients with 5p deletion: genotype-phenotype correlation. *J Med Genet* **38**, 151–8.

176. Belloni E, Martucciello G, Verderio D, Ponti E, Seri M, Jasonni V, Torre M, Ferrari M, Tsui LC, Scherer SW. (2000) Involvement of the *HLXB9* homeobox gene in Currarino syndrome. *Am J Hum Genet* **66**, 312–9

177. Kohlhase, J., Heinrich, M., Schubert, L., Liebers, M., Kispert, A., Laccone, F., Turnpenny, P., Winter, R.M., Reardon, W. (2002) Okihiro syndrome is caused by *SALL4* mutations. *Hum Mol Genet* **11**, 2979–87.

178. den Dunnen, J.T., Grootscholten, P.M., Bakker, E., Blonden, L.A.J., Ginjaar, H.B., Wapenaar, M.C., van Paassen, H.M.B., van Broeckhoven, C., Pearson, P.L., van Ommen, G.J.B. (1989) Topography of the Duchenne muscular dystrophy (DMD) gene: FIGE and cDNA analysis of 194 cases reveals 115 deletions and 13 duplications. *Am J Hum Genet* **45**, 835–47.

179. Erez, A., Patel, A. J., Wang, X., Xia, Z., Bhatt, S. S., Craigen, W., Cheung, S. W., Lewis, R. A., Fang, P., Davenport, S. L. H., Stankiewicz, P., Lalani, S. R. (2009) Alu-specific microhomology-mediated deletions in *CDKL5* in females with early-onset seizure disorder. *Neurogenetics* **10**, 363–9.

180. Wang, X., Reid Sutton, V., Omar Peraza-Llanes, J., Yu, Z., Rosetta, R., Kou, Y.C., Eble, T.N., Patel, A., Thaller, C., Fang, P., Van den Veyver, I.B. (2007) Mutations in X-linked *PORCN*, a putative regulator of Wnt signaling, cause focal dermal hypoplasia. *Nat Genet* **39**, 836–8.

181. Grzeschik, K.H., Bornholdt, D., Oeffner, F., König, A., del Carmen Boente, M., Enders, H., Fritz, B., Hertl, M., Grasshoff, U., Höfling, K., Oji, V., Paradisi, M., Schuchardt, C., Szalai, Z., Tadini, G., Traupe, H., Happle, R. (2007) Deficiency of PORCN, a regulator of Wnt signaling, is associated with focal dermal hypoplasia. *Nat Genet* **39**, 833–5.

182. Vortkamp, A., Gessler, M., Grzeschik, K.-H. (1991) *GLI3* zinc-finger gene interrupted by translocations in Greig syndrome families. *Nature* **352**, 539–40.

183. Basson, C. T., Bachinsky, D. R., Lin, R. C., Levi, T., Elkins, J. A., Soults, J., Grayzel, D., Kroumpouzou, E., Traill, T. A., Leblanc-Straceski, J., Renault, B., Kucherlapati, R., Seidman, J. G., Seidman, C. E. (1997) Mutations in human cause limb and cardiac malformation in Holt-Oram syndrome. *Nat Genet* **15**, 30–5.

184. Van Esch, H., Groenen, P., Nesbit, M.A., Schuffenhauer, S., Lichtner, P., Vanderlinden, G., Harding, B., Beetz, R., Bilous, R.W., Holdaway, I., Shaw, N.J., Fryns, J.P., Van de Ven, W., Thakker, R.V., Devriendt, K. (2000) *GATA3* haplo-insufficiency causes human HDR syndrome. *Nature* **406**, 419–22.

185. Coldren, C.D., Lai, Z., Shragg, P., Rossi, E., Glidewell, S.C., Zuffardi, O., Mattina, T., Ivy, D.D., Curfs, L.M., Mattson, S.N., Riley, E.P., Treier, M., Grossfeld, P.D. (2009) Chromosomal microarray mapping suggests a role for BSX and neurogranin in neurocognitive and behavioral defects in the 11q terminal deletion disorder (Jacobsen syndrome). *Neurogenetics* **10**, 89–95.

186. Ballabio, A., Parenti, G., Tippett, P., Mondello, C., Di Maio, S., Tenore, A., Andria, G. (1986) X-linked ichthyosis, due to steroid sulphatase deficiency, associated with Kallmann syndrome (hypogonadotropic hypogonadism and anosmia): linkage relationships with Xg and cloned DNA sequences from the distal short arm of the X chromosome. *Hum Genet* **72**, 237–40.

187. Ludecke, H. J., Schaper, J., Meinecke, P., Momeni, P., Gross, S., von Holtum, D., Hirche, H., Abramowicz, M. J., Albrecht, B., Apacik, C., Christen, H. J., Claussen, U Devriendt, K., Fastnacht, E., Forderer, A., Friedrich, U., Goodship, T.H., Greiwe, M., Hamm, H., Hennekam, R.C., Hinkel, G.K., Hoeltzenbein, M., Kayserili, H., Majewski, F., Mathieu, M., McLeod, R., Midro, A.T., Moog, U., Nagai, T., Niikawa, N., Orstavik, K.H., Plöchl, E., Seitz, C., Schmidtke, J., Tranebjaerg, L., Tsukahara, M., Wittwer, B., Zabel, B., Gillessen-Kaesbach, G., Horsthemke, B. (2001) Genotypic and phenotypic spectrum in tricho-rhino-phalangeal syndrome types I and III. *Am J Hum Genet* **68**, 81–91.

188. Rao, E., Weiss, B., Fukami, M., Rump, A., Niesler, B., Mertz, A., Muroya, K., Binder, G., Kirsch, S., Winkelmann, M., Nordsiek, G., Heinrich, U., Breuning, M.H., Ranke, M.B., Rosenthal, A., Ogata, T., Rappold, G.A. (1997) Pseudoautosomal deletions encompassing a novel homeobox gene cause growth failure in idiopathic short stature and Turner syndrome. *Nat Genet* **16**, 54–63.

189. Benito-Sanz, S., Gorbenko del Blanco, D., Huber, C., Thomas, N.S., Aza-Carmona, M., Bunyan, D., Maloney, V., Argente, J., Cormier-Daire, V., Campos-Barros, A., Heath, K.E. (2006) Characterization of *SHOX* deletions in Leri-Weill Dyschondrosteosis (LWD) reveals genetic heterogeneity and no recombination hotspots. *Am J Hum Genet* **79**, 409–14.
190. des Portes, V., Pinard, J.M., Billuart, P., Vinet, M.C., Koulakoff, A., Carrie, A., Gelot, A., Dupuis, E., Motte, J., Berwald-Netter, Y., Catala, M., Kahn, A., Beldjord, C., Chelly, J. (1998) A novel CNS gene required for neuronal migration and involved in X-linked subcortical laminar heterotopia and lissencephaly syndrome. *Cell* **92**, 51–61.
191. Gleeson, J.G., Allen, K.M., Fox, J.W., Lamperti, E.D., Berkovic, S., Scheffer, I., Cooper, E.C., Dobyns, W.B., Minnerath, S.R., Ross, M.E., Walsh, C.A. (1998) Doublecortin, a brain-specific gene mutated in human X-linked lissencephaly and double cortex syndrome, encodes a putative signaling protein. *Cell* **92**, 63–72.
192. Lubs, H., Abidi, F., Bier, J. A. B., Abuelo, D., Ouzts, L., Voeller, K., Fennell, E., Stevenson, R. E., Schwartz, C. E., Arena, F. (1999) XLMR syndrome characterized by multiple respiratory infections, hypertelorism, severe CNS deterioration and early death localizes to distal Xq28. *Am J Med Genet* **85**, 243–8.
193. Schinzel, A. (1988) Microdeletion syndromes, balanced translocations, and gene mapping. *J Med Genet* **25**, 454–62.
194. Amiel, J., Espinosa-Parrilla, Y., Steffann, J., Gosset, P., Pelet, A., Prieur, M., Boute, O., Choiset, A., Lacombe, D., Philip, N., Le Merrer, M., Tanaka, H., Till, M., Touraine, R., Toutain, A., Vekemans, M., Munnich, A., Lyonnet, S. (2001) Large-scale deletions and *SMADIP1* truncating mutations in syndromic Hirschsprung disease with involvement of midline structures. *Am J Hum Genet* **69**, 1370–7.
195. Dreyer, S. D., Zhou, G., Baldini, A., Winterpacht, A., Zabel, B., Cole, W., Johnson, R. L., Lee, B. (1998) Mutations in *LMX1B* cause abnormal skeletal patterning and renal dysplasia in nail patella syndrome. *Nat Genet* **19**, 47–50.
196. Asthagiri, A.R., Parry, D.M., Butman, J.A., Kim, H.J., Tsilou, E.T., Zhuang, Z., Lonser, R.R. 2009. Neurofibromatosis type 2. *Lancet* **373**, 1974–86.
197. Rodriguez-Revenga, L., Madrigal, I., Alkhalidi, L.S., Armengol, L., González, E., Badenas, C., Estivill, X., Milà, M. (2007) Contiguous deletion of the *NDP*, *MAOA*, *MAOB*, and *EFHC2* genes in a patient with Norrie disease, severe psychomotor retardation and myoclonic epilepsy. *Am J Med Genet A* **143**, 916–20.
198. Ferrentino, R., Bassi, M.T., Chitayat, D., Tabolacci, E., Meroni, G. (2007) MID1 mutation screening in a large cohort of Opitz G/BBB syndrome patients: twenty-nine novel mutations identified. *Hum Mutat* **28**, 206–7.
199. Tuchman, M., Plante, R.J., Garcia-Perez, M.A., Rubio, V. (1996) Relative frequency of mutations causing ornithine transcarbamylase deficiency in 78 families. *Hum Genet* **97**, 274–6.
200. Schinzel, A. (1991) Tetrasomy 12p (Pallister-Killian syndrome). *J Med Genet* **28**, 122–5.
201. Polymeropoulos, M. H., Higgins, J. J., Golbe, L. I., Johnson, W. G., Ide, S. E., Di Iorio, G., Sanges, G., Stenroos, E. S., Pho, L. T., Schaffer, A. A., Lazzarini, A. M., Nussbaum, R. L., Duvoisin, R. C. (1996) Mapping of a gene for Parkinson's disease to chromosome 4q21-q23. *Science* **274**, 1197–8.
202. Inoue K. (2005) *PLP1*-related inherited dysmyelinating disorders: Pelizaeus-Merzbacher disease and spastic paraplegia type 2. *Neurogenetics* **6**, 1–16.
203. Zweier, C., Peippo, M. M., Hoyer, J., Sousa, S., Bottani, A., Clayton-Smith, J., Reardon, W., Saraiva, J., Cabral, A., Gohring, I., Devriendt, K., de Ravel, T., Bijlsma, E. K., Hennekam, R. C., M., Orrico, A., Cohen, M., Dreweke, A., Reis, A., Nurnberg, P., Rauch, A. (2007) Haploinsufficiency of *TCF4* causes syndromal mental retardation with intermittent hyperventilation (Pitt-Hopkins syndrome). *Am J Hum Genet* **80**, 994–1001.
204. Potocki, L., Shaffer, L. G. (1996) Interstitial deletion of 11(p11.2p12): a newly described contiguous gene deletion syndrome involving the gene for hereditary multiple exostoses (*EXT2*). *Am J Med Genet* **62**, 319–25.
205. Zhu, X., Dunn, J. M., Phillips, R. A., Goddard, A. D., Paton, K. E., Becker, A., Gallie, B. L. (1989) Preferential germline mutation of the paternal allele in retinoblastoma. *Nature* **340**, 312–3.
206. Moog, U., Smeets, E.E.J., van Roozendaal, K.E.P., Schoenmakers, S., Herbergs, J., Schoonbrood-Lenssen, A.M.J., Schrander-Stumpel, C.T.R.M. (2003) Neurodevelopmental disorders in males related to the gene causing Rett syndrome in females (MECP2). *Eur J Paediat Neurol* **7**, 5–12.
207. Petrij, F., Dauwerse, H. G., Blough, R. I., Giles, R. H., van der Smagt, J. J., Wallerstein,

R., Maaswinkel-Mooy, P. D., van Karnebeek, C. D., van Ommen, G.-J. B., van Haeringen, A., Rubinstein, J. H., Saal, H. M., Hennekam, R. C. M., Peters, D. J. M., Breuning, M. H. (2000) Diagnostic analysis of the Rubinstein-Taybi syndrome: five cosmids should be used for microdeletion detection and low number of protein truncating mutations. *J Med Genet* **37**, 168–76.

208. El Ghouzzi, V., Le Merrer, M., Perrin-Schmitt, F., Lajeunie, E., Benit, P., Renier, D., Bourgeois, P., Bolcato-Bellemin, A.-L., Munnich, A., Bonaventure, J. (1997) Mutations of the *TWIST* gene in the Saethre-Chotzen syndrome. *Nat Genet* **15**, 42–46.

209. Scherer, S.W., Poorkaj, P., Allen, T., Kim, J., Geshuri, D., Nunes, M., Soder, S., Stephens, K., Pagon, R.A., Patton, M.A., et al. (1994) Fine mapping of the autosomal dominant split hand/split foot locus on chromosome 7, band q21.3-q22.1. *Am J Hum Genet* **55**, 12–20.

210. de Mollerat, X.J., Gurrieri, F., Morgan, C.T., Sangiorgi, E., Everman, D.B., Gaspari, P., Amiel, J., Bamshad, M.J., Lyle, R., Blouin, J.L., Allanson, J.E., Le Marec, B., Wilson, M., Braverman, N.E., Radhakrishna, U., Delozier-Blanchet, C., Abbott, A., Elghouzzi, V., Antonarakis, S., Stevenson, R.E., Munnich, A., Neri, G., Schwartz, C.E. (2003) A genomic rearrangement resulting in a tandem duplication is associated with split hand-split foot malformation 3 (SHFM3) at 10q24. *Hum Mol Genet* **12**, 1959–71.

211. Kohlhase, J., Wischermann, A., Reichenbach, H., Froster, U., Engel, W. (1998) Mutations in the *SALL1* putative transcription factor gene cause Townes-Brocks syndrome. *Nat Genet* **18**, 81–3.

212. Brook-Carter, P.T., Peral, B., Ward, C.J., Thompson, P., Hughes, J., Maheshwar, M.M., Nellist, M., Gamble, V., Harris, P.C., Sampson, J.R. (1994) Deletion of the TSC2 and PKD1 genes associated with severe infantile polycystic kidney disease: a contiguous gene syndrome. *Nat Genet* **8**, 328–32.

213. Jones, A. C., Daniells, C. E., Snell, R.G., Tachataki, M., Idziaszczyk, S.A., Krawczak, M., Sampson, J.R., Cheadle, J.P. (1997) Molecular genetic and phenotypic analysis reveals differences between TSC1 and TSC2 associated familial and sporadic tuberous sclerosis. *Hum Molec Genet* **6**, 2155–61.

214. Klopocki, E., Neumann, L. M., Tonnies, H., Ropers, H.-H., Mundlos, S., Ullmann, R. (2006) Ulnar-mammary syndrome with dysmorphic facies and mental retardation caused by a novel 1.28 Mb deletion encompassing the *TBX3* gene. *Eur J Hum Genet* **14**, 1274–9.

215. Kondo, S., Schutte, B.C., Richardson, R.J., Bjork, B. C., Knight, A.S., Watanabe, Y., Howard, E., Ferreira de Lima, R.L.L., Daack-Hirsch, S., Sander, A., McDonald-McGinn, D. M., Zackai, E.H., Lammer, E.J., Aylsworth, A.S., Ardinger, H.H., Lidral, A.C., Pober, B.R., Moreno, L., Arcos-Burgos, M., Valencia, C., Houdayer, C., Bahuau, M., Moretti-Ferreira, D., Richieri-Costa, A., Dixon, M.J., Murray, J.C. (2002) Mutations in *IRF6* cause Van der Woude and popliteal pterygium syndromes. *Nat Genet* **32**, 285–9.

216. Zollino, M., Lecce, R., Fischetto, R., Murdolo, M., Faravelli, F., Selicorni, A., Buttè, C., Memo, L., Capovilla, G., Neri, G. (2003) Mapping the Wolf-Hirschhorn syndrome phenotype outside the currently accepted WHS critical region and defining a new critical region, WHSCR-2. *Am J Hum Genet* **72**, 590–7.

217. Kato, M., Das, S., Petras, K., Kitamura, K., Morohashi, K., Abuelo, D.N., Barr, M., Bonneau, D.,,Brady, A.F., Carpenter, N.J., Cipero, K.L., Frisone, F., Fukuda, T., Guerrini, R., Iida, E., Itoh, M., Lewanda, A.F., Nanba, Y., Oka, A., Proud, V.K., Saugier-Veber, P., Schelley, S.L., Selicorni, A., Shaner, R., Silengo, M., Stewart, F., Sugiyama, N., Toyama, J., Toutain, A., Vargas, A.L., Yanazawa, M., Zackai, E.H., Dobyns, W.B. (2004) Mutations of *ARX* are associated with striking pleiotropy and consistent genotype-phenotype correlation. *Hum Mutat* **23**, 147–59.

218. Sumegi, J., Huang, D., Lanyi, A., Davis, J.D., Seemayer, T.A., Maeda, A., Klein, G., Seri, M., Wakiguchi, H., Purtilo, D.T., Gross, T.G. (2000) Correlation of mutations of the *SH2D1A* gene and epstein-barr virus infection with clinical phenotype and outcome in X-linked lymphoproliferative disease. *Blood* **96**, 3118–25.

219. Zhang, Y.H., Huang, B.L., Niakan, K.K., McCabe, L.L., McCabe, E.R., Dipple, K.M. (2004) *IL1RAPL1* is associated with mental retardation in patients with complex glycerol kinase deficiency who have deletions extending telomeric of DAX1. *Hum Mutat* **24**, 273.

220. Laumonnier, F., Ronce, N., Hamel, B.C., Thomas, P., Lespinasse, J., Raynaud, M., Paringaux, C., Van Bokhoven, H., Kalscheuer, V., Fryns, J.P., Chelly, J., Moraine, C., Briault, S. (2002) Transcription factor *SOX3* is involved in X-linked mental retardation with growth hormone deficiency. *Am J Hum Genet* **71**, 1450–5.

221. Najm, J., Horn, D., Wimplinger, I., Golden, J.A., Chizhikov, V.V., Sudi, J., Christian, S.L.,

Ullmann, R., Kuechler, A., Haas, C.A., Flubacher, A., Charnas, L.R., Uyanik, G., Frank, U., Klopocki, E., Dobyns, W.B., Kutsche, K. (2008) Mutations of *CASK* cause an X-linked brain malformation phenotype with microcephaly and hypoplasia of the brainstem and cerebellum. *Nat Genet* **40**, 1065–7.

222. Girirajan, S., Campbell, C.D., Eichler, E. E. (2011) Human copy number variation and complex genetic disease. *Annu Rev Genet* **45**, 203–26.

223. Ramocki, M..B., Bartnik, M., Szafranski, P., Kołodziejska, K. E., Xia, Z., Bravo, J., Miller, G. S., Rodriguez, D.L., Williams, C.A., Bader, P.I., Szczepanik, E., Mazurczak, T., Antczak-Marach, D., Coldwell, J.G., Akman, C.I., McAlmon, K., Cohen, M.P., McGrath, J., Roeder, E., Mueller, J., Kang, S.-H., Bacino, C. A., Patel, A., Bocian, E., Shaw, C. A., Cheung, S. W., Mazurczak, T., Stankiewicz, P. (2010) Recurrent Distal 7q11.23 Deletion Including HIP1 and YWHAG Identified in Patients with Intellectual Disabilities, Epilepsy, and Neurobehavioral Problems. *Am J Hum Genet* **87**, 857–65.

224. Stankiewicz, P., Kulkarni, S., Dharmadhikari, A. V., Sampath, S., Bhatt, S. S., Shaikh, T. H., Xia, Z., Pursley, A.N., Cooper, M. L., Shinawi, M., Paciorkowski, A. R., Grange, D. K., Noetzel, M. J., Saunders, S., Simons, P., Summar, M., Lee, B., Scaglia, F., Fellmann, F., Martinet, D., Beckmann, J. S., Asamoah, A., Platky, K., Sparks, S., Martin, A. S., Madan-Khetarpal, S., Hoover, J., Medne, L., Bonnemann, C.G., Moeschler, J. B., Vallee, S. E., Parikh, S., Irwin, P., Dalzell, V. P., Smith, W. E., Banks, V. C., Flannery, D. B., Lovell, C. M., Bellus, G. A., Golden-Grant, K., Gorski, J. L., Kussmann, J. L., McGregor, T. L., Hamid, R., Pfotenhauer, J., Ballif, B.C., Shaw, C. A., Kang, S.-H., Bacino, C. A., Patel, A., Rosenfeld, J. A., Cheung, S. W., Shaffer, L. G. (2011) Recurrent deletions and reciprocal duplications of 10q11.21q11.23 including CHAT and SLC18A3 are likely mediated by complex low-copy repeats. *Hum Mutat In press.*

225. Dimitrov, B., Balikova, I., de Ravel, T., Van Esch, H., De Smedt, M., Baten, E., Vermeesch, J. R., Bradinova, I., Simeonov, E., Devriendt, K., Fryns, J. P., Debeer, P. (2011) 2q31.1 microdeletion syndrome: redefining the associated clinical phenotype. *J Med Genet* **48**, 98–104.

226. Palomares, M., Delicado, A., Mansilla, E., de Torres, M. L., Vallespín, E., Fernandez, L., Martinez-Glez, V., García-Miñaur, S., Nevado, J., Simarro, F. S., Ruiz-Perez, V. L., Lynch, S. A., Sharkey, F. H., Thuresson, A. C., Annerén, G., Belligni, E. F., Martínez-Fernández, M. L., Bermejo, E., Nowakowska, B., Kutkowska-Kazmierczak, A., Bocian, E., Obersztyn, E., Martínez-Frías, M. L., Hennekam, R. C., Lapunzina, P. (2011) Characterization of a 8q21.11 microdeletion syndrome associated with intellectual disability and a recognizable phenotype. *Am J Hum Genet* **89**, 295–301.

227. de Kovel, C.G., Trucks, H., Helbig, I., Mefford, H.C., Baker, C., Leu, C., Kluck, C., Muhle, H., von Spiczak, S., Ostertag, P., Obermeier, T., Kleefuss-Lie, A.A., Hallmann, K., Steffens, M., Gaus, V., Klein, K.M., Hamer, H.M., Rosenow, F., Brilstra, E.H., Trenité, D.K., Swinkels, M.E., Weber, Y.G., Unterberger, I., Zimprich, F., Urak, L., Feucht, M., Fuchs, K., Møller, R.S., Hjalgrim, H., De Jonghe, P., Suls, A., Rückert, I.M., Wichmann, H.E., Franke, A., Schreiber, S., Nürnberg, P., Elger, C.E., Lerche, H., Stephani, U., Koeleman, B.P., Lindhout, D., Eichler, E.E., Sander, T. (2010) Recurrent microdeletions at 15q11.2 and 16p13.11 predispose to idiopathic generalized epilepsies. *Brain* **133**, 23–32.

228. Nagamani, S.C., Erez, A., Bay, C., Pettigrew, A., Lalani, S.R., Herman, K., Graham, B.H., Nowaczyk, M.J., Proud, M., Craigen, W.J., Hopkins, B., Kozel, B., Plunkett, K., Hixson, P., Stankiewicz, P., Patel, A., Cheung, S.W. (2011) Delineation of a deletion region critical for corpus callosal abnormalities in chromosome 1q43–q44. *Eur J Hum Genet* in press.

229. Ballif, B.C., Rosenfeld, J.A., Traylor, R., Theisen, A., Bader, P.I., Ladda, R.L., Sell, S.L., Steinraths, M., Surti, U., McGuire, M., Williams, S., Farrell, S.A., Filiano, J., Schnur, R.E., Coffey, L.B., Tervo, R.C., Stroud, T., Marble, M., Netzloff, M., Hanson, K., Aylsworth, A.S., Bamforth, J.S., Babu, D., Niyazov, D.M., Ravnan, J.B., Schultz, R.A., Lamb, A.N., Torchia, B.S., Bejjani, B.A., Shaffer, L.G. (2011) High-resolution array CGH defines critical regions and candidate genes for microcephaly, abnormalities of the corpus callosum, and seizure phenotypes in patients with microdeletions of 1q43q44. *Hum Genet* in press.

230. Cho, T.-J., Kim, O.-H., Choi, I. H., Nishimura, G., Superti-Furga, A., Kim, K. S., Lee, Y.-J., Park, W.-Y. (2010) A dominant mesomelic dysplasia associated with a 1.0-Mb microduplication of HOXD gene cluster at 2q31.1. *J Med Genet* **47**, 638–9.

Chapter 3

Structural Genomic Variation in Intellectual Disability

Rolph Pfundt and Joris A. Veltman

Abstract

The genetic causes of mental retardation are highly heterogeneous and for a large proportion unknown. Mutations as well as large chromosomal abnormalities are known to contribute to mental retardation, and recently more subtle structural genomic variations have been shown to contribute significantly to this common and complex disorder. Genomic microarrays with increasing resolution levels have revealed the presence of rare *de novo* CNVs in approximately 15% of all mentally retarded patients. Microarray-based CNV screening is rapidly replacing conventional karyotyping in the diagnostic workflow, resulting in an increased diagnostic yield as well as biological insight into this disorder. In this chapter, an overview is given of the detection and interpretation of copy number variations in mental retardation, with a focus on diagnostic applications. In addition, a detailed protocol is provided for the diagnostic interpretation of copy-number variations in mental retardation.

Key words: Copy-number variation, Intellectual disability, Genomic microarrays, Genetic diagnosis

1. Introduction

1.1. Mental Retardation: A Common Complex Disorder

Mental retardation, also termed learning disability or developmental delay, is a complex disorder affecting 2–3% of the general population (1). It is defined by significant limitations both in intellectual functioning and in adaptive behavior originating before the age of 18 years (2). It often involves slow learning of basic motor and language skills during childhood, and a significantly below-normal global intellectual capacity as an adult. With regard to the intellectual capacities, mental retardation is represented by an intelligence quotient (IQ) test score of 70 or below. Based on this IQ score, mental retardation can be classified into three categories: mild (70–50), moderate (50–30), and severe (<30) mental retardation. Mental retardation occurs either in isolation or in combination with other malformations and/or dysmorphisms.

Lars Feuk (ed.), *Genomic Structural Variants: Methods and Protocols*, Methods in Molecular Biology, vol. 838,
DOI 10.1007/978-1-61779-507-7_3, © Springer Science+Business Media, LLC 2012

The prevalence, the lifelong severity, and poor curability emphasize the impact of mental retardation and the need for research. Patients live lifelong in institutes and in special housing facilities, and require special schools as well as lifelong support from parents, health care, and society.

1.2. The Many Causes of Mental Retardation

The causes of mental retardation are highly heterogeneous. Apart from genetic causes, prenatal causes include intrauterine infections, excess maternal alcohol consumption during pregnancy, and premature birth. In addition, perinatal factors, such as hypoxia, and postnatal factors, such as brain trauma, can also influence the mental ability. In clinical practice, even after extensive investigations, about half of the individuals with mental retardation remain without a diagnosis (3–5). Genetic anomalies include structural genomic variations and single-gene mutations. To date, genes that influence cognitive function have predominantly been found on the X chromosome. This is mainly due to the fact that it is easy to identify and subsequently map the gene in X-linked families (6). Already, more than 90 genes on the X chromosome are known to be involved in mental retardation (7), whereas the majority of autosomal mental retardation genes await identification. Chromosome abnormalities on the other hand have been recognized for a long time as an important cause of mental retardation (8, 9). Microscopically detectable chromosome aberrations, including numerical and structural chromosome aberrations, generally cause mental retardation in conjunction with physical anomalies because they are associated with large genetic imbalances affecting many genes. In addition, fluorescence in situ hybridization (FISH) analysis is frequently applied in the diagnostics of mental retardation mainly because this technology has a much higher resolution as compared to karyotyping. This technology is either applied to confirm a clinical suspicion by screening for well-known microdeletion syndromes associated with mental retardation (10) or for the analysis of all subtelomeric regions of the genome, as these are known to be frequently affected in mental retardation (11, 12). The combined diagnostic yield of karyotyping and FISH analysis in the general population of patients with unexplained mental retardation is approximately 7–10%.

1.3. Structural Genomic Variation in Mental Retardation

Microarray-based genome profiling technologies have dramatically changed the nature of human genome analysis by combining the targeted high-resolution approach of the FISH technology and the whole-genome approach of the karyotyping technology. In recent years, the use of genomic microarrays has revolutionized our insight in the role of structural genomic variation in mental retardation. Genomic microarray technology has enhanced the genome-wide resolving power to detect genomic deletions and duplications, the so-called copy-number variations (CNVs), from the megabase to the kilobase (kb) level. Eight years ago, this technology was used for the first time in the field of clinical genetics, when we described microarray-based copy number analysis of all human telomeres in patients with mental retardation (13). At that time, we suggested that "the robustness and simplicity of this array-based telomere copy number screening makes it highly suited for introduction into the clinic as a rapid and sensitive automated diagnostic procedure." Indeed microarrays have found their way into the clinic setting, although the majority of applications now target the entire genome instead of the telomeres only.

From its introduction using low-resolution, clone-based genomic microarrays, microarray-based CNV detection showed a lot of potential by detecting causative microdeletions and/or duplications in patients, where no chromosome abnormality was detected by conventional chromosome analysis (14, 15). Pilot studies provided insight into the quality and reproducibility of the procedure, the need for validation of the microarray data by independent technologies, such as FISH or QPCR, as well as the way to translate these data into clinical practice (16, 17). The clinical usefulness of microarray-based CNV detection or molecular karyotyping was substantiated in larger, less-selected cohorts of individuals with mental retardation using arrays with one genomic clone per mega base (18, 19) or arrays with one genomic clone per 100 kb (20). In the latter study, CNVs were detected in 97 of 100 patients tested. The majority of these CNVs were inherited from phenotypically normal parents, reflecting normal genomic variation in the human population. In contrast, 10 patients contained CNVs that were not present in their parents, nor in any of the other parents studied. These so-called rare *de novo* CNVs occurred throughout the genome and included seven deletions and three duplications, varying in size from 540 kb to 12 Mb. Many similar studies have been published since on the detection of CNVs in mental retardation (reviewed by 18, 21–23). When taking all studies together, three main conclusions can be drawn: (1) rare *de novo* CNVs are responsible for a considerable proportion of cases with mental retardation, (2) these rare *de novo* CNVs occur all over the genome, and (3) these rare *de novo* CNVs can be very small. This argues for the use of the highest resolution genome-wide microarray approaches in the detection of the genetic causes of mental retardation.

1.4. Commercial Microarrays in Mental Retardation Diagnostics

Genomic microarray technology has greatly matured over the past decade, and many diagnostic genetics laboratories have started to implement this technology into a routine diagnostic setting. Commercially available oligonucleotide arrays are increasingly being used as they can be produced according to industrial quality standards, offer very high genomic coverage and resolution, and are available to all laboratories. Moreover, genome-wide oligonucleotide

a

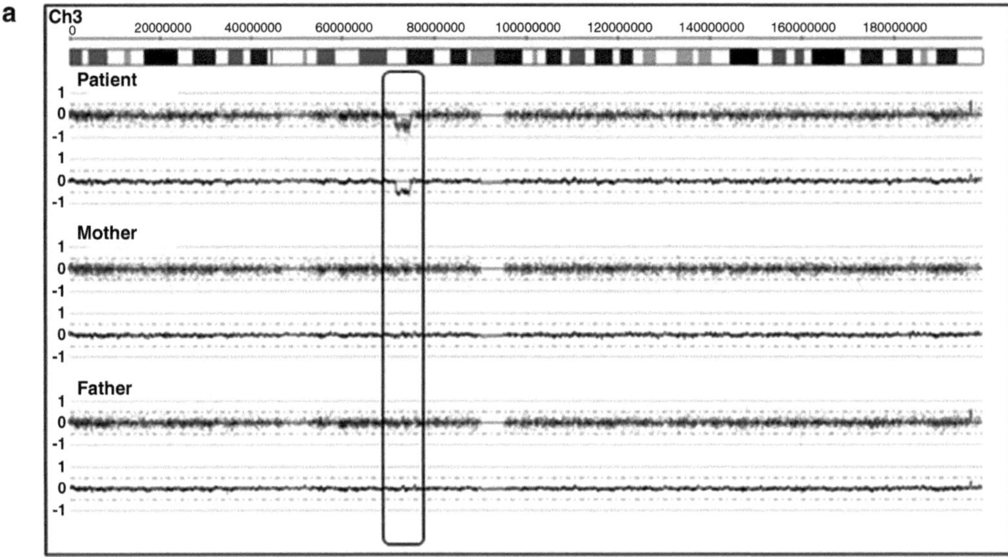

b

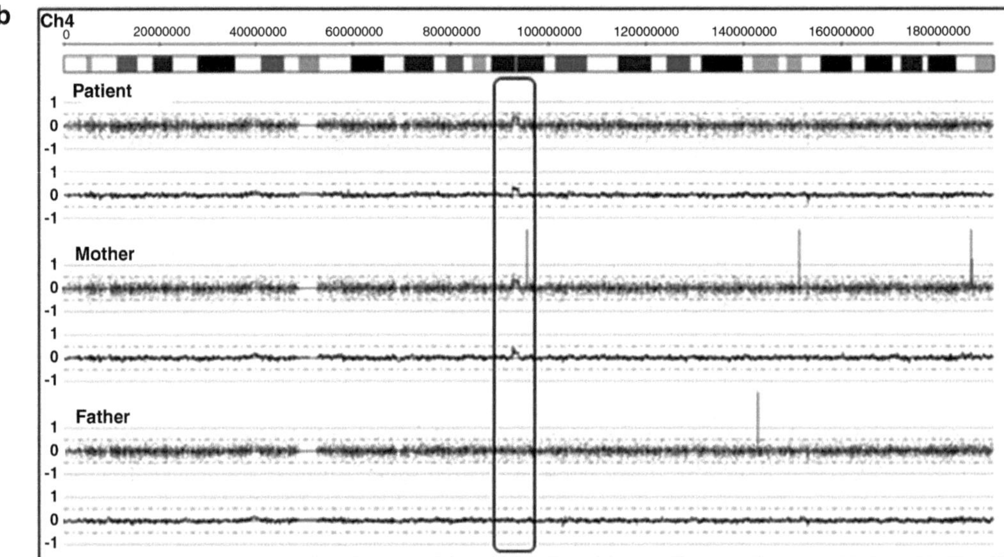

c

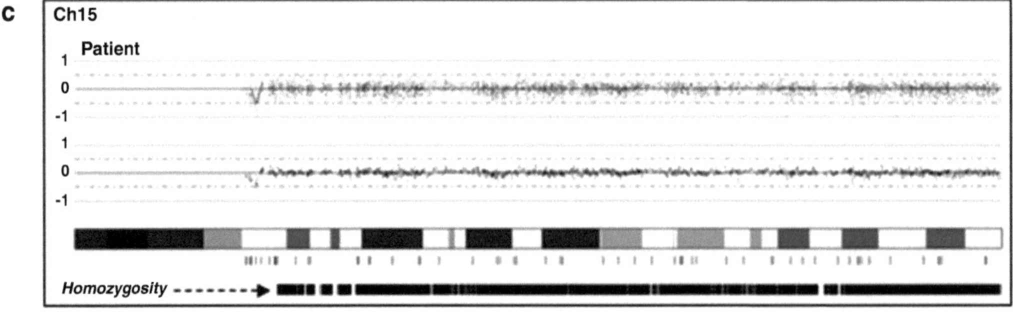

Fig. 1. Example of copy number profiles obtained using Affymetrix GeneChip® 250 K and 500 K Arrays. These chromosome plots were obtained using the CNAG software package. Below each chromosome drawing with indicated banding pattern, the normalized and log2-transformed patient-over-reference intensity ratios for each SNP are displayed (*dots*) according to their chromosomal position. This plot also shows the result of the statistical HMM algorithm, with a normal copy at the 0-line, a deletion going down, and a duplication going up. Below this plot, a moving average of these intensity ratios from 10

arrays targeting single-nucleotide polymorphisms (SNPs) offer an additional advantage by providing high-resolution genotype next to CNV information in a single experiment. This genotype information can be used to check for potential sample mix-up, test for paternity, and to determine whether a *de novo* CNV occurred on the paternal or maternal chromosome. In addition, it can be used to study copy number-neutral genomic variations, such as uniparental disomy (UPD) or homozygosity. See Fig. 1 for examples of *de novo* CNVs, inherited CNVs, as well as a UPD detected on the Affymetrix SNP array platform.

Because of these advantages and after a detailed microarray platform comparison (24), we decided with two other European diagnostic laboratories to validate the use of Affymetrix 500 K SNP Arrays for diagnostic applications (25). First, we demonstrated that these arrays containing more than 500,000 SNP probes could reliably detect CNVs previously associated to mental retardation, such as CNVs in subtelomeric regions and CNVs causing known mental retardation syndromes. In addition, we used this array format to analyze a total of 120 patients with unexplained mental retardation and their unaffected parents. This analysis detected 362 CNVs larger than 100 kb, of which CNVs in 18 patients were shown to have occurred *de novo* and did not show more than 50% overlap with CNVs detected in any of the other parents tested. This diagnostic yield of 15% was slightly higher than that reported by previous genome-wide microarray studies (23), which could be explained by the higher resolution of our microarray platform; five of the rare *de novo* CNVs were smaller than 1 Mb. Importantly, in this aspect, all of these patients underwent conventional karyotyping prior to microarray analysis, and large chromosome abnormalities were therefore already excluded as a cause of mental retardation in this cohort.

Recently, the first diagnostic laboratories decided that it was no longer needed to perform both conventional chromosome analysis and microarrays in mental retardation diagnostics. Instead, these laboratories now only perform microarray analysis in the diagnostic workflow of this patient cohort. Balanced chromosome abnormalities, such as balanced inversions or translocations, however, are not detected by genomic microarrays, and this is why many cytogeneticists still hesitate to give up karyotyping. Approximately 40% of the chromosome abnormalities that appeared to be balanced by chromosome analysis, however, turn out to be unbalanced

Fig. 1. (continued) adjacent SNPs is displayed (*line*). (**a**) In the chromosome 3 plot of patient A, a *de novo* 2.3 Mb deletion CNV can be observed in the *black rectangle*. (**b**) In the chromosome 4 plot of patient B, a maternally inherited CNV, this time a 1.3 Mb duplication, can be observed in the *black rectangle*. (**c**) In the chromosome 15 plot of patient C, no CNVs are detected. However, the *black bar* at the bottom of this graph indicates (nearly) complete homozygosity of chromosome 15, indicative for a uniparental isodisomy of this chromosome. The *small vertical lines* above this bar represent heterozygous SNPs.

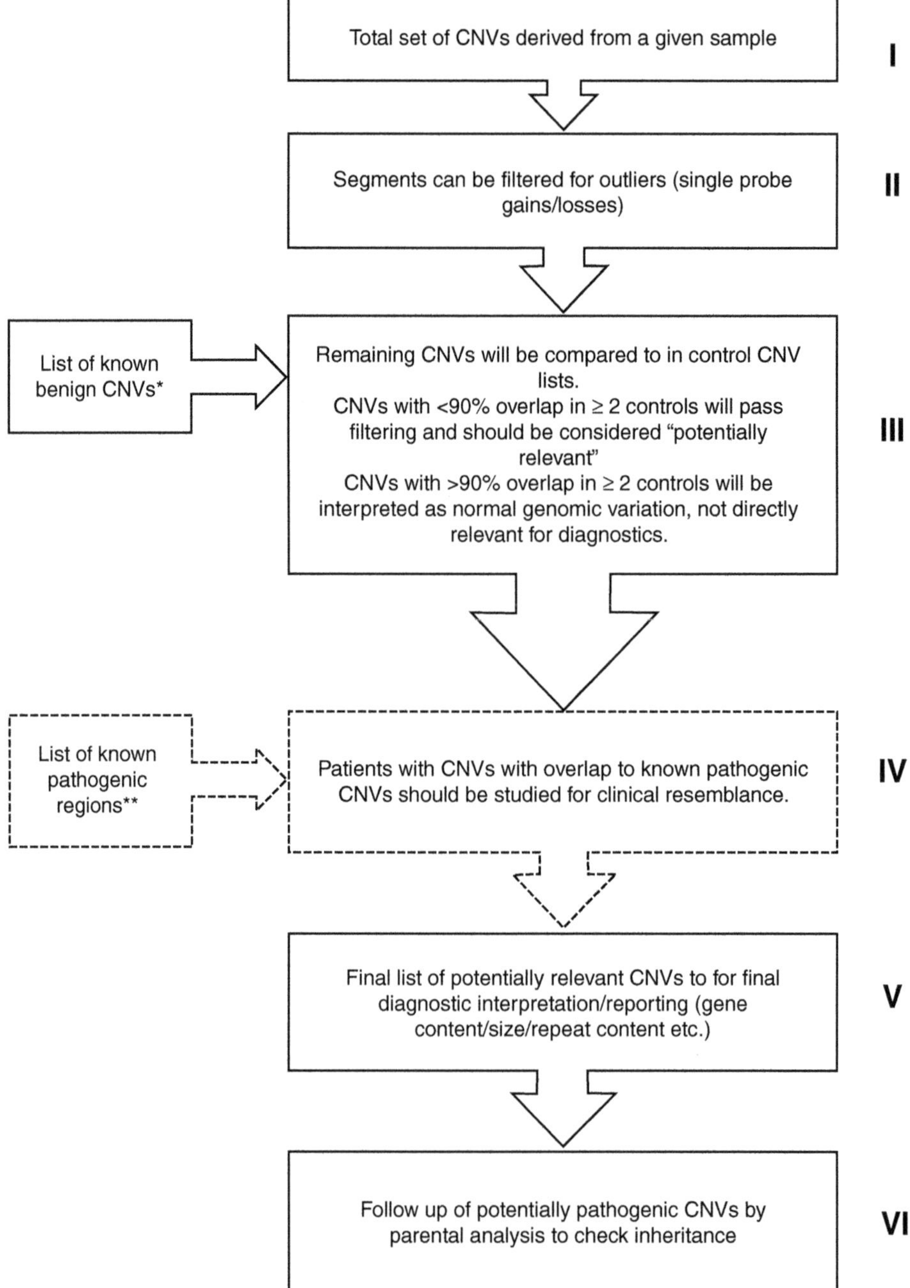

Fig. 2. Flow scheme for CNV classification: All CNVs that are identified in a single sample (I) should be interpreted with respect to their potential pathogenicity. In step II, the probe values that represent outliers (single clones with a very high/low T/R ration) are removed from the list of potential CNVs. In step II, the remaining CNVs are compared with a CNV list derived from normal healthy controls. When a CNV in the sample has a >90% overlap with a CNV that has been detected in at least two healthy individuals, it is considered to be a benign CNV. NOTE that for this comparison the CNV should have the identical copy number state (gain/loss) in sample and controls. If a CNV has less than 90% overlap with a CNV found in healthy controls, it is considered to be possibly pathogenic. As an option (so not required for pathogenicity) in step IV, the possibly pathogenic CNVs could be cross-checked with (a) list(s) of known pathogenic CNVs. If any overlap with known pathogenic CNVs is found, phenotypes of the patient and the patient with the reported pathogenic CNV could be compared.

by high-resolution microarray analysis (26). It has been estimated that truly balanced chromosomal rearrangements cause disease in less than 1% of the patients with mental retardation (27, 28). This percentage is relatively small, especially when you estimate how many causative CNVs are still missed by the microarrays now used in most diagnostic laboratories. Most microarrays do not cover the genome evenly, and many are not capable of detecting CNVs smaller than 100,000 bases. Ideally, one would like to screen the entire genome for the presence of CNVs, especially in the exome, the coding region of the genome, since a single-exon deletion or duplication can already disrupt a gene and result in a disorder like mental retardation. Arrays containing millions of oligonucleotides have recently been introduced in a research setting and undoubtedly enter the diagnostic arena as soon as they are affordable, reliable, and their interpretation is straightforward.

1.5. Differentiating Between CNVs Causing Mental Retardation and Benign CNVs

The use of genomic microarrays has only recently shown us that all of our genomes are full of CNVs: high-resolution microarrays and whole-genome sequencing approaches are able to identify 600–900 CNVs in a single individual (29–31). Current clinical interpretation relies on information about the frequencies of a CNV in affected versus unaffected individuals, as well as on inheritance information (23, 32). Information about this frequency in healthy individuals can be found in the Toronto Database of Genomic Variants (DGV). In addition, clinically relevant CNVs are categorized in the DatabasE of Chromosomal Imbalance and Phenotype in Humans using Ensembl Resources (DECIPHER) (33) as well as in the European Cytogeneticists Association Register of Unbalanced Chromosome Aberrations (ECARUCA) (34). See Fig. 2 for a practical workflow of CNV interpretation in mental retardation diagnostics.

Although information about CNV frequency is useful, the latest generation of CNV arrays with millions of genomic probes identify an enormous amount of relatively small CNVs per individual, and it is unclear how large a control cohort should be before one can conclude that a CNV is not known to vary in the normal population, let alone the observation that different ethnical populations show a different CNV distribution (31, 35). In addition, an increasing

Fig. 2. (continued) Step V leaves a (list of) CNV(s) that could be pathogenic and should be reported on. To evaluate the pathogenicity of these CNVs, the inheritance pattern should be studied by analyzing parental samples. *Lists of known benign controls can be generated by hybridizing a set of normal healthy controls to the identical array platform that is used for sample analyses. CNVs that occur in multiple (>2) healthy control individuals can be considered as normal genomic variants. As an alternative, publically available databases of benign CNVs can be used (such as the database of genomic variants http://projects.tcag.ca/variation/). **Lists of known pathogenic regions can consist of either an in-house list of aberrations that are proven to be pathogenic or derived from publically available lists of pathogenic CNVs, such as the DECIPHER (https://decipher.sanger.ac.uk/) or ECARUCA (http://www.ecaruca.net) database.

number of CNVs with variable inheritance and penetrance have been described in mental retardation as well as in other neurological disorders. Rare CNVs at 1q21.1 (36, 37), 15q13.3 (38, 39), and 16p13.11 (40, 41) have been shown to be significantly enriched in these disorders, but they do occur sporadically in unaffected individuals and are sometimes inherited from unaffected or mildly affected parents. These findings demonstrate that there are limitations in considering CNVs as benign when common and inherited or causal when rare and *de novo.*

An alternative approach to differentiate benign from disease-causing CNVs may be found by studying their genomic content in a systematic and statistically robust manner. In 2009, we demonstrated that genes present in mental retardation-associated CNVs are different from genes in benign CNVs or from the average human gene. For this study, we selected a set of 148 rare *de novo* CNVs identified in patients with mental retardation as well as a set of 26,472 CNVs form the general population. In contrast to benign CNVs, we noticed that mental retardation-associated CNVs contain a greater number of genes that give a specific nervous system phenotype when disrupted in the mouse. In addition, these CNVs contain an enrichment of genes that show specific expression in the brain as well as an enrichment of genes involved in neurodegenerative pathways (42). These results are of great value for two reasons. Firstly, we can use this functional enrichment to predict which of the genes located in a mental retardation CNV is the dosage-sensitive gene that most likely caused the phenotype. In this study, we used this information to predict 55 candidate mental retardation genes, the majority of which were not associated to mental retardation before. Large-scale CNV as well as sequencing studies are now required to validate the role of these individual genes in mental retardation. Secondly, we can use this functional difference between genes in benign and pathogenic CNVs to assist in the clinical decision-making process. Recently, we developed a computational tool called GEnomic Classification of CNVs Objectively (GECCO) that measures each CNV for the presence and frequency of 13 genomic features. Based on this analysis, the CNV receives a probability score of being a mental retardation CNV, varying from 0 (not likely to cause mental retardation) to 1 (highly likely to cause mental retardation). We validated this tool by showing that it can reliably classify CNVs causing MR syndromes and achieve a high accuracy (94%) and negative predicted value (99%) on a blinded set of CNVs from a large cohort of patients with mental retardation (43). These results indicate that this classification method is of value for objectively identifying CNVs associated to mental retardation. We should, however, realize that the phenotypic consequences of genetic variations like CNVs depend in part on the rest of the genome as well as on environmental influences. In this aspect, Girirajan and colleagues recently described an enrichment of the so-called second genomic hits in patients with developmental delay (44). These patients all carried a CNV on chromosome 16p12.1 that was mostly inherited from unaffected parents, but a quarter of the patients contained a second-large CNV that likely contributed to the phenotype. These kind of detailed and complex genotype–phenotype analyses are required in order to fully understand the genotype–phenotype correlations in individual patients with mental retardation.

1.6. Conclusion

The introduction of genomic microarray technology in the study of mental retardation has dramatically increased both the diagnostic yield as well as the biological understanding of this common and complex neurological disorder. Genomic microarrays have largely replaced karyotyping in the routine diagnostic workflow, and hundreds of CNVs can robustly be detected in each patient. Now that microarray-based CNV identification has become robust and reliable, the major challenge shifts toward the interpretation of all of these CNVs. This interpretation is until now mostly based on inheritance information as well as information about the frequency of a particular CNV in the affected as well as the unaffected population. In practice, this means that CNVs can only be linked to disease in an individual patient when the CNV is rare and occurred *de novo* in the patient or is segregating with disease in the family. This works well for relatively large CNVs, but more and more complex associations between CNVs are being described that complicate clinical decision making. The interpretation of these CNVs can be improved by the introduction of computational tools that can objectively identify functional differences between benign and pathogenic CNVs. In addition, the clinical relevance of a CNV should be interpreted in the context of the remainder of the genome, as genetic modifiers can influence the phenotypic expression of these CNVs. This will soon become a possibility, now that whole genome sequencing identifies all genomic variations present in an individual genome in a single experiment.

2. Materials

2.1. Affymetrix Genechip 250/500 K SNP Array Hybridization for CNV Analysis

Laboratory procedures are performed essentially as described by the manufacturer (http://www.affymetrix.com/support/technical/index.affx) consisting of the following steps.

All chemicals are of "p.a. quality" unless indicated otherwise.

- GeneChip Mapping 250 K Nsp Assay Kit (Affymetrix 900766).

Including:	Adaptor Nsp I
	PCR primer 002
	Fragmentation reagent (DNase I)
	10× Fragmentation buffer
	Labeling reagent
	5× TdT buffer
	TdT (30 U/μl)
	OCR

- GeneChip Human Mapping 250 K Array Nsp (Affymetrix 900768).
- Nsp I enzyme (10,000 U/ml) (New England Biolabs (NE R0602L)).

Including:	NE buffer 2 10×
	BSA 10 mg/ml

- TMACL (Sigma T3411).
- MES Hydrate SigmaUltra (Sigma, M5287).
- MES Sodium Salt (Sigma M5057).
- DMSO (Sigma D5879).
- Denhardt's Solution (50× concentrate) (Sigma D2532).
- DNA amplification kit (Clontech 639240).

Including:	Titanium Taq DNA polymerase
	G-C Melt (5 M)
	dNTP mixture (2, 5 mM each)

- Recovery Buffer (RB) (Clontech 636976).
- Human Cot-1 (Invitrogen 15279-011).
- PBS pH 7.4 (Invitrogen 10010-056).
- 1 M Tris–HCl, pH 8.0 (Invitrogen 15568-025).
- T4 DNA Ligase (400,000 U/ml) (New England Biolabs M0202L).
- Including: 10× T4 DNA ligase buffer.
- 20× SSPE (Invitrogen 15591043).
- 0.5 M EDTA, pH 8.0 (Ambion 9260G).
- Surfact-Amps 20 (10% Tween-20) (Pierce 28320).
- Herring Sperm DNA (HSDNA, 10 mg/ml, sonicated) (Promega D1815).

- Streptavidin, R-phycoerythrin (SAPE) conjugate (Molecular Probes S866).
- Ab (Biotinylated Anti-Streptavidin) (Vector BA-0500).
- Bleach (4% sodium hypochlorite).
- QIAamp DNA purification kit (Qiagen 51306).
- RNAse A (Qiagen 19101).
- 100 bp marker.
- Ethanol 100%.
- 5 M NaCl (Ambion 9759).
- Molecular biology-grade water.

2.1.1. Solutions

Wash A: Nonstringent Wash Buffer (6× SSPE, 0.01% Tween 20)

For 1,000 ml:

- 300 ml 20× SSPE.
- 1.0 ml 10% Tween 20.
- 699 ml of water.
- Filter through a 0.2-μm filter.
- Store at room temperature.

Wash B: Stringent Wash Buffer (0.6× SSPE, 0.01% Tween 20)

For 1,000 ml:

- 30 ml 20× SSPE.
- 1.0 ml 10% Tween 20.
- 969 ml of water.
- Filter through a 0.2-μm filter.
- Store at room temperature.

0.5 mg/ml Anti-streptavidin Antibody

Resuspend 0.5 mg in 1 ml of water. Aliquot in 100 μl in Eppendorf tubes. Store at −20°C (stock solution). Store work solution at 4°C.

12× MES Stock Buffer (1.22 M MES, 0.89 M [Na^+])

For 1,000 ml:

- 70.4 g MES hydrate.
- 193.3 g MES sodium salt.
- 800 ml molecular-grade water.
- Mix and adjust volume to 1,000 ml.
- The pH should be between 6.5 and 6.7.
- Filter through a 0.2-μm filter.

Do not autoclave. Store at 2–8°C, and shield from light. Discard solution if yellow.

1× Array Holding Buffer (100 mM MES, 1 M [Na^+], 0.01% Tween 20)

For 100 ml:

- 8.3 ml of 12× MES stock buffer.
- 18.5 ml of 5 M NaCl.
- 0.1 ml of 10% Tween 20.
- 73.1 ml of water.
- Store at 2–8°C, and shield from light.

TE Buffer (0.1 mM EDTA, 10 mM Tris-HCl, pH 8.0)

For 500 ml:

- 5 ml 1 M Tris–HCl, pH 8.
- 0.1 ml 0.5 M EDTA, pH 8.
- 494.9 ml of water.
- Filter through a 0.2-μm filter.

3. Methods

3.1. Affymetrix Genechip 250/500 K SNP Array Hybridization for CNV Analysis

Laboratory procedures are performed essentially as described by the manufacturer (http://www.affymetrix.com/support/technical/index.affx) consisting of the following steps (see Notes 1 and 2).

3.1.1. Step 1: Restriction Enzyme Digestion

- 250 ng of genomic DNA is digested by the array type-matched restriction enzyme (Nsp-I or Sty-I) (see Notes 3 and 4).
  - Samples can be stored at −20°C if not proceeding to the next step.

3.1.2. Step 2: Ligation

- Adapters that contain primer sequences are ligated to the digested genomic DNA. The digestions enzyme that is used in step 1 determines the types of adapters that is used in step 2.
- After ligation, the sample is diluted with 75 μl of water (total volume = 100 μl) (see Notes 5 and 6).
  - Samples can be stored at −20°C if not proceeding to the next step.

3.1.3. Step 3: PCR

- Three 100 μl PCR reactions (30 cycles) are performed for every initial sample by using 10 μl of diluted ligation product as template (leaving 70 μl).
- 10. Dilute 4 μl of fragmented PCR product with loading buffer and analyze on agarose gel.
  - On agarose gel, DNA fragment length should be between 200–1100 bp. Some bands can be visible (see Notes 7–9).
  - Samples can be stored at −20°C if not proceeding to the next step.

3.1.4. Step 4: PCR Product Purification and Elution

- The PCR products of every patient are purified and pooled for further processing by using a 96-well filter plate system.
- The purified pooled product is eluted in 40 μl recovery buffer (see Notes 9–12).

3.1.5. Step 5: Quantification of Purified PCR Product

- The purified PCR products are quantified by OD 260/280.
- For next step, 90 μg of each of the purified DNA samples is collected (see Notes 13–16).
  - Samples can be stored at −20°C if not proceeding to the next step.

3.1.6. Step 6: Fragmentation

- The normalized samples are fragmented by DNase treatment.
- Fragmentation of PCR product before hybridization onto the SNP arrays has been shown to be critical in obtaining optimal assay performance. Due to the sensitive nature of the fragmentation reagent (DNase I), these general rules need to be followed to ensure the success of the step (see Notes 17–26):
- After incubation, 4 μl of fragmented PCR product is checked on agarose gel.
- Proceed *immediately* to step 7 if DNA fragments are of good length.
- On agarose gel, all fragments should be smaller than 300 bp (preferably, <200 bp) (see Notes 27–29).
  - At this stage, products cannot be stored at −20°C, and immediately proceed to next step.

3.1.7. Step 7: Labeling

- Fragmented samples are "labeled" with biotinylated nucleotides.
  - Samples can be stored at −20°C if not proceeding to the next step.

3.1.8. Step 8: Target Hybridization

- Labeled samples are hybridized onto 250K SNP arrays overnight (>16 h) at 49°C (see Notes 30–33).

3.1.9. Step 9: Washing, Staining, and Scanning

- After 16–20 h of hybridization, the arrays are extensively washed, stained with fluorescence, and finally scanned (see Note 34).

3.1.10. Step 10: Processing of Array Files

- The data files derived from the scanning are processed in multiple steps.
  - The array data is analyzed with respect to the SNP calling (we normally use the Affymetrix software tool "GTYPE" for this analysis).
  - The array data is analyzed with respect to the CNV calling by calculating the ^{2}Log(test/reference) ratio for

every probe of the array. We normally use the freeware software package, Copy Number Analyzer for Affymetrix GeneChip® (CNAG) published by Nannya et al. (*Cancer Research* 2005).

- The CNVs that are detected in each sample are classified into "likely benign" (not clinically relevant) and "possibly pathogenic." This latter group requires follow-up like parental analyses. A flow scheme for the discrimination between "likely benign" and "possibly pathogenic" is depicted in Fig. 2 (see Notes 35–38).

4. Notes

1. For practical reasons, it is advised to perform all incubations in a PCR machine.
2. For preparation of master mixes, a 5% excess is advised relative to volumes indicated.

Restriction Enzyme Digestion

3. The sample needs to be provided in a maximum of 14.75 μl, which means a minimum sample concentration of 17 ng/μl. Lower concentrations (down to ~10 ng/μl) can be used but might require an extra PCR reaction (step 4).
4. A negative control (water) should be included in the experiment to exclude DNA contamination.

Ligation

5. Ligase buffer contains ATP and should be defrosted and kept at 4°C. Mix ligase buffer thoroughly before use to ensure that precipitate is resuspended. Avoid multiple freeze–thaw cycles.
6. To prevent self-ligation of the adaptor, the T4 DNA ligase should be added to the digested DNA as the last component.

PCR

7. Careful pipetting of viscous fluids is very important.
8. According to the manufacturer's protocol, three PCR reactions are required to produce sufficient product for hybridization to one array (each PCR reaction = 100 μl).
9. 90 μg of purified PCR product is needed for fragmentation. This yield is normally derived from these three PCRs. Yields down to 70 μg can be subsequently processed without major consequences. For DNA samples with very low concentrations, extra PCR reactions can be used to generate extra product.

PCR Product Purification and Elution

10. Cover wells that are not needed with PCR plate cover.
11. When the solution is pipetted into the well of the plate, be careful not to pierce the membrane.
12. For easier recovery of the eluates, the plate can be held at a slight angle.

Quantification of Purified PCR Product

13. To reach 90 μg of product, the concentration should be >2 μg/μl.
14. Use the elution buffer as a blank for OD260/280.
15. Samples with yields down to 70 μg can proceed.
16. Samples with lower yields should be repeated (starting at step 1 and 3.

Fragmentation

17. Store the fragmentation reagent stock at –20°C until ready for use.
18. Make sure that the purified PCR products are in RB buffer with proper volume (45 μl).
19. Preheat the thermal cycler at 37°C.
20. Prepare the fragmentation reagent dilution just prior to use.
21. Prepare diluted fragmentation reagent in 10% excess.
22. Make the fragmentation reagent dilution ON ICE.
23. Perform all the steps AS QUICKLY AS POSSIBLE.
24. Discard remaining fragmentation reagent after use.
25. Proceed *immediately* to step 7 if DNA fragments are of good length!!
26. The concentration of stock fragmentation reagent (U/μl) may vary from lot to lot!!!
27. Varying fragment length within one series results in varying array data quality.
28. Samples with fragment length > 300 bp should be repeated.
29. Samples with fragment length <50–100 base pairs result in high SNP calls but poor CNV data quality.

Target Hybridization

30. Equilibrate the array to hybridization temperature (49°C) before usage.
31. Specifically, if the rubber septa are not equilibrated, they may be prone to cracking.
32. If the pellet is visible, pipette briefly to resuspend before putting the sample at 49°C.

33. When processing multiple samples, leave samples at 49°C until ready to load onto the array.

Washing, Staining, and Scanning

34. Remove the hybridization cocktail from the probe array and set it aside in a microcentrifuge vial for potential rehybridization (*do not discard*). Store at –20°C. This hybridization cocktail can be reused if necessary (see Notes 32 and 34).

Processing of Array Files

35. The SNP call percentage can be used as a quality parameter (e.g., >85% with the dynamic models (DM) algorithm). If samples have an SNP call <85%, the hybridization cocktail (step 9) can be rehybridized onto a new array. This normally leads to higher SNP calls.
36. Note that different SNP calling algorithms are available (e.g., DM, Bayesian Robust Linear Model with Mahalanobis distance classifier (BRLMM)).
37. The average ratio of the X chromosome after sex-mismatch CNV analysis can be used as a quality parameter (e.g., X-ratio should be <–0.3 (for male samples) or >0.3 (for female samples)). If samples have an X-ratio between –0.3 and 0.3, the hybridization cocktail (step 9) can be rehybridized onto a new array. This normally leads to a better X-ratio.
38. The standard deviation (SD) of all the autosomal ^{2}Log(T/R) ratios can be used as a quality parameter (e.g., <0.2). If samples have an SD > 0.2, the entire procedure should be repeated after the initial DNA samples have been purified (preferably by column purification).

References

1. Leonard H and Wen X (2002) The epidemiology of mental retardation: challenges and opportunities in the new millennium. *Ment Retard Dev Disabil Res Rev* **8**, 117–4.
2. Luckasson R, Reeve A. (2001) Naming, defining, and classifying in mental retardation. *Ment Retard.* **39**, 47–52.
3. de Vries BB, van den Ouweland AM, Mohkamsing S, Duivenvoorden HJ, Mol E, Gelsema K, van RM, Halley DJ, Sandkuijl LA, Oostra BA, Tibben A, and Niermeijer MF (1997) Screening and diagnosis for the fragile X syndrome among the mentally retarded: an epidemiological and psychological survey. Collaborative Fragile X Study Group. *Am J Hum Genet* **61**, 660–7.
4. Moog U (2005) The outcome of diagnostic studies on the etiology of mental retardation: considerations on the classification of the causes. *Am J Med Genet* **A 137**, 228–31.
5. Rauch A, Hoyer J, Guth S, Zweier C, Kraus C, Becker C, Zenker M, Huffmeier U, Thiel C, Ruschendorf F, Nurnberg P, Reis A, and Trautmann U (2006) Diagnostic yield of various genetic approaches in patients with unexplained developmental delay or mental retardation. *Am J Med Genet* **A 140**, 2063–74.
6. Ropers HH and Hamel BC (2005) X-linked mental retardation. *Nat Rev Genet* **6**, 46–57.
7. Gécz J, Shoubridge C, Corbett M (2009) The genetic landscape of intellectual disability arising from chromosome X. *Trends Genet* **25**, 308–16.
8. Graham SM, Selikowitz M (1993) Chromosome testing in children with developmental delay in

whom the aetiology is not evident clinically. *J Paediatr Child Health* **29**, 360–2.

9. van Karnebeek CD, Jansweijer MC, Leenders AG, Offringa M, Hennekam RC (2005) Diagnostic investigations in individuals with mental retardation: a systematic literature review of their usefulness. *Eur J Hum Genet* **13,** 6–25.
10. Shapira SK (1998) An update on chromosome deletion and microdeletion syndromes. *Curr Opin Pediatr* **10,** 622–7.
11. Flint J, Wilkie AO, Buckle VJ, Winter RM, Holland AJ, McDermid HE (1995) The detection of subtelomeric chromosomal rearrangements in idiopathic mental retardation. *Nat Genet* **9,** 132–40.
12. De Vries BB, Winter R, Schinzel A, van Ravenswaaij-Arts C. (2003) Telomeres: a diagnosis at the end of the chromosomes. *J Med Genet* **40,** 385–98.
13. Veltman JA, Schoenmakers EF, Eussen BH, Janssen I, Merkx G, van Cleef B, van Ravenswaaij CM, Brunner HG, Smeets D, Geurts van Kessel AG (2002) High-throughput analysis of subtelomeric chromosome rearrangements by use of array-based comparative genomic hybridization. *Am J Hum Genet* **70,** 1269–76.
14. Vissers LE, de Vries BB, Osoegawa K, Janssen IM, Feuth T, Choy CO, Straatman H, van der Vliet W, Huys EH, van Rijk A, Smeets D, van Ravenswaaij-Arts CM, Knoers NV, van der Burgt I, de Jong PJ, Brunner HG, Geurts van Kessel A, Schoenmakers EF, Veltman JA (2003) Array-based comparative genomic hybridization for the genomewide detection of submicroscopic chromosomal abnormalities. *Am J Hum Genet* **73,** 1261–70.
15. Shaw-Smith C, Redon R, Rickman L, Rio M, Willatt L, Fiegler H, Firth H, Sanlaville D, Winter R, Colleaux L, Bobrow M, Carter NP (2004) Microarray based comparative genomic hybridisation (array-CGH) detects submicroscopic chromosomal deletions and duplications in patients with learning disability/mental retardation and dysmorphic features. *J Med Genet* **41**, 241–8.
16. Schoumans J, Ruivenkamp C, Holmberg E, Kyllerman M, Anderlid BM, Nordenskjöld M (2005) Detection of chromosomal imbalances in children with idiopathic mental retardation by array based comparative genomic hybridisation (array-CGH). *J Med Genet* **42,** 699–705.
17. Tyson C, Harvard C, Locker R, Friedman JM, Langlois S, Lewis ME, Van Allen M, Somerville M, Arbour L, Clarke L, McGilivray B, Yong SL, Siegel-Bartel J, Rajcan-Separovic E (2005) Submicroscopic deletions and duplications in individuals with intellectual disability detected by array-CGH. *Am J Med Genet* **A 139,** 173–85.
18. Menten B, Maas N, Thienpont B, Buysse K, Vandesompele J, Melotte C, de Ravel T, Van Vooren S, Balikova I, Backx L, Janssens S, De Paepe A, De Moor B, Moreau Y, Marynen P, Fryns JP, Mortier G, Devriendt K, Speleman F, Vermeesch JR (2006) Emerging patterns of cryptic chromosomal imbalance in patients with idiopathic mental retardation and multiple congenital anomalies: a new series of 140 patients and review of published reports. *J Med Genet* **43,** 625–33.
19. Rosenberg C, Knijnenburg J, Bakker E, Vianna-Morgante AM, Sloos W, Otto PA, Kriek M, Hansson K, Krepischi-Santos AC, Fiegler H, Carter NP, Bijlsma EK, van Haeringen A, Szuhai K, Tanke HJ (2006) Array-CGH detection of micro rearrangements in mentally retarded individuals: clinical significance of imbalances present both in affected children and normal parents. *J Med Genet* **43,** 180–6.
20. de Vries BB, Pfundt R, Leisink M, Koolen DA, Vissers LE, Janssen IM, Reijmersdal S, Nillesen WM, Huys EH, Leeuw N, Smeets D, Sistermans EA, Feuth T, van Ravenswaaij-Arts CM, Geurts van Kessel A, Schoenmakers EF, Brunner HG, Veltman JA (2005) Diagnostic genome profiling in mental retardation. *Am J Hum Genet* **77**, 606–16.
21. Veltman JA. Genomic microarrays in clinical diagnosis (2006) *Curr Opin Pediatr* **18,** 598–603.
22. Knight SJ, Regan R (2006) Idiopathic learning disability and genome imbalance. *Cytogenet Genome Res* **115,** 215–24.
23. Koolen DA, Pfundt R, de Leeuw N, Hehir-Kwa JY, Nillesen WM, Neefs I, Scheltinga I, Sistermans E, Smeets D, Brunner HG, Geurts van Kessel A, Veltman JA, de Vries BB (2009) Genomic microarrays in mental retardation: a practical workflow for diagnostic applications. *Hum Mutation* **30,** 283–92.
24. Hehir-Kwa JY, Egmont-Petersen M, Janssen IM, Smeets D, Geurts van Kessel A, Veltman JA (2007) Genome-wide copy number profiling on high-density BAC, SNP and oligonucleotide microarrays: a platform comparison based on statistical power analysis. *DNA research* **14,** 1–11.
25. McMullan DJ, Bonin M, Hehir-Kwa JY, de Vries BB, Dufke A, Rattenberry E, Steehouwer M, Moruz L, Pfundt R, de Leeuw N, Riess A, Altug-Teber O, Enders H, Singer S, Grasshoff U, Walter M, Walker JM, Lamb CV, Davison EV, Brueton L, Riess O, Veltman JA (2009) Molecular karyotyping of patients with unexplained mental retardation by SNP arrays: A multicenter study. *Hum Mutation* **30,** 1082–92.
26. De Gregori M, Ciccone R, Magini P, Pramparo T, Gimelli S, Messa J, Novara F, Vetro A, Rossi

E, Maraschio P, Bonaglia MC, Anichini C, Ferrero GB, Silengo M, Fazzi E, Zatterale A, Fischetto R, Previderé C, Belli S, Turci A, Calabrese G, Bernardi F, Meneghelli E, Riegel M, Rocchi M, Guerneri S, Lalatta F, Zelante L, Romano C, Fichera M, Mattina T, Arrigo G, Zollino M, Giglio S, Lonardo F, Bonfante A, Ferlini A, Cifuentes F, Van Esch H, Backx L, Schinzel A, Vermeesch JR, Zuffardi O (2007). Cryptic deletions are a common finding in "balanced" reciprocal and complex chromosome rearrangements: a study of 59 patients. *J Med Genet* **44,** 750–62.

27. Hochstenbach R, van Binsbergen E, Engelen J, Nieuwint A, Polstra A, Poddighe P, Ruivenkamp C, Sikkema-Raddatz B, Smeets D, Poot M (2009). Array analysis and karyotyping: workflow consequences based on a retrospective study of 36,325 patients with idiopathic developmental delay in the Netherlands. *Eur J Med Genet* **52(4)**, 161–9.

28. Schluth-Bolard C, Delobel B, Sanlaville D, Boute O, Cuisset JM, Sukno S, Labalme A, Duban-Bedu B, Plessis G, Jaillard S, Dubourg C, Henry C, Lucas J, Odent S, Pasquier L, Copin H, Latour P, Cordier MP, Nadeau G, Till M, Edery P, Andrieux J (2009) Cryptic genomic imbalances in *de novo* and inherited apparently balanced chromosomal rearrangements: array CGH study of 47 unrelated cases. *Eur J Med Genet* **52,** 291–6.

29. Levy S, Sutton G, Ng PC, Feuk L, Halpern AL, Walenz BP, Axelrod N, Huang J,Kirkness EF, Denisov G, Lin Y, MacDonald JR, Pang AW, Shago M, Stockwell TB, Tsiamouri A, Bafna V, Bansal V, Kravitz SA, Busam DA, Beeson KY, McIntosh TC, Remington KA, Abril JF, Gill J, Borman J, Rogers YH, Frazier ME, Scherer SW, Strausberg RL, Venter JC (2007) The Diploid Genome Sequence of an Individual Human. *PLoS Biol* **5,** e254.

30. Korbel JO, Urban AE, Affourtit JP, Godwin B, Grubert F, Simons JF, Kim PM, Palejev D, Carriero NJ, Du L, Taillon BE, Chen Z, Tanzer A, Saunders AC, Chi J, Yang F, Carter NP, Hurles ME, Weissman SM, Harkins TT, Gerstein MB, Egholm M, Snyder M (2007) Paired-End Mapping Reveals Extensive Structural Variation in the Human Genome. *Science* **318,** 420–6.

31. Conrad DF, Pinto D, Redon R, Feuk L, Gokcumen O, Zhang Y, Aerts J, Andrews TD, Barnes C, Campbell P, Fitzgerald T, Hu M, Ihm CH, Kristiansson K, Macarthur DG, Macdonald JR, Onyiah I, Pang AW, Robson S, Stirrups K, Valsesia A, Walter K, Wei J; Wellcome Trust Case Control Consortium, Tyler-Smith C, Carter NP, Lee C, Scherer SW, Hurles ME (2010) Origins and functional impact of copy number variation in the human genome. *Nature* **464,** 704–12.

32. Lee C, Iafrate AJ, Brothman AR (2007) Copy number variations and clinical cytogenetic diagnosis of constitutional disorders. *Nat Genet* **39**, S48–54.

33. Firth HV, Richards SM, Bevan AP, Clayton S, Corpas M, Rajan D, Van Vooren S, Moreau Y, Pettett RM, Carter NP (2009) DECIPHER: Database of Chromosomal Imbalance and Phenotype in Humans Using Ensembl Resources. *Am J Hum Genet* **84,** 524–33.

34. Feenstra I, Fang J, Koolen DA, Siezen A, Evans C, Winter RM, Lees MM, Riegel M, de Vries BB, Van Ravenswaaij CM, Schinzel A (2006) European Cytogeneticists Association Register of Unbalanced Chromosome Aberrations (ECARUCA); an online database for rare chromosome abnormalities. *Eur J Med Genet* **49**, 279–91.

35. Redon R, Ishikawa S, Fitch KR, Feuk L, Perry GH, Andrews TD, Fiegler H, Shapero MH, Carson AR, Chen W, Cho EK, Dallaire S, Freeman JL, González JR, Gratacòs M, Huang J, Kalaitzopoulos D, Komura D, MacDonald JR, Marshall CR, Mei R, Montgomery L, Nishimura K, Okamura K, Shen F, Somerville MJ, Tchinda J, Valsesia A, Woodwark C, Yang F, Zhang J, Zerjal T, Zhang J, Armengol L, Conrad DF, Estivill X, Tyler-Smith C, Carter NP, Aburatani H, Lee C, Jones KW, Scherer SW, Hurles ME (2006) Global variation in copy number in the human genome. *Nature* **444,** 444–54.

36. Brunetti-Pierri N, Berg JS, Scaglia F, Belmont J, Bacino CA, Sahoo T, Lalani SR, Graham B, Lee B, Shinawi M, Shen J, Kang SH, Pursley A, Lotze T, Kennedy G, Lansky-Shafer S, Weaver C, Roeder ER, Grebe TA, Arnold GL, Hutchison T, Reimschisel T, Amato S, Geragthy MT, Innis JW, Obersztyn E, Nowakowska B, Rosengren SS, Bader PI, Grange DK, Naqvi S, Garnica AD, Bernes SM, Fong CT, Summers A, Walters WD, Lupski JR, Stankiewicz P, Cheung SW, Patel A. (2008) Recurrent reciprocal 1q21.1 deletions and duplications associated with microcephaly or macrocephaly and developmental and behavioral abnormalities. *Nat Genet* **40,** 1466–71.

37. Mefford HC, Sharp AJ, Baker C, Itsara A, Jiang Z, Buysse K, Huang S, Maloney VK, Crolla JA, Baralle D, Collins A, Mercer C, Norga K, de Ravel T, Devriendt K, Bongers EM, de Leeuw N, Reardon W, Gimelli S, Bena F, Hennekam RC, Male A, Gaunt L, Clayton-Smith J, Simonic I, Park SM, Mehta SG, Nik-Zainal S, Woods CG, Firth HV, Parkin G, Fichera M, Reitano S, Lo Giudice M, Li KE, Casuga I, Broomer A,

Conrad B, Schwerzmann M, Räber L, Gallati S, Striano P, Coppola A, Tolmie JL, Tobias ES, Lilley C, Armengol L, Spysschaert Y, Verloo P, De Coene A, Goossens L, Mortier G, Speleman F, van Binsbergen E, Nelen MR, Hochstenbach R, Poot M, Gallagher L, Gill M, McClellan J, King MC, Regan R, Skinner C, Stevenson RE, Antonarakis SE, Chen C, Estivill X, Menten B, Gimelli G, Gribble S, Schwartz S, Sutcliffe JS, Walsh T, Knight SJ, Sebat J, Romano C, Schwartz CE, Veltman JA, de Vries BB, Vermeesch JR, Barber JC, Willatt L, Tassabehji M, Eichler EE. (2008) Recurrent Rearrangements of Chromosome 1q21.1 and Variable Pediatric Phenotypes. *N Engl J Med* **359,** 1685–99.

38. Sharp AJ, Mefford HC, Li K, Baker C, Skinner C, Stevenson RE, Schroer RJ, Novara F, De Gregori M, Ciccone R, Broomer A, Casuga I, Wang Y, Xiao C, Barbacioru C, Gimelli G, Bernardina BD, Torniero C, Giorda R, Regan R, Murday V, Mansour S, Fichera M, Castiglia L, Failla P, Ventura M, Jiang Z, Cooper GM, Knight SJ, Romano C, Zuffardi O, Chen C, Schwartz CE, Eichler EE (2008) A recurrent 15q13.3 microdeletion syndrome associated with mental retardation and seizures. *Nat Genet* **40,** 322–328.
39. van Bon BW, Mefford HC, Menten B, Koolen DA, Sharp AJ, Nillesen WM, Innis JW, de Ravel TJ, Mercer CL, Fichera M, Stewart H, Connell LE, Ounap K, Lachlan K, Castle B, Van der Aa N, van Ravenswaaij C, Nobrega MA, Serra-Juhé C, Simonic I, de Leeuw N, Pfundt R, Bongers EM, Baker C, Finnemore P, Huang S, Maloney VK, Crolla JA, van Kalmthout M, Elia M, Vandeweyer G, Fryns JP, Janssens S, Foulds N, Reitano S, Smith K, Parkel S, Loeys B, Woods CG, Oostra A, Speleman F, Pereira AC, Kurg A, Willatt L, Knight SJ, Vermeesch JR, Romano C, Barber JC, Mortier G, Pérez-Jurado LA, Kooy F, Brunner HG, Eichler EE, Kleefstra T, de Vries BB (2009) Further delineation of the 15q13 microdeletion and duplication syndromes: a clinical spectrum varying from non-pathogenic to a severe outcome. *J Med Genet* **46**: 511–23.
40. Ullmann R, Turner G, Kirchhoff M, Chen W, Tonge B, Rosenberg C, Field M, Vianna-Morgante AM, Christie L, Krepischi-Santos AC, Banna L, Brereton AV, Hill A, Bisgaard AM, Müller I, Hultschig C, Erdogan F, Wieczorek G, Ropers HH (2007) Array CGH identifies reciprocal 16p13.1 duplications and deletions that predispose to autism and/or mental retardation. *Human Mutation* **28,** 674–82.
41. Hannes FD, Sharp AJ, Mefford HC, de Ravel T, Ruivenkamp CA, Breuning MH, Fryns JP, Devriendt K, Van Buggenhout G, Vogels A, Stewart H, Hennekam RC, Cooper GM, Regan R, Knight SJ, Eichler EE, Vermeesch JR (2008) Recurrent reciprocal deletions and duplications of 16p13.11: The deletion is a risk factor for MR/MCA while the duplication may be a rare benign variant. *J Med Genet* **46,** 223–32.
42. Webber C, Hehir-Kwa JY, Nguyen DQ, de Vries BBA, Veltman JA, Ponting CP (2009) Forging links between human mental retardation-associated CNVs and mouse gene knockout models. *PLoS Genetics* **5,** e1000531.
43. Hehir-Kwa JY, Wieskamp N, Webber C, Pfundt R, Brunner HG, Gilissen C, de Vries BBA, Ponting CP, Veltman JA (2010) Accurate distinction of pathogenic from benign CNVs in Mental Retardation. *PLoS Computational Biology* **6,** e1000752.
44. Girirajan S, Rosenfeld JA, Cooper GM, Antonacci F, Siswara P, Itsara A, Vives L, Walsh T, McCarthy SE, Baker C, Mefford HC, Kidd JM, Browning SR, Browning BL, Dickel DE, Levy DL, Ballif BC, Platky K, Farber DM, Gowans GC, Wetherbee JJ, Asamoah A, Weaver DD, Mark PR, Dickerson J, Garg BP, Ellingwood SA, Smith R, Banks VC, Smith W, McDonald MT, Hoo JJ, French BN, Hudson C, Johnson JP, Ozmore JR, Moeschler JB, Surti U, Escobar LF, El-Khechen D, Gorski JL, Kussmann J, Salbert B, Lacassie Y, Biser A, McDonald-McGinn DM, Zackai EH, Deardorff MA, Shaikh TH, Haan E, Friend KL, Fichera M, Romano C, Gécz J, DeLisi LE, Sebat J, King MC, Shaffer LG, Eichler EE (2010) A recurrent 16p12.1 microdeletion supports a two-hit model for severe developmental delay. *Nat Genet* **42,** 203–9.

Web Resources

Database of Genomic Variation: http://projects.tcag.ca/variation

DatabasE of Chromosomal Imbalance and Phenotype in Humans using Ensembl Resources (DECIPHER): https://decipher.sanger.ac.uk/information

European Cytogeneticists Association Register of Unbalanced Chromosome Aberrations (ECARUCA): www.ecaruca.net

Genomic Classification of CNVs Objectively (GECCO) http://genomegecco.sourceforge.net

Chapter 4

Copy Number Variation and Psychiatric Disease Risk

Rebecca J. Levy, Bin Xu, Joseph A. Gogos, and Maria Karayiorgou

Abstract

Psychiatric disorders are multifactorial in nature with complex genetic architecture. A number of recent studies, building upon earlier findings of copy number variants (CNVs) at the 22q11.2 locus, suggest that rare CNVs represent an important component of genetic heterogeneity in the etiology of complex psychiatric diseases, such as schizophrenia. De novo CNVs are found with higher frequency among sporadic cases, whereas inherited CNVs are enriched among familial cases. Despite substantial progress, a number of challenges remain, such as pinpointing causative relationships between specific gene(s) affected by CNVs and disease phenotypes as well as distinguishing abnormal structural mutations from neutral polymorphisms and establishing a clear association between individual pathogenic CNV and disease phenotypes.

Key words: Schizophrenia, Copy number variants, 22q11.2 Microdeletion, Rare allele

1. Introduction

Schizophrenia is a devastating neuropsychiatric disorder characterized by impaired cognition, positive psychotic symptoms, such as hallucinations, delusions, and disorganized behavior, as well as negative symptoms, such as social withdrawal and apathy. Early onset, poor response to treatment, frequent relapses, and a chronic course make treatment complicated; considering that the lifetime prevalence is high at 1%, this disorder poses a significant burden on society. Genetic epidemiological studies revealed a recurrence risk with heritability of schizophrenia of about 80% (1). Although, with few exceptions, genetic risk factors for schizophrenia remain largely unidentified despite decades of research into the etiology of the disease, recent studies have provided important new insights into the genetic architecture of the disease as well as promising new leads that will almost certainly lead to the unequivocal identification of additional genetic risk factors in the near future.

Lars Feuk (ed.), *Genomic Structural Variants: Methods and Protocols*, Methods in Molecular Biology, vol. 838, DOI 10.1007/978-1-61779-507-7_4, © Springer Science+Business Media, LLC 2012

1.1. Genetic Architecture of Psychiatric Disorders

Psychiatric disorders are multifactorial in nature with complex genetic etiologies. Both risk allele frequency and penetrance contribute to disease susceptibility. The common disease–common variant (CDCV) hypothesis emphasizes the importance of relatively common alleles, each of small effect, acting together to increase disease risk. The common disease–rare variant (CDRV) hypothesis, conversely, emphasizes the impact of individually rare yet highly penetrant alleles. Psychiatric disorders are undoubtedly caused by both common and rare alleles, yet the relative contribution of each type of allele is unknown.

Association studies to examine the CDCV hypothesis can either investigate candidate genes based on a priori functional or positional evidence or identify candidates using unbiased genome-wide association studies (GWAS) (2). After nearly 1,500 association studies on almost 800 candidate genes for schizophrenia (http://www.szgene.org), no genes have unequivocal statistical support (3, 4). Recently, unbiased GWAS identified additional putative loci for schizophrenia, as well as bipolar disorder and autism (5–7). It is interesting to note that these loci do not include any of the previously implicated top candidate genes.

Because few families transmit psychiatric disorders in a Mendelian pattern and technology was inefficient to recognize rare alleles, the CDRV hypothesis had not received much attention. As technology to detect rare mutations improves, it is becoming increasingly clear that rare alleles may contribute substantially to both familial and sporadic cases of schizophrenia (8–10). Ironically, while the limitations of linkage studies spurred GWAS to elucidate common risk alleles, it was the unintended repurposing of GWAS data for copy number variation detection that provided strong supporting evidence for the rare variant hypothesis. With the discovery of copy number variants (CNVs) as a form of genetic variation, there has been a wave of research demonstrating that rare, occasionally recurrent CNVs contribute to schizophrenia's genetic heterogeneity and etiology. This represents the final stage of a conceptual shift in our understanding of the genetic architecture of schizophrenia. This shift started 15 years ago with the description of an association between 22q11.2 microdeletions with schizophrenia (11).

2. Copy Number Variation and Schizophrenia

2.1. The 22q11.2 Microdeletion and Susceptibility to Schizophrenia

Microdeletion of 22q11.2 was the first CNV described in schizophrenia (11). Since its initial discovery, a strong and specific relationship has been established between 22q11.2 microdeletion and psychosis (9, 11). As they enter adulthood, up to one-third of all individuals carrying this deletion develop psychosis, diagnosed as schizophrenia or schizoaffective disorder (12–14). A number of studies, including CNV arrays, indicate that 22q11.2 deletions account for as many as 1–2% of sporadic (i.e., nonfamilial) schizophrenia cases (8, 9, 11, 15) representing the only confirmed, recurrent structural mutation responsible for introducing new sporadic cases of schizophrenia.

There are no major clinical differences in the core schizophrenia phenotype between individuals with schizophrenia who are 22q11.2 microdeletion carriers and those who are not (16, 17). Moreover, many of these individuals have neither obvious congenital abnormalities nor serious intellectual disabilities, making them phenotypically impossible to differentiate from other schizophrenia patients. The neurocognitive and biological features distinguishing 22q11.2 carriers who do and do not develop schizophrenia remain unknown, although results from a number of preliminary cross-sectional studies suggest that these features align with signs and findings generally associated with schizophrenia. These include increased lateral and third ventricle volumes, generalized decrease in gray and white matter volumes particularly in the frontal and temporal lobes (18, 19), plus a number of cognitive impairments that implicate differences in the development and function of frontal brain regions (20, 21). Overall, the 22q11.2 microdeletion is one of the highest known risk factors for schizophrenia with only a greater risk conferred by being the child of two parents with schizophrenia or the monozygotic co-twin of an affected individual. Yet much is still unknown about how this CNV contributes to schizophrenia etiology.

2.2. Recent CNV Discoveries in Schizophrenia

As CNVs gained acceptance as a new form of genetic variation, the original finding at the 22q11.2 locus hinted that the role of rare CNVs in the genetic makeup of complex psychiatric disorders could be widespread. Following rapid developments in high-density microarray technologies designed to screen for structural variants throughout the whole genome and advancements in statistical analysis methods, several groups in rapid succession provided evidence that rare CNVs contribute to the genetic etiology of schizophrenia.

A 30-Kb resolution SNP array analysis performed on 1,077 individuals (159 schizophrenia and 200 control trios (i.e., affected individual and both biological parents)) determined that de novo (i.e., noninherited) CNVs are 8 times more common in sporadic schizophrenia cases than controls, a 10% rate compared to a 1.3%, respectively (9). In contrast with the de novo CNVs, inherited rare CNVs were only slightly (1.5 times) more common in sporadic cases than controls, representing a smaller but significant increase from 20 to 30%. Analysis of the genes affected by the de novo CNVs demonstrated enrichment for pathways participating in neural development and RNA processing. A parallel scan that focused on familial schizophrenia (defined as cases with at least one additional first- or second-degree relative with schizophrenia) showed a converse picture, where rare inherited CNVs had a prominent role, being almost twice as common in familial cases of schizophrenia compared to sporadic cases or controls (22), while there was no enrichment of de novo CNVs. This discovery resulted from a high-resolution SNP array scan on 48 schizophrenia cases plus their parents and affected relatives compared to more than 300 controls and sporadic schizophrenia cases. The familial cases were 2.7 times more likely than controls to have genic CNVs (i.e., CNVs interrupting or containing genes). Nine of twelve families with inherited CNVs showed evidence of CNV and disease cosegregation. These studies present strong empirical evidence supporting the hypothesis that multiple rare CNVs contribute to the genetic risk of schizophrenia, including private CNVs that in total affect diverse genes. These studies also suggest a divergence in the genetic architecture of familial and nonfamilial (i.e., sporadic) schizophrenia (Fig. 1). While the overall frequency of rare structural variants is equivalent between the familial and sporadic cases (~40%), the heritable nature of variants is markedly different. Sporadic schizophrenia is characterized by a significant enrichment of rare de novo mutations and only a modest increase in the rate of rare *inherited* CNVs, which do not appear to preferentially affect genes. By contrast, familial schizophrenia is characterized by enrichment in rare, *inherited* genic CNVs that hypothetically have higher penetrance while de novo mutations are found at lower frequency.

Three case/control studies (8, 15, 23) also provided evidence for a collective enrichment of rare CNVs in schizophrenia, although they could not assess CNV heritability. Initially, Walsh et al. used ROMA on 150 American schizophrenia cases plus controls followed by SNP-based validation (23). They reported a significant threefold increase in the frequency of rare genic CNVs among

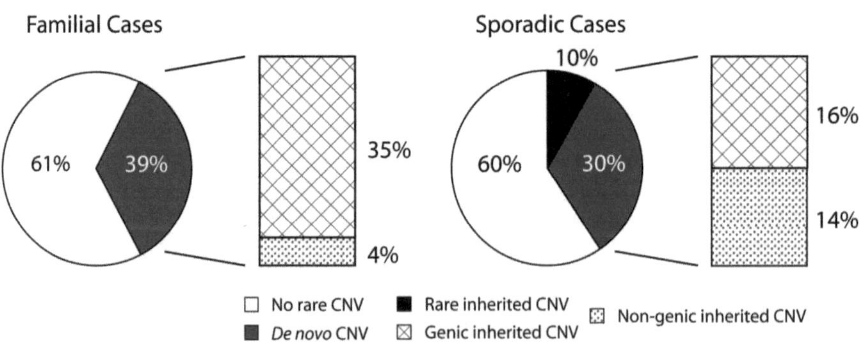

Fig. 1. Divergence in the genetic architecture of familial and nonfamilial schizophrenia. Frequency distribution of rare CNVs identified in familial and sporadic cases of schizophrenia at ~30-kb resolution. There is an ~20% basal rate of inherited CNVs in unaffected controls while the overall frequency of carriers of all rare CNVs is the same between familial and sporadic cases (~40%). Modified from Xu et al. (22).

cases (15% compared to 5% in controls), and a fourfold increase when only early-onset cases were considered (20% compared to 5% in controls). In an independent sample of 83 early-onset trios employing three CNV assays, they determined that rare genic CNVs were present in 28% of cases compared to 13% of nontransmitted parental chromosomes, a significant increase. One recurrent CNV was identified at 16p11.2. Genes affected in the cases were mostly involved in brain development and neural function and included *ERBB4* and *Neurexin1* (*NRXN1*), discussed further below.

The International Schizophrenia Consortium (ISC) analyzed 3,391 European schizophrenia cases and controls using SNP genotype arrays to identify large (>100 kb) CNVs (8). When rare CNVs with <1% overall frequency were considered, cases had 1.15 times as many CNVs as controls. Furthermore, cases had 1.41 times as many genes affected by CNVs. This study confirmed that 22q11.2 deletions were significantly associated with schizophrenia and identified two other recurrent CNVs at 15q13.3 and 1q21.1 that had significantly increased frequency among cases.

Additionally, Stefansson et al. employed a three-step strategy that partially took advantage of a family design to provide additional evidence for the role of rare CNVs in schizophrenia (15). First, they identified 66 de novo CNVs in over 7,500 control European and Chinese families. Next, they tested these 66 CNVs for association with schizophrenia in 1,433 cases and over 33,000 controls. Three recurrent deletions at 1q21.1, 15q11.2, and 15q13.3 were nominally significant among cases with schizophrenia or psychosis, two of which were also reported in the ISC scan (8). Finally, they performed association analysis in 3,285 additional cases and controls, and reported that all three CNVs were significantly associated with psychosis with high odds ratios. Only 1q21.1, however, was significantly associated with a strict definition of schizophrenia. Moreover, the controls included individuals with other psychiatric disorders, thus possibly masking nonpsychotic neuropsychiatric phenotypes associated with these CNVs.

Subsequent to these reports, Need et al. performed both a GWAS and a CNV scan in 871 cases plus controls, followed by independent replication in over 1,400 cases plus controls (24). The GWAS yielded no results that remained significant upon independent replication. Collective CNV analysis, however, indicated an enrichment of large structural variants (>2 Mb) in the cases. CNVs within *NRXN1*, as well as at 1q21.1 and 22q11.2, replicated earlier findings.

Three additional, smaller-scale studies also investigated the importance of rare CNVs in the etiology of schizophrenia. A BAC-CGH scan of 93 Bulgarian schizophrenia trios followed by SNP array verification identified 13 rare CNVs in cases, which were absent from controls or the CNV database (25). The most interesting findings in light of the results discussed above were a deletion at 2p16.3 involving *NRXN1* inherited by affected siblings from their unaffected mother, a de novo duplication at 15q13.1, and a deletion at 16p12.2 inherited from a parent with a nonpsychotic affective disorder. Another SNP array analysis of 54 Dutch cases discovered CNVs affecting *NRXN1*, as well as candidate genes *MYT1L*, *CTNND2*, and *ASTN2* (26). Subsequent targeted screening for CNVs affecting these 4 loci in 752 additional cases and 706 controls demonstrated that CNVs in *MYT1L*, *ASTN2*, and *NRXN1* were only found in cases. A SNP array and CGH validation study in 471 UK schizophrenia cases investigated how CNV frequency, size, and copy number were associated with schizophrenia (27). It was determined that CNVs with frequency <1% and length >1 Mb were 2.3 times more frequent in cases than controls. After combining their data with the Stefansson et al. (15) and ISC (8) studies in a preliminary meta-analysis, 1q21.1, 15q11.2, 15q13.3, and 17p12 demonstrated significant association with a broad schizophrenia diagnosis.

Four groups examined candidate regions of structural variation in schizophrenia. Ingason et al. examined 4,345 European schizophrenia cases and 35,079 controls for copy number variation at the 16p13.1 locus using microarray data (28). They found a threefold increase in duplications and deletions in schizophrenia cases compared with controls, with duplications present in 0.3% of cases versus 0.09% of controls and deletions in 0.12% of cases and 0.04% of controls. Interestingly, in an Icelandic family, one duplication cosegregated in two cases of schizophrenia, as well as one case each of alcoholism, attention deficit hyperactivity disorder, and dyslexia. Additionally, CNVs within 16p11.2 have been implicated in schizophrenia, as well as autism, bipolar disorder, and developmental delay. McCarthy et al. reported an association between 16p11.2 microduplications and schizophrenia in two large cohorts (29). The microduplication was detected in 0.63% cases and 0.03% controls from the initial cohort, and in 0.34% cases and 0.04% controls from the replication cohort, resulting in a 14.5-fold enrichment in the combined sample. Analysis of family cosegregation demonstrates variable penetrance, as well as phenotypic pleiotropy. Next, a screen of 28 candidate loci for neuropsychiatric disorders replicated around 40% of reported CNVs, although no CNVs reached statistically significant association with schizophrenia (30). The conclusion from that study was that CNVs, and thus hypothetically genetic risk factors, are shared between schizophrenia, autism, and mental retardation. An effort to screen 140 CNV loci from the schizophrenia literature yielded only two significant results at 1p13.3 and 22q11.23, both of which contain glutathione transferase genes (31). The results from targeted screens have, therefore, been mixed to date.

Finally, there is one negative report in the literature of genome-wide scans for CNVs. Analysis of 155 Han Chinese cases plus controls on a SNP array did not demonstrate any significant increase in large (>100 kb), rare CNVs among cases (32). The lack of significance was also true among a subset of cases with early-onset schizophrenia. Shi et al. note that population stratification or genetic disease heterogeneity could play a role in this sample set. A dearth of thorough CNV mapping in a large cohort of controls across multiple populations is a significant challenge to data analysis, as discussed below.

Overall, these studies suggest that rare structural rearrangements collectively contribute significantly to schizophrenia risk. De novo mutations appear to contribute more to sporadic than familial schizophrenia. Thus, rare CNVs represent an important source of genetic heterogeneity in the etiology of complex psychiatric diseases, such as schizophrenia.

3. Candidate Schizophrenia Risk Genes within CNVs

Thorough meta-analysis data has not yet been reported, but there are regions of recurrent CNV reports as well as results from preliminary meta-analyses of a subset of studies (27, 33, 34), which are summarized in Table 1. Within these CNVs are many possible candidate genes for neuropsychiatric processes (Table 2). Although the CNVs appear to be quite disparate, it is possible that the genes

Table 1
Loci overlapping in CNV reports

Locus of interest	Study	Candidate genes	Odds ratio (95% CI)[a]
1q21.1	(8, 15, 24, 27)	*GJP8*	9.1 (4.2–19.4)
2p16.3	(8, 23–26, 39)	*NRXN1*	4.78 (2.44–9.37)
15q11.2	(15, 27)	*CYFIP1*	2.8 (2.0–3.9)
15q13.1–q13.3	(8, 15, 25, 27)	*CHRNA7, APBA2, NDNL2, TJP1*	11.4 (4.8–27)
16p11.2	(23, 29)	*MAPK3, DOC2A, SEZ6L2*	8.4 (2.8–25.4)
16p12.2–p12.1	(8, 25)	*EEF2K, CDR2*	
16p12.4–p13.1	(27, 28)	*NDE1, NXPH2, NTAN1*	2.98 (nonsignificant)
22q11.2	(8, 9, 15, 24)	*DGCR8, ZDHHC8, PRODH, COMT, TBX1*	~30

[a]Estimates from incomplete meta-analysis in ref. 34

Table 2
Candidate genes from CNV reports

Study	Genes of interest	Pathways of interest
Walsh et al. (23)	*ERBB4, NRXN1, SLC1A3, GRM7, PRKCD, SKP2, MAGI2, CAV1, PRKAG2, PTK2, DLG2, LAMA1, PTPRM*	Cell adhesion, glutamate receptors, cell cycle, cell growth and extension
Kirov et al. (25)	*NRXN1, APBA2, NDNL2, TJP1, EEF2K, CDR2*	Cell adhesion, amyloid processing, calmodulin signaling
ISC (8)	*CHRNA7, NRXN1, CNTNAP2, NOTCH1, PAK7, GJA8*	Cell adhesion, gap junctions, acetylcholine receptors
Xu et al. (9)	*RAPGEF6, EphB1, DICER1*	Cell signaling, Ephrin signaling, RNA processing
Stefansson et al. (15)	*GJA8, CYFIP1, CHRNA7*	Gap junctions, translation, acetylcholine receptors
Vrijenhoek et al. (26)	*NRXN1, MYT1L, ASTN2*	Cell/synaptic adhesion
Kirov et al., 2009 (27)	*PMP22, NDE1, NXPH2, GJA8, CYFIP1, CHRNA7*	Myelin sheath, cell adhesion
Xu et al. (22)	*NRG3, RAPGEF2, PEX13, KIAA1841, AHSA2, USP34, C4orf45, PTPRN2, CSMD1, MACROD2, A26B3, LOC441956*	Neural differentiation, peroxisomal targeting, ubiquitination
Ingason et al. (28)	*NTAN1, NDE1*	Disc1 signaling, neuronal proliferation, memory regulation

affected by these rare CNVs will converge on a set of neural pathways central to schizophrenia etiology.

The 22q11.2 deletion locus has the highest risk for schizophrenia with an odds ratio (OR) around 30. The minimal 1.5-Mb region found rearranged in patients contains approximately 27 genes of which *COMT*, *PRODH*, *ZDHH8*, *DGCR8*, and *TBX1* have already yielded animal models with behavioral and neuronal deficits (35–38).

Three studies in schizophrenia cohorts reported CNVs in 16p11.2–p13.1, a locus previously found in autism, bipolar disorder, and mental retardation. Possible candidate genes in this region include *NDE1*, which binds *DISC1*, a well-known schizophrenia susceptibility gene; *NXPH2*, which binds neurexins; *EEF2K*, a key kinase downstream of calmodulin; and *CDR2*, which is a target of autoantibodies in cerebellar diseases. Additionally, 1q21.1 is a region previously reported deleted in cases of mental retardation and autism that now has also been reported in schizophrenia.

Many of the regions and candidate genes affected by CNVs demonstrate phenotypic heterogeneity; this could indicate an underlying neuropsychiatric genetic risk or poor clinical differentiation.

Neurexin1 (*NRXN1*) was disrupted by CNVs at 2p16.3 in several schizophrenia genome-wide scans. A targeted screen in 2,977 European schizophrenia patients and 33,746 controls of all 3 neurexin genes found CNVs that affect exons of *NRXN1*, but not *NRXN2* or *3* (39). *NRXN1* could affect the overall brain structure and connectivity since it is a cell-surface receptor involved in synapse formation in the central nervous system (40).

Another interesting set of candidate genes involves microRNA biogenesis and microRNA-mediated translation control. Beyond *DGCR8* within 22q11.2, a candidate gene within the recurrent 15q11.2 CNV region is *CYFIP1*, which binds the fragile X mental retardation protein FMR1 and translation initiation factor eIF4E (41). Both FMR1 and eIF4E have been implicated in microRNA-mediated translational control machinery (42, 43). Moreover, *DICER1*, another microRNA-processing enzyme, has also been affected by a de novo CNV at 14q32 (9).

Additionally, there are also excellent candidate genes located within rare CNVs implicating a number of neural pathways, including glutamate, dopamine, and acetylcholine-signaling pathways, as well as aspects of cell–cell and cell–matrix adhesion and signaling (Table 2). Finally, CNVs may disrupt regulatory elements, transcription factors, or microRNAs that modulate genes at a distance; thus, the directly implicated genes may not represent the full extent of the genetic information affected by these genomic rearrangements.

4. Challenges in Neuropsychiatric CNV Research

Although rare CNVs have been repeatedly shown collectively connected to psychiatric conditions including schizophrenia and autism, it is a challenging task to distinguish abnormal structural mutations from neutral polymorphisms and establish a clear association between individual pathogenic CNV and the disease phenotypes. The main difficulties arise from both our limited knowledge of the nature of CNVs and limitations in current methodology.

Given a set of validated CNVs in the genome of a patient, the degree of difficulty in determining their pathogenicity is dependent on some important properties of the CNVs. Establishing causality may be easier for the CNVs that recur relatively frequently in the population. When a sufficient number of patients is available, a bidirectional association between CNV and disease phenotypes, as exemplified by the 22q11.2 microdeletion situation described above, can be established and used to facilitate further development of diagnostic assays. Specifically, CNV enrichment has to be seen among cases while demonstrating a relatively high disease penetrance among CNV carriers (higher than the baseline rate of psychosis in mental retardation), as well as an absence of major clinical differences in the core schizophrenia phenotype between individuals with schizophrenia who are CNV carriers and those who are not. So far, this has been accomplished only for the 22q11.2 microdeletions.

Establishing association is harder for very rare CNVs or private CNVs found only in a single individual or family, which represents the majority of CNVs described to date in case cohorts. In such instances, a number of features related to the inheritance pattern of the CNV in the family are used to determine a possible causal connection between the CNV locus and disease phenotypes. For example, a rare de novo CNV is more likely to be pathogenic compared to a CNV inherited by an unaffected parent. For inherited CNVs, cosegregation with disease is a strong indicator of causal connection between disease and the CNV. In particular, observation that all affected members of a family carry a rare CNV is a strong indication of pathogenicity (it is not detrimental if unaffected members also carry it since there may be incomplete penetrance). The issue of whether a potentially pathogenic CNV segregates to all affected members within a family has not received much attention in the current literature, which is heavily based on case/control studies. This is a source of concern as it may lead to false findings. In addition to their inheritance pattern, the gene content of a CNV is also an important indicator of its pathogenicity. For example, CNVs that affect the coding region or the splicing pattern of a given gene are more likely to have a detrimental effect than intronic or intergenic CNVs. CNVs that contain genes that are expressed in the brain are more likely to have a negative impact on the nervous system than genes expressed in other organs or tissues. Overall, support for a pathogenic role can be offered by the location of a CNV in a given gene as well as by observation of recurrent incidence of independent CNVs in more than one exon of the same gene or members of the same gene family.

Establishment of causal relationship is complicated by additional methodological issues. First, sample collection design determines whether the desirable CNV properties described above can be reliably evaluated. Case/control design is much easier for sample collection and pooling than a family-based design, but the inheritance pattern information is missing from case/control studies. Second, different assay platforms and analysis algorithms have different sensitivity and specificity in CNV detection. This variation can affect the CNV properties reported and prevent reliable evaluation of causality. It also increases the difficulties in cross-sample comparison and follow-up validation (44). To address this issue, a joint estimation method was proposed recently to estimate copy number from multiple platforms (45). Third, the rarity of CNVs that are collectively associated with neuropsychiatric disorders forces many investigators to pool large numbers of samples collected from different sources to increase the power in study design. These large, collaborative studies identified several recurrent CNVs as described above and provided important insights into the etiology of schizophrenia. However, such design faces the same problems as association studies, including population stratification and inconsistency in diagnostic criteria and control selection (46, 47). The pooling of samples also makes it difficult to evaluate the independence of follow-up validation sample cohorts in relation to the original cohorts.

To reduce the false-positive rate due to these methodological limits, continuously improving and cross-validated reference CNV maps across the whole genome can be used as a filter to remove CNVs that are present in normal populations and, therefore, less likely to associate with disease phenotypes. The Database of Genomic Variants (DGV, http://projects.tcag.ca/variation) is such an accumulation of data, which as of this writing holds reports on over 8,400 CNV loci. It also provides public access to the Genome Structural Variation Consortium results from high-resolution CNV scans on 40 individuals with European or African ancestry. By offering a compendium of current findings, this database is especially valuable to determine whether newly discovered CNVs are unique to one's case population or found at relatively high frequency among controls in prior reports. However, because the CNV annotations in this database are also generated by different labs via different platforms and analysis algorithms, the false-positive detection rate, CNV frequency, CNV size, and CNV breakpoints might also vary. Nevertheless, CNVs that are repeatedly confirmed by independent experiments should be reliable. Caution is needed when considering CNVs reported only once, which could be potentially false-positive findings. Another database, the Wellcome Trust Case–Control Consortium (WTCCC, http://www.wtccc.org.uk), is a collaborative effort among 50 groups in the UK originally assembled to accumulate SNP genotype information across a variety of diseases for over 17,000 individuals. As this information can also be used to identify CNVs, this represents an important collection for genomic rearrangements. However, it should be kept in mind that because this dataset was initially used for SNP genotyping it has suboptimal CNV signal/noise ratio due to its design and low coverage for the CNV regions that have few SNPs. More recently, Conrad et al. provided a population-based CNV map by using tiling oligonucleotide microarray of 42 million probes on 800 individuals from different ethnic groups (48). Another recently launched international research effort, the 1000 Genomes Project (http://www.1000genomes.org/), aims to establish a complete and detailed catalogue of human genetic variations, which in turn can be used for association studies relating genetic variation to disease. Once completed, the project will provide a map with >95% of the variants (e.g., SNPs, CNVs, insertions/deletions) with minor allele frequencies as low as 1% across the genome, 0.1–0.5% in gene regions and a high-resolution reference map for future CNV-based association studies. Integration of these reference CNV databases can provide a relatively solid background in evaluating the pathogenicity of a CNV.

Another way to increase our confidence in establishing CNV–phenotype connections is to explore broader clinical phenotypes that are discovered in different cytogenetic analyses. It is possible (although not always true) that a single CNV can produce multiple spectrum disorders and one disorder might involve multiple CNVs. In that case, a cross-phenotype comparison might help to establish the contribution of a CNV to a specific endophenotype shared among various diseases. Several databases have been established to collect pathogenic CNVs observed in various diseases: the Database of Chromosomal Imbalance and Phenotype in Human using Ensembl Resources (DECIPHER), the Chromosome Abnormality Database (CAD), the Mendelian Cytogenetics Network Online Database, and the European Cytogeneticists Association Register of Unbalanced Chromosome Aberrations (ECARUCA). The accumulation of recurrent CNVs in certain phenotypes provides information on the potential connections between related diseases and might help to dissect some aspects of psychiatric diseases by establishing the causal connection between endophenotypes and underlying molecular mechanisms.

In addition to the challenges in establishing causality between CNVs and psychiatric diseases, another difficulty lies in pinpointing causative relationships between specific gene(s) affected by CNVs and disease phenotypes. An initial challenge is to determine the exact boundaries of CNVs identified in both patient genomes and control populations. The precise information on CNV boundaries offers some insight on what sets of genes are affected and how individual genes are affected, such as whether the structure of an affected gene has been directly disrupted or changes only occur in copy number or at potential regulatory elements. This information can help to determine to what degree the affected genes may contribute to the severity of phenotypes, which is an important issue in clarifying gene contribution to disease phenotypes. Considering the set of genes within a given CNV, there are at least two ways in which the removal or duplication of genes from a region deleted or duplicated by a CNV could be acting: a single dosage-sensitive gene may exert a major effect or alternatively the CNV-associated neurobehavioral phenotypes may stem from the cumulative effect of a number of gene products, acting either additively or synergistically. According to the latter model, although one or a few loci may have a greater phenotypic impact, it is the cumulative effect of the imbalance of several genes within the CNV that determines the overall phenotype. This scenario is likely to complicate gene identification efforts using traditional human genetic approaches. Under this condition, several complementary approaches can potentially provide insights. First, animal models can be very informative in evaluating the impact of genes within CNVs (35, 37, 38, 49). Generation of induced pluripotent stem cells (iPSCs), a newly developed cellular reprogramming technology, also provides opportunities to determine the potential deficit associated with a specific CNV by directly working on patient-derived neurons (50). Finally, higher density arrays and deep sequencing technology are now providing improved resolution to fully determine the CNV boundaries and mutational load of resident genes. These complementary approaches can provide us with important insights into how one or more genes within a CNV can contribute to some aspects of psychiatric disorders.

Finally, the contribution of CNVs has to be evaluated in the context of the genetic architecture of schizophrenia. As discussed above, CDCV and CDRV hypotheses propose two distinct, but not mutually exclusive, classes of variants for the etiology of schizophrenia. It is likely that both classes of variants are involved, but a key issue is to evaluate their relative importance. That is, how the genetic architecture of schizophrenia is built from these different variants, their interactions with each other, and their interactions with environmental and epigenetic factors. Current studies in both rare CNVs and common variants have identified several variants that are strongly associated with the disease, but these variants only account for a small portion of schizophrenia cases. Perhaps, there are more rare variants with lower frequencies, many more common variants with smaller effects, or, more likely, both. It is also possible that interactions among these different variants might provide additional phenotypes that cannot be explained by individual mutations. There are several scenarios on how mutations and their polymorphism modifiers combine to influence the variable expressivity of phenotypes. Two CNVs, or a CNV and a common variant, do not show significant contribution to a phenotype on their own, but their interaction increases the risk of developing an abnormal phenotype. One deletion CNV may expose a common risk variant that is normally masked by diploidy. Another important aspect in this integrated picture is how these genetic variants are influenced by environmental or other epigenetic factors. The impact of nongenetic factors is evident, for example, in the fact that there is less than 50% concordance for schizophrenia in genetically identical twins.

References

1. Sullivan, P. F., Kendler, K. S., and Neale, M. C. (2003) Schizophrenia as a complex trait: evidence from a meta-analysis of twin studies, *Arch. Gen. Psychiatry.* **60**, 1187–1192.
2. Altshuler, D., Daly, M. J., and Lander, E. S. (2008) Genetic mapping in human disease, *Science* **322**, 881–888.
3. Need, A. C. and Goldstein, D. B. (2009) Next generation disparities in human genomics: concerns and remedies, *Trends Genet.* **25**, 489–494.
4. Sanders, A. R., Duan, J., Levinson, D. F., Shi, J., He, D., Hou, C., Burrell, G. J., Rice, J. P., Nertney, D. A., Olincy, A., Rozic, P., Vinogradov, S., Buccola, N. G., Mowry, B. J., Freedman, R., Amin, F., Black, D. W., Silverman, J. M., Byerley, W. F., Crowe, R. R., Cloninger, C. R., Martinez, M., and Gejman, P. V. (2008) No significant association of 14 candidate genes with schizophrenia in a large European ancestry sample: implications for psychiatric genetics, *Am. J. Psychiatry* **165**, 497–506.
5. Ferreira, M. A., O'Donovan, M. C., Meng, Y. A., Jones, I. R., Ruderfer, D. M., Jones, L., Fan, J., Kirov, G., Perlis, R. H., Green, E. K., Smoller, J. W., Grozeva, D., Stone, J., Nikolov, I., Chambert, K., Hamshere, M. L., Nimgaonkar, V. L., Moskvina, V., Thase, M. E., Caesar, S., Sachs, G. S., Franklin, J., Gordon-Smith, K., Ardlie, K. G., Gabriel, S. B., Fraser, C., Blumenstiel, B., Defelice, M., Breen, G., Gill, M., Morris, D. W., Elkin, A., Muir, W. J., McGhee, K. A., Williamson, R., MacIntyre, D. J., MacLean, A. W., St, C. D., Robinson, M., Van, B. M., Pereira, A. C., Kandaswamy, R., McQuillin, A., Collier, D. A., Bass, N. J., Young, A. H., Lawrence, J., Ferrier, I. N., Anjorin, A., Farmer, A., Curtis, D., Scolnick, E. M., McGuffin, P., Daly, M. J., Corvin, A. P., Holmans, P. A., Blackwood, D. H., Gurling, H. M., Owen, M. J., Purcell, S. M., Sklar, P., and Craddock, N. (2008) Collaborative genome-wide association analysis supports a role for ANK3 and CACNA1C in bipolar disorder, *Nat. Genet.* **40**, 1056–1058.
6. O'Donovan, M. C., Craddock, N., Norton, N., Williams, H., Peirce, T., Moskvina, V., Nikolov, I., Hamshere, M., Carroll, L., Georgieva, L., Dwyer, S., Holmans, P., Marchini, J. L., Spencer, C. C., Howie, B., Leung, H. T., Hartmann, A. M., Moller, H. J., Morris, D. W., Shi, Y., Feng, G., Hoffmann, P., Propping, P., Vasilescu, C., Maier, W., Rietschel, M., Zammit, S., Schumacher, J., Quinn, E. M., Schulze, T. G., Williams, N. M., Giegling, I., Iwata, N., Ikeda, M., Darvasi, A., Shifman, S., He, L., Duan, J., Sanders, A. R., Levinson, D. F., Gejman, P. V., Cichon, S., Nothen, M. M., Gill, M., Corvin, A., Rujescu, D., Kirov, G., Owen, M. J., Buccola, N. G., Mowry, B. J., Freedman, R., Amin, F., Black, D. W., Silverman, J. M., Byerley, W. F., and Cloninger, C. R. (2008) Identification of loci associated with schizophrenia by genome-wide association and follow-up, *Nat. Genet.* **40**, 1053–1055.
7. Wang, K., Zhang, H., Ma, D., Bucan, M., Glessner, J. T., Abrahams, B. S., Salyakina, D., Imielinski, M., Bradfield, J. P., Sleiman, P. M., Kim, C. E., Hou, C., Frackelton, E., Chiavacci, R., Takahashi, N., Sakurai, T., Rappaport, E., Lajonchere, C. M., Munson, J., Estes, A., Korvatska, O., Piven, J., Sonnenblick, L. I., varez Retuerto, A. I., Herman, E. I., Dong, H., Hutman, T., Sigman, M., Ozonoff, S., Klin, A., Owley, T., Sweeney, J. A., Brune, C. W., Cantor, R. M., Bernier, R., Gilbert, J. R., Cuccaro, M. L., McMahon, W. M., Miller, J., State MW, Wassink, T. H., Coon, H., Levy, S. E., Schultz, R. T., Nurnberger, J. I., Haines, J. L., Sutcliffe, J. S., Cook, E. H., Minshew, N. J., Buxbaum, J. D., Dawson, G., Grant, S. F., Geschwind, D. H., Pericak-Vance, M. A., Schellenberg, G. D., and Hakonarson, H. (2009) Common genetic variants on 5p14.1 associate with autism spectrum disorders, *Nature* **459**, 528–533.
8. ISC (2008) Rare chromosomal deletions and duplications increase risk of schizophrenia, *Nature* **455**, 237–241.
9. Xu, B., Roos, J. L., Levy, S., van Rensburg, E. J., Gogos, J. A., and Karayiorgou, M. (2008) Strong association of de novo copy number mutations with sporadic schizophrenia, *Nat. Genet.* **40**, 880–885.
10. Bodmer, W. and Bonilla, C. (2008) Common and rare variants in multifactorial susceptibility to common diseases, *Nat. Genet.* **40**, 695–701.
11. Karayiorgou, M., Morris, M. A., Morrow, B., Shprintzen, R. J., Goldberg, R., Borrow, J., Gos, A., Nestadt, G., Wolyniec, P. S., Lasseter, V. K., and . (1995) Schizophrenia susceptibility associated with interstitial deletions of chromosome 22q11, *Proc. Natl. Acad. Sci. USA* **92**, 7612–7616.
12. Pulver, A. E., Nestadt, G., Goldberg, R., Shprintzen, R. J., Lamacz, M., Wolyniec, P. S., Morrow, B., Karayiorgou, M., Antonarakis, S. E., Housman, D., and . (1994) Psychotic illness in patients diagnosed with velo-cardio-facial syndrome and their relatives, *J. Nerv. Ment. Dis.* **182**, 476–478.

13. Murphy, K. C., Jones, L. A., and Owen, M. J. (1999) High rates of schizophrenia in adults with velo-cardio-facial syndrome, *Arch. Gen. Psychiatry* **56**, 940–945.
14. Gothelf, D., Feinstein, C., Thompson, T., Gu, E., Penniman, L., Van, S. E., Kwon, H., Eliez, S., and Reiss, A. L. (2007) Risk factors for the emergence of psychotic disorders in adolescents with 22q11.2 deletion syndrome, *Am. J. Psychiatry* **164**, 663–669.
15. Stefansson, H., Rujescu, D., Cichon, S., Pietilainen, O. P., Ingason, A., Steinberg, S., Fossdal, R., Sigurdsson, E., Sigmundsson, T., Buizer-Voskamp, J. E., Hansen, T., Jakobsen, K. D., Muglia, P., Francks, C., Matthews, P. M., Gylfason, A., Halldorsson, B. V., Gudbjartsson, D., Thorgeirsson, T. E., Sigurdsson, A., Jonasdottir, A., Jonasdottir, A., Bjornsson, A., Mattiasdottir, S., Blondal, T., Haraldsson, M., Magnusdottir, B. B., Giegling, I., Moller, H. J., Hartmann, A., Shianna, K. V., Ge, D., Need, A. C., Crombie, C., Fraser, G., Walker, N., Lonnqvist, J., Suvisaari, J., Tuulio-Henriksson, A., Paunio, T., Toulopoulou, T., Bramon, E., Di, F. M., Murray, R., Ruggeri, M., Vassos, E., Tosato, S., Walshe, M., Li, T., Vasilescu, C., Muhleisen, T. W., Wang, A. G., Ullum, H., Djurovic, S., Melle, I., Olesen, J., Kiemeney, L. A., Franke, B., Sabatti, C., Freimer, N. B., Gulcher, J. R., Thorsteinsdottir, U., Kong, A., Andreassen, O. A., Ophoff, R. A., Georgi, A., Rietschel, M., Werge, T., Petursson, H., Goldstein, D. B., Nothen, M. M., Peltonen, L., Collier, D. A., St, C. D., and Stefansson, K. (2008) Large recurrent microdeletions associated with schizophrenia, *Nature* **455**, 232–236.
16. Bassett, A. S., Hodgkinson, K., Chow, E. W., Correia, S., Scutt, L. E., and Weksberg, R. (1998) 22q11 deletion syndrome in adults with schizophrenia, *Am. J. Med. Genet.* **81**, 328–337.
17. Bassett, A. S., Chow, E. W., AbdelMalik, P., Gheorghiu, M., Husted, J., and Weksberg, R. (2003) The schizophrenia phenotype in 22q11 deletion syndrome, *Am. J. Psychiatry* **160**, 1580–1586.
18. Chow, E. W., Zipursky, R. B., Mikulis, D. J., and Bassett, A. S. (2002) Structural brain abnormalities in patients with schizophrenia and 22q11 deletion syndrome, *Biol. Psychiatry* **51**, 208–215.
19. vanAmelsvoort, T., Daly, E., Henry, J., Robertson, D., Ng, V., Owen, M., Murphy, K. C., and Murphy, D. G. (2004) Brain anatomy in adults with velocardiofacial syndrome with and without schizophrenia: preliminary results of a structural magnetic resonance imaging study, *Arch. Gen. Psychiatry* **61**, 1085–1096.
20. vanAmelsvoort, T., Henry, J., Morris, R., Owen, M., Linszen, D., Murphy, K., and Murphy, D. (2004) Cognitive deficits associated with schizophrenia in velo-cardio-facial syndrome, *Schizophr. Res.* **70**, 223–232.
21. Chow, E. W., Watson, M., Young, D. A., and Bassett, A. S. (2006) Neurocognitive profile in 22q11 deletion syndrome and schizophrenia, *Schizophr. Res.* **87**, 270–278.
22. Xu, B., Woodroffe, A., Rodriguez-Murillo, L., Roos, J. L., van Rensburg, E. J., Abecasis, G. R., Gogos, J. A., and Karayiorgou, M. (2009) Elucidating the genetic architecture of familial schizophrenia using rare copy number variant and linkage scans, *Proc. Natl. Acad. Sci. USA* **106**, 16746–16751.
23. Walsh, T., McClellan, J. M., McCarthy, S. E., Addington, A. M., Pierce, S. B., Cooper, G. M., Nord, A. S., Kusenda, M., Malhotra, D., Bhandari, A., Stray, S. M., Rippey, C. F., Roccanova, P., Makarov, V., Lakshmi, B., Findling, R. L., Sikich, L., Stromberg, T., Merriman, B., Gogtay, N., Butler, P., Eckstrand, K., Noory, L., Gochman, P., Long, R., Chen, Z., Davis, S., Baker, C., Eichler, E. E., Meltzer, P. S., Nelson, S. F., Singleton, A. B., Lee, M. K., Rapoport, J. L., King, M. C., and Sebat, J. (2008) Rare structural variants disrupt multiple genes in neurodevelopmental pathways in schizophrenia, *Science* **320**, 539–543.
24. Need, A. C., Ge, D., Weale, M. E., Maia, J., Feng, S., Heinzen, E. L., Shianna, K. V., Yoon, W., Kasperaviciute, D., Gennarelli, M., Strittmatter, W. J., Bonvicini, C., Rossi, G., Jayathilake, K., Cola, P. A., McEvoy, J. P., Keefe, R. S., Fisher, E. M., St Jean, P. L., Giegling, I., Hartmann, A. M., Moller, H. J., Ruppert, A., Fraser, G., Crombie, C., Middleton, L. T., St, C. D., Roses, A. D., Muglia, P., Francks, C., Rujescu, D., Meltzer, H. Y., and Goldstein, D. B. (2009) A genome-wide investigation of SNPs and CNVs in schizophrenia, *PLoS. Genet.* **5**, e1000373.
25. Kirov, G., Gumus, D., Chen, W., Norton, N., Georgieva, L., Sari, M., O'Donovan, M. C., Erdogan, F., Owen, M. J., Ropers, H. H., and Ullmann, R. (2008) Comparative genome hybridization suggests a role for NRXN1 and APBA2 in schizophrenia, *Hum. Mol. Genet.* **17**, 458–465.
26. Vrijenhoek, T., Buizer-Voskamp, J. E., van, d. S., I, Strengman, E., Sabatti, C., Geurts van, K. A., Brunner, H. G., Ophoff, R. A., and Veltman, J. A. (2008) Recurrent CNVs disrupt three candidate genes in schizophrenia patients, *Am. J. Hum. Genet.* **83**, 504–510.
27. Kirov, G., Grozeva, D., Norton, N., Ivanov, D., Mantripragada, K. K., Holmans, P.,

Craddock, N., Owen, M. J., and O'Donovan, M. C. (2009) Support for the involvement of large copy number variants in the pathogenesis of schizophrenia, *Hum. Mol. Genet.* **18**, 1497–1503.

28. Ingason, A., Rujescu, D., Cichon, S., Sigurdsson, E., Sigmundsson, T., Pietilainen, O. P., Buizer-Voskamp, J. E., Strengman, E., Francks, C., Muglia, P., Gylfason, A., Gustafsson, O., Olason, P. I., Steinberg, S., Hansen, T., Jakobsen, K. D., Rasmussen, H. B., Giegling, I., Moller, H. J., Hartmann, A., Crombie, C., Fraser, G., Walker, N., Lonnqvist, J., Suvisaari, J., Tuulio-Henriksson, A., Bramon, E., Kiemeney, L. A., Franke, B., Murray, R., Vassos, E., Toulopoulou, T., Muhleisen, T. W., Tosato, S., Ruggeri, M., Djurovic, S., Andreassen, O. A., Zhang, Z., Werge, T., Ophoff, R. A., Rietschel, M., Nothen, M. M., Petursson, H., Stefansson, H., Peltonen, L., Collier, D., Stefansson, K., and Clair, D. M. (2009) Copy number variations of chromosome 16p13.1 region associated with schizophrenia, *Mol. Psychiatry*, Sept 29 (Epub ahead of print).

29. McCarthy, S. E., Makarov, V., Kirov, G., Addington, A. M., McClellan, J., Yoon, S., Perkins, D. O., Dickel, D. E., Kusenda, M., Krastoshevsky, O., Krause, V., Kumar, R. A., Grozeva, D., Malhotra, D., Walsh, T., Zackai, E. H., Kaplan, P., Ganesh, J., Krantz, I. D., Spinner, N. B., Roccanova, P., Bhandari, A., Pavon, K., Lakshmi, B., Leotta, A., Kendall, J., Lee, Y. H., Vacic, V., Gary, S., Iakoucheva, L. M., Crow, T. J., Christian, S. L., Lieberman, J. A., Stroup, T. S., Lehtimaki, T., Puura, K., Haldeman-Englert, C., Pearl, J., Goodell, M., Willour, V. L., Derosse, P., Steele, J., Kassem, L., Wolff, J., Chitkara, N., McMahon, F. J., Malhotra, A. K., Potash, J. B., Schulze, T. G., Nothen, M. M., Cichon, S., Rietschel, M., Leibenluft, E., Kustanovich, V., Lajonchere, C. M., Sutcliffe, J. S., Skuse, D., Gill, M., Gallagher, L., Mendell, N. R., Craddock, N., Owen, M. J., O'Donovan, M. C., Shaikh, T. H., Susser, E., Delisi, L. E., Sullivan, P. F., Deutsch, C. K., Rapoport, J., Levy, D. L., King, M. C., and Sebat, J. (2009) Microduplications of 16p11.2 are associated with schizophrenia, *Nat. Genet.* **41**, 1223–1227.

30. Guilmatre, A., Dubourg, C., Mosca, A. L., Legallic, S., Goldenberg, A., Drouin-Garraud, V., Layet, V., Rosier, A., Briault, S., Bonnet-Brilhault, F., Laumonnier, F., Odent, S., Le, V. G., Joly-Helas, G., David, V., Bendavid, C., Pinoit, J. M., Henry, C., Impallomeni, C., Germano, E., Tortorella, G., Di, R. G., Barthelemy, C., Andres, C., Faivre, L., Frebourg, T., Saugier, V. P., and Campion, D. (2009) Recurrent rearrangements in synaptic and neurodevelopmental genes and shared biologic pathways in schizophrenia, autism, and mental retardation, *Arch. Gen. Psychiatry* **66**, 947–956.

31. Rodriguez-Santiago, B., Brunet, A., Sobrino, B., Serra-Juhe, C., Flores, R., Armengol, L., Vilella, E., Gabau, E., Guitart, M., Guillamat, R., Martorell, L., Valero, J., Gutierrez-Zotes, A., Labad, A., Carracedo, A., Estivill, X., and Perez-Jurado, L. A. (2009) Association of common copy number variants at the glutathione S-transferase genes and rare novel genomic changes with schizophrenia, *Mol. Psychiatry*, June 16 (Epub ahead of print).

32. Shi, Y. Y., He, G., Zhang, Z., Tang, W., Zhang, J., Jr., Zhao, Q., Zhang, J., Sr., Li, X. W., Xi, Z. R., Fang, C., Zhao, X. Z., Feng, G. Y., and He, L. (2008) A study of rare structural variants in schizophrenia patients and normal controls from Chinese Han population, *Mol. Psychiatry* **13**, 911–913.

33. Itsara, A., Cooper, G. M., Baker, C., Girirajan, S., Li, J., Absher, D., Krauss, R. M., Myers, R. M., Ridker, P. M., Chasman, D. I., Mefford, H., Ying, P., Nickerson, D. A., and Eichler, E. E. (2009) Population analysis of large copy number variants and hotspots of human genetic disease, *Am. J. Hum. Genet.* **84**, 148–161.

34. Kirov, G. (2010) The role of copy number variation in schizophrenia, *Expert. Rev. Neurother.* **10**, 25–32.

35. Paterlini, M., Zakharenko, S. S., Lai, W. S., Qin, J., Zhang, H., Mukai, J., Westphal, K. G., Olivier, B., Sulzer, D., Pavlidis, P., Siegelbaum, S. A., Karayiorgou, M., and Gogos, J. A. (2005) Transcriptional and behavioral interaction between 22q11.2 orthologs modulates schizophrenia-related phenotypes in mice, *Nat. Neurosci.* **8**, 1586–1594.

36. Paylor, R., Glaser, B., Mupo, A., Ataliotis, P., Spencer, C., Sobotka, A., Sparks, C., Choi, C. H., Oghalai, J., Curran, S., Murphy, K. C., Monks, S., Williams, N., O'Donovan, M. C., Owen, M. J., Scambler, P. J., and Lindsay, E. (2006) Tbx1 haploinsufficiency is linked to behavioral disorders in mice and humans: implications for 22q11 deletion syndrome, *Proc. Natl. Acad. Sci. USA* **103**, 7729–7734.

37. Mukai, J., Dhilla, A., Drew, L. J., Stark, K. L., Cao, L., MacDermott, A. B., Karayiorgou, M., and Gogos, J. A. (2008) Palmitoylation-dependent neurodevelopmental deficits in a mouse model of 22q11 microdeletion, *Nat. Neurosci.* **11**, 1302–1310.

38. Stark, K. L., Xu, B., Bagchi, A., Lai, W. S., Liu, H., Hsu, R., Wan, X., Pavlidis, P., Mills, A. A., Karayiorgou, M., and Gogos, J. A. (2008) Altered brain microRNA biogenesis contributes to phenotypic deficits in a 22q11-deletion mouse model, *Nat. Genet.* **40**, 751–760.

39. Rujescu, D., Ingason, A., Cichon, S., Pietilainen, O. P., Barnes, M. R., Toulopoulou, T., Picchioni, M., Vassos, E., Ettinger, U., Bramon, E., Murray, R., Ruggeri, M., Tosato, S., Bonetto, C., Steinberg, S., Sigurdsson, E., Sigmundsson, T., Petursson, H., Gylfason, A., Olason, P. I., Hardarsson, G., Jonsdottir, G. A., Gustafsson, O., Fossdal, R., Giegling, I., Moller, H. J., Hartmann, A. M., Hoffmann, P., Crombie, C., Fraser, G., Walker, N., Lonnqvist, J., Suvisaari, J., Tuulio-Henriksson, A., Djurovic, S., Melle, I., Andreassen, O. A., Hansen, T., Werge, T., Kiemeney, L. A., Franke, B., Veltman, J., Buizer-Voskamp, J. E., Sabatti, C., Ophoff, R. A., Rietschel, M., Nothen, M. M., Stefansson, K., Peltonen, L., St, C. D., Stefansson, H., and Collier, D. A. (2009) Disruption of the neurexin 1 gene is associated with schizophrenia, *Hum. Mol. Genet.* **18**, 988–996.
40. Reissner, C., Klose, M., Fairless, R., and Missler, M. (2008) Mutational analysis of the neurexin/neuroligin complex reveals essential and regulatory components, *Proc. Natl. Acad. Sci. USA* **105**, 15124–15129.
41. Napoli, I., Mercaldo, V., Boyl, P. P., Eleuteri, B., Zalfa, F., De, R. S., Di, M. D., Mohr, E., Massimi, M., Falconi, M., Witke, W., Costa-Mattioli, M., Sonenberg, N., Achsel, T., and Bagni, C. (2008) The fragile X syndrome protein represses activity-dependent translation through CYFIP1, a new 4E-BP, *Cell* **134**, 1042–1054.
42. Jin, P., Zarnescu, D. C., Ceman, S., Nakamoto, M., Mowrey, J., Jongens, T. A., Nelson, D. L., Moses, K., and Warren, S. T. (2004) Biochemical and genetic interaction between the fragile X mental retardation protein and the microRNA pathway, *Nat. Neurosci.* **7**, 113–117.
43. Pillai, R. S., Bhattacharyya, S. N., Artus, C. G., Zoller, T., Cougot, N., Basyuk, E., Bertrand, E., and Filipowicz, W. (2005) Inhibition of translational initiation by Let-7 MicroRNA in human cells, *Science* **309**, 1573–1576.
44. Curtis, C., Lynch, A. G., Dunning, M. J., Spiteri, I., Marioni, J. C., Hadfield, J., Chin, S. F., Brenton, J. D., Tavare, S., and Caldas, C. (2009) The pitfalls of platform comparison: DNA copy number array technologies assessed, *BMC. Genomics* **10**, 588.
45. Zhang, N. R., Senbabaoglu, Y., and Li, J. Z. (2010) Joint estimation of DNA copy number from multiple platforms, *Bioinformatics.* **26**, 153–160.
46. McCarroll, S. A., Kuruvilla, F. G., Korn, J. M., Cawley, S., Nemesh, J., Wysoker, A., Shapero, M. H., de Bakker, P. I., Maller, J. B., Kirby, A., Elliott, A. L., Parkin, M., Hubbell, E., Webster, T., Mei, R., Veitch, J., Collins, P. J., Handsaker, R., Lincoln, S., Nizzari, M., Blume, J., Jones, K. W., Rava, R., Daly, M. J., Gabriel, S. B., and Altshuler, D. (2008) Integrated detection and population-genetic analysis of SNPs and copy number variation, *Nat. Genet.* **40**, 1166–1174.
47. Ionita-Laza, I., Rogers, A. J., Lange, C., Raby, B. A., and Lee, C. (2009) Genetic association analysis of copy-number variation (CNV) in human disease pathogenesis, *Genomics* **93**, 22–26.
48. Conrad, D. F., Pinto, D., Redon, R., Feuk, L., Gokcumen, O., Zhang, Y., Aerts, J., Andrews, T. D., Barnes, C., Campbell, P., Fitzgerald, T., Hu, M., Ihm, C. H., Kristiansson, K., Macarthur, D. G., MacDonald, J. R., Onyiah, I., Pang, A. W., Robson, S., Stirrups, K., Valsesia, A., Walter, K., Wei, J., Tyler-Smith, C., Carter, N. P., Lee, C., Scherer, S. W., and Hurles, M. E. (2009) Origins and functional impact of copy number variation in the human genome, *Nature*, Oct. 7 (Epub ahead of print).
49. Sigurdsson, T., Stark, K. L., Karayiorgou, M., Gogos, J. A., and Gordon, J. A. (2010) Impaired hippocampal-prefrontal synchrony in a genetic mouse model of schizophrenia. *Nature*, in press.
50. Raff, M. (2009) New routes into the human brain, *Cell* **139**, 1209–1211.

Chapter 5

Detection and Characterization of Copy Number Variation in Autism Spectrum Disorder

Christian R. Marshall and Stephen W. Scherer

Abstract

There now exist multiple lines of evidence pointing to a significant genetic component underlying the aetiology of autism spectrum disorders (ASDs). The advent of methodologies for scanning the human genome at high resolution, coupled with the recognition of copy number variation (CNV) as a prevalent source of genomic variation, has led to new strategies in the identification of clinically relevant loci. Balanced genomic changes, such as translocations and inversions, also contribute to ASD, but current studies have shown that screening with microarrays has up to fivefold increase in diagnostic yield. Recent work by our group and others has shown unbalanced genomic alterations that are likely pathogenic in upwards of 10% of cases, highlighting an important role for CNVs in the genetic aetiology of ASD. A trend in our empirical data has shifted focus for discovery of candidate loci towards individually rare but highly penetrant CNVs instead of looking for common variants of low penetrance. This strategy has proven largely successful in identifying ASD-susceptibility candidate loci, including gains and losses at 16p11.2, SHANK2, NRXN1, and PTCHD1. Another emerging and intriguing trend is the identification of the same genes implicated by rare CNVs across neurodevelopmental disorders, including schizophrenia, attention deficit hyperactivity disorder, and intellectual disability. These observations indicate that similar pathways may be involved in phenotypically distinct outcomes. Although interrogation of the genome at high resolution has led to these novel discoveries, it has also made cataloguing, characterization, and clinical interpretation of the increasing amount of CNV data difficult. Herein, we describe the history of genomic structural variation in ASD and how CNV discovery has been used to pinpoint novel ASD-susceptibility loci. We also discuss the overlap of CNVs across neurodevelopmental disorders and comment on the current challenges of understanding the relationship between CNVs and associated phenotypes in a clinical context.

Key words: Autism spectrum disorder, Copy number variation, Genetics, Genome variation, Microarray, Neurodevelopmental disorders

1. Introduction

Autism spectrum disorders (ASDs) are lifelong neurological conditions that affect a person's development and how their brain processes information. Autism is the prototypic form of the ASDs

Lars Feuk (ed.), *Genomic Structural Variants: Methods and Protocols*, Methods in Molecular Biology, vol. 838,
DOI 10.1007/978-1-61779-507-7_5, © Springer Science+Business Media, LLC 2012

while other subtypes include Asperger syndrome and pervasive developmental disorder not otherwise specified (PDD-NOS). The ASD phenotype is characterized by challenges in communication, social interaction, and learning, as well as by unusual behaviour, interests, and activities (1, 2). The incidence of ASD is currently observed to be as high as 1 in 150 (3) with diagnosis in males four times more common than in females. Autism conditions are spectrum disorders, which mean that the symptoms can be present in a variety of combinations with a range of severity. Adding to the phenotypic complexity in ASD is a 40% prevalence of developmental disability/intellectual disability (DD/ID), as well as clinically significant symptoms consistent with schizophrenia (SZ), attention deficit hyperactivity disorder (ADHD), and obsessive compulsive disorder (OCD) being observed in 5, 59–75, and 60% of ASD cases, respectively.

Although first identified by Leo Kanner in 1943 (1), it took more than 30 years for the importance of a genetic contribution to the aetiology of ASD to be recognized (reviewed in ref. 4). From family studies, the heritability of ASD is estimated to be as high as 90%, making it the neuropsychiatric disorder most strongly influenced by genetics. Due to rapid technical advances in the last 10 years, there has been significant progress towards unravelling the genetic causes associated with the ASD phenotype. One strategy that has proven particularly useful for discovery of ASD candidate genes is genome-wide screening for copy number variants (CNVs), defined as segments of DNA present in varying number of copies in different individuals. CNVs are now recognized as a prevalent source of genomic variation that account for both normal phenotypic variation between individuals and that associated with risk for disease. The latter observation has resulted in a paradigm shift away from the previously held "common disease–common variant" hypothesis to a "common disease–rare variant" model for the genetic architecture of ASD. The central tenet of this model suggests a role for multiple, rare, highly penetrant, genetic risk factors for ASD, many of which are in the form of CNVs. Perhaps not surprisingly, given the phenotypic overlap between conditions, many of the ASD-associated CNVs and the genes they implicate have also been associated with other conditions, such as DD/ID, SZ, and ADHD. In this review, we summarize the latest CNV findings associated with ASD, with particular emphasis on the strategies used to detect, characterize, and interpret CNVs in a clinical context.

1.1. The Genetics of ASD

There is strong evidence from twin and family studies for the importance of complex genetic factors in the development of idiopathic autism (5, 6). Family studies have shown that a recurrence rate of autism in siblings of affected probands is as high as 8–10% (5, 7). Thus, the recurrence risk in siblings is roughly

100 times higher than found in the general population. The substantial degree of familial clustering in ASD could reflect shared environmental factors, but twin studies strongly point to genetics. Several epidemiological same-sex twin studies have clearly demonstrated significant differences in the monozygotic (MZ) and dizygotic (DZ) twin concordance rates. The largest of these studies (8) found that 60% of the MZ pairs were concordant for autism compared with none of the DZ pairs, suggesting a heritability estimate of >90% assuming a multifactorial threshold model. When concordance was examined for a phenotype that consists of autism or milder cognitive and social deficits, 82% of MZ twins, compared with ~10% DZ twin pairs, were concordant (9). Strikingly discordant forms of ASD can occur in the same family (10), although affected siblings in general tend to be similar with respect to the level of function and other domains (11, 12). The marked difference in pairwise concordance between MZ and DZ twins and the rapid fall-off in recurrence risks with decreasing genetic relatedness both suggest that autism is a multi-locus disorder (reviewed in refs. 6, 13–15). Through analysis of recurrence rates, current genetic models suggest that sporadic autism in the low-risk families would be mainly caused by spontaneous mutations of high penetrance in males (16). The same mutations may be of relatively low penetrance in females that could then be passed on to affected male offspring (16).

Approximately 10–15% of ASD individuals have an identifiable Mendelian condition or genetic syndrome (including chromosomal anomalies; see below) (6, 13, 15, 17). Fragile X syndrome (0–20%), tuberous sclerosis (TSC; 0.4–2.9%), and neurofibromatosis (NF1; 0.2–14%) are the most frequently cited Mendelian conditions. Other rare microdeletion or single-gene defects have been associated with ASD, including those found in Williams, Sotos, Cowden, Moebius, and Timothy syndrome. ASD may also occur in some mitochondrial diseases and untreated phenylketonuria. In addition, cytogenetically visible anomalies on all chromosomes are observed in ASD (see the Autism Chromosome Rearrangement Database or *ACRD* hosted by our group; http://www.projects.tcag.ca/autism/; ref. 18). Recent studies, which employ standardized diagnostic assessment tools and exclude children with profound intellectual disabilities, tend to identify a lower prevalence of cytogenetic abnormalities compared to other studies using less-rigorous diagnostic standards (18). Estimates from some studies may also be biased upwards by selective testing based on the presence of dysmorphic features and/or severe disability. Nevertheless, in our survey of 15 worldwide studies (including our own data), the mean rate observed was 7.4% (129/1,749) with a range from 0 to 54% (18, 19). Among the most frequent findings are fra(X)(q27)(3.1%; 28/899) and anomalies involving maternally inherited proximal 15q11-q13 (0.97%; 17/1,749).

Microarray technologies can now detect all of the unbalanced chromosomal changes traditionally observed by karyotyping. When we applied both of these methods to over 400 ASD probands, we found that of the ~7.4% of cases that carried de novo structural variants 7.1% were unbalanced changes (or CNVs) all found by microarrays and 0.4% were balanced translocations or inversions only seen by karyotyping (20). Other studies have reported a similar trend, with a higher diagnostic yield in unexplained DD/ID and/or ASD (20 vs. 3%) (21) and ASD (7 vs. 2%) (22) for microarray testing versus conventional G-banded karyotype. Low-level mosaicism and balanced rearrangements that cannot be detected by microarrays are generally found to be infrequent causes (up to 1%) of abnormal phenotype suggesting that microarray screens should be used as a first-tier cytogenetic diagnostic test in these cohorts (21).

Although not discussed in detail here (and reviewed extensively in ref. 6), it should be noted that other strategies are being used in the search for ASD candidate loci. Candidate gene exon re-sequencing has identified causative mutations in some genes, most notably *NLGN3* and *NLGN4* (23), but these are mostly extremely rare and sequencing in general has failed to uncover a common explanation for ASD. Much effort has also been invested in linkage and genome-wide association (GWA) studies with only moderate success and little replication thus far. The first common single-nucleotide polymorphism (SNP) variant identified to reach genome-wide significance mapped 1 Mb away from the cadherin 9 (CDH9) and cadherin 10 (CDH10) genes of chromosome 5p14.1, two genes that encode neuronal cell-adhesion molecules (24). Other genes with promising common associations include *SEMA5A* (25) and CNTNAP2 (26), but the overall lack of genome-wide significance for any common variant is indicative of a small effect size attributable to any particular gene.

1.2. Rare De Novo and Inherited CNVs Identify ASD-Susceptibility Genes

Taken together, the strong genetic contribution shown in family studies and the association of cytogenetic changes, but apparent lack of common risk factors in ASD, led to the hypothesis that rare sub-microscopic unbalanced changes in the form of CNVs likely contribute to the ASD phenotype. With the development of microarrays capable of scanning the genome at sub-microscopic resolution, there is accumulating evidence that multiple, rare de novo and some inherited CNVs contribute to the genetic vulnerability to ASD (reviewed in ref. 27). For example, some karyotypic and sub-microscopic CNV abnormalities involving many genes have been known for some time, including the maternal 15q11-q13 duplications in 1–3% of ASD (28). In some syndromes, such as Potocki-Lupski syndrome/dup17p11.2 also involving many genes, over 90% of cases exhibit ASD features (29). CNVs can also be small involving a single gene and act more similar to a sequence-level mutation. In either instance, screening for CNVs has proven to be a rapid method to identify candidate ASD-susceptibility loci.

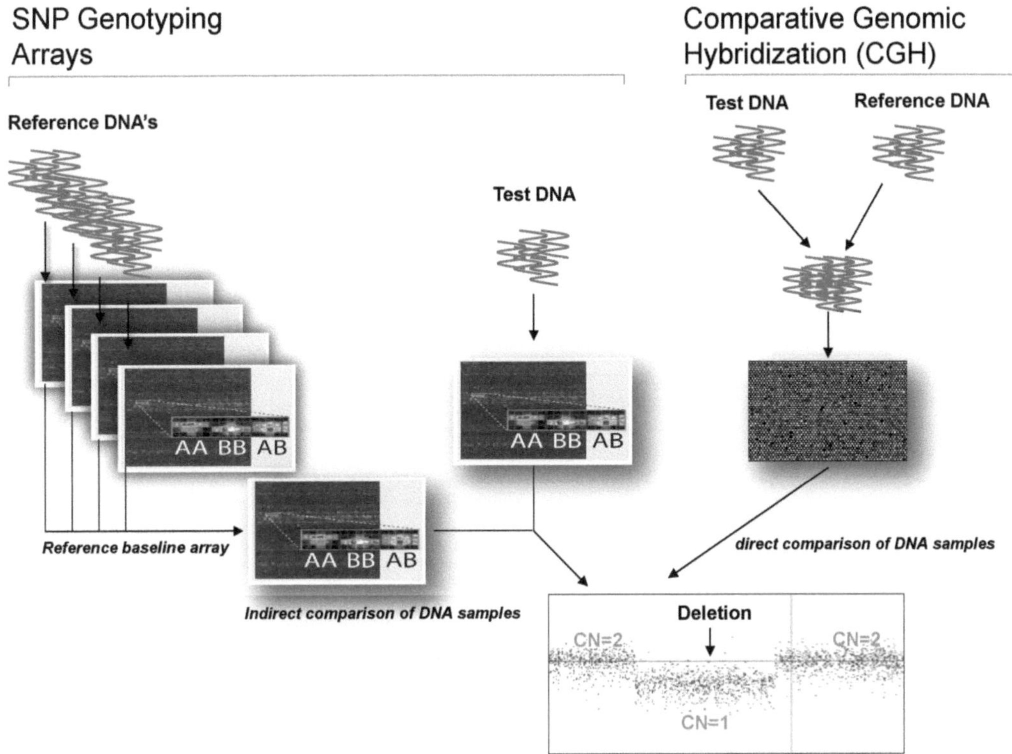

Fig. 1. Comparison of SNP- and CGH-based platforms for CNV calling. Both SNP and CGH microarray platforms can be used for CNV analysis. For SNP arrays, fluorescently labelled genomic DNA from one individual (test) is hybridized to an array. CNVs can be detected in the test individual by comparing the intensities of the test array to a reference baseline made of several other experiments. The *lower panel* shows an example of a deletion detected in the test individual. Array probe intensities are denoted as *black dots* centred around a *horizontal line* denoting a copy number (CN) of two (diploid sequence). The *black arrow* denotes a downward deflection of intensities of the test sample compared to the reference baseline indicating CN = 1 (deletion). For the CGH array, the test and reference DNA are differentially labelled before hybridization to the array, thus creating a direct comparison of DNA samples with no need to create a reference baseline. The difference in fluorescence between the test and reference is then plotted in a similar manner to that described above for detection of CNVs. Refer to Note 1 for additional details.

To quantify the role of CNV in ASD, a series of large studies (20, 30–38) have used different microarray platforms examining different ASD cohorts (there are also smaller studies and case reports). The general study design is typically family based, whereby trios from simplex or multiplex families meeting minimal standard criteria for an ASD are genotyped. Most groups have used SNP-based platforms at various resolutions from either Affymetrix or Illumina and employed a wide range of tools for CNV detection. A more detailed description of the microarray technologies and an example of our CNV analysis pipeline are described in Note 1 (also see Fig. 1).

Our strategy for determining the significance of a CNV in ASD probands involves several steps and is outlined in Fig. 2 (also see Note 2). Using these strategies, relevant findings considering all of the data from the larger CNV screening studies include the

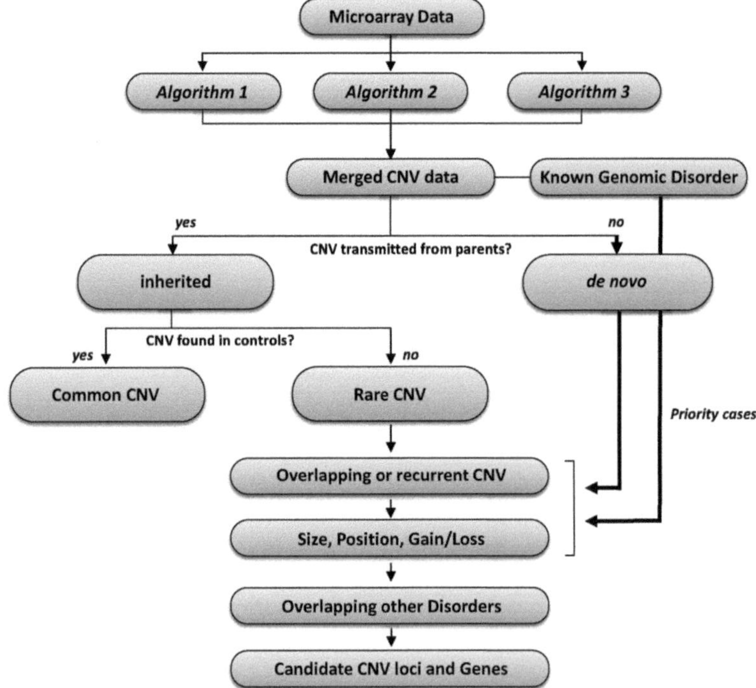

Fig. 2. General analyses and prioritization workflow for rare CNVs. Microarray data is analyzed for CNV using three different algorithms to maximize CNV discovery resulting in a merged CNV dataset at the sample level. Priority is given to those CNVs detected by more than one algorithm (validation >95% using qPCR) in order to decease potential false CNV calls (also see Note 1). CNVs for follow-up are prioritized based on several criteria as described in Note 2. Those CNVs that are already defined as a genomic disorder and those that are de novo are given the highest priority for follow-up. For rare-inherited CNVs not found in controls, parameters, such as genomic size, position, recurrence, and overlap with other neurodevelopmental disorders, are used to assess potential significance as candidate loci for follow-up.

following: (1) The proportion of de novo CNVs differed among simplex (7–10%), multiplex (2–3%), and non-ASD controls (~1%), similar to other complex inherited disorders that display evidence for different mechanisms underlying presumed sporadic and familial cases (20, 31). (2) Some individuals with ASD had two or more de novo CNVs (emphasizing the need for high-resolution microarrays) (20) and 27% of families with syndromic forms of ASD carried de novo events (35). (3) Up to 40% of family-specific CNVs were found to be inherited from an apparently non-ASD parent (20). (4) In several examples, unrelated individuals carried de novo or inherited CNVs at the same locus, and some of these coincided with known genomic disorders, while others involved genes associated with DD/ID (20, 35). (5) There is an overall enrichment of rare CNVs overlapping genes in ASD cases compared to controls, especially for those loci previously implicated in ASD and/or ID (38).

(6) Comprehensive pathway analysis shows significant enrichment of particular gene sets in ASD, including those involved in cellular proliferation, projection and motility, GTPase/Ras signalling (38), as well as neuronal synaptic complex genes, notably, *SHANK2*, *SHANK3*, *NLGN4*, and *NRXN1* (reviewed in ref. 6, 27) and ubiquitin degradation genes (37). Suggested CNV interpretation and further analysis are discussed in Note 3.

One of the most significant findings was highlighted in three studies published concurrently with identification of an ~600-kb microdeletion/microduplication of chromosome 16p11.2 (~1%) (20, 32, 33). These 16p11.2 CNVs were also observed in ASD cases with additional dysmorphology, in non-ASD cases having DD (32), and in some controls. In our latest work, we found that the 16p11.2 deletions are more penetrant (100% for either ASD or DD/ID in our cohort and controls) than the duplications (~50%) (39). Another study with consanguineous ASD families identified rare homozygous CNVs deleting both copies of genes, such as *PCDH10*, *DIA1*, and *NHE926*, all genes regulated by neuronal activity (36). Other novel ASD genes recently implicated by CNVs include SHANK2 (40), SYNGAP1, and DLGAP2 (38).

A recent observation that provides potential explanation of the skewed male ASD diagnosis is the finding of CNVs and sequence mutations at the X-linked DDX53-PTCHD1 locus (38, 41). Initially, a 167-kb microdeletion spanning exon 1 was found in two ASD brothers (20), and in a second family a 90-kb microdeletion spanning the entire *PTCHD1* gene was found in three males with MR. In additional males with ASD, we identified deletions in the 5′-flanking region of PTCHD1 disrupting a complex non-coding RNA and potential regulatory elements; equivalent changes were not found in male control individuals. Using conventional Sanger-based sequencing, we tested 900 ASD and 208 ID male probands and identified 7 different missense changes in eight probands. All of these X-linked "mutations" were inherited from unaffected mothers and not found in 1,000 matched controls. Interestingly, two of the ASD individuals with missense changes also carried a de novo deletion at another ASD-susceptibility locus (*DPYD* and *DPP6*) identified through our CNV studies (20), suggesting complex genetic contributions. Systematic screening at PTCHD1 and 5′-flanking regions suggests involvement of this locus in ~1% of ASD and ID individuals.

CNVs can rapidly identify candidate ASD genes, and DNA sequencing can then be applied to search in other (or the same) patients for subtler sequence-level mutations. As with the PTCHD1 locus described above, this approach has been used effectively to implicate *SHANK3* (42), *NLGN3* and *NLGN4* (43), and *SHANK2* (40) as rare-highly penetrant potentially monogenic genes in ASD. In a related example, we have found de novo CNV deletions in the *ANKRD11* gene at 16q24.3 in multiple unrelated families with either ASD or DD/ID (44). In this instance, the phenotype seems more severe when the CNV encompasses both the *ANKRD11* gene and *CDH15*, a gene just recently shown to be involved in intellectual disability (45). These observations also emphasize the importance of performing CNV and mutation screening on the same samples so as to increase the likelihood of finding multi-genic contributors.

To make sense of the contribution of CNVs to ASD, a "threshold" model has recently been proposed (27). The model posits that different CNVs exhibit different penetrance depending on the dosage sensitivity and function (relative to ASD) of the gene(s) they affect. Some CNVs have a large impact on ASD susceptibility and these are typically de novo in origin, cause more severe ASD symptoms, are more prevalent among sporadic forms of ASD, and are less influenced by other factors like gender and parent of origin. Other CNVs have moderate or mild effects that probably require other genetic (or non-genetic) factors to take the phenotype across the ASD threshold. Some of these CNVs demonstrate variable phenotypic expression, are found in other disorders (see below), and are observed in non-ASD family members and also in some population controls. Features of this model will likely also apply to DD/ID (as discussed above for 16p11.2, *PTCHD1*, and *ANKRD11*), SZ, ADHD, and OCD.

1.3. Neurodevelopmental Disorders with Common CNVs and Phenotypes to ASD

Perhaps, the most intriguing CNV data emerging is that common genes/pathways are being identified across neurodevelopmental disorders. In Table 1, based on an extensive literature search and our unpublished data, we summarize >20 gene/CNV loci from across the genome associated with two or more neuropsychiatric phenotypes. Below, we describe phenotypic overlaps between ASD and SZ, ADHD, and ID/DD and touch on some of the more interesting CNVs found across disorders.

Other than ASD, the disorder most commonly screened for CNVs is SZ. The clinical overlap between ASDs and childhood-onset schizophrenia is already well-documented as early editions of the DSM-placed autism and childhood-onset SZ in the same diagnostic category (46). It is reported that 12% of adults with ASD met criteria for a psychotic disorder (47) and, conversely, that autistic symptoms are exceedingly common in children who later develop SZ (48). Evidence is accumulating that, like in ASD, CNVs contribute to the aetiology of SZ. The best known example is the 22q11.2 microdeletion, which has been reported in some ASD subjects (frequency not well-studied) (6, 20, 28, 30) and 1% of SZ studies (49). Conversely, in clinical studies of children with known 22q11.2 deletion syndrome, 20–50% have an ASD, and in adults 25% have diagnosable psychotic disorder (49, 50). One case–control study (51) found more novel rare CNVs in SZ cases with adult (15%) or childhood onset (20%) compared with controls (5%).

Table 1
Genomic loci associated with multiple neuropsychiatric disorders

#	Cytoband	Locus	Disorder[a]	References[b]
1	1q21.1	1q21.1 locus	ASD, SZ	(68, 70, 76)
2	1q44	KIF26B	SZ, ADHD	(58, 69, 72)
3	2p16.3	NRXN1	ASD, SZ	(20, 37, 66, 69, 72, 75, 78)
4	5p15.2	CTNND2	SZ, ADHD	(58, 78)
5	5p15.3	TPPP	ASD, SZ	(66, 70)
6	6q25.2–q27	PARK2	ASD, ADHD	(37, 58)
7	7q11.22	AUTS2	ASD, ADHD	(37, 58)
8	7q35-q36	CNTNAP2	ASD, SZ, ADHD	(58, 65, 67, 74)
9	7q36.1	PRKAG2	ASD, SZ	(51, 77)
10	7q36.2	DPP6	ASD, ADHD	(20, 58)
11	8p22	NAT2	SZ, ADHD	(58, 72)
12	9q33.1	ASTN2	ASD, SZ, ADHD	(25, 37, 72, 78)
13	15q11.2–q12	GABRA5	ASD, SZ	(66, 72)
14	15q13.1	NDNL2	ASD, SZ	(30, 69, 73)
15	15q21.1	DUOXA1	ASD, ADHD	(58, 77)
16	16p11.2	16p11.2 locus	ASD, SZ, ADHD	(20, 25, 32, 33, 51)
17	16p13.3	A2BP1	ASD, ADHD	(36, 58, 71)
18	19p12	ZNF676	ASD, ADHD	(58, 66)
19	20p12	PAK7	ASD, SZ, ADHD	(37, 58, 72)
20	22q11.21	22q11.2 locus	ASD, SZ	(30, 37, 68, 70, 72, 73, 75)
21	22q12.3	LARGE	ASD, SZ	(51, 77)
22	Xp22.11	PTCHD1 locus	ASD, ADHD	un, (38, 41)

[a] ASD - autism spectrum disorder; SZ - schizophrenia; ADHD - attention deficit hyperactivity disorder; OCD - obsessive compulsive disorder.
[b] un - unpublished observations

Another study (52) tested for association of de novo CNVs with SZ, finding 10% in sporadic cases, 1.3% in unaffected controls, and 0% in familial cases (noting that the 0% may reflect small sample size). Similar to findings for ASD (20, 34), ~1% of schizophrenia cases carried two or more de novo CNV events (52). Notable candidate genes for SZ from CNV studies include NRXN1, the 16p11.2 locus, and the 1q21.1 locus (51), which are also found in ASD cohorts.

Attention deficits and hyperactivity are clinically significant problems in up to 75% of ASD individuals (47, 53), but DSM-IV discourages the diagnosis of ADHD in a child who is already diagnosed with an ASD. Conversely, surveys of ADHD populations have found a higher-than-expected rate of co-morbid autistic symptoms (54, 55). Like ASD, there is strong evidence for a genetic aetiology of ADHD. GWASs studies have identified susceptibility loci for ADHD that overlap with regions implicated in ASD, for example 16p13, 17p11, and 5p13 (56, 57). Recently (including our unpublished data), overlapping rare CNVs between ASD and ADHD have been found at nine different loci, including the 16p11.2 locus, ASTN2, and DPP6 (see Table 1). Our preliminary data suggests that the de novo CNV mutation rate in ADHD is ~1%, not significantly different than that in controls. Some studies (58), including our own observations, suggest a higher frequency of rare-inherited CNVs present in cases compared to controls, and in many instances they overlap other neuropsychiatric CNV regions.

DD/ID is operationally defined as having an IQ score less than 70 and impairments in adaptive functioning. Individuals with ASDs can have a wide range of IQ scores, but DD/ID is a common co-morbidity in 40% of cases (59). As already discussed, studies of CNVs in ASD have revealed deletions and duplications that seem to manifest as ASD in some individuals, but as non-specific or syndromic DD/ID in others.

The overlaps of CNVs described above and in Table 1 may be due to (1) co-occurrence by chance, (2) misdiagnosis, (3) overlap in symptoms across psychiatric diagnostic categories, or (4) the likely pleiotropy and/or variable expressivity of the genes involved. A working hypothesis states that human disease-phenotype groups (phenome connections) might reflect the different functional domains of a single protein, interaction between different proteins (such as ligand and receptor), interaction between proteins in a multi-protein complex, and different steps in a cellular pathway (60–62).

2. Implications and Conclusions

Although much has been accomplished in finding some genetic causes of ASD, there remains a lot of work to do. The development of the microarray technologies has made whole-genome analysis increasingly fast, cost-effective, and routine, allowing reliable detection of much of the larger unbalanced genetic variation in an individual DNA sample. The detection and characterization of CNVs in ASD cases has proven a powerful tool for discovery and has led to the identification of numerous candidate genes and loci, including NRXN1, SHANK3, SHANK2, PTCHD1, and the 16p11.2 locus. Although many of these loci are individually rare, representing causation in no more than 1–2% of a case cohort, collectively one may expect to find a putative pathogenic CNV (either de novo or large) in upwards of 10–20% of cases at the current resolution of microarray screens. Re-sequencing to identify additional mutations of those candidate genes pinpointed through CNV studies can increase yields as is the case for SHANK isoforms. As high-resolution technologies are being used to scan the genome, it is becoming increasingly apparent that the genetic architecture responsible for the expression of the ASD phenotype is made up of many, individually rare (familial) variants. With this increasing yield per experiment, the challenge is correct interpretation of the clinical significance of each variant. For the vast majority of these variants, it is not possible to assign causality until larger numbers of cases and controls are run. With next-generation sequencing technologies now capable of exome sequencing at a reasonable cost, the number of rare variants needing interpretation increase substantially.

The other trend emerging from genome scanning studies is the identification of the same CNVs or genes interrupted by CNVs across multiple neurodevelopmental disorders, like SZ, ADHD, and ID. The fact that many of these pathogenic variants are pleiotropic has broad implications for diagnostics and highlights the need for well-phenotyped cases. The preliminary success in identifying CNVs, genes, and mutations associated with ASD has already raised expectations both in the clinical community and in the public. Considering all data, it is imperative that research discoveries are introduced into clinical testing even though there may be incomplete evidence of their clinical utility. The translation of research findings into clinical practice must be a continuing priority with the ultimate goal of early detection and treatment of ASDs.

3. Notes

1. There are several commercially available high-resolution microarrays available for CNV analysis, falling into two general categories. The first is SNP-genotyping microarrays, which can interrogate over a million SNP genotypes in a single individual and are provided mainly by the companies Affymetrix and Illumina. The second type of microarray, provided by companies such as Agilent and Roche Nimblegen, is known as comparative genomic hybridization (CGH) arrays. Both platforms are made up of oligonucleotide probes that bind to fluorescently labelled genomic DNA for which the intensities can be used to interrogate copy number. The main difference is that for CGH arrays the reference DNA is hybridized to the array along with the test DNA allowing a direct comparison of copy number. For SNP-genotyping arrays, only one sample is hybridized to

the array and the reference must be built from other experiments in order to analyze for copy number (see Fig. 1). Both platforms are capable of providing excellent CNV data and whenever possible both approaches should ideally be employed since in our experience the platforms are complementary for CNV calling. When only using one platform, we generally prefer using an SNP-based platform. Despite the disadvantage of potential decreased sensitivity for CNV calling, SNP-based platforms have the advantage over CGH arrays of providing genotype information that can be used in a variety of ways including: (1) association analysis with SNPs in or near CNV regions; (2) determination of population substructure of samples for case control studies; (3) quality control checks to ensure that there has been no sample mix-up and to determine parentage if the study is family based; (4) determination of allele-specific copy number changes (e.g. amplification of a specific allele or copy-neutral loss of heterozygosity); and (5) using genotype calls (homozygosity) as confirmation of deletion calls. The following is a description of our analysis methods for Affymetrix SNP-genotyping arrays, but the same principles and strategies can be used for analysis of CGH arrays.

The Affymetrix® Genome-Wide Human SNP Array 6.0 includes more than 906,600 SNPs and more than 946,000 probes for the detection of copy number variation. The array also contains probes specifically designed for detection of 5,677 known copy number variable regions from the Database of Genomic Variants (http://www.projects.tcag.ca/variation/) housed at The Centre for Applied Genomics (TCAG). Greater than 200,000 probes are used to interrogate copy number at these known CNVs for an average of 61 probes per loci. In addition, more than 744,000 probes are evenly spaced along the genome to enable the detection of novel and putative disease-associated CNV. All 1.8 million probes can be used to detect CNVs enabling routine detection of CNVs as small as 2,000 bases.

For quality control, array files are loaded into Affymetrix® Genotyping Console (GTC). Those meeting a contrast QC > 0.4 were SNP genotyped using the built-in algorithm Birdseed version 2. For CNV analysis, we used three different CNV analysis tools (see Fig. 2) in order to obtain the most comprehensive CNV dataset possible: (1) Birdsuite, (2) iPattern, and (3) Affymetrix GTC. Analysis using Birdsuite is done using a pool of patients and controls as previously described (63) using a confidence score cut-off of 10. iPattern (64) implements a non-parametric density-based clustering model that integrates intensity data across samples to assign distinct copy number states to genomic loci. A two-stage analytical framework is used with an optimal moving window-based approach followed by secondary boundary refinement. For Affymetrix GTC, experiments were compared to a baseline constructed of controls run in the same facility and a built-in HMM was used for CNV detection. CNVs were merged using the outside probe boundaries if they were detected in the same individual by more than one algorithm. Using these three tools provides a unique merged CNV dataset averaging ~150 CNVs per sample, whereas the stringent calling criteria (regions found by two or more analysis tools) yield ~50–60 CNVs per sample. In our experience, the combination of using multiple calling algorithms and strict filters maximizes CNV discovery while allowing for a simple prioritization scheme for CNV for follow-up. We have found that CNVs detected by multiple analysis tools have a low false-positive rate, validating at a rate of >95% using laboratory techniques, such as quantitative PCR (qPCR). If further size cut-offs are used, we expect the sensitivity and specificity for CNVs >500 kb in size to be approaching 100% with a slight drop to 98% for CNVs >100 kb in size. We also note that many CNVs called by only one of the above methods (upwards of 50%) could be true positives and should not be discounted.

2. There are several criteria to prioritize CNVs for follow-up with the general overall scheme we use shown in Fig. 2. To identify putative pathogenic variants, CNVs from ASD cases should be compared to large (>1,000 s) control datasets. Ideally, these controls should be run on the same platforms and analyzed in a similar manner, but we have also taken advantage of large control datasets run on different platforms to determine if a variant is rare or not found in the population. CNVs in the case dataset that are at least 50% unique (by genomic length) compared with the control CNV datasets are considered unique variants (different breakpoints). With a high-resolution array (>1 million probes), we can conservatively expect that ~15% of cases will have a significant CNV that warrants follow-up in the families. As shown in the Fig. 2 flow chart, we prioritize based on the following.

   (a) *CNVs known to be associated with a genomic disorder*. Priority is given to those known to be associated with neuropsychiatric disorders (e.g. 1q21.1, 16p11.2, 22q13.3, etc.).

   (b) De novo *CNV*. If using a family-based study with trios, we identify those CNVs that are in probands and not parents as high priority for validation and follow-up. Significance for follow-up is also prioritized based on the scheme for rare CNVs (see below).

   (c) *Rare CNV*. This category of CNVs is defined by frequency and with respect to the degree of overlap with those identified in large control populations. Rare CNVs are those present at a frequency of <1%; however, we give priority to those not present in control populations (noting that it is important to use ancestry-matched cases and controls).

We use the following additional factors to prioritize the evaluation of CNVs.

- *Genomic position – genic CNVs.* We prioritize CNVs that overlap genes with those interrupting exons given priority over those that are intronic.
- *Size – genomic extent.* Large (>500 kb) CNVs are generally rare in the population and are more likely to overlap multiple genes (also having a decreased false-positive rate).
- *Loss vs. gain.* All other factors being equal, CNV losses (deletion) are more likely to significantly affect gene expression than gains (duplication). Also note that some gains, if the breakpoint interrupts the gene, may have an effect similar to that of a deletion.
- *Known candidate genes.* Based on the evidence available, genes already considered candidates for ASD and other neuropsychiatric disorders (i.e. NRXN1, CNTNAP2) as well ID (i.e. ILRAPL1) are prioritized for follow-up.
- *Recurrence.* If more than one of the subjects in our cohort has the same (recurrent) or overlapping CNV that is not in controls, the region will be given some priority within the context of the above criteria. This also applies to CNVs that are in controls but at a significantly higher frequency in cases. Also note that this definition can apply to a locus (e.g. PTCHD1 locus), so non-overlapping CNVs interrupting the same genomic feature (gene, ncRNA, etc.) are also considered.

3. Analysis of primary CNV and phenotypic data can be used to address the question of clinical validity of the observations. Here are some strategies that should be employed when looking at any CNV dataset involving neuropsychiatric cohorts.

   (a) *Prevalence of de novo putative pathogenic CNVs.* For any CNV dataset, a de novo rate should be established with the knowledge that non-disease trios show a de novo CNV prevalence in <1% of trios. A true de novo rate would best be calculated in consecutive sampling of patients as de novo rate varies depending on patient ascertainment and inclusion/exclusion criteria. Those cases carrying multiple or exceptionally large (i.e. >1Mb) CNVs with many genes involvedshould be flagged for further phenotypic assessment. Ultimately, this information should be tested in genotype and phenotype correlations as those with de novo CNVs may have more severe and complicated phenotypes.

(b) *Correlation of clinical features with particular CNV(s).* There has been no systematic attempt in a large cohort to phenotypically characterize individuals with ASD who carry presumed aetiologic CNVs, compared to those who do not. Our work on chromosome 16p11.2 microdeletions, for example (20, 39), suggests that these individuals have a more complex presentation than is typically observed in ASD. In del(16)p11.2 ASD individuals, there appears to be a higher prevalence than expected of growth abnormalities, epilepsy, congenital abnormalities, lower IQ, medical problems, and birth defects, as well as a more equal sex ratio. To substantiate this type of anecdotal evidence, a random sample of ASD subjects without a pathogenic CNV (a comparison group) should be blindly assessed using the same instruments. In addition to diagnostic information, assessment of CNV association with parameters, such as adaptive functioning, cognitive and language skills, minor physiological anomalies, social skills, impulsivity, hyperactivity, and developmental delay, is necessary. CNVs are also classified based on the functional characteristics of the genes they contain (using ontology, pathway, and network predictions) to determine if there is clustering of clinical features with the CNVs/genes involved. For proper correlation of CNVs with phenotypes, it is imperative that the diagnosis of these disorders is representative of a proper clinical (and not research based) diagnosis performed by paediatricians, child and adult psychiatrists, and clinical psychologists.

(c) *Aetiologic role of rare-inherited CNVs.* It is well-established that inherited mechanisms contribute in the aetiology of ASD. For example, the maternally inherited chromosome 15q11-q13 is observed in 1–3% of ASD cases. Also mother–son CNV transmission has been observed in X-linked genes, such as *NLGN3* and *NLGN4*, and most recently in our newly discovered *PTCHD1* ASD-susceptibility gene (see above). There are also examples of the inherited CNVs at 1q21.1, 15q13.2-q13.3, 16p11.2, and 22q13. Moreover, the MZ twin concordance data of close to 100% and observation that ~20% of parents and siblings of children with ASD have mild social communication impairments that fall below the threshold for a diagnosis of ASD also re-enforce inherited mechanisms. It is possible that these broader autism phenotypes (BAPs) are the result of single or combinations of inherited CNVs. If true, the parents and siblings who carry such CNVs might be more likely to have features of the BAP than parents and siblings who do not carry these rare CNVs. Phenotypic features in probands, parents, and controls should be identified to see if they co-segregate with the CNV in family members.

(d) *Rare CNVs shared across disorders.* Our preliminary observations indicate that some CNVs are pleiotropic, meaning the same CNV can cause several different outcomes (see Table 1). It is possible that a CNV at a particular locus will affect IQ in probands with ASD and inflexible behaviour in those with OCD. It does appear that across disorders those with CNVs that we attribute to being pathogenic appear to be more severely affected and to have a greater burden of medical problems, growth abnormalities, and congenital anomalies. In addition, if the "families of phenotypes concept" is true for neuropsychiatric conditions, it is possible that when two conditions share a CNV they also share a component phenotype. This may include a neuropsychiatric phenotype that is present to some degree in DD/ID, SZ, ADHD, or OCD. The aforementioned issues will be extremely important for future diagnostics of neuropsychiatric disorders.

References

1. Kanner L (1943). Autistic disturbances of affected contact. Nervous Child, 2, 217–250.
2. Asperger H (1944). Die autistischen psychopathen im kindesalter. Arch fur Psychiatrie und Nerenkrankheiten, 117, 76–136.
3. Fombonne E (2001). Is there an epidemic of autism? Pediatrics, 107, 411–2.
4. Folstein SE and Rosen-Sheidley B (2001). Genetics of autism: complex aetiology for a heterogeneous disorder. Nature Reviews Genetics, 12:943–55.
5. Szatmari P, Jones MB, Zwaigenbaum L, MacLean JE (1998). Genetics of autism: overview and new directions. Journal of Autism and Developmental Disorders, 28, 351–68.
6. Abrahams BS, Geschwind DH. (2008). Advances in autism genetics: on the threshold of a new neurobiology. Nature Reviews Genetics, 9, 341–355.
7. Zwaigenbaum L, Bryson S, Roberts W, Brian J, Szatmari P (2005). Behavioral markers of autism in the first year of life. International Journal of Developmental Neurosciences, 23, 143–152.
8. Bailey A, Le Couteur A, Gottesman I, Bolton P, Simonoff E, Yuzda E, Rutter M (1995). Autism as a strongly genetic disorder: Evidence from a British twin study. Psychological Medicine, 25, 63–77.
9. Folstein S, Rutter M (1977). Infantile autism: a genetic study of 21 twin pairs. Journal of Child Psychology and Psychiatry, 18, 297–321.
10. Zwaigenbaum L, Szatmari P, Mahoney WJ, Bryson SE, Bartolucci G, MacLean JE (2000). High functioning autism and childhood disintegrative disorder in halfbrothers. Journal of Autism and Developmental Disorders, 30, 121–26.
11. MacLean JE, Szatmari P, Jones MB, Bryson SE, Mahoney WJ, Bartolucci G, Tuff L (1999). Familial factors influence level of functioning in pervasive developmental disorder. Journal of the American Academy of Child and Adolescent Psychiatry, 38, 746–53.
12. Szatmari P, Bryson SE, Boyle MH, Streiner DL, Duku E (2003). Predictors of outcome among high functioning children with autism and Asperger syndrome. Journal of Child Psychology and Psychiatry, and Allied Disciplines, 44, 520–8.
13. Veenstra-VanderWeele J, Cook EH (2004). Molecular genetics of autism spectrum disorder. Molecular Psychiatry, 9, 819–32.
14. Pickles A, Bolton P, Macdonald H, Bailey A, Le Couteur A, Sim CH, Rutter M (1995). Latent-class analysis of recurrence risks for complex phenotypes with selection and measurement error: a twin and family history study of autism. American Journal of Human Genetics, 3, 717–26.
15. Risch N, Spiker D, Lotspeich L, Nouri N, Hinds D, Hallmayer J, Kalaydjieva L, McCague P, Dimiceli S, Pitts T, Nguyen L, Yang J,

Harper C, Thorpe D, Vermeer S, Young H, Hebert J, Lin A, Ferguson J, Chiotti C, Wiese-Slater S, Rogers T, Salmon B, Nicholas P, Myers RM, et al (1999). A genomic screen of autism:evidence for a multilocus etiology. American Journal of Human Genetics, 65, 493–507.

16. Zhao X, Leotta A, Kustanovich V, Lajonchere C, Geschwind DH, Law K, Law P, Qiu S, Lord C, Sebat J, Ye K, Wigler M. (2007). A unified genetic theory for sporadic and inherited autism. Proceedings of the National Academy of Sciences USA, 104, 12831–12836.
17. Lord C, Cook EH, Leventhal B, Amaral, DG (2000). Autism Spectrum Disorders. Cell, 28, 355–63.
18. Xu, J, Zwaigenbaum, L Szatmari, P and Scherer SW (2004). Molecular cytogenetics of autism. Current Genomics, 5, 347–64.
19. Buchanan JA, Scherer SW. (2008). Contemplating effects of genomic structural variation. Genetics inMedicine, 10, 639–647.
20. Marshall CR, Noor A, Vincent JB, Lionel AC, Feuk L, Skaug J, Shago M, Moessner R, Pinto D, Ren Y, Thiruvahindrapduram B, Fiebig A, Schreiber S, Friedman J, Ketelaars CE, Vos YJ, Ficicioglu C, Kirkpatrick S, Nicolson R, Sloman L, Summers A, Gibbons CA, Teebi A, Chitayat D, Weksberg R, Thompson A, Vardy C, Crosbie V, Luscombe S, Baatjes R, Zwaigenbaum L, Roberts W, Fernandez B, Szatmari P, Scherer SW. (2008). Structural variation of chromosomes in autism spectrum disorder. American Journal of Human Genetics, 82, 477–488.
21. Miller DT, Adam MP, Aradhya S, Biesecker LG, Brothman AR, Carter NP, Church DM, Crolla JA, Eichler EE, Epstein CJ, Faucett WA, Feuk L, Friedman JM, Hamosh A, Jackson L, Kaminsky EB, Kok K, Krantz ID, Kuhn RM, Lee C, Ostell JM, Rosenberg C, Scherer SW, Spinner NB, Stavropoulos DJ, Tepperberg JH, Thorland EC, Vermeesch JR, Waggoner DJ, Watson MS, Martin CL, Ledbetter DH (2010). Consensus statement: chromosomal microarray is a first-tier clinical diagnostic test for individuals with developmental disabilities or congenital anomalies. American Journal of Human Genetics, 86, 749–64.
22. Shen Y, Dies KA, Holm IA, Bridgemohan C, Sobeih MM, Caronna EB, Miller KJ, Frazier JA, Silverstein I, Picker J, Weissman L, Raffalli P, Jeste S, Demmer LA, Peters HK, Brewster SJ, Kowalczyk SJ, Rosen-Sheidley B, McGowan C, Duda AW 3rd, Lincoln SA, Lowe KR, Schonwald A, Robbins M, Hisama F, Wolff R, Becker R, Nasir R, Urion DK, Milunsky JM, Rappaport L, Gusella JF, Walsh CA, Wu BL, Miller DT; Autism Consortium Clinical Genetics/DNA Diagnostics Collaboration (2010). Clinical genetic testing for patients with autism spectrum disorders. Pediatrics, 125, e727–35.
23. Jamain S, Quach H, Betancur C, Råstam M, Colineaux C, Gillberg IC, Soderstrom H, Giros B, Leboyer M, Gillberg C, Bourgeron T; Paris Autism Research International Sibpair Study. (2003). Mutations of the X-linked genes encoding neuroligins NLGN3 and NLGN4 are associated with autism. Nature Genetics, 34, 27–29.
24. Wang K, Zhang H, Ma D, Bucan M, Glessner JT, Abrahams BS, Salyakina D, Imielinski M, Bradfield JP, Sleiman PM, Kim CE, Hou C, Frackelton E, Chiavacci R, Takahashi N, Sakurai T, Rappaport E, Lajonchere CM, Munson J, Estes A, Korvatska O, Piven J, Sonnenblick LI, Alvarez Retuerto AI, Herman EI, Dong H, Hutman T, Sigman M, Ozonoff S, Klin A, Owley T, Sweeney JA, Brune CW, Cantor RM, Bernier R, Gilbert JR, Cuccaro ML, McMahon WM, Miller J, State MW, Wassink TH, Coon H, Levy SE, Schultz RT, Nurnberger JI, Haines JL, Sutcliffe JS, Cook EH, Minshew NJ, Buxbaum JD, Dawson G, Grant SF, Geschwind DH, Pericak-Vance MA, Schellenberg GD, Hakonarson H. (2009). Common genetic variants on 5p14.1 associate with autism spectrum disorders. Nature, 459, 528–533.
25. Weiss LA, Arking DE; Gene Discovery Project of Johns Hopkins & the Autism Consortium, Daly MJ, Chakravarti A (2009) A genome-wide linkage and association scan reveals novel loci for autism. Nature, 461, 802–8.
26. Arking DE, Cutler DJ, Brune CW, Teslovich TM, West K, Ikeda M, Rea A, Guy M, Lin S, Cook EH, Chakravarti A (2008). A common genetic variant in the neurexin superfamily member CNTNAP2 increases familial risk of autism. American Journal of Human Genetics, 82, 160–4.
27. Cook EH Jr, Scherer SW. (2008). Copy-number variations associated with neuropsychiatric conditions. Nature, 16, 919–923.
28. Cook EH Jr, Lindgren V, Leventhal BL, Courchesne R, Lincoln A, Shulman C, Lord C, Courchesne E.(1997). Autism or atypical autism in maternally but not paternally derived proximal 15q duplication. American Journal of Human Genetics, 60, 928–934.
29. Potocki L, Bi W, Treadwell-Deering D, Carvalho CM, Eifert A, Friedman EM, Glaze D, Krull K, Lee JA, Lewis RA, Mendoza-Londono R, Robbins-Furman P, Shaw C, Shi X, Weissenberger G, Withers M, Yatsenko SA, Zackai EH, Stankiewicz P, Lupski JR. (2007). Characterization of Potocki-Lupski syndrome

(dup(17)(p11.2p11.2)) and delineation of a dosage-sensitive critical interval that can conveyan autism phenotype. American Journal of Human Genetics, 80, 633–649.

30. Autism Genome Project Consortium, Szatmari P, Paterson AD, Zwaigenbaum L, Roberts W, Brian J, Liu XQ, Vincent JB, Skaug JL, Thompson AP, Senman L, Feuk L, Qian C, Bryson SE, Jones MB, Marshall CR, Scherer SW, Vieland VJ, Bartlett C, Mangin LV, Goedken R, Segre A, Pericak-Vance MA, Cuccaro ML, Gilbert JR, Wright HH, Abramson RK, Betancur C, Bourgeron T, Gillberg C, Leboyer M, Buxbaum JD, Davis KL, Hollander E, Silverman JM, Hallmayer J, Lotspeich L, Sutcliffe JS, Haines JL, Folstein SE, Piven J, Wassink TH, Sheffield V, Geschwind DH, Bucan M, Brown WT, Cantor RM, Constantino JN, Gilliam TC, Herbert M, Lajonchere C, Ledbetter DH, Lese-Martin C, Miller J, Nelson S, Samango-Sprouse CA, Spence S, State M, Tanzi RE, Coon H, Dawson G, Devlin B, Estes A, Flodman P, Klei L, McMahon WM, Minshew N, Munson J, Korvatska E, Rodier PM, Schellenberg GD, Smith M, Spence MA, Stodgell C, Tepper PG, Wijsman EM, Yu CE, Rogé B, Mantoulan C, Wittemeyer K, Poustka A, Felder B, Klauck SM, Schuster C, Poustka F, Bölte S, Feineis-Matthews S, Herbrecht E, Schmötzer G, Tsiantis J, Papanikolaou K, Maestrini E, Bacchelli E, Blasi F, Carone S, Toma C, Van Engeland H, de Jonge M, Kemner C, Koop F, Langemeijer M, Hijmans C, Staal WG, Baird G, Bolton PF, Rutter ML, Weisblatt E, Green J, Aldred C, Wilkinson JA, Pickles A, Le Couteur A, Berney T, McConachie H, Bailey AJ, Francis K, Honeyman G, Hutchinson A, Parr JR, Wallace S, Monaco AP, Barnby G, Kobayashi K, Lamb JA, Sousa I, Sykes N, Cook EH, Guter SJ, Leventhal BL, Salt J, Lord C, Corsello C, Hus V, Weeks DE, Volkmar F, Tauber M, Fombonne E, Shih A, Meyer KJ. (2007). Mapping autism risk loci using genetic linkage and chromosomal rearrangements. Nature Genetics, 39, 319–328.

31. Sebat J, Lakshmi B, Malhotra D, Troge J, Lese-Martin C, Walsh T, Yamrom B, Yoon S, Krasnitz A, Kendall J, Leotta A, Pai D, Zhang R, Lee YH, Hicks J, Spence SJ, Lee AT, Puura K, Lehtimäki T, Ledbetter D, Gregersen PK, Bregman J, Sutcliffe JS, Jobanputra V, Chung W, Warburton D, King MC, Skuse D, Geschwind DH, Gilliam TC, Ye K, Wigler M. (2007) Strong association of de novo copy number mutations with autism. Science, 316, 445–449.

32. Weiss LA, Shen Y, Korn JM, Arking DE, Miller DT, Fossdal R, Saemundsen E, Stefansson H, Ferreira MA, Green T, Platt OS, Ruderfer DM, Walsh CA, Altshuler D, Chakravarti A, Tanzi RE, Stefansson K, Santangelo SL, Gusella JF, Sklar P, Wu BL, Daly MJ; Autism Consortium. (2008). Association between microdeletion and microduplication at 16p11.2 and autism. New England Journal of Medicine, 358, 667–675.

33. Kumar RA, KaraMohamed S, Sudi J, Conrad DF, Brune C, Badner JA, Gilliam TC, Nowak NJ, Cook EH Jr, Dobyns WB, Christian SL. (2008). Recurrent 16p11.2 microdeletions in autism. Human Molecular Genetics, 17, 628–638.

34. Christian SL, Brune CW, Sudi J, Kumar RA, Liu S, Karamohamed S, Badner JA, Matsui S, Conroy J, McQuaid D, Gergel J, Hatchwell E, Gilliam TC, Gershon ES, Nowak NJ, Dobyns WB, Cook EH Jr. (2008). Novel submicroscopic chromosomal abnormalities detected in autism spectrum disorder. Biological Psychiatry, 63, 1111–1117.

35. Jacquemont ML, Sanlaville D, Redon R, Raoul O, Cormier-Daire V, Lyonnet S, Amiel J, Le Merrer M, Heron D, de Blois MC, Prieur M, Vekemans M, Carter NP, Munnich A, Colleaux L, Philippe A. (2006). Array-based comparative genomic hybridisation identifies high frequency of cryptic chromosomal rearrangements in patients with syndromic autism spectrum disorders. Journal of Medical Genetics, 43, 843–849.

36. Morrow EM, Yoo SY, Flavell SW, Kim TK, Lin Y, Hill RS, Mukaddes NM, Balkhy S, Gascon G, Hashmi A, Al-Saad S, Ware J, Joseph RM, Greenblatt R, Gleason D, Ertelt JA, Apse KA, Bodell A, Partlow JN, Barry B, Yao H, Markianos K, Ferland RJ, Greenberg ME, Walsh CA. (2008). Identifying autism loci and genes by tracing recent shared ancestry. Science, 321, 218–223.

37. Glessner JT, Wang K, Cai G, Korvatska O, Kim CE, Wood S, Zhang H, Estes A, Brune CW, Bradfield JP, Imielinski M, Frackelton EC, Reichert J, Crawford EL, Munson J, Sleiman PM, Chiavacci R, Annaiah K, Thomas K, Hou C, Glaberson W, Flory J, Otieno F, Garris M, Soorya L, Klei L, Piven J, Meyer KJ, Anagnostou E, Sakurai T, Game RM, Rudd DS, Zurawiecki D, McDougle CJ, Davis LK, Miller J, Posey DJ, Michaels S, Kolevzon A, Silverman JM, Bernier R, Levy SE, Schultz RT, Dawson G, Owley T, McMahon WM, Wassink TH, Sweeney JA, Nurnberger JI, Coon H, Sutcliffe JS, Minshew NJ, Grant SF, Bucan M, Cook EH, Buxbaum JD, Devlin B, Schellenberg GD, Hakonarson H. (2009). Autism genome-wide copy number variation reveals ubiquitin and neuronal genes. Nature, 459, 569–573.

38. Pinto D et al. Autism Genome Project Consortium (2010). Functional impact of global rare copy number variation in autism spectrum disorder. Nature, In press.
39. Fernandez B, Roberts W, Chung B, Weksberg R, Meyn S, Szatmari P, Joseph-George AM, MacKay S, Whitten K, Noble B, Vardy C, Crosbie V, Luscombe S, Tucker E, Turner L, Marshall CR, Scherer SW. (2010). Phenotypic spectrum associated with de novo and inherited deletions and duplications at 16p11.2 in individuals ascertained for diagnosis of autism spectrum disorder. Journal of Medical Genetics, 47, 195–203
40. Berkel S, Marshall CR, Weiss B, Howe J, Roeth R, Moog U, Endris V, Roberts W, Szatmari P, Pinto D, Bonin M, Riess A, Engels H, Sprengel R, Scherer SW, Rappold GA (2010). Mutations in the SHANK2 synaptic scaffolding gene in autism spectrum disorder and mental retardation. Nature Genetics, 42, 489–91.
41. Noor A, Whibley A, Marshall CR, Gianakopoulos PJ, Piton A, Orlic M, Fernandez B, Pinto D, Baatjes-Young R, Zhang X, Mo R, Gauthier J, Roberts R, Szatmari P, Gallagher L, Stratton M, Gecz J, Brady A, Schwartz CE, Monaco AP, Rouleau GA, Hui C-C, Raymond FL, Scherer SW and Vincent JB.(2009). Disruption at the PTCHD1 locus on Xp22.11 in autism spectrum disorder and intellectual disability. Science Translational Medicine, In review.
42. Durand CM, Betancur C, Boeckers TM, Bockmann J, Chaste P, Fauchereau F, Nygren G, Rastam M, Gillberg IC, Anckarsäter H, Sponheim E, Goubran-Botros H, Delorme R, Chabane N, Mouren-Simeoni MC, de Mas P, Bieth E, Rogé B, Héron D, Burglen L, Gillberg C, Leboyer M, Bourgeron T. (2007). Mutations in the gene encoding the synaptic scaffolding protein SHANK3 are associated with autism spectrum disorders. Nature Genetics, 39, 25–27.
43. Jamain S, Quach H, Betancur C, Råstam M, Colineaux C, Gillberg IC, Soderstrom H, Giros B, Leboyer M, Gillberg C, Bourgeron T; Paris Autism Research International Sibpair Study. (2003). Mutations of the X-linked genes encoding neuroligins NLGN3 and NLGN4 are associated with autism. Nature Genetics, 34, 27–29.
44. Willemsen MH, Fernandez BA, Bacino C, Gerkes E, de Brouwer APM, Pfundt R, Sikkema-Raddatz B, Scherer SW, Marshall CR, Potocki L, van Bokhoven H, Kleefstra T. (2010). Identification of ANKRD11 and ZNF778 as candidate genes for autism and variable cognitive impairment in the 16q24.3 microdeletion syndrome. European Journal of Human Genetics, 18, 429–35.
45. Bhalla K, Luo Y, Buchan T, Beachem MA, Guzauskas GF, Ladd S, Bratcher SJ, Schroer RJ, Balsamo J, DuPont BR, Lilien J, Srivastava AK. (2008). Alterations in CDH15 and KIRREL3 in patients with mild to severe intellectual disability. American Journal of Human Genetics, 83, 703–713
46. Rapoport J, Chavez A, Greenstein D, Addington A, Gogtay N. (2009). Autism spectrum disorders and childhood-onset schizophrenia: clinical and biological contributions to a relation revisited. Journal of the American Academy of Child and Adolescent Psychiatry, 48, 10–18.
47. Hofvander B, Delorme R, Chaste P, Nydén A, Wentz E, Ståhlberg O, Herbrecht E, Stopin A, Anckarsäter H, Gillberg C, Råstam M, Leboyer M. (2009). Psychiatric and psychosocial problems in adults with normal-intelligence autism spectrum disorders. BMC Psychiatry, 9, 35–44.
48. Sporn AL, Addington AM, Gogtay N, Ordoñez AE, Gornick M, Clasen L, Greenstein D, Tossell JW, Gochman P, Lenane M, Sharp WS, Straub RE, Rapoport JL. (2004). Pervasive developmental disorder and childhood-onset schizophrenia: comorbid disorder or a phenotypic variant of a very early onset illness? Biological Psychiatry, 55, 989–994.
49. Bassett AS, Chow EW. (2008). Schizophrenia and 22q11.2 deletion syndrome. Current Psychiatry Reports, 10, 148–157.
50. Fine SE, Weissman A, Gerdes M, Pinto-Martin J, Zackai EH, McDonald-McGinn DM, Emanuel BS. (2005). Autism spectrum disorders and symptoms in children with molecularly confirmed 22q11.2 deletion syndrome. Journal of Autism and Developmental Disorders, 35, 461–470.
51. Walsh T, McClellan JM, McCarthy SE, Addington AM, Pierce SB, Cooper GM, Nord AS, Kusenda M,Malhotra D, Bhandari A, Stray SM, Rippey CF, Roccanova P, Makarov V, Lakshmi B, Findling RL, Sikich L, Stromberg T, Merriman B, Gogtay N, Butler P, Eckstrand K, Noory L, Gochman P, Long R, Chen Z, Davis S, Baker C, Eichler EE, Meltzer PS, Nelson SF, Singleton AB, Lee MK, Rapoport JL, King MC, Sebat J. (2008). Rare structural variants disrupt multiple genes in neurodevelopmental pathways in schizophrenia. Science, 320, 539–543.
52. Xu B, Roos JL, Levy S, van Rensburg EJ, Gogos JA, Karayiorgou M. (2008). Strong association of de novo copy number mutations with sporadic schizophrenia. Nature Genetics, 40, 880–885.
53. Goldstein S, Schwebach AJ. (2004). The comorbidity of Pervasive Developmental

Disorder and Attention Deficit Hyperactivity Disorder: results of a retrospective chart review. Journal of Autism and Developmental Disorders, 34, 329–339.

54. Clark T, Feehan C, Tinline C, Vostanis P. (1999) Autistic symptoms in children with attention deficithyperactivity disorder. European Child & Adolescent Psychiatry, 8, 50–55.

55. Mulligan A, Anney RJ, O'Regan M, Chen W, Butler L, Fitzgerald M, Buitelaar J, Steinhausen HC, Rothenberger A, Minderaa R, Nijmeijer J, Hoekstra PJ, Oades RD, Roeyers H, Buschgens C, Christiansen H, Franke B, Gabriels I, Hartman C, Kuntsi J, Marco R, Meidad S, Mueller U, Psychogiou L, Rommelse N, Thompson M, Uebel H, Banaschewski T, Ebstein R, Eisenberg J, Manor I, Miranda A, Mulas F, Sergeant J, Sonuga-Barke E, Asherson P, Faraone SV, Gill M. (2009). Autism symptoms in Attention-Deficit/Hyperactivity Disorder: a familial trait which correlates with conduct, oppositional defiant, language and motor disorders. Journal of Autism and Developmental Disorders, 39, 197–209.

56. Smalley SL, Kustanovich V, Minassian SL, Stone JL, Ogdie MN, McGough JJ, McCracken JT, MacPhie IL, Francks C, Fisher SE, Cantor RM, Monaco AP, Nelson SF. (2002). Genetic linkage of attention-deficit/hyperactivity disorder on chromosome 16p13, in a region implicated in autism. American Journal of Human Genetics, 71, 959–963.

57. Ogdie MN, Fisher SE, Yang M, Ishii J, Francks C, Loo SK, Cantor RM, McCracken JT, McGough JJ, Smalley SL, Nelson SF. (2004). Attention deficit hyperactivity disorder: fine mapping supports linkage to 5p13, 6q12, 16p13, and 17p11. American Journal of Human Genetics, 75, 661–668.

58. Elia J, Gai X, Xie HM, Perin JC, Geiger E, Glessner JT, D'arcy M, Deberardinis R, Frackelton E, Kim C, Lantieri F, Muganga BM, Wang L, Takeda T, Rappaport EF, Grant SF, Berrettini W, Devoto M, Shaikh TH, Hakonarson H, White PS. (2010). Rare structural variants found in attention-deficit hyperactivity disorder are preferentially associated with neurodevelopmental genes. Molecular Psychiatry, 15, 637–46.

59. O'Brien G, Pearson J. (2004). Autism and learning disability. Autism, 8, 125–140.

60. Oti M, Huynen MA, Brunner HG (2008). Phenome connections. Trends in Genetics, 24, 103–106.

61. Brunner HG, van Driel MA. (2004). From syndrome families to functional genomics. Nature Reviews Genetics, 5, 545–551.

62. Rzhetsky A, Wajngurt D, Park N, Zheng T. (2007). Probing genetic overlap among complex human phenotypes. Proceedings of the National Academy of Sciences of the USA, 104, 11694–11699.

63. Korn JM, Kuruvilla FG, McCarroll SA, Wysoker A, Nemesh J, Cawley S, Hubbell E, Veitch J, Collins PJ, Darvishi K, Lee C, Nizzari MM, Gabriel SB, Purcell S, Daly MJ, Altshuler D (2008). Integrated genotype calling and association analysis of SNPs, common copy number polymorphisms and rare CNVs. Nature Genetics, 40, 1253–60.

64. Zhang J, Pinto D, Thiruvahindrapduram B, Wang Z, Prasad A, Marshall CR, Lionel A, Hu P, Greenwood CM, Feuk L, Wintle RF, Scherer SW (2010). iPattern: a cross-sample copy number variation discovery method for multiple array platforms. Nucleic Acid Research, In Review.

65. Alarcón M, Abrahams BS, Stone JL, Duvall JA, Perederiy JV, Bomar JM, Sebat J, Wigler M, Martin CL, Ledbetter DH, Nelson SF, Cantor RM, Geschwind DH (2008). Linkage, association, and gene-expression analyses identify CNTNAP2 as an autism-susceptibility gene. American Journal of Human Genetics, 82, 150–9.

66. Bucan M, Abrahams BS, Wang K, Glessner JT, Herman EI, Sonnenblick LI, Alvarez, Retuerto AI, Imielinski M, Hadley D, Bradfield JP, Kim C, Gidaya NB, Lindquist I, Hutman T, Sigman M, Kustanovich V, Lajonchere CM, Singleton A, Kim J, Wassink TH, McMahon WM, Owley T, Sweeney JA, Coon H, Nurnberger JI, Li M, Cantor RM, Minshew NJ, Sutcliffe JS, Cook EH, Dawson G, Buxbaum JD, Grant SF, Schellenberg GD, Geschwind DH, Hakonarson H (2009). Genome-wide analyses of exonic copy number variants in a family-based study point to novel autism susceptibility genes. PLoS Genetics, 6, e1000536.

67. Friedman JI, Vrijenhoek T, Markx S, Janssen IM, van der Vliet WA, Faas BH, Knoers NV, Cahn W, Kahn RS, Edelmann L, Davis KL, Silverman JM, Brunner HG, van Kessel AG, Wijmenga C, Ophoff RA, Veltman JA (2008). CNTNAP2 gene dosage variation is associated with schizophrenia and epilepsy. Molecular Psychiatry, 3, 261–6.

68. International Schizophrenia Consortium (2008). Rare chromosomal deletions and duplications increase risk of schizophrenia. Nature, 455, 237–41.

69. Kirov G, Gumus D, Chen W, Norton N, Georgieva L, Sari M, O'Donovan MC, Erdogan F, Owen MJ, Ropers HH, Ullmann R (2008). Comparative genome hybridization suggests a

role for NRXN1 and APBA2 in schizophrenia. Human Molecular Genetics, 17, 458–65.

70. Kirov G, Grozeva D, Norton N, Ivanov D, Mantripragada KK, Holmans P; International Schizophrenia Consortium; Wellcome Trust Case Control Consortium, Craddock N, Owen MJ, O'Donovan MC (2009). Support for the involvement of large copy number variants in the pathogenesis of schizophrenia. Human Molecular Genetics, 18, 1497–503.

71. Martin CL, Duvall JA, Ilkin Y, Simon JS, Arreaza MG, Wilkes K, Alvarez-Retuerto A, Whichello A, Powell CM, Rao K, Cook E, Geschwind DH (2007). Cytogenetic and molecular characterization of A2BP1/FOX1 as a candidate gene for autism. American Journal of Medical Genetics B Neuropsychiatric Genetics, 144B, 869–76.

72. Need AC, Ge D, Weale ME, Maia J, Feng S, Heinzen EL, Shianna KV, Yoon W, Kasperavičiūte D, Gennarelli M, Strittmatter WJ, Bonvicini C, Rossi G, Jayathilake K, Cola PA, McEvoy JP, Keefe RS, Fisher EM, St Jean PL, Giegling I, Hartmann AM, Möller HJ, Ruppert A, Fraser G, Crombie C, Middleton LT, St Clair D, Roses AD, Muglia P, Francks C, Rujescu D, Meltzer HY, Goldstein DB (2009). A genome-wide investigation of SNPs and CNVs in schizophrenia. PLoS Genetics, 5, e1000373.

73. Rodríguez-Santiago B, Brunet A, Sobrino B, Serra-Juhé C, Flores R, Armengol L, Vilella E, Gabau E, Guitart M, Guillamat R, Martorell L, Valero J, Gutiérrez-Zotes A, Labad A, Carracedo A, Estivill X, Pérez-Jurado LA (2009). Association of common copy number variants at the glutathione S-transferase genes and rare novel genomic changes with schizophrenia. Molecular Psychiatry. June 16 epub.

74. Rossi E, Verri AP, Patricelli MG, Destefani V, Ricca I, Vetro A, Ciccone R, Giorda R, Toniolo D, Maraschio P, Zuffardi O (2008). A 12Mb deletion at 7q33-q35 associated with autism spectrum disorders and primary amenorrhea. European Journal of Medical Genetics, 51, 631–8.

75. Rujescu D, Ingason A, Cichon S, Pietiläinen OP, Barnes MR, Toulopoulou T, Picchioni M, Vassos E, Ettinger U, Bramon E, Murray R, Ruggeri M, Tosato S, Bonetto C, Steinberg S, Sigurdsson E, Sigmundsson T, Petursson H, Gylfason A, Olason PI, Hardarsson G, Jonsdottir GA, Gustafsson O, Fossdal R, Giegling I, Möller HJ, Hartmann AM, Hoffmann P, Crombie C, Fraser G, Walker N, Lonnqvist J, Suvisaari J, Tuulio-Henriksson A, Djurovic S, Melle I, Andreassen OA, Hansen T, Werge T, Kiemeney LA, Franke B, Veltman J, Buizer-Voskamp JE; GROUP Investigators, Sabatti C, Ophoff RA, Rietschel M, Nöthen MM, Stefansson K, Peltonen L, St Clair D, Stefansson H, Collier DA (2009). Disruption of the neurexin 1 gene is associated with schizophrenia. Human Molecular Genetics, 18, 988–96.

76. Stefansson H, Rujescu D, Cichon S, Pietiläinen OP, Ingason A, Steinberg S, Fossdal R, Sigurdsson E, Sigmundsson T, Buizer-Voskamp JE, Hansen T, Jakobsen KD, Muglia P, Francks C, Matthews PM, Gylfason A, Halldorsson BV, Gudbjartsson D, Thorgeirsson TE, Sigurdsson A, Jonasdottir A, Jonasdottir A, Bjornsson A, Mattiasdottir S, Blondal T, Haraldsson M, Magnusdottir BB, Giegling I, Möller HJ, Hartmann A, Shianna KV, Ge D, Need AC, Crombie C, Fraser G, Walker N, Lonnqvist J, Suvisaari J, Tuulio-Henriksson A, Paunio T, Toulopoulou T, Bramon E, Di Forti M, Murray R, Ruggeri M, Vassos E, Tosato S, Walshe M, Li T, Vasilescu C, Mühleisen TW, Wang AG, Ullum H, Djurovic S, Melle I, Olesen J, Kiemeney LA, Franke B; GROUP, Sabatti C, Freimer NB, Gulcher JR, Thorsteinsdottir U, Kong A, Andreassen OA, Ophoff RA, Georgi A, Rietschel M, Werge T, Petursson H, Goldstein DB, Nöthen MM, Peltonen L, Collier DA, St Clair D, Stefansson K (2008). Large recurrent microdeletions associated with schizophrenia. Nature, 455, 232–6.

77. van der Zwaag B, Franke L, Poot M, Hochstenbach R, Spierenburg HA, Vorstman JA, van Daalen E, de Jonge MV, Verbeek NE, Brilstra EH, van 't Slot R, Ophoff RA, van Es MA, Blauw HM, Veldink JH, Buizer-Voskamp JE, Beemer FA, van den Berg LH, Wijmenga C, van Amstel HK, van Engeland H, Burbach JP, Staal WG (2009). Gene-network analysis identifies susceptibility genes related to glycobiology in autism. PLoS One, 4, e5324.

78. Vrijenhoek T, Buizer-Voskamp JE, van der Stelt I, Strengman E, Genetic Risk and Outcome in Psychosis (GROUP) Consortium, Sabatti C, Geurts van Kessel A, Brunner HG, Ophoff RA, Veltman JA (2008). Recurrent CNVs disrupt three candidate genes in schizophrenia patients. American Journal of Human Genetics, 83, 504–10.

Chapter 6

Structural Variation in Subtelomeres

M. Katharine Rudd

Abstract

Subtelomeres are an incredibly dynamic part of the human genome located at the ends of chromosomes just proximal to telomere repeats. Although subtelomeric variation contributes to normal polymorphism in the human genome and is a by-product of rapid evolution in these regions, rearrangements in subtelomeres can also cause intellectual disabilities and birth defects, making robust methods of detecting copy number variation in chromosome ends a must for cytogenetics labs. In recent years, methods for detecting structural variation in subtelomeres have moved from fluorescence in situ hybridization (FISH) to array technology; however, FISH is still necessary to determine the chromosomal structure of subtelomeric gains and losses identified by arrays.

Key words: Subtelomere, Chromosome, Array CGH, FISH, Structural variation, Polymorphism, CNV

1. Subtelomeres: *The Parts*

Human subtelomeres can be divided into two distinct zones: a terminal region consisting of segmental duplications and an adjacent region of chromosome-specific sequences (Fig. 1). Variation in the segmental duplication region represents normal polymorphism, whereas variation in the more internal chromosome-specific region can lead to intellectual disabilities and birth defects. Therefore, we must consider the two regions separately when analyzing subtelomeric CNV.

Segmental duplications make up more than 5% of the human genome and are preferentially located at pericentromeres and subtelomeres (1). Human subtelomeres include clustered segmental duplications located just proximal to telomere repeats. Each duplicated segment is shared between a subset of chromosome ends and ranges from 3 to 50 kilobases (kb) in size, constituting a duplication zone of 5–300 kb at a given chromosome end. Copies of the same duplication are 88–99.9% identical and are found on from

Lars Feuk (ed.), *Genomic Structural Variants: Methods and Protocols*, Methods in Molecular Biology, vol. 838,
DOI 10.1007/978-1-61779-507-7_6,

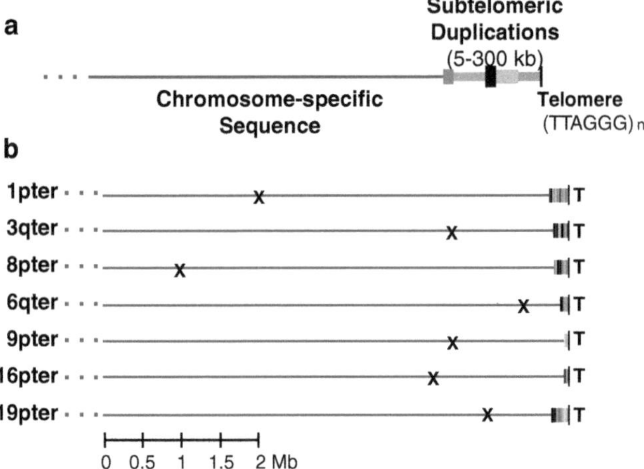

Fig. 1. Organization of human subtelomeres. (**a**) Schematic of chromosome ends. Telomere repeats (TTAGGG)n at the very end of the chromosome are indicated by a *black vertical line*. Subtelomeric segmental duplications are shown as rectangles. Chromosome-specific sequences are depicted as a gray line. (**b**) Distal 5 Mb of selected chromosome ends with sites of rearrangement breakpoints indicated by X (6). Chromosome-specific sequences (*horizontal gray lines*), subtelomeric duplications (*thin, vertical shaded lines*), and telomere repeats (T) are shown.

2 to 18 different chromosome ends (2). Subtelomeric segmental duplications are almost always in the same orientation and relative order, suggesting a translocation mechanism of sequence transfer (2). Though the human genome assembly does not capture the spectrum of subtelomeric polymorphism, fluorescence in situ hybridization (FISH) studies show that every subtelomeric segmental duplication tested varies in copy number and chromosomal location among individuals (2, 3).

Just proximal to subtelomeric duplications begins a region of chromosome-specific DNA (Fig. 1) that was once believed to be a more static region of the genome, but recently parts of this zone have proved to be polymorphic (4–6). Copy number gains and losses in this region have been detected in phenotypically normal individuals, suggesting that, as in the duplicated zones, some alterations in terminal chromosome-specific regions are tolerated.

Diagnostic testing of subtelomeric zones has also uncovered a number of benign CNVs in the chromosome-specific region. Deletions, duplications, and translocations inherited from a phenotypically normal parent with the same structural complement are assumed to be variants that do not contribute to the abnormalities of the proband. However, it is possible that some inherited CNVs exhibit incomplete penetrance, complicating the clinical interpretation. Many clinical cytogenetics labs have described inherited variants with the same apparent subtelomeric breakpoints, suggesting that these are common CNVs segregating in the normal population

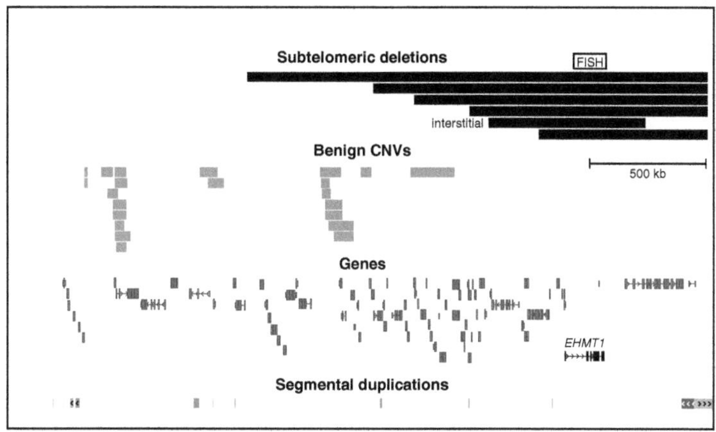

Fig. 2. Pathogenic subtelomeric deletions vary in size. Five disease-causing terminal deletions and one interstitial deletion of 9q are shown relative to benign CNVs (*light gray*) (33), genes (*gray*), and segmental duplications (1). The benign CNVs include genomic gains and losses. All six pathogenic deletions include the *EHMT1* gene (*black*) and are detectable using the FISH probe (*open box*), which does not distinguish larger or smaller deletions

(5–11). As with other benign CNVs in the human genome, these regions include genes, but do not affect normal development.

Genomic changes in the chromosome-specific region of subtelomeres can also cause intellectual disabilities and birth defects. In general, these imbalances are larger and contain more genes than benign CNVs in the same genomic neighborhood (Fig. 2). Between 3 and 6% of children with previously idiopathic mental retardation are diagnosed with a pathogenic subtelomeric rearrangement, making subtelomeric assays a standard clinical test (5, 11–13).

Subtelomeric rearrangements identified by diagnostic testing have led to the discovery of several recognizable syndromes. Studies of patients with common phenotypic features and overlapping genomic regions of loss/gain have pinpointed critical regions associated with disease and revealed causative genes in a given subtelomere. The recurrent 3q29 microdeletion and microduplication syndromes are caused by the loss/gain of an ~1.5-Mb region containing 22 genes (14, 15). The reciprocal deletions and duplications in 3q29 are caused by nonallelic homologous recombination (NAHR) between highly homologous segmental duplications bordering the site of imbalance. The 9q subtelomere deletion syndrome was first identified in patients with overlapping deletions of several Mb in size (16, 17). Here, the critical region of loss includes the *EHMT1* gene, which is responsible for the phenotype, as de novo *EHMT1* mutations have a typical 9q phenotype (18) (Fig. 2). Individuals with the 22q13 deletion syndrome have terminal deletions 130 kb to 9 Mb in size, all with a common critical region that includes the *SHANK3* gene. Mutations in *SHANK3* also cause the language and social disorders associated with the syndrome (19–21). Overall, the discovery of pathogenic subtelomeric rearrangements has linked certain critical regions with specific syndromes, uncovered novel disease-causing genes, and yielded careful genotype–phenotype correlations to better predict patient outcomes.

Despite these advances, distinguishing disease-causing subtelomeric CNVs from normal genomic variants can still be a challenge and requires careful characterization of subtelomeric imbalances. Genomic assays that accurately size the subtelomeric CNV and determine the gene content of the region are, therefore, essential.

2. Subtelomeres: *The Tools*

There are two major tools for detecting and characterizing subtelomeric rearrangements: FISH and arrays. Whereas array technology typically assesses copy number throughout the genome, FISH only detects gains or losses in the particular region(s) included in the FISH probe. Another advantage of arrays is that they size the region of gain/loss, identifying genes within the region. Yet despite the limitations of FISH, it provides critical data on the chromosomal structure of subtelomeric imbalances, which is necessary both to determine the mechanism of rearrangement and to interpret recurrence risks for families with pathogenic subtelomeric rearrangements. Hence, a combination of array and FISH technologies is the most comprehensive approach for analyzing structural variation in subtelomeres.

Systematic detection of subtelomeric rearrangements began ~15 years ago with the introduction of commercially available subtelomeric FISH probes (22, 23). Panels of FISH probes corresponding to the chromosome-specific regions of subtelomeres paved the way for detection of disease-causing deletions, duplications, and translocations conventional chromosome banding could not pick up (8, 24). Subtelomeric deletions and translocations are detected via FISH analysis of metaphase chromosomes. A missing subtelomeric FISH signal is indicative of a deletion, and the transfer of one or more FISH signals to another chromosome represents a translocation. Duplications of subtelomeric segments are visible using interphase FISH, as long as the duplication is large enough to distinguish two separate signals. Subtelomeric FISH studies are incredibly time consuming as 41 chromosome ends are evaluated, typically in mixes of FISH probe sets. Chromosome-specific FISH probes are commercially available and hybridize to the ends of the short and long arms of all chromosomes, except the acrocentric p arms (chromosomes 13, 14, 15, 21, and 22). Subtelomeric FISH probes for the sex chromosomes do not distinguish between the ends of Xp and Yp or Xq and Yq due to the shared homology between the pseudoautosomal regions (8). Subtelomeric FISH is a vast improvement over chromosome banding for the detection of subtelomeric imbalances; however, it does not accurately size the genomic imbalance or determine breakpoints. As shown in Fig. 2, heterozygous subtelomeric deletions of various sizes all have the same FISH output: a loss of probe signal on one chromosome. Further, subtelomeric FISH studies do not distinguish between terminal and interstitial gains or losses.

Terminal chromosome-specific sequences have since been moved to microarrays to monitor all subtelomeres in one experiment. First developed as bacterial artificial chromosome (BAC)-based arrays (5, 6, 11), subtelomeric copy number assays are now typically performed via oligonucleotide array comparative genome hybridization (CGH) or single-nucleotide polymorphism (SNP) platforms. Array technology has improved the resolution of subtelomeric gains and losses, refining breakpoints to tens or hundreds of kilobases, depending on probe density. As described above, sizing subtelomeric rearrangements via array provides critical information on the genes involved in phenotype. Moreover, array technologies cost less and take less time than performing FISH assays for all chromosome ends, making subtelomeric FISH panels virtually obsolete. In clinical cytogenetics diagnostic labs, subtelomeric copy number detection is performed as part of whole-genome array analysis. These assays are particularly useful for detecting imbalances in subtelomeres *and* other disease-causing loci, which is ideal for unearthing genomic abnormalities in children with intellectual disabilities and birth defects that could be due to a number of genetic causes. This approach is extremely effective, as ~16% of patients with previously idiopathic mental retardation carry a pathogenic copy number abnormality that ultimately proves responsible for their phenotype (25).

With improvements in technology and the collection of more data, subtelomeric imbalances have emerged as quite a diverse group of rearrangements. Subtelomeric rearrangements are found on every chromosome end, and the locations of breakpoints are rarely recurrent (5, 6, 11). To determine the causes of subtelomeric rearrangements, a handful of labs have performed high-resolution array analysis coupled with sequencing of subtelomeric breakpoint junctions. These studies reveal that the predominant mechanism of subtelomeric repair is nonhomologous end joining (NHEJ), requiring little or no homology between rearranging segments (2, 26–28); however, the causes of chromosome breakage and the mechanisms of genomic instability at chromosome ends have yet to be discovered.

Detecting variation in the segmental duplications at the very ends of subtelomeres poses more of a challenge than detecting imbalances in the chromosome-specific regions. Subtelomeric segmental duplications are incredibly polymorphic, present at from 2 to 18 different chromosome ends among individuals, making the detection of copy number variation (CNV) in these regions via array virtually impossible. Imagine comparing the genomes of two

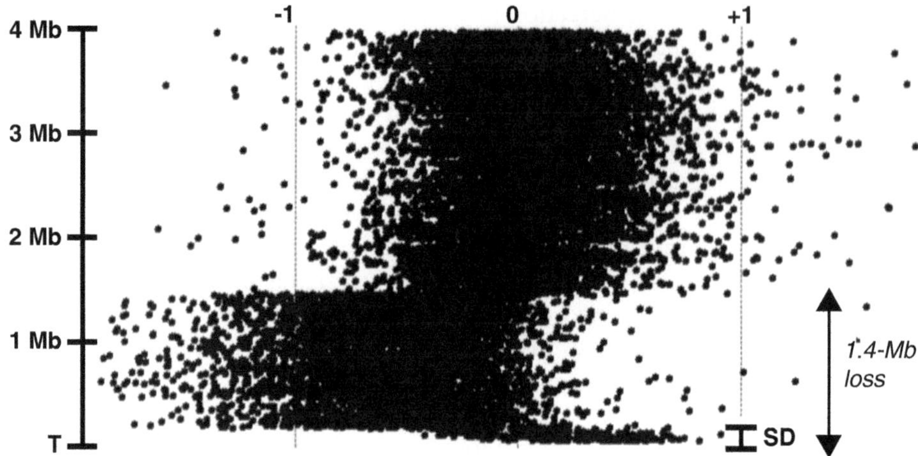

Fig. 3. Terminal deletion. Array CGH reveals the 1.4-Mb terminal deletion. Averaged $\log_2$ ratios of probe signal intensities are shown (*black dots*). *Dashed lines* indicate $\log_2$ ratios of −1, 0, and +1. The vertical scale indicates distance in megabases (Mb) from the end of telomeric sequence in the human genome assembly (T). Probes located in distal segmental duplications (SD) show no copy number loss.

individuals with nine and ten copies of a segmental duplication via array CGH. The ratio of signal intensities for probes in the segmental duplication (9:10 copies) is much more subtle than the ratio of a heterozygous deletion (1:2 copies) present in the chromosome-specific region. For example, a terminal deletion that encompasses the terminal 1.4 megabases (Mb) including the segmental duplications appears as an obvious loss in signal ratio in the chromosome-specific region, whereas the segmental duplication region appears as normal copy number (Fig. 3).

Studies of subtelomeric segmental duplications have relied on FISH assays to determine copy number and chromosomal location. Originally, cosmids or BACs containing multiple segmental duplications were used as FISH probes to query polymorphism among individuals (3, 29, 30). More targeted approaches to analyze individual segmental duplications push the limits of FISH because subtelomeric segmental duplications are 3–50 kb in size. Particular subtelomeric segmental duplications have been subcloned from BACs and amplified via long-range PCR to generate FISH probes as small as 20 kb (2, 31, 32). Though these probes represent the lower limits of FISH detection, they provide valuable information on the location and transfer of independent segmental duplications.

3. Subtelomeres: *The Caveats*

Given the repetitive nature and polymorphism of human subtelomeres, interpreting copy number data can be a challenge. As mentioned above, the best way to analyze subtelomeric variation is via a combination of FISH and array technology. This section explains the caveats to subtelomeric analyses, highlighting examples of complex rearrangements.

4. Terminal Deletions

Terminal deletions are the most common type of pathogenic subtelomeric rearrangement (5, 6, 11, 13). As shown in Fig. 3, multicopy segmental duplications complicate array data from terminal deletions, making them appear interstitial. Thus, most clinical cytogenetics labs exclude segmental duplications from array designs. However, there are other polymorphisms that can also confuse array interpretation.

Array CGH detects net copy number differences between two genomes, but it does not distinguish alleles. Therefore, the loss of one allele can be masked by an overlapping gain in the other allele. Such is the case for the terminal deletion shown in Fig. 4.

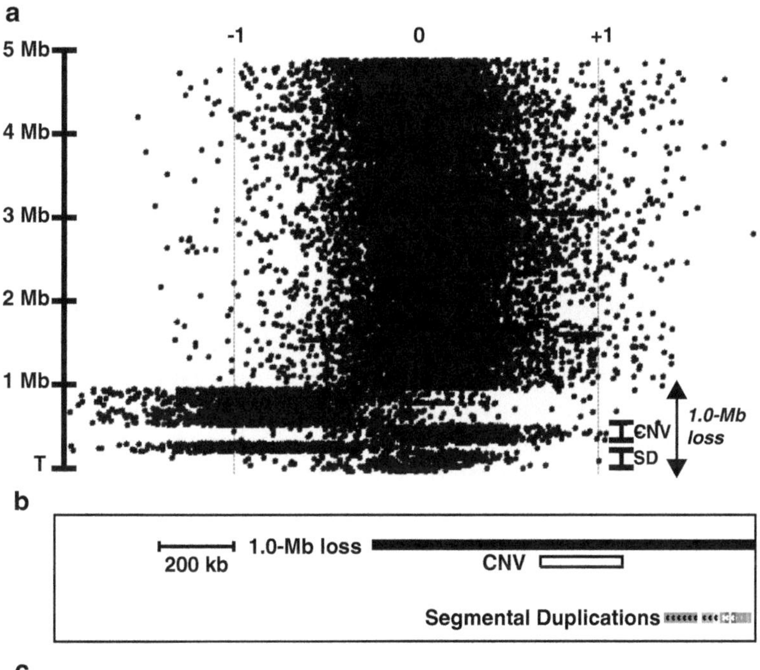

c

TCCTGCAGGTTAAACGAATTCGCCCTTCTACACATGGACGTGGGACACGGAGTTTGAAACCATGACAAATACCAAATGACT
TTTAGCTACTGGGCCAAACCCAATAACTCAAGGTGTAACTACTTCATTTAACAGCCTAATCTCAAGACACTCTACAAGTAC
CACAGGTGTTTGAACAAAATGCAAAAGGAAAGTACAATTATTACGATTTTGCACATGAGAAAACTAAGTCCCTAAATACTT
GATACTTCCTCGAGTAAGGAAGAATATCCTG***TTAGGGTTAGGGTTAGGGTTAGGGTTAGGGTTAGGGTTAGGGTTAAGGG***

Fig. 4. A terminal deletion with an overlapping duplication on the opposite allele. (**a**) Array CGH analysis of the 1.0-Mb terminal deletion reveals an overlapping gain (CNV) on the opposite allele that results in a net zero copy number difference between the patient and the reference genomes. (**b**) Genomic organization of the terminal deletion (*black rectangle*) relative to the CNV gain (*open rectangle*) and segmental duplications (*gray*). (**c**) Sequenced breakpoint junction from the intact (*nondeleted*) region to the telomere repeat sequence $(TTAGGG)_n$ (*bold italics*) (M.K.R. and K.E.H., unpublished data).

The terminal deletion is 1.0 Mb in size, but appears to be interrupted by a region of normal copy number. FISH analysis of the "normal" copy number region demonstrated that it is a product of an ~250-kb duplication on the nondeleted allele. This is a common CNV present in databases of genomic variation in normal individuals (4, 33). The structure of the terminal deletion was further confirmed by sequencing across the deletion breakpoint (Fig. 4c). Sequencing captures the junction between the proximal breakpoint identified by array CGH and the telomere repeat, $(TTAGGG)_n$. A SNP array could distinguish the alleles gained and lost, provided there are enough informative SNPs in the particular subtelomere (34, 35).

5. Translocations

Unbalanced translocations are detectable by array CGH, which reveals a terminal loss of one chromosome end and a terminal gain of another chromosome end (Fig. 5). Such array findings, however, should always be confirmed by FISH to rule out the possibility that they represent independent copy number alterations: a terminal deletion and a terminal duplication. Unbalanced translocations can also appear as an abnormality on only one chromosome end, if array coverage is lacking in the other gain/loss region. Such is the case for a gain detected by array CGH, but determined by FISH to reside on a different chromosome end. Alternatively, a translocation could appear as a loss by array CGH, but the participating gain could include only segmental duplications from another chromosome end, typically undetectable by arrays.

Truly balanced translocations, on the other hand, are not detectable by array, since there is no net gain or loss of genetic material. However, apparently balanced translocations with a concomitant gain or loss at the translocation junction may show a copy number alteration at that site (36). Balanced translocations are identified more often via family studies after an unbalanced translocation has been diagnosed in an affected proband. Once an unbalanced translocation is found in a family, parental FISH studies are essential to predict recurrence risks for the unbalanced form of the rearrangement. Carriers of the balanced translocation may be detected by FISH analysis of the participating chromosome ends.

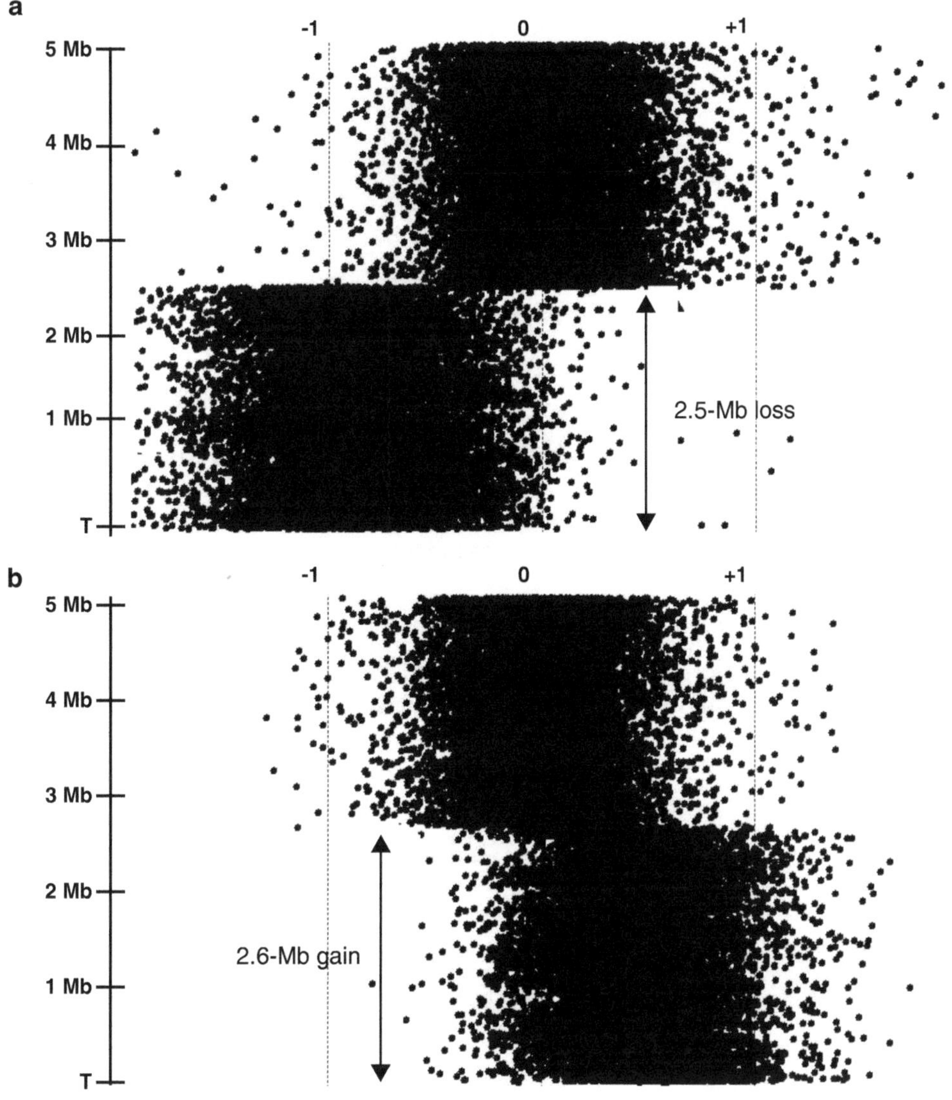

Fig. 5. Unbalanced translocation detected by array CGH. The 2.5-Mb loss (**a**) and the 2.6-Mb gain (**b**) are visible as terminal shifts in signal intensities.

6. Terminal Deletions Adjacent to Inverted Duplications

First described in maize by Barbara McClintock (37), terminal deletions adjacent to inverted duplications represent one of the most complex subtelomeric rearrangements. Using array CGH, these rearrangements appear as a genomic loss adjacent to a gain (Fig. 6); however, array CGH cannot determine the orientation of the duplicated segment. In cases where FISH studies have been performed, the duplications are usually found to be in an inverted

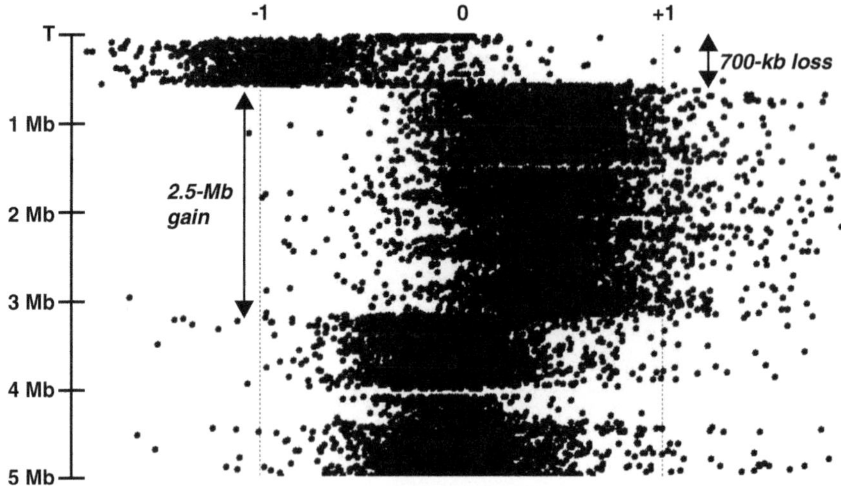

Fig. 6. A terminal deletion adjacent to an inverted duplication has a characteristic array pattern. The 700-kb terminal deletion is adjacent to a 2.5-Mb duplication that proves to be inverted by FISH and breakpoint sequencing.

orientation (38). Terminal deletions adjacent to inverted duplications have been reported in many different chromosome ends, with varying sizes of deletion and duplication (38–42). Nevertheless, the common genomic structure of the rearrangement likely represents a particular DNA repair mechanism.

7. Structural Variation in Subtelomeres: *The Future*

Variation in human subtelomeres is a double-edged sword; although small rearrangements are a source of normal polymorphism and signify rapidly evolving regions of the genome, larger gains and losses involving dosage-sensitive genes can cause intellectual disabilities and birth defects, making these regions particularly relevant to studies of human disease and diversity.

FISH and copy number technologies are the current methods we use to unravel subtelomeric rearrangements; however, as sequencing technologies become cheaper and better, we could see high-throughput sequencing platforms move into the subtelomeric variation field. If so, sequencing data must be analyzed carefully, with an eye on the precise genomic structure, to interpret the organization of subtelomeric imbalances. Such technology would not only identify subtelomeric imbalances, but would also capture subtelomeric breakpoint junctions, shedding light on the mechanisms of DNA breakage and repair.

Acknowledgment

We thank Karen E. Hermetz for providing array images and Cheryl T. Strauss for editorial assistance.

References

1. Bailey, J. A., Gu, Z., Clark, R. A., Reinert, K., Samonte, R. V., Schwartz, S., Adams, M. D., Myers, E. W., Li, P. W., and Eichler, E. E. (2002) Recent segmental duplications in the human genome. *Science 297*, 1003–7.
2. Linardopoulou, E. V., Williams, E. M., Fan, Y., Friedman, C., Young, J. M., and Trask, B. J. (2005) Human subtelomeres are hot spots of interchromosomal recombination and segmental duplication. *Nature 437*, 94–100.
3. Trask, B. J., Friedman, C., Martin-Gallardo, A., Rowen, L., Akinbami, C., Blankenship, J., Collins, C., Giorgi, D., Iadonato, S., Johnson, F., Kuo, W. L., Massa, H., Morrish, T., Naylor, S., Nguyen, O. T., Rouquier, S., Smith, T., Wong, D. J., Youngblom, J., and van den Engh, G. (1998) Members of the olfactory receptor gene family are contained in large blocks of DNA duplicated polymorphically near the ends of human chromosomes. *Hum Mol Genet 7*, 13–26.
4. Redon, R., Ishikawa, S., Fitch, K. R., Feuk, L., Perry, G. H., Andrews, T. D., Fiegler, H., Shapero, M. H., Carson, A. R., Chen, W., Cho, E. K., Dallaire, S., Freeman, J. L., Gonzalez, J. R., Gratacos, M., Huang, J., Kalaitzopoulos, D., Komura, D., MacDonald, J. R., Marshall, C. R., Mei, R., Montgomery, L., Nishimura, K., Okamura, K., Shen, F., Somerville, M. J., Tchinda, J., Valsesia, A., Woodwark, C., Yang, F., Zhang, J., Zerjal, T., Zhang, J., Armengol, L., Conrad, D. F., Estivill, X., Tyler-Smith, C., Carter, N. P., Aburatani, H., Lee, C., Jones, K. W., Scherer, S. W., and Hurles, M. E. (2006) Global variation in copy number in the human genome. *Nature 444*, 444–54.
5. Ballif, B. C., Sulpizio, S. G., Lloyd, R. M., Minier, S. L., Theisen, A., Bejjani, B. A., and Shaffer, L. G. (2007) The clinical utility of enhanced subtelomeric coverage in array CGH. *Am J Med Genet A 143*, 1850–7.
6. Martin, C. L., Nawaz, Z., Baldwin, E. L., Wallace, E. J., Justice, A. N., and Ledbetter, D. H. (2007) The evolution of molecular ruler analysis for characterizing telomere imbalances: from fluorescence in situ hybridization to array comparative genomic hybridization. *Genet Med 9*, 566–73.
7. Ballif, B. C., Kashork, C. D., and Shaffer, L. G. (2000) The promise and pitfalls of telomere region-specific probes. *Am J Hum Genet 67*, 1356–9.
8. Knight, S. J., Lese, C. M., Precht, K. S., Kuc, J., Ning, Y., Lucas, S., Regan, R., Brenan, M., Nicod, A., Lawrie, N. M., Cardy, D. L., Nguyen, H., Hudson, T. J., Riethman, H. C., Ledbetter, D. H., and Flint, J. (2000) An optimized set of human telomere clones for studying telomere integrity and architecture. *Am J Hum Genet 67*, 320–32.
9. Adeyinka, A., Adams, S. A., Lorentz, C. P., Van Dyke, D. L., and Jalal, S. M. (2005) Subtelomere deletions and translocations are frequently familial. *Am J Med Genet A 135*, 28–35.
10. Barber, J. C. (2005) Directly transmitted unbalanced chromosome abnormalities and euchromatic variants. *J Med Genet 42*, 609–29.
11. Ravnan, J. B., Tepperberg, J. H., Papenhausen, P., Lamb, A. N., Hedrick, J., Eash, D., Ledbetter, D. H., and Martin, C. L. (2006) Subtelomere FISH analysis of 11 688 cases: an evaluation of the frequency and pattern of subtelomere rearrangements in individuals with developmental disabilities. *J Med Genet 43*, 478–89.
12. Biesecker, L. G. (2002) The end of the beginning of chromosome ends. *Am J Med Genet 107*, 263–6.
13. Shao, L., Shaw, C. A., Lu, X. Y., Sahoo, T., Bacino, C. A., Lalani, S. R., Stankiewicz, P., Yatsenko, S. A., Li, Y., Neill, S., Pursley, A. N., Chinault, A. C., Patel, A., Beaudet, A. L., Lupski, J. R., and Cheung, S. W. (2008) Identification of chromosome abnormalities in subtelomeric regions by microarray analysis: a study of 5,380 cases. *Am J Med Genet A 146A*, 2242–51.
14. Willatt, L., Cox, J., Barber, J., Cabanas, E. D., Collins, A., Donnai, D., FitzPatrick, D. R., Maher, E., Martin, H., Parnau, J., Pindar, L., Ramsay, J., Shaw-Smith, C., Sistermans, E. A., Tettenborn, M., Trump, D., de Vries, B. B., Walker, K., and Raymond, F. L. (2005) 3q29 microdeletion syndrome: clinical and molecular characterization of a new syndrome. *Am J Hum Genet 77*, 154–60.

15. Lisi, E. C., Hamosh, A., Doheny, K. F., Squibb, E., Jackson, B., Galczynski, R., Thomas, G. H., and Batista, D. A. (2008) 3q29 interstitial microduplication: a new syndrome in a three-generation family. *Am J Med Genet A 146A*, 601–9.
16. Harada, N., Visser, R., Dawson, A., Fukamachi, M., Iwakoshi, M., Okamoto, N., Kishino, T., Niikawa, N., and Matsumoto, N. (2004) A 1-Mb critical region in six patients with 9q34.3 terminal deletion syndrome. *J Hum Genet 49*, 440–4.
17. Stewart, D. R., Huang, A., Faravelli, F., Anderlid, B. M., Medne, L., Ciprero, K., Kaur, M., Rossi, E., Tenconi, R., Nordenskjold, M., Gripp, K. W., Nicholson, L., Meschino, W. S., Capua, E., Quarrell, O. W., Flint, J., Irons, M., Giampietro, P. F., Schowalter, D. B., Zaleski, C. A., Malacarne, M., Zackai, E. H., Spinner, N. B., and Krantz, I. D. (2004) Subtelomeric deletions of chromosome 9q: a novel microdeletion syndrome. *Am J Med Genet A 128A*, 340–51.
18. Kleefstra, T., Brunner, H. G., Amiel, J., Oudakker, A. R., Nillesen, W. M., Magee, A., Genevieve, D., Cormier-Daire, V., van Esch, H., Fryns, J. P., Hamel, B. C., Sistermans, E. A., de Vries, B. B., and van Bokhoven, H. (2006) Loss-of-function mutations in euchromatin histone methyl transferase 1 (EHMT1) cause the 9q34 subtelomeric deletion syndrome. *Am J Hum Genet 79*, 370–7.
19. Phelan, M. C., Rogers, R. C., Saul, R. A., Stapleton, G. A., Sweet, K., McDermid, H., Shaw, S. R., Claytor, J., Willis, J., and Kelly, D. P. (2001) 22q13 deletion syndrome. *Am J Med Genet 101*, 91–9.
20. Wilson, H. L., Wong, A. C., Shaw, S. R., Tse, W. Y., Stapleton, G. A., Phelan, M. C., Hu, S., Marshall, J., and McDermid, H. E. (2003) Molecular characterisation of the 22q13 deletion syndrome supports the role of haploinsufficiency of SHANK3/PROSAP2 in the major neurological symptoms. *J Med Genet 40*, 575–84.
21. Durand, C. M., Betancur, C., Boeckers, T. M., Bockmann, J., Chaste, P., Fauchereau, F., Nygren, G., Rastam, M., Gillberg, I. C., Anckarsater, H., Sponheim, E., Goubran-Botros, H., Delorme, R., Chabane, N., Mouren-Simeoni, M. C., de Mas, P., Bieth, E., Roge, B., Heron, D., Burglen, L., Gillberg, C., Leboyer, M., and Bourgeron, T. (2007) Mutations in the gene encoding the synaptic scaffolding protein SHANK3 are associated with autism spectrum disorders. *Nat Genet 39*, 25–7.
22. National Institutes of Health and Institute of Molecular Medicine Collaboration (1996) A complete set of human telomeric probes and their clinical application. *Nat Genet 14*, 86–9.
23. Knight, S. J., Horsley, S. W., Regan, R., Lawrie, N. M., Maher, E. J., Cardy, D. L., Flint, J., and Kearney, L. (1997) Development and clinical application of an innovative fluorescence in situ hybridization technique which detects submicroscopic rearrangements involving telomeres. *Eur J Hum Genet 5*, 1–8.
24. Knight, S. J., Regan, R., Nicod, A., Horsley, S. W., Kearney, L., Homfray, T., Winter, R. M., Bolton, P., and Flint, J. (1999) Subtle chromosomal rearrangements in children with unexplained mental retardation. *Lancet 354*, 1676–81.
25. Baldwin, E. L., Lee, J. Y., Blake, D. M., Bunke, B. P., Alexander, C. R., Kogan, A. L., Ledbetter, D. H., and Martin, C. L. (2008) Enhanced detection of clinically relevant genomic imbalances using a targeted plus whole genome oligonucleotide microarray. *Genet Med 10*, 415–29.
26. Gajecka, M., Pavlicek, A., Glotzbach, C. D., Ballif, B. C., Jarmuz, M., Jurka, J., and Shaffer, L. G. (2006) Identification of sequence motifs at the breakpoint junctions in three t(1;9)(p36.3;q34) and delineation of mechanisms involved in generating balanced translocations. *Hum Genet 120*, 519–26.
27. Gajecka, M., Gentles, A. J., Tsai, A., Chitayat, D., Mackay, K. L., Glotzbach, C. D., Lieber, M. R., and Shaffer, L. G. (2008) Unexpected complexity at breakpoint junctions in phenotypically normal individuals and mechanisms involved in generating balanced translocations t(1;22)(p36;q13). *Genome Res 18*, 1733–42.
28. Yatsenko, S. A., Brundage, E. K., Roney, E. K., Cheung, S. W., Chinault, A. C., and Lupski, J. R. (2009) Molecular mechanisms for subtelomeric rearrangements associated with the 9q34.3 microdeletion syndrome. *Hum Mol Genet 18*, 1924–36.
29. Monfouilloux, S., Avet-Loiseau, H., Amarger, V., Balazs, I., Pourcel, C., and Vergnaud, G. (1998) Recent human-specific spreading of a subtelomeric domain. *Genomics 51*, 165–76.
30. Martin, C. L., Wong, A., Gross, A., Chung, J., Fantes, J. A., and Ledbetter, D. H. (2002) The evolutionary origin of human subtelomeric homologies – or where the ends begin. *Am J Hum Genet 70*, 972–84.
31. Rudd, M. K., Friedman, C., Parghi, S. S., Linardopoulou, E. V., Hsu, L., and Trask, B. J. (2007) Elevated rates of sister chromatid exchange at chromosome ends. *PLoS Genet 3*, e32.
32. Rudd, M. K., Endicott, R. M., Friedman, C., Walker, M., Young, J. M., Osoegawa, K., de Jong, P. J., Green, E. D., and Trask, B. J. (2009) Comparative sequence analysis of primate subtelomeres originating from a chromosome fission event. *Genome Res 19*, 33–41.

33. Itsara, A., Cooper, G. M., Baker, C., Girirajan, S., Li, J., Absher, D., Krauss, R. M., Myers, R. M., Ridker, P. M., Chasman, D. I., Mefford, H., Ying, P., Nickerson, D. A., and Eichler, E. E. (2009) Population analysis of large copy number variants and hotspots of human genetic disease. *Am J Hum Genet 84*, 148–61.
34. Wang, K., Chen, Z., Tadesse, M. G., Glessner, J., Grant, S. F., Hakonarson, H., Bucan, M., and Li, M. (2008) Modeling genetic inheritance of copy number variations. *Nucleic Acids Res 36*, e138.
35. Wang, K., Li, M., Hadley, D., Liu, R., Glessner, J., Grant, S. F., Hakonarson, H., and Bucan, M. (2007) PennCNV: an integrated hidden Markov model designed for high-resolution copy number variation detection in whole-genome SNP genotyping data. *Genome Res 17*, 1665–74.
36. Baptista, J., Mercer, C., Prigmore, E., Gribble, S. M., Carter, N. P., Maloney, V., Thomas, N. S., Jacobs, P. A., and Crolla, J. A. (2008) Breakpoint mapping and array CGH in translocations: comparison of a phenotypically normal and an abnormal cohort. *Am J Hum Genet 82*, 927–36.
37. McClintock, B. (1941) The Stability of Broken Ends of Chromosomes in Zea Mays. *Genetics 26*, 234–82.
38. Ballif, B. C., Yu, W., Shaw, C. A., Kashork, C. D., and Shaffer, L. G. (2003) Monosomy 1p36 breakpoint junctions suggest pre-meiotic breakage-fusion-bridge cycles are involved in generating terminal deletions. *Hum Mol Genet 12*, 2153–65.
39. Stetten, G., Charity, L. L., Kasch, L. M., Scott, A. F., Berman, C. L., Pressman, E., and Blakemore, K. J. (1997) A paternally derived inverted duplication of 7q with evidence of a telomeric deletion. *Am J Med Genet 68*, 76–81.
40. Jenderny, J., Poetsch, M., Hoeltzenbein, M., Friedrich, U., and Jauch, A. (1998) Detection of a concomitant distal deletion in an inverted duplication of chromosome 3. Is there an overall mechanism for the origin of such duplications/deficiencies? *Eur J Hum Genet 6*, 439–44.
41. Bonaglia, M. C., Giorda, R., Poggi, G., Raggi, M. E., Rossi, E., Baroncini, A., Giglio, S., Borgatti, R., and Zuffardi, O. (2000) Inverted duplications are recurrent rearrangements always associated with a distal deletion: description of a new case involving 2q. *Eur J Hum Genet 8*, 597–603.
42. Cotter, P. D., Kaffe, S., Li, L., Gershin, I. F., and Hirschhorn, K. (2001) Loss of subtelomeric sequence associated with a terminal inversion duplication of the short arm of chromosome 4. *Am J Med Genet 102*, 76–80.

Chapter 7

Array-Based Approaches in Prenatal Diagnosis

Paul D. Brady, Koenraad Devriendt, Jan Deprest, and Joris R. Vermeesch

Abstract

The diagnostic benefits of array comparative genomic hybridisation (CGH) have been demonstrated, with this technique now being applied as the first-line test for patients with intellectual disabilities and/or multiple congenital anomalies in numerous laboratories. There are no technical barriers preventing the introduction of array CGH to prenatal diagnosis. The question is rather how this is best implemented, and for whom. The challenges lie in the interpretation of copy number variations, particularly those which exhibit reduced penetrance or variable expression, and how to deal with incidental findings, which are not related to the observed foetal anomalies, or unclassified variants which are currently of uncertain clinical significance. Recently, applications of array technologies to the field of pre-implantation genetic diagnosis have also been demonstrated. It is important to address the ethical questions raised concerning the genome-wide analysis of prenatal samples to ensure the maximum benefit for patients. We provide an overview of the recent developments on the use of array CGH in the prenatal setting, and address the challenges posed.

Key words: Prenatal diagnosis, Array CGH, PGD, Copy number variation, CNV, Incidental findings, Unclassified variants, Miscarriage, POC

Abbreviations

cffDNA	Cell-free foetal DNA
UCV	Unclassified variants
WGA	Whole-genome amplification
CNV	Copy number variation
POC	Product of conception

1. Introduction

Initially, chromosome studies were performed using simple staining techniques which only allowed for the detection of entire groups of chromosomes. In 1966, Steel and Breg (1) demonstrated that the chromosomal constitution of the foetus could be determined by the

Lars Feuk (ed.), *Genomic Structural Variants: Methods and Protocols*, Methods in Molecular Biology, vol. 838,
DOI 10.1007/978-1-61779-507-7_7, © Springer Science+Business Media, LLC 2012

analysis of cultured amniotic fluid (AF) cells. One year later, Jacobson and Barter (2) performed the first prenatal diagnosis of a chromosomal abnormality. In the following years, several series of prenatal diagnoses with diverse chromosomal abnormalities were reported (3, 4). The degree of precision was increased in the 1970s with the introduction of chromosome-banding techniques. These enabled the detection of individual chromosomes and segments (bands) within chromosomes. Although chromosomal karyotyping allows for the genome-wide detection of large chromosomal abnormalities and translocations, it has a number of inherent limitations, namely, the time required to culture cells and the limited resolution.

Prenatal genetic diagnosis is carried out with time constraints. Since the vast majority of prenatal diagnosis is aimed at the detection of well-known precise anomalies, such as trisomy 21 (Down syndrome), trisomy 18 (Edward syndrome), trisomy 13 (Patau syndrome), or sex chromosome aneuploidies (Turner syndrome, Klinefelter syndrome), the technique of FISH using probes targeted to loci on chromosomes 21, 18, 13, X, and Y became the method of choice due to the ability to provide a rapid result (2–3 days) for these conditions (5–8). More recently, novel molecular biology techniques have also been applied to rapid aneuploidy detection, notably, quantitative fluorescent PCR (qf-PCR) (9, 10) and multiplex ligation-dependent probe amplification (MLPA) (11, 12) which are able to provide a result in 1–2 days. These techniques apply DNA-based methods to determine genomic copy number changes, but have the same locus-specific disadvantages as with FISH. Rapid methods of prenatal diagnosis for additional specific indications also have the advantage of fast reporting times, and relative low cost, for example FISH as rapid test for deletion of 22q11 (cleft palate, cardiac defects—DiGeorge/velocardiofacial syndrome, DG/VCFS) or tetrasomy 12p (congenital diaphragmatic hernia—Pallister–Killian syndrome).

However, the above techniques target only specific loci, and are unable to provide a rapid genome-wide screen. Until recently, G-banded karyotyping has been the gold standard for the genome-wide detection of genomic imbalances in prenatal diagnosis. Today, a new technology that is overcoming the resolution, locus specificity, and time limitations of the aforementioned techniques is molecular karyotyping using genome-wide array comparative genomic hybridisation (CGH).

1.1. Prenatal Diagnosis by Array Comparative Genomic Hybridisation

There are no technical barriers to performing array CGH as a prenatal test. In a proof of principle experiment, Rickman et al. (13) demonstrated the feasibility of performing array CGH for prenatal diagnosis on DNA extracted from uncultured AF cells. With the exception of a triploidy, 29/30 results were in complete concordance with the karyotype. The feasibility of performing array CGH for the prenatal detection of genomic imbalances using BAC, oligo, and SNP arrays on DNA extracted from uncultured amniocytes has been further demonstrated over recent years (14–17). A point to note is that the quality of DNA obtained from AF is often suboptimal due to the presence of dead cells, small, degraded DNA fragments, and other inhibiting factors, as well as often being of limited quantity. Cell-free foetal DNA in the AF supernatant has also been shown to be suitable for performing array CGH (18–20). In addition, whole-genome amplification (WGA) has been applied to small quantities of DNA to provide accurate prenatal results from limited amounts of starting material (16). An overview of the reports using array CGH for prenatal diagnosis and the types of arrays used is presented in Table 1.

The application of array CGH in the prenatal diagnostic setting is an attractive alternative to karyotyping, providing a genome-wide screen for genomic imbalances at a higher resolution, allowing for the detection of all known recurrent microdeletion/microduplication syndromes when using a minimum resolution of approximately 400 kb. Further advantages include a higher degree of automation, and with no cell culture required array CGH provides a more rapid result than karyotyping while overcoming the issue of culture failure. It is, therefore, not surprising that several diagnostic centres have implemented this technology (21, 22), and a number of reports have suggested that the technique is now ready for mainstream use (23–25). Array CGH allows a more precise characterisation of imbalances, providing a better prediction of the phenotype, or the severity of the disease. The more precise delineation of an imbalance allows for comparison of the phenotype with other patients displaying imbalances in the same region, and to correlate the phenotype with genes in the region.

1.2. Copy Number Variations and Polymorphisms

Aside from the identification of pathogenic copy number variations (CNVs), array CGH has also revealed large numbers of CNVs in normal individuals. Up to five years ago, SNPs were generally considered to be the main source of genetic variation. Hence, the discovery of an unexpectedly large number of apparently benign CNVs, in unrelated "normal" individuals, was rightly described by *Science* as the discovery of the year in 2007. A number of array CGH studies have now demonstrated the presence of large numbers of polymorphic copy number variants (26–31). In the first large-scale systematic study, Redon et al. (32) mapped all CNVs using both array CGH and SNP genotyping arrays on the 270 individuals of the HapMap collection with ancestry from Europe, Africa, and Asia, which revealed 1,447 submicroscopic copy number variable regions in the human genome. These non-pathogenic variations, initially estimated at covering 12% of the genome, includes deletions, duplications, insertions, and complex multisite variants. Furthermore, these CNVs encompass hundreds of genes,

Table 1
Summarises the published research into the use of array CGH for prenatal diagnosis

References	Number of samples tested	Type of array platform	Sample details	Additional comments
Coppinger et al. (37)	244; 182 (genome wide), 62 (targeted)	Genome-wide 161 BAC and 21 oligo arrays; targeted 62 BAC arrays	Cultured and uncultured AF and CVS	Genome-wide vs. targeted arrays; clinically significant 3.8 vs. 0.9%, UCV 0.5 vs. 0.5%, benign CNV 8.8 vs. 8.0%
Kleeman et al. (73)	50	26 targeted BAC arrays; 24 genome-wide BAC arrays	47 AF and 3 CVS	2% clinically significant
Tyreman et al. (17)	106	Affymetrix GeneChip 6.0 SNP array	87 AF; 15 CVS; 4 POC	10% clinically significant, triploidy detected
Vialard et al. (74)	39	Targeted BAC array	28 AF; 11 CVS (after TOP)	10.8% clinically significant (unidentified on karyotype)
Van den Veyver et al. (21)	300	Targeted BAC arrays and oligo arrays	AF cultured (65), uncultured (189); CVS cultured (20), uncultured (33); other (2)	5% clinically significant, 1.0% UCV. WGA performed for some AF and CVS samples
Bi et al. (16)	15	Targeted oligo array	Uncultured AF and WGA	Proof of principle for uncultured AF and oligo array, and for WGA

Shaffer et al. (22)	151	Targeted BAC array	Cultured AF and CVS	
Lapaire et al. (20)	10	Targeted BAC array	Uncultured AF and cffDNA from AF supernatant	Proof of principle for extraction of cffDNA from fresh/frozen AF
Miura et al. (19)	13	Custom BAC array (chr.13, 18, 21, *X* and *Y* only)	Uncultured AF and cffDNA from AF supernatant	Proof of principle for DNA extracted from AF
Rickman et al. (13)	30	Targeted BAC array	30 uncultured prenatal samples	Proof of principle for uncultured prenatal arrays using as little as 1-ml AF
Sahoo et al. (14)	98	Targeted BAC array	56 AF and 42 CVS	
Larrabee et al. (18)	36	Targeted BAC array	28 cffDNA from frozen AF supernatant, 8 cultured AF	Proof of principle for cffDNA, higher noise with cffDNA, foetal sex and aneuploidy detected

including a large number of genes known to be involved in genetic disorders and registered in OMIM, as well as disease loci and functional elements. In the most recent and comprehensive CNV study to date, Conrad et al. (33) used tiling path oligonucleotide arrays comprising 42 million probe features, spread over 20 arrays, to generate a comprehensive map of 11,700 CNV loci, after merging the 51,997 putative CNV segments greater than 443 bp in size detected in 40 HapMap individuals (vs. 1 reference sample). The authors estimate that their study detects 80–90% of common CNVs greater than 1 kb in length, and furthermore the results identify an average of 1,098 CNVs, comprising 24 Mb or 0.78% of the genome, when comparing any two individual genomes by array CGH.

Recent fine-mapping studies have revealed that apparently benign CNVs can cause intragenic variation, resulting in different splice variants and the use of different exons and even new gene products (34). Gene ontology analysis has also shown an enrichment of genes involved in extracellular processes, such as cell adhesion, recognition, and communication occurring in CNV regions, while genes involved in intracellular processes, such as biosynthetic and metabolic pathways, are relatively under-represented in CNV regions.

2. The Challenge of Interpretation of CNVs in the Prenatal Diagnostic Setting

A major challenge in the use of array CGH is in the interpretation of CNVs of unknown clinical significance. It is currently impossible to predict the phenotypic effect of these so-called unclassified variants. Targeted screening for specific imbalances by dedicated arrays offers an attractive alternative to conventional karyotyping given their speed and the increase in resolution. However, there is much debate as to whether there is a best resolution for diagnostic purposes, and whether to use an array strictly targeted at “clinically relevant” loci or an array with genome-wide coverage (35). A compromise is to use an array with variable resolution, providing higher resolution for target regions than the genomic backbone (36). It is hoped that this type of array design will provide maximum detection of clinically significant imbalances, and good definition of breakpoints, while minimising the detection of large numbers of CNVs of uncertain clinical significance. In a recent study, however, Coppinger et al. (37) demonstrated that genome-wide array CGH for prenatal diagnosis identified clinically significant submicroscopic chromosome abnormalities without an increase in unclassified variants or benign CNVs when compared to a targeted array platform, at the resolution provided by these platforms.

While the identification of a CNV known to be associated with a specific phenotype can generally be considered causal for the phenotype observed in a particular patient post-natally, it is much more difficult to predict the outcome, or severity, of the future phenotype in the prenatal setting. Indeed, many well-known pathogenic CNVs can be tolerated by apparently normal individuals. A classic example is that of deletion of 22q11 causing DG/VCFS, and the recently reported reciprocal microduplication which has been reported to often be inherited and of a highly variable phenotypic outcome (38, 39).

A major reason for the slow uptake of prenatal diagnosis by array CGH is that the interpretation of submicroscopic imbalances post-natally remains challenging, since the phenotypic consequences may be of reduced penetrance or variable expression. These challenges, and the extent to which they occur when performing array CGH at high resolutions, create the necessity for parental samples to be made available along with the prenatal sample at the time of array CGH analysis to aid the interpretation of the prenatal result. Without the availability of parental DNA, it may be impossible to interpret the clinical consequences of rare CNVs detected by array CGH. At present, the following reasoning is one approach used in the clinical interpretation of array CGH results (40).

- If the imbalance in a patient involves a known microdeletion or microduplication syndrome, which also fits with the observed phenotypic features, then this is considered pathogenic.
- If the imbalance has occurred de novo in the patient, and particularly if it contains genes whose effects are compatible with the clinical findings in the patient, this supports the hypothesis of pathogenicity, but is not absolute proof.
- If the imbalance is familial and is not a known benign CNV, then the phenotypic relationship is difficult to interpret.

Additional means of determining causality include investigating whether a particular imbalance is segregating with the phenotype within a family, creating a need for additional family members to be tested to aid the interpretation. For rare variants, the use of databases, such as *D*atabas*e* of *C*hromosomal *I*mbalance and *P*henotype in *H*umans using *E*nsemble *R*esources (DECIPHER) (41) (https://decipher.sanger.ac.uk/) and *E*uropean *C*ytogeneticists *A*ssociation *R*egister of *U*nbalanced *C*hromosome *A*berrations (ECARUCA) (42) (https://www.ecaruca.net), allows the comparison of array results and patient phenotype with those obtained from other centres around the world. Chapter 9 provides further information on the interpretation of array data in clinical genetics.

2.1. Incidental Findings

One of the ethical issues raised by high-resolution, genome-wide array CGH, and one of the main objections (43), is the detection of “unwanted” information, which has been referred to as “the incidentalome” (44, 45). Chromosome imbalances with clinical consequences but which are not relevant to the reason for referral and the observed phenotype may be identified, such as late onset disorders, infertility, or cancer predisposition loci (46). These incidental findings create challenges for the laboratory and clinicians in the interpretation of results, and for the provision of genetic counselling (47, 48). Though not entirely new, the extent to which these events occur is more frequent when using genome-wide array CGH at high resolution in comparison to karyotyping. There are several forms of incidental findings.

- Those of importance to the patients themselves, but which are not related to the phenotype for which the initial investigation was undertaken: For example, increased risk of late-onset diseases, some of which may be of high risk, or of severe disease, e.g. tumour regulatory genes (46), BRCA1, p53, APC. In addition, there are a number of CNVs associated with late-onset disorders which are less severe or of relatively low increased risk, e.g. TCF2 (MODY5), PMP22 (HNPP/CMT), DAZ genes (infertility).
- Those of importance to other family members: In the assessment of the status of de novo or inherited, which requires analysis of parents, similar incidental information may be obtained on the parents themselves.
- Those not of importance to the individual, for which there is no health risk, but which may pose an increased risk of a genetic disorder for their children or other relatives, e.g. carrier for recessive disorders.
- The issue of false paternity.

These issues are not unusual in the classical setting of genetic diagnosis. Understandably, the opinions about the procedures to be followed are different, and nowadays are seen as a spectrum with two extremes, one in which all information should be provided and another in which only the information of medical benefit is provided.

An example of such an incidental finding is displayed in Fig. 1. In this case, a prenatal sample referred due to an isolated congenital diaphragmatic hernia (CDH) being detected on ultrasound was found to carry a duplication of 22q11 by array CGH analysis. Following parental analysis, this duplication was shown to have been inherited from the apparently normal father. There is a lack of evidence to associate this finding with the observed foetal abnormalities; however, there may be future post-natal consequences which are impossible to predict. This finding is also of clinical relevance for the father and for future pregnancies. This example highlights the importance of well-defined procedures for pre-test counselling, and of the challenges raised in genome-wide array CGH for prenatal diagnosis.

An alternative to the shielded approach, which is somewhat subjective, is to provide the option during the pre-test counselling of an

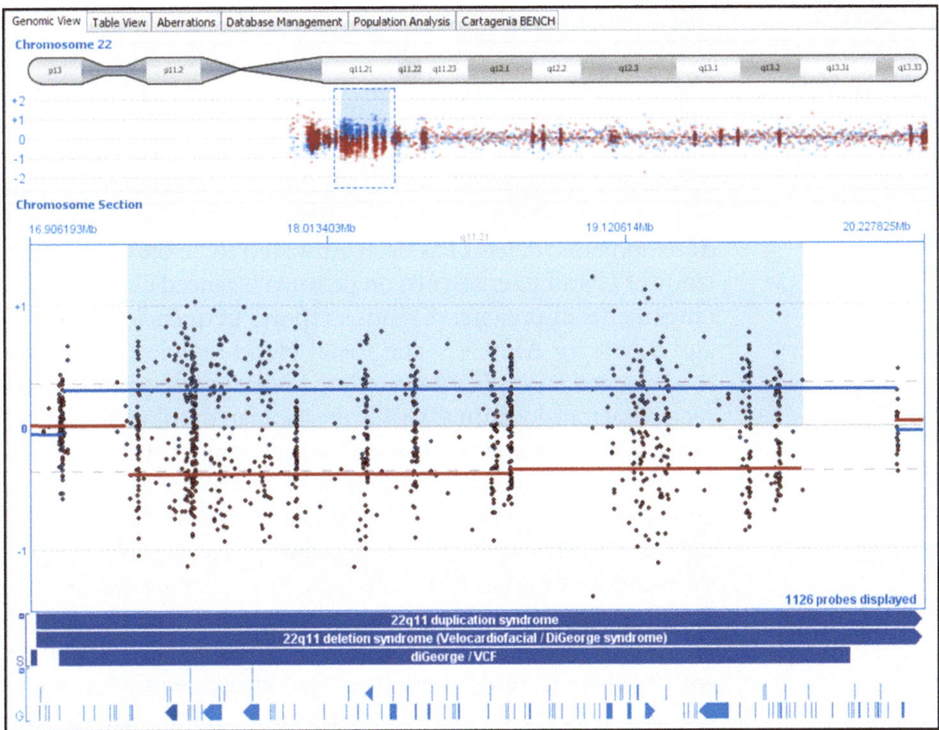

Fig. 1. Displays a duplication of 22q11 detected in a prenatal sample using an array comprising 105,000 (105 K) oligonucleotide probes with genome-wide and targeted coverage. The *blue* dataset shows the array result with patient labelled in Cy5, and the *red* dataset the result from the dye swap with patient labelled in Cy3, hybridised against reference DNA in Cy3 and Cy5, respectively. The minimal duplicated region is highlighted in *blue* in the main Chromosome Section view, and by the *dashed box* above. The Log_2 signal intensity ratios of the individual probes are plotted on the *y*-axis, with the base-pair position shown in Mb on the *x*-axis. At the top of the figure, an image of chromosome 22 is displayed, with cytogenetic bands displayed. The tracks below indicate known syndromes (S) and genes (G).

informed consent in which the patient(s) themselves decide whether to be informed, or not, of the unexpected findings that genome-wide array CGH at high resolution may reveal. The implications, and uncertainties, of these findings are addressed during the traditional post-test counselling session. However, experience has taught us that in the prenatal setting, in which the parents are in an anxious and often distressful situation, this decision can be cumbersome.

A related question is whether the information obtained from array CGH should be made available at a later date, or perhaps after birth, particularly if there are new clinical questions. Do we have a duty to store this information, since the large amount of information obtained by high-resolution array CGH and from SNP arrays concerning susceptibility to complex diseases or important pharmacogenetic information may be revealed? As the number of available array platforms increases, their resolution increases and our understanding of the influence of CNVs also increases; this is an important issue that should be addressed and is likely to be of more importance in the future. More research in this area is certainly needed.

2.2. Chromosomal Rearrangements not Detectable by Array CGH

Inherent to the technique, balanced chromosomal rearrangements (balanced translocations and inversions) are not identified by array CGH. When a balanced rearrangement is detected prenatally on karyotype, the parents are usually tested and if the same rearrangement is present in the "normal" parent then the translocation is considered benign. However, if the rearrangement is de novo, genetic counselling is more challenging, and the empiric risk for developmental defects has been estimated to be 6% (49). The recent study of Giardino et al. (50) on de novo balanced chromosome rearrangements in prenatal diagnosis reports frequencies of 0.09, 0.08, and 0.05% for AF, CVS, and foetal blood samples, respectively. Of the balanced anomalies detected, the reported frequencies were 72% reciprocal translocations, 18% Robertsonian translocations, 7% inversions, and 3% complex chromosome rearrangements.

However, array CGH analysis of patients with developmental anomalies and apparently balanced translocations reveals microdeletions and microduplications which disrupt genes at the translocation sites, and in some cases an imbalance in a chromosome not involved in the translocation is responsible for the observed phenotype (51–53). The study of De Gregori et al. (51) found that 40% of patients with a "chromosomal phenotype" and an apparently balanced translocation were in fact unbalanced, and a similar study by Schluth-Bolard et al. (53) confirms the figure of 40% of apparently balanced chromosome rearrangements being in fact unbalanced. In light of these findings, genome-wide array CGH can provide important information in cases of apparently balanced translocations, particularly where abnormalities are detected on ultrasound scan. Considering that de novo translocations occur about 1/1,000 births, with 6% having an abnormal phenotype and 40% of these detectable by array CGH, this would leave only 0.0036% of de novo pathogenic translocations undetected if no karyotype were to be performed.

In addition to truly balanced rearrangements, triploidies (69,XXX and 69,XXY) and tetraploidies are not detectable by array CGH. However, the use of DNA from a patient with Klinefelter syndrome (47,XXY) results in aberrant X and Y chromosome intensity ratios, enabling the detection of XXX triploidies and all tetraploidies (54). An exception to this is the use of SNP arrays which do allow for the detection of polyploidy (17).

A further issue is that of mosaicism, and the ability of array CGH to detect mosaicism at the same level or better than with current methods. It has been shown that array CGH does have the ability to detect mosaicism at a level comparable to conventional karyotyping with the use of BAC arrays (55); however, it is advisable for individual laboratories to determine the level of detection and resolution at which an accurate result can be achieved. More research is needed to establish the level of detection with oligo arrays, which are rapidly replacing BAC arrays in the diagnostic setting. Figure 2 demonstrates the detection of mosaicism in a

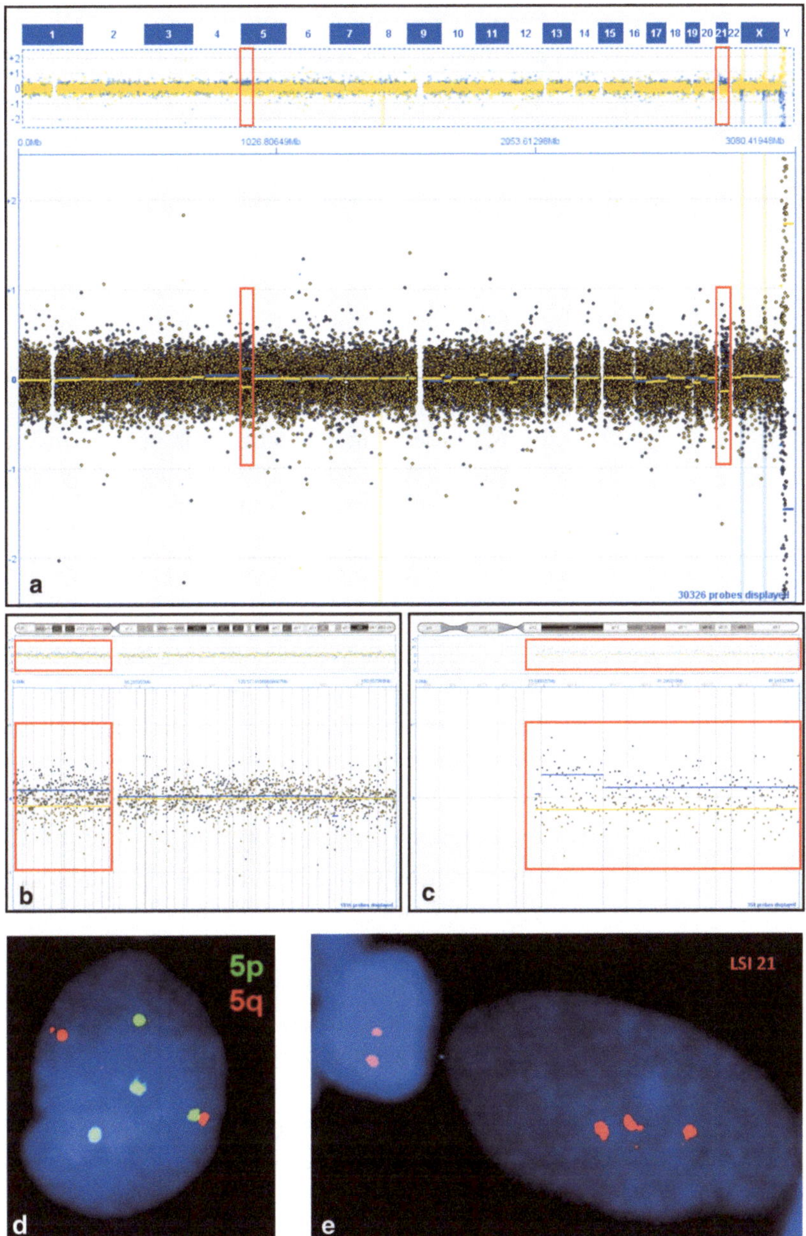

Fig. 2. Displays the array result for a female POC sample labelled in a dye swap experiment against an XXY reference sample using an array comprising 15,000 (15 K) oligonucleotide probes. Patient sample is labelled in Cy5 (*blue* dataset) and Cy3 (*yellow* dataset). Also shown are images from the confirmatory FISH analysis. (**a**) At the *top* of the image are chromosome numbers from 1 to 22, *X* and *Y* (*left* to *right*). The Log_2 signal intensity ratios of the individual probes are plotted on the *y*-axis in whole genome view in both windows. The *red boxes* highlight chromosome arm 5p and chromosome 21, both of which show a deviation from zero of the Log_2 signal intensity ratio indicating mosaicism for tetrasomy 5p and trisomy 21 in this sample. (**b**) Displays a view of chromosome 5, with the deviation from zero of the Log_2 signal intensity ratios highlighted in the *red box* for the 5p arm. (**c**) Displays a view of chromosome 21, with the deviation from zero of the Log_2 signal intensity ratios highlighted in the *red box*. There are no probes present for the 21p microsatellite region. (**d**) Shows an image of a cell with tetrasomy 5p indicated by the four *green* signals, with the normal 5q arm indicated by the two *red* signals. FISH analysis confirmed mosaicism for tetrasomy 5p in 30% of the cells examined. (**e**) Shows an image of a cell with trisomy 21 and a normal cell with 2 copies of chromosome 21, indicated by the three *red* signals and two *red* signals, respectively. FISH analysis-confirmed mosaicism for trisomy 21 in 50% of the cells examined.

sample from a spontaneous abortion of an IVF pregnancy at 9 weeks gestation, in which mosaicisms for trisomy 21 and tetrasomy 5p were both observed and subsequently confirmed by FISH.

3. Application of Array CGH

3.1. Indications for Array CGH

In order of clinical importance, it is possible to distinguish between a number of indications for prenatal genetic testing.

1. Multiple congenital abnormalities detected in a foetus: For a foetus with multiple abnormalities and without an etiological diagnosis, there is an indication for genetic investigation. In this situation, array CGH offers a higher quality of genetic research. The a priori probability of a significant imbalance is relatively high, and therefore imbalances are likely easier to interpret. The definition of congenital malformations is based on a similar classification post-natally:
   (a) Abnormalities in two or more organs/systems (e.g. heart, kidney, face, brain)
   (b) A single major abnormality as defined above, along with minor abnormalities or additional soft marker(s) (e.g. increased nuchal translucency, low-set ears)
2. Isolated defects: Where it is possible to distinguish between:
   (a) Abnormalities with a high risk of a chromosome imbalance (e.g. omphalocoele) or with important therapeutic consequences (e.g. diaphragmatic hernia, where antenatal intervention is contemplated)
   (b) Abnormalities with less serious clinical implications or of low risk of a chromosome imbalance (e.g. cardiopathy, cleft lip/palate)
3. Prenatal screening: Array CGH may be implemented as an alternative to traditional karyotyping, which is currently undertaken regardless of the indication. There are several different situations as follows.
   (a) Abnormal first-trimester biochemical screening, mainly an indication for trisomy 21.
   (b) Maternal age: The largest referral group for prenatal diagnosis, where an increased risk due to advanced maternal age is the only indication, mainly for aneuploidy.
   (c) Psychological: A small minority of patients who wish to undergo prenatal diagnosis, although there is no medical indication.

   Due to the aforementioned challenges in interpretation, during the first phase of introduction of array CGH in prenatal diagnosis, most laboratories currently limit testing to those groups defined in 1 and 2(a). As our knowledge of the influence of CNVs increases, it is likely that the indications for which prenatal array CGH is offered will broaden, but at this time routine prenatal screening by array CGH without an increased risk of a serious genetic disorder may undermine the rationale and purpose of prenatal screening.

3.2. The Use of Array Comparative Genomic Hybridisation for Miscarriages

Spontaneous abortions are a common occurrence, with 10–15% of all clinically recognised pregnancies ending in early pregnancy loss. Cytogenetic analysis has shown that around 50% of first-trimester miscarriages are caused by foetal chromosome abnormalities, most of which consist of numerical abnormalities (86%), including trisomies, monosomies, and polyploidies. Structural abnormalities represent a further 6% of observed anomalies (56, 57). Identification of the cause of spontaneous abortion helps to assess recurrence risks for future pregnancies, and can provide some comfort to parents when an anomaly has been identified. Traditionally, routine analysis of products of conception (POC) has been performed by karyotyping of metaphase spreads following tissue culture. However, due to failure of culture growth, suboptimal chromosome preparations, or maternal cell contamination (58, 59), on occasion either no result or an erroneous result is obtained. This might be overcome by sampling by embryofetoscopy prior to expulsion of the POC.

The concept of array CGH has been applied successfully for the detection of chromosome abnormalities in POCs. In an early proof of principle study, Schaeffer et al. (60) demonstrated that all abnormalities detected by G-banding were detected by array CGH, as well as additional abnormalities in 9.8% of cases, thereby enhancing the detection of foetal chromosome aberrations. The studies of Benkhalifa et al. (61) and Shimokawa et al. (62) have also confirmed the ability to detect genomic imbalances in POC samples which either fail to grow or have normal karyotypes. Given the high rate of chromosome abnormalities in miscarriage samples and the added value array CGH has been shown to offer, a number of laboratories are now applying this technique to the diagnostic workup of these samples (63).

3.3. The Use of Array Comparative Genomic Hybridisation for PGD

Pre-implantation genetic diagnosis (PGD) was first introduced almost 20 years ago, providing genetic testing prior to pregnancy for the purpose of identifying only those embryos which are unaffected, thereby avoiding the possibility of subsequent pregnancy loss or termination. In one of the first reports by Handyside et al. (64), the gender of embryos in couples at risk of X-linked diseases was determined by PCR, allowing for unaffected female embryos to be selected for. Since PGD requires the removal of one or more single cells from the early embryo, karyotyping is unreliable as a genome-wide diagnostic test. The technique of FISH is, therefore, routinely used for the detection of aneuploidy, termed pre-implantation genetic screening (PGS), and for certain structural chromosome abnormalities while PCR is the method of choice for known single-gene defects. However, FISH has the limitation of the number of loci that can be investigated, with typically only 7–10 chromosomes or loci being interrogated in a single cell or blastomere. In addition, for the screening of chromosomal imbalances in translocation carriers, novel probes may require development for each PGD cycle and its accuracy tested, which can be costly and time consuming.

A number of reports have demonstrated the feasibility of WGA by a variety of methods from single cells, as well as single blastomeres, allowing for genome-wide analysis by conventional CGH and more recently by array CGH (65–68), thus overcoming the limitations of FISH. The genome-wide analysis of blastomeres by CGH revealed the frequency of chromosome abnormalities in the early embryo, confirming that FISH is not an accurate screen for viable embryos. More recently, Vanneste et al. (69) used array CGH techniques to reveal not only mosaicism for whole-chromosome aneuploidies and uniparental disomies in most cleavage-stage embryos, but also frequent segmental deletions, duplications, and amplifications that were reciprocal in sister blastomeres, as shown in Fig. 3, implying the occurrence of breakage–fusion–bridge cycles. This finding explains the low human fecundity and identifies post-zygotic chromosome instability as a leading cause of constitutional chromosomal disorders, indicating that screening only a single blastomere may result in the discarding of potentially viable embryos or, conversely, selection of a non-viable embryo (70). The consequences of this observation that during early human embryogenesis chromosomal instability is common, a feature thus far only observed in tumours, are that pre-implantation genetic aneuploidy screening of cleavage-stage embryos may be of limited use as a diagnostic tool.

In 2008, Hellani et al. (71) reported the successful clinical application of array CGH for aneuploidy screening in a clinical setting resulting in successful pregnancies. More recently, Handyside et al. (72) report on the potential application of "karyomapping" for PGD, in which SNP arrays are used for the genome-wide linkage-based analysis of inheritance and detection of chromosome imbalances. Aside from using this technique to screen for aneuploidy, through the analysis of SNP genotypes for parents and a sibling or appropriate family member the allelic inheritance can also be determined, and therefore the affected status, in two families segregating mutant CFTR alleles. This promising technique has the ability to improve the pregnancy rate for couples undergoing PGD in IVF pregnancies, a factor which has remained elusive with current methods. It is likely that array technology will become increasingly available in the clinical setting of PGD/PGS in the near future.

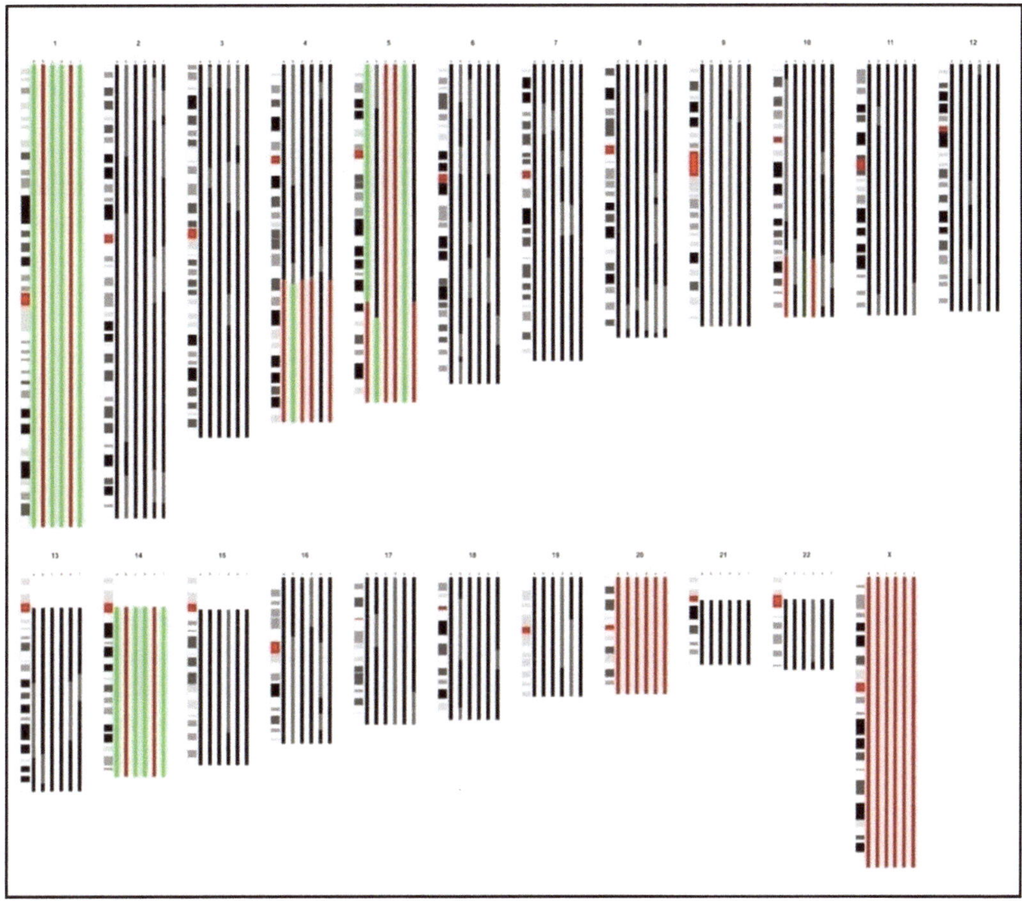

Fig. 3. Displays a karyogram of six individual blastomeres within a single embryo. Chromosomes are numbered, with the *coloured bars* representing the copy number state of the specific blastomeres, based upon BAC array results and SNP array results of copy number and genotype in each individual blastomere. *Black bars* represent a region of normal diploid copy number, *red* a hemizygous deletion, *green* a duplication, *dark green* an amplification, and *grey* for discordance between analyses or unreliable results. This male embryo displays whole chromosome imbalances of mitotic origin in chromosomes 1 and 14, whereas chromosome 20 displays monosomy for all sister blastomeres suggestive of meiotic non-disjunction. In addition, 4q and 10q terminal deletions with the reciprocal 4q duplication and 10q amplification, respectively, were detected in a number of sister blastomeres. Finally, a 5q terminal deletion and the reciprocal 5q duplication were detected in two blastomeres, whereas the remaining part of the chromosome located proximal to the 5q deletion was trisomic. Two sister blastomeres contained a monosomy for chromosome 5 and a remaining sister blastomere contained a trisomy 5, whereas another lacked a 5q terminal portion for which the size was equal to the partial deletion and duplication in its sister blastomeres (69).

4. The Future of Prenatal Diagnosis

The use of array CGH is now challenging conventional karyotyping as the gold standard in prenatal diagnosis. The successful implementation of array CGH in the clinical setting requires excellent collaboration between obstetricians/gynaecologists, clinical geneticists, genetic counsellors, and cytogeneticists and technical staff, with well-defined ethically approved protocols for indications for prenatal array CGH, pre- and post-test counselling, analysis and interpretation of results, and reporting of results. The general introduction of genome-wide screening tools for all indications largely exceeds current practice, and raises a number of ethical questions. It is advisable that in the initial stages of introduction of array CGH in the clinic to offer this only to those patients for whom abnormalities are observed on ultrasound scan, and to limit those specialists offering, and counselling for, array CGH analysis in prenatal diagnosis.

The use of expression arrays for prenatal samples may also provide relevant information related to developmental abnormalities; however, more research is needed in this field to evaluate the diagnostic benefits. Non-invasive prenatal diagnosis using cell-free foetal DNA or isolated foetal cells in maternal blood has been slow in reaching cost-effective clinical applications. However, the research in this field is promising and could replace invasive procedures if accurate, cost-effective techniques become available.

As whole-genome sequencing becomes more readily available and the costs reduce, should this technology be applied to prenatal diagnosis, since whole-genome sequencing is an ethical minefield for the purpose of prenatal diagnosis? The concept of resequencing using dedicated prenatal capture arrays for all relevant developmental genes may reduce the ethical issues raised with respect to a prenatal whole-genome sequence and the information contained within while allowing the benefits of sequencing technology to be applied, in the appropriate circumstances, to prenatal diagnosis. This approach may bypass the main ethical objections and provide a clinically relevant mutation screen for all known prenatal disorders.

Array-based technologies are shown to have various applications in many aspects of prenatal genetic diagnosis. Their success in this field is dependent on the cautious introduction into the clinical setting. The further demonstration of the benefits that this technique offers and constructive discussion into the circumstances in which this should be applied allow for the successful transition of array CGH from a research tool into a valuable asset in the clinical setting of prenatal genetic diagnosis.

Acknowledgements

Paul Brady is funded by a Marie-Curie Early Stage Research Fellowship (MEST-CT-2005-019707). Part of the foetal therapy programme is funded by the European Commission in the 6th framework programme (LSHC-CT-2006-037409) and the Flemish Regional Government (IWT/070715). JDP is a clinical researcher for the Fonds voor Wetenschappelijk Onderzoek Vlaanderen

(1.8.012.07.N.02). KD is a senior clinical investigator of the F.W.O. Flanders. Part of this work was made possible by Grants from the IWT (SBO-60848), FWO (GOA/2006/12), and Center of Excellence SymBioSys (Research Council K.U.Leuven EF/05/007), JRV.

References

1. Steel, M. W. and Breg, W. R. (1966) Chromosome analysis of human amniotic fluid cells. *Lancet* **i**, 383-
2. Jacobson, C. B. and Barter, R. H. (1967) Intrauterine diagnosis and management of genetic defects. *Am J Obstet Gynecol* **99**, 796-
3. Jacobson, C. B. and Barter, R. H. (1967) Some cytogenetic aspects of habitual abortion. *Am J Obstet Gynecol* **97**, 666-
4. Nadler, H. L. (1968) Antenatal detection of heriditary disorders. *Pediatrics* **42**, 912-
5. Philip, J., Bryndorf, T., and Christensen, B. (1994) Prenatal aneuploidy detection in interphase cells by fluorescence in situ hybridization (FISH) *Prenat Diagn* **14**, 1203–1215.
6. Bryndorf, T., Christensen, B., Vad, M., Parner, J., Carelli, M. P., Ward, B. E., Klinger, K. W., Bang, J., and Philip, J. (1996) Prenatal detection of chromosome aneuploidies in uncultured chorionic villus samples by FISH *Am J Hum Genet* **59**, 918–926.
7. Bryndorf, T., Lundsteen, C., Lamb, A., Christensen, B., and Philip, J. (2000) Rapid prenatal diagnosis of chromosome aneuploidies by interphase fluorescence in situ hybridization: a one-year clinical experience with high-risk and urgent fetal and postnatal samples *Acta Obstet Gynecol Scand* **79**, 8–14.
8. Tepperberg, J., Pettenati, M. J., Rao, P. N., Lese, C. M., Rita, D., Wyandt, H., Gersen, S., White, B., and Schoonmaker, M. M. (2001) Prenatal diagnosis using interphase fluorescence in situ hybridization (FISH): 2-year multi-center retrospective study and review of the literature *Prenat Diagn* **21**, 293–301.
9. Adinolfi, M., Pertl, B., and Sherlock, J. (1997) Rapid detection of aneuploidies by microsatellite and the quantitative fluorescent polymerase chain reaction *Prenat Diagn* **17**, 1299–1311.
10. Mann, K., Fox, S. P., Abbs, S. J., Yau, S. C., Scriven, P. N., Docherty, Z., and Ogilvie, C. M. (2001) Development and implementation of a new rapid aneuploidy diagnostic service within the UK National Health Service and implications for the future of prenatal diagnosis *Lancet* **358**, 1057–1061.
11. Schouten, J. P., McElgunn, C. J., Waaijer, R., Zwijnenburg, D., Diepvens, F., and Pals, G. (2002) Relative quantification of 40 nucleic acid sequences by multiplex ligation-dependent probe amplification *Nucleic Acids Res* **30**, e57-
12. Schouten, J. and Galjaard, R. J. (2008) MLPA for prenatal diagnosis of commonly occurring aneuploidies *Methods Mol Biol* **444**, 111–122.
13. Rickman, L., Fiegler, H., Shaw-Smith, C., Nash, R., Cirigliano, V., Voglino, G., Ng, B. L., Scott, C., Whittaker, J., Adinolfi, M., Carter, N. P., and Bobrow, M. (2006) Prenatal detection of unbalanced chromosomal rearrangements by array CGH *J Med Genet* **43**, 353–361.
14. Sahoo, T., Cheung, S. W., Ward, P., Darilek, S., Patel, A., del, G. D., Kang, S. H., Lalani, S. R., Li, J., McAdoo, S., Burke, A., Shaw, C. A., Stankiewicz, P., Chinault, A. C., Van den Veyver, I. B., Roa, B. B., Beaudet, A. L., and Eng, C. M. (2006) Prenatal diagnosis of chromosomal abnormalities using array-based comparative genomic hybridization *Genet Med* **8**, 719–727.
15. Lapierre, J. M., Cacheux, V., Luton, D., Collot, N., Oury, J. F., Aurias, A., and Tachdjian, G. (2000) Analysis of uncultured amniocytes by comparative genomic hybridization: a prospective prenatal study *Prenat Diagn* **20**, 123–131.
16. Bi, W., Breman, A. M., Venable, S. F., Eng, P. A., Sahoo, T., Lu, X. Y., Patel, A., Beaudet, A. L., Cheung, S. W., and White, L. D. (2008) Rapid prenatal diagnosis using uncultured amniocytes and oligonucleotide array CGH *Prenat Diagn* **28**, 943–949.
17. Tyreman, M., Abbott, K. M., Willatt, L. R., Nash, R., Lees, C., Whittaker, J., and Simonic, I. (2009) High resolution array analysis: diagnosing pregnancies with abnormal ultrasound findings *J Med Genet* **46**, 531–541.
18. Larrabee, P. B., Johnson, K. L., Pestova, E., Lucas, M., Wilber, K., LeShane, E. S., Tantravahi, U., Cowan, J. M., and Bianchi, D. W. (2004) Microarray analysis of cell-free fetal DNA in amniotic fluid: a prenatal molecular karyotype *Am J Hum Genet* **75**, 485–491.
19. Miura, S., Miura, K., Masuzaki, H., Miyake, N., Yoshiura, K., Sosonkina, N., Harada, N.,

Shimokawa, O., Nakayama, D., Yoshimura, S., Matsumoto, N., Niikawa, N., and Ishimaru, T. (2006) Microarray comparative genomic hybridization (CGH)-based prenatal diagnosis for chromosome abnormalities using cell-free fetal DNA in amniotic fluid *J Hum Genet* **51**, 412–417.

20. Lapaire, O., Lu, X. Y., Johnson, K. L., Jarrah, Z., Stroh, H., Cowan, J. M., Tantravahi, U., and Bianchi, D. W. (2007) Array-CGH analysis of cell-free fetal DNA in 10 mL of amniotic fluid supernatant *Prenat Diagn* **27**, 616–621.

21. Van den Veyver, I. B., Patel, A., Shaw, C. A., Pursley, A. N., Kang, S. H., Simovich, M. J., Ward, P. A., Darilek, S., Johnson, A., Neill, S. E., Bi, W., White, L. D., Eng, C. M., Lupski, J. R., Cheung, S. W., and Beaudet, A. L. (2009) Clinical use of array comparative genomic hybridization (aCGH) for prenatal diagnosis in 300 cases *Prenat Diagn* **29**, 29–39.

22. Shaffer, L. G., Coppinger, J., Alliman, S., Torchia, B. A., Theisen, A., Ballif, B. C., and Bejjani, B. A. (2008) Comparison of microarray-based detection rates for cytogenetic abnormalities in prenatal and neonatal specimens *Prenat Diagn* **28**, 789–795.

23. Van den Veyver, I. B. and Beaudet, A. L. (2006) Comparative genomic hybridization and prenatal diagnosis *Curr Opin Obstet Gynecol* **18**, 185–191.

24. Kashork, C. D., Theisen, A., and Shaffer, L. G. (2008) Prenatal diagnosis using array CGH *Methods Mol Biol* **444**, 59–69.

25. Breman, A. M., Bi, W. M., and Cheung, S. W. (2009) Prenatal diagnosis by array-based comparative genomic hybridization in the clinical laboratory setting *Beijing Da Xue Xue Bao* **41**, 500–504.

26. Iafrate, A. J., Feuk, L., Rivera, M. N., Listewnik, M. L., Donahoe, P. K., Qi, Y., Scherer, S. W., and Lee, C. (2004) Detection of large-scale variation in the human genome *Nat Genet* **36**, 949–951.

27. Sebat, J., Lakshmi, B., Troge, J., Alexander, J., Young, J., Lundin, P., Maner, S., Massa, H., Walker, M., Chi, M., Navin, N., Lucito, R., Healy, J., Hicks, J., Ye, K., Reiner, A., Gilliam, T. C., Trask, B., Patterson, N., Zetterberg, A., and Wigler, M. (2004) Large-scale copy number polymorphism in the human genome *Science* **305**, 525–528.

28. Conrad, D. F., Andrews, T. D., Carter, N. P., Hurles, M. E., and Pritchard, J. K. (2006) A high-resolution survey of deletion polymorphism in the human genome *Nat Genet* **38**, 75–81.

29. Hinds, D. A., Kloek, A. P., Jen, M., Chen, X., and Frazer, K. A. (2006) Common deletions and SNPs are in linkage disequilibrium in the human genome *Nat Genet* **38**, 82–85.

30. McCarroll, S. A. and Altshuler, D. M. (2007) Copy-number variation and association studies of human disease *Nat Genet* **39**, S37–S42.

31. Feuk, L., Carson, A. R., and Scherer, S. W. (2006) Structural variation in the human genome *Nat Rev Genet* 7, 85–97.

32. Redon, R., Ishikawa, S., Fitch, K. R., Feuk, L., Perry, G. H., Andrews, T. D., Fiegler, H., Shapero, M. H., Carson, A. R., Chen, W., Cho, E. K., Dallaire, S., Freeman, J. L., Gonzalez, J. R., Gratacos, M., Huang, J., Kalaitzopoulos, D., Komura, D., MacDonald, J. R., Marshall, C. R., Mei, R., Montgomery, L., Nishimura, K., Okamura, K., Shen, F., Somerville, M. J., Tchinda, J., Valsesia, A., Woodwark, C., Yang, F., Zhang, J., Zerjal, T., Zhang, J., Armengol, L., Conrad, D. F., Estivill, X., Tyler-Smith, C., Carter, N. P., Aburatani, H., Lee, C., Jones, K. W., Scherer, S. W., and Hurles, M. E. (2006) Global variation in copy number in the human genome *Nature* **444**, 444–454.

33. Conrad, D. F., Pinto, D., Redon, R., Feuk, L., Gokcumen, O., Zhang, Y., Aerts, J., Andrews, T. D., Barnes, C., Campbell, P., Fitzgerald, T., Hu, M., Ihm, C. H., Kristiansson, K., Macarthur, D. G., MacDonald, J. R., Onyiah, I., Pang, A. W., Robson, S., Stirrups, K., Valsesia, A., Walter, K., Wei, J., Tyler-Smith, C., Carter, N. P., Lee, C., Scherer, S. W., and Hurles, M. E. (2009) Origins and functional impact of copy number variation in the human genome *Nature*

34. Perry, G. H., Ben-Dor, A., Tsalenko, A., Sampas, N., Rodriguez-Revenga, L., Tran, C. W., Scheffer, A., Steinfeld, I., Tsang, P., Yamada, N. A., Park, H. S., Kim, J. I., Seo, J. S., Yakhini, Z., Laderman, S., Bruhn, L., and Lee, C. (2008) The fine-scale and complex architecture of human copy-number variation *Am J Hum Genet* **82**, 685–695.

35. Bejjani, B. A., Saleki, R., Ballif, B. C., Rorem, E. A., Sundin, K., Theisen, A., Kashork, C. D., and Shaffer, L. G. (2005) Use of targeted array-based CGH for the clinical diagnosis of chromosomal imbalance: is less more? *Am J Med Genet A* **134**, 259–267.

36. Baldwin, E. L., Lee, J. Y., Blake, D. M., Bunke, B. P., Alexander, C. R., Kogan, A. L., Ledbetter, D. H., and Martin, C. L. (2008) Enhanced detection of clinically relevant genomic imbalances using a targeted plus whole genome oligonucleotide microarray *Genet Med* **10**, 415–429.

37. Coppinger, J., Alliman, S., Lamb, A. N., Torchia, B. S., Bejjani, B. A., and Shaffer, L. G. (2009) Whole-genome microarray analysis in prenatal specimens identifies clinically significant chromosome alterations without increase

in results of unclear significance compared to targeted microarray *Prenat Diagn*

38. Yobb, T. M., Somerville, M. J., Willatt, L., Firth, H. V., Harrison, K., MacKenzie, J., Gallo, N., Morrow, B. E., Shaffer, L. G., Babcock, M., Chernos, J., Bernier, F., Sprysak, K., Christiansen, J., Haase, S., Elyas, B., Lilley, M., Bamforth, S., and McDermid, H. E. (2005) Microduplication and triplication of 22q11.2: a highly variable syndrome *Am J Hum Genet* 76, 865–876.
39. Ou, Z., Berg, J. S., Yonath, H., Enciso, V. B., Miller, D. T., Picker, J., Lenzi, T., Keegan, C. E., Sutton, V. R., Belmont, J., Chinault, A. C., Lupski, J. R., Cheung, S. W., Roeder, E., and Patel, A. (2008) Microduplications of 22q11.2 are frequently inherited and are associated with variable phenotypes *Genet Med* **10**, 267–277.
40. de Ravel, T. J., Devriendt, K., Fryns, J. P., and Vermeesch, J. R. (2007) What's new in karyotyping? The move towards array comparative genomic hybridisation (CGH) *Eur J Pediatr* **166**, 637–643.
41. Firth, H. V., Richards, S. M., Bevan, A. P., Clayton, S., Corpas, M., Rajan, D., Van Vooren, S., Moreau, Y., Pettett, R. M., and Carter, N. P. (2009) DECIPHER: Database of Chromosomal Imbalance and Phenotype in Humans Using Ensembl Resources *Am J Hum Genet* **84**, 524–533.
42. Feenstra, I., Fang, J., Koolen, D. A., Siezen, A., Evans, C., Winter, R. M., Lees, M. M., Riegel, M., de Vries, B. B., Van Ravenswaaij, C. M., and Schinzel, A. (2006) European Cytogeneticists Association Register of Unbalanced Chromosome Aberrations (ECARUCA); an online database for rare chromosome abnormalities *Eur J Med Genet* **49**, 279–291.
43. Shuster, E. (2007) Microarray genetic screening: a prenatal roadblock for life? *Lancet* **369**, 526–529.
44. Kohane, I. S., Masys, D. R., and Altman, R. B. (2006) The incidentalome: a threat to genomic medicine *JAMA* **296**, 212–215.
45. Pergament, E. (2007) Controversies and challenges of array comparative genomic hybridization in prenatal genetic diagnosis *Genet Med* **9**, 596–599.
46. Adams, S. A., Coppinger, J., Saitta, S. C., Stroud, T., Kandamurugu, M., Fan, Z., Ballif, B. C., Shaffer, L. G., and Bejjani, B. A. (2009) Impact of genotype-first diagnosis: the detection of microdeletion and microduplication syndromes with cancer predisposition by aCGH *Genet Med* **11**, 314–322.
47. Darilek, S., Ward, P., Pursley, A., Plunkett, K., Furman, P., Magoulas, P., Patel, A., Cheung, S. W., and Eng, C. M. (2008) Pre- and postnatal genetic testing by array-comparative genomic hybridization: genetic counseling perspectives *Genet Med* **10**, 13–18.
48. Mencarelli, M. A., Katzaki, E., Papa, F. T., Sampieri, K., Caselli, R., Uliana, V., Pollazzon, M., Canitano, R., Mostardini, R., Grosso, S., Longo, I., Ariani, F., Meloni, I., Hayek, J., Balestri, P., Mari, F., and Renieri, A. (2008) Private inherited microdeletion/microduplications: implications in clinical practice *Eur J Med Genet* **51**, 409–416.
49. Warburton, D. (1991) De novo balanced chromosome rearrangements and extra marker chromosomes identified at prenatal diagnosis: clinical significance and distribution of breakpoints *Am J Hum Genet* **49**, 995–1013.
50. Giardino, D., Corti, C., Ballarati, L., Colombo, D., Sala, E., Villa, N., Piombo, G., Pierluigi, M., Faravelli, F., Guerneri, S., Coviello, D., Lalatta, F., Cavallari, U., Bellotti, D., Barlati, S., Croci, G., Franchi, F., Savin, E., Nocera, G., Amico, F. P., Granata, P., Casalone, R., Nutini, L., Lisi, E., Torricelli, F., Giussani, U., Facchinetti, B., Guanti, G., Di, G. M., Susca, F. P., Pecile, V., Romitti, L., Cardarelli, L., Racalbuto, E., Police, M. A., Chiodo, F., Rodeschini, O., Falcone, P., Donti, E., Grimoldi, M. G., Martinoli, E., Stioui, S., Caufin, D., Lauricella, S. A., Tanzariello, S. A., Voglino, G., Lenzini, E., Besozzi, M., Larizza, L., and Dalpra, L. (2009) De novo balanced chromosome rearrangements in prenatal diagnosis *Prenat Diagn* **29**, 257–265.
51. De Gregori, M., Ciccone, R., Magini, P., Pramparo, T., Gimelli, S., Messa, J., Novara, F., Vetro, A., Rossi, E., Maraschio, P., Bonaglia, M. C., Anichini, C., Ferrero, G. B., Silengo, M., Fazzi, E., Zatterale, A., Fischetto, R., Previdere, C., Belli, S., Turci, A., Calabrese, G., Bernardi, F., Meneghelli, E., Riegel, M., Rocchi, M., Guerneri, S., Lalatta, F., Zelante, L., Romano, C., Fichera, M., Mattina, T., Arrigo, G., Zollino, M., Giglio, S., Lonardo, F., Bonfante, A., Ferlini, A., Cifuentes, F., Van, E. H., Backx, L., Schinzel, A., Vermeesch, J. R., and Zuffardi, O. (2007) Cryptic deletions are a common finding in "balanced" reciprocal and complex chromosome rearrangements: a study of 59 patients *J Med Genet* **44**, 750–762.
52. Baptista, J., Mercer, C., Prigmore, E., Gribble, S. M., Carter, N. P., Maloney, V., Thomas, N. S., Jacobs, P. A., and Crolla, J. A. (2008) Breakpoint mapping and array CGH in translocations: comparison of a phenotypically normal and an abnormal cohort *Am J Hum Genet* **82**, 927–936.
53. Schluth-Bolard, C., Delobel, B., Sanlaville, D., Boute, O., Cuisset, J. M., Sukno, S., Labalme,

A., Duban-Bedu, B., Plessis, G., Jaillard, S., Dubourg, C., Henry, C., Lucas, J., Odent, S., Pasquier, L., Copin, H., Latour, P., Cordier, M. P., Nadeau, G., Till, M., Edery, P., and Andrieux, J. (2009) Cryptic genomic imbalances in de novo and inherited apparently balanced chromosomal rearrangements: array CGH study of 47 unrelated cases *Eur J Med Genet* **52**, 291–296.

54. Ballif, B. C., Kashork, C. D., Saleki, R., Rorem, E., Sundin, K., Bejjani, B. A., and Shaffer, L. G. (2006) Detecting sex chromosome anomalies and common triploidies in products of conception by array-based comparative genomic hybridization *Prenat Diagn* **26**, 333–339.
55. Ballif, B. C., Rorem, E. A., Sundin, K., Lincicum, M., Gaskin, S., Coppinger, J., Kashork, C. D., Shaffer, L. G., and Bejjani, B. A. (2006) Detection of low-level mosaicism by array CGH in routine diagnostic specimens *Am J Med Genet A* **140**, 2757–2767.
56. Hassold, T., Chen, N., Funkhouser, J., Jooss, T., Manuel, B., Matsuura, J., Matsuyama, A., Wilson, C., Yamane, J. A., and Jacobs, P. A. (1980) A cytogenetic study of 1000 spontaneous abortions *Ann Hum Genet* **44**, 151–178.
57. Hassold, T. J. (1980) A cytogenetic study of repeated spontaneous abortions *Am J Hum Genet* **32**, 723–730.
58. Goddijn, M. and Leschot, N. J. (2000) Genetic aspects of miscarriage *Baillieres Best Pract Res Clin Obstet Gynaecol* **14**, 855–865.
59. Bell, K. A., Van Deerlin, P. G., Haddad, B. R., and Feinberg, R. F. (1999) Cytogenetic diagnosis of "normal 46,XX" karyotypes in spontaneous abortions frequently may be misleading *Fertil Steril* **71**, 334–341.
60. Schaeffer, A. J., Chung, J., Heretis, K., Wong, A., Ledbetter, D. H., and Lese, M. C. (2004) Comparative genomic hybridization-array analysis enhances the detection of aneuploidies and submicroscopic imbalances in spontaneous miscarriages *Am J Hum Genet* **74**, 1168–1174.
61. Benkhalifa, M., Kasakyan, S., Clement, P., Baldi, M., Tachdjian, G., Demirol, A., Gurgan, T., Fiorentino, F., Mohammed, M., and Qumsiyeh, M. B. (2005) Array comparative genomic hybridization profiling of first-trimester spontaneous abortions that fail to grow in vitro *Prenat Diagn* **25**, 894–900.
62. Shimokawa, O., Harada, N., Miyake, N., Satoh, K., Mizuguchi, T., Niikawa, N., and Matsumoto, N. (2006) Array comparative genomic hybridization analysis in first-trimester spontaneous abortions with 'normal' karyotypes *Am J Med Genet A* **140**, 1931–1935.
63. Robberecht, C., Schuddinck, V., Fryns, J. P., and Vermeesch, J. R. (2009) Diagnosis of miscarriages by molecular karyotyping: benefits and pitfalls *Genet Med* **11**, 646–654.
64. Handyside, A. H., Kontogianni, E. H., Hardy, K., and Winston, R. M. (1990) Pregnancies from biopsied human preimplantation embryos sexed by Y-specific DNA amplification *Nature* **344**, 768–770.
65. Wells, D., Sherlock, J. K., Handyside, A. H., and Delhanty, J. D. (1999) Detailed chromosomal and molecular genetic analysis of single cells by whole genome amplification and comparative genomic hybridisation *Nucleic Acids Res* **27**, 1214–1218.
66. Le Caignec, C., Spits, C., Sermon, K., De Rycke, M., Thienpont, B., Debrock, S., Staessen, C., Moreau, Y., Fryns, J. P., Van Steirteghem, A., Liebaers, I., and Vermeesch, J. R. (2006) Single-cell chromosomal imbalances detection by array CGH *Nucleic Acids Res* **34**, e68-
67. Fiegler, H., Geigl, J. B., Langer, S., Rigler, D., Porter, K., Unger, K., Carter, N. P., and Speicher, M. R. (2007) High resolution array-CGH analysis of single cells *Nucleic Acids Res* **35**, e15-
68. Handyside, A. H., Robinson, M. D., Simpson, R. J., Omar, M. B., Shaw, M. A., Grudzinskas, J. G., and Rutherford, A. (2004) Isothermal whole genome amplification from single and small numbers of cells: a new era for preimplantation genetic diagnosis of inherited disease *Mol Hum Reprod* **10**, 767–772.
69. Vanneste, E., Voet, T., Le Caignec, C., Ampe, M., Konings, P., Melotte, C., Debrock, S., Amyere, M., Vikkula, M., Schuit, F., Fryns, J. P., Verbeke, G., D'Hooghe, T., Moreau, Y., and Vermeesch, J. R. (2009) Chromosome instability is common in human cleavage-stage embryos *Nat Med* **15**, 577–583.
70. Vanneste, E., Voet, T., Melotte, C., Debrock, S., Sermon, K., Staessen, C., Liebaers, I., Fryns, J. P., D'Hooghe, T., and Vermeesch, J. R. (2009) What next for preimplantation genetic screening? High mitotic chromosome instability rate provides the biological basis for the low success rate *Hum Reprod* **24**, 2679–2682.
71. Hellani, A., Abu-Amero, K., Azouri, J., and El-Akoum, S. (2008) Successful pregnancies after application of array-comparative genomic hybridization in PGS-aneuploidy screening *Reprod Biomed Online* **17**, 841–847.
72. Handyside, A. H., Harton, G. L., Mariani, B., Thornhill, A. R., Affara, N. A., Shaw, M. A., and Griffin, D. K. (2009) Karyomapping: a Universal Method for Genome Wide Analysis of Genetic Disease based on Mapping Crossovers between Parental Haplotypes *J Med Genet*

73. Kleeman, L., Bianchi, D., Shaffer, L. G., Rorem, E., Cowan, J., Craigo, S. D., Tighiouart, H., and Wilkins-Haug, L. E. (2009) Use of array comparative genomic hybridization for prenatal diagnosis of fetuses with sonographic anomalies and normal metaphase karyotype *Prenat Diagn*

74. Vialard, F., Molina Gomes, D., Leroy, B., Quarello, E., Escalona, A., Le Sciellour, C., Serazin, V., Roume, J., Ville, Y., de Mazancourt, P., and Selva, J. (2009) Array comparative genomic hybridization in prenatal diagnosis: another experience *Fetal Diagn Ther* **25**, 277–284.

Chapter 8

Structural Variation and Its Effect on Expression

Louise Harewood, Evelyne Chaignat, and Alexandre Reymond

Abstract

Structural variation, whether it is caused by copy number variants or present in a balanced form, such as reciprocal translocations and inversions, can have a profound and dramatic effect on the expression of genes mapping within and close to the rearrangement, as well as affecting others genome wide. These effects can be caused by altering the copy number of one or more genes or regulatory elements (dosage effect) or from physical disruption of links between regulatory elements and their associated gene or genes, resulting in perturbation of expression. Similarly, large-scale structural variants can result in genome-wide expression changes by altering the positions that chromosomes occupy within the nucleus, potentially disrupting not only local *cis* interactions, but also *trans* interactions that occur throughout the genome. Structural variation is, therefore, a significant factor in the study of gene expression and is discussed here in more detail.

Key words: Structural variation, CNV, Translocation, Inversion, Gene expression, Position effect

1. Structural Variation

Whole-genome catalogues of copy number variants (CNVs) have now been established for multiple model species (mouse (1–4), rat (5), and *Drosophila melanogaster* (6, 7)), diverse primates (human (8–10), chimpanzee (11, 12), and rhesus macaque (13)), and domestic animals (dog (14, 15) and cow (16)). The Database of Genomic Variants (http://projects.tcag.ca/variation/) contains around 58,000 human CNVs, defining close to 14,500 CNV regions. In other words, approximately 0.8% of the length of the human genome differs between two individuals (17). A certain synteny has been observed among related species. For example, 20–25% of chimpanzee and macaque CNVs overlap with those found in human (11, 13). In addition, CNVs identified in multiple macaques were frequently observed in multiple human samples, suggesting the existence of hotspots for copy number variation (13).

Lars Feuk (ed.), *Genomic Structural Variants: Methods and Protocols*, Methods in Molecular Biology, vol. 838,
DOI 10.1007/978-1-61779-507-7_8,

Besides CNVs, genomes also harbour balanced structural rearrangements, such as reciprocal translocations and inversions. These variants are the product of reorganizations that elicit no gain or loss of genetic material. Around 850 inversions are reported in the aforementioned Database of human Genomic Variants while balanced translocations are estimated to occur in around 1 in 500 individuals in the general population (18). A study analyzing amniocenteses results from over a 10-year period found the incidence of *de novo* reciprocal translocations to be 1 in 2,000, that of Robertsonian translocations (or centric-fusion translocations—those that occur between the centromeres of acrocentric chromosomes, chromosomes 13, 14, 15, 21, and 22 in human) to be 1 in 9,000, and that of *de novo* inversions to be 1 in 10,000 (19). The proportion of these rearrangements that are associated with a clinical phenotype is unknown, but the risk of a serious congenital anomaly occurring in the presence of a chromosomal rearrangement has been estimated to be around 6.1% for reciprocal translocations and 9.4% for inversions (19).

2. CNVs and Gene Expression

Around 1 in 20 human genes are altered in dosage by CNVs (17) and 1 in 8 genes in the mouse genome map entirely to CNVs (3). Human CNV-mapping genes are functionally enriched for cell adhesion, sensory perception, and neurophysiological processes (8) while genes important for MHC class II antigen presentation and neurotransmission are likely to map to mouse CNVs, and those for immunity, lactation, reproduction, and rumination likely to map to CNVs in cattle (3, 16). These deletions or duplications of genomic intervals can result in gene expression changes or clinical phenotypes via several molecular mechanisms, the most commonly recognized of which is copy number alteration of a dosage-sensitive gene (or genes) (e.g. duplication of *PMP22* is associated with Charcot-Marie-Tooth type 1A (CMT1A) (20)). Alternatively, CNVs that involve deletions can produce or alter a phenotype by unmasking recessive mutations or functional polymorphisms on the "remaining" allele. Such cases of compound heterozygosity have been reported in disorders, such as Cohen syndrome (21), spastic ataxia of Charlevoix–Saguenay (ARSACS) (22), and Peters Plus syndrome (23).

Recent reports have attempted to assess the effects of CNVs on tissue transcriptomes of rats and mice (3–5, 24), and studies have reported an overrepresentation of differentially expressed genes among CNV-mapping transcripts (3, 5). Globally, a weak, yet significant, positive correlation has been found between relative expression levels and gene dosage. This correlation is driven by a restricted number of genes, as only a fraction of genes (reported to be between

5 and 18% dependent on the tissue type and rodent species analyzed (3–5)) show a strong correlation between number of copies and relative expression levels. There are many potential explanations for these findings. Firstly, there may be additional transcription factors or regulatory elements, not included in the CNV, that have a *trans*-acting effect on CNV-mapping genes. Conversely, the expression of normal copy number genes may be disrupted if their regulatory elements are affected by a CNV. There are many suggested examples of these, although further work is required to elucidate the exact mechanisms by which these CNV-mapping regulatory elements elicit an effect on gene expression. For example, a microduplication was found in two families with the autosomal dominant limb disorder Brachydactyly type A2 (BDA2, OMIM 112600). This 5.5-kb duplication was located around 110 kb downstream of the *BMP2* gene and was shown, with the use of transgenic mice, to contain a limb-specific enhancer of the gene (25). A 186-kb duplication on chromosome 17p12 has also recently been found in 11 patients with CMT1A (OMIM 118220), a disorder known to be caused by duplication of the *PMP22* gene on 17p12. However, in these 11 cases, the gene itself did not vary in copy number, but the duplication encompassed a 3-kb region situated upstream that presumably contains *PMP22* regulatory elements (26). Similarly, a duplication of less than 150 kb located approximately 136 kb downstream of the *PLP1* gene has been shown to potentially silence the normal copy number gene in individuals with spastic paraplegia type 2 (SPG2) (OMIM 312920) (27).

As well as duplications, deletions of regulatory elements can also have an effect on gene expression. A deletion around one of the breakpoints of a chromosomal inversion, inv(7)(q21.3q35), was found to be present in a family with hearing loss and craniofacial defects (28). This approximately 5-kb deletion was suggested to remove a tissue-specific enhancer of the *DLX5* and/or *DLX6* gene, situated around 100 kb away from the CNV, most likely resulting in dysregulation of the gene(s) in the inner ear and developing bones and production of the associated phenotype.

Genetic imprinting may also modulate the expression at a CNV locus in a parental allele-dependent manner (29) and the possibility of compensatory mechanisms within the cell to counteract the copy number change (for example, an increase/decrease in transcription or altered degradation of mRNA molecules) cannot be excluded. A case in which genetic compensation occurs due to a deletion on one allele of a locus and duplication on the other has also been reported in the unaffected father of a clinically affected child (30). Multiple copies of a gene may also not occur in tandem and may, therefore, map to different chromatin environments and be under differential regulatory control, thereby causing some copies to show expression changes while others do not or producing counteracting expression changes. Indeed, array comparative genomic hybridization (aCGH) does not allow discrimination between tandem and non-tandem duplications. Another potential explanation involves the reliability of CNV detection, particularly in defining CNV boundaries with confidence when aCGH data are used (8). Unreliable boundary calling can lead to genes being erroneously included or excluded from CNVs, which can artificially skew expression profiles and affect correlation results. Finally, results can be largely dependent on tissue type, as analyses suggest tissue-specific expression, with a greater number of genes showing expression that is concordant with their copy number being found in largely homogenous tissues, such as adipose tissue, and a lower degree of concordance being seen in highly specialized tissues, such as brain (31).

Although expression is generally increased in genes found in regions of higher copy number and decreased in genes found in lower copy CNVs, this is not always the case as many genes show either inverse correlation between copy number and gene expression or show no differential expression at all. For 2–15% of the genes, relative expression levels were significantly inversely correlated with copy number (3–5). The mechanism of this effect is still poorly understood, but may be explained by two models: in the first, the negative correlation between number of copies and relative expression is explained by immediate early genes (IEGs). These genes are initially expressed at levels proportional to their copy number, but subsequently induce either directly or indirectly the expression of a repressor that controls expression of the CNV gene via a negative feedback loop, thus reducing or even abolishing it. In the second hypothesis, the extra copies of a gene impair, through steric hindrance, its access to a specific transcription factory (discrete foci within the nucleus containing high levels of RNA polymerase II at which actively transcribed alleles have been shown to be positioned (32, 33)) at which that particular locus is preferentially transcribed, thereby affecting its transcription (34).

Large-scale CNVs have an effect on expression that is greater than that caused merely by altering gene dosage. The deletion associated with Williams–Beuren syndrome (WBS, OMIM 194050) was shown to modify the expression of some neighbouring normal-copy-number genes in human lymphoblastoid and skin fibroblast cell lines (35). Consistently, Stranger et al. (36) observed that although CNVs capture 18% of the detected genetic variation in lymphoblastoid cell lines, more than half of the identified associations between copy number and expression levels involved genes mapping outside CNV intervals. Likewise, an engineered duplication of a segment of mouse chromosome 11, which models the rearrangement present in Potocki–Lupski syndrome (PTLS) patients (OMIM 610883), was shown to affect the expression of both the genes mapping within the duplicated interval and those at its flanks (37). Thus, CNVs also profoundly affect the expression of genes located in their vicinity. This effect extends over half a megabase into their genomic neighbourhoods (3, 5). But how may changes in copy number of CNV regions alter the expression of genes in their vicinity?

One way is via physical dissociation of the transcription unit from its *cis*-acting regulators. There are many examples of this caused by both CNVs and balanced chromosomal rearrangements (see below). Other CNV-induced mechanisms that affect gene expression include modification of transcriptional control through alteration of chromatin structure (38–41) and modification of the positioning of chromatin within the nucleus and/or within a chromosome territory of a genomic region (42, 43). All may play a role, either individually or in combination. Copy number changes may also influence gene expression through perturbation of transcript structure (44–46). Similar mechanisms may explain how some pathways are perturbed without correlation to a particular number of copies of a CNV (see above). Detailed investigations of the different mechanisms by which CNVs influence gene expression are warranted to shed light on how CNVs alter the architecture of chromosomal segments and thus influence the expression of genes.

As well as their association with clinical disease phenotypes, studies of expression quantitative trait loci (eQTL) in mice have demonstrated that CNV regions can also be used as genetic markers to monitor expression profiles associated with quantitative traits (31, 47). Several genes mapping to CNVs have been shown to be highly correlated to metabolic traits, such as body weight and adiposity. Likewise, clinical quantitative trait loci (cQTL) for these traits map to CNVs. This suggests that these clinical traits may, at least in part, be caused by genes or regulatory elements that are located within CNVs (31). However, only a small proportion of CNV-mapping eQTL are estimated to be due to gene-dosage effects (7.3% in 47) with the remaining presumably reflecting expression variation resulting from alteration of regulatory control mechanisms and/or changes of the local chromatin structure.

3. Balanced Structural Rearrangements and Gene Expression

While CNVs can affect gene expression and be the basis of a clinical phenotype, balanced structural rearrangements, in which there is no net gain or loss of genetic material, can also elicit an effect. The association of a balanced chromosomal rearrangement, such as a reciprocal translocation or an inversion, with a clinical phenotype may have many underlying causes, the simplest of which is direct disruption of a dosage-dependent gene (or genes) by one or more breakpoints. These disease-associated balanced chromosomal rearrangements (DBCRs) have proved to be instrumental in the mapping of disease loci and the positional cloning of disease genes. They have been used to identify a growing number of disease loci, and subsequently causative genes, for a variety of different conditions. A few examples are outlined below. The first Mendelian disorder to be mapped on the basis of DBCR information was Duchenne Muscular Dystrophy (DMD, OMIM 310200), an X-linked disorder characterized by rapid progressive muscle degeneration. A number of reciprocal X; autosome translocations in females allowed the DMD locus to be assigned to chromosome Xp21 and breakpoint mapping identified the causative gene (48–50). In these cases, the normal X chromosome is inactivated and the rearrangement directly disrupts the only copy of the causative gene. While X inactivation in these individuals is still random, if the der(X) is inactivated, the cells will be monosomic for part of X and whichever autosome is involved in the translocation, most likely causing the cell to be selected against or to undergo apoptosis. The presence of X; autosome translocations in females has since allowed the causative genes for other disorders, such as choroideremia (OMIM 303100) (51), Lowe syndrome (OMIM 309000) (52), and lissencephaly (OMIM 300067) (53) to be similarly mapped.

Autosomal dominant conditions constitute a large proportion of the disorders in which causative genes have been mapped using DBCRs, as would be expected as only one copy of a gene needs to be disrupted to produce a phenotype. However, at least one autosomal recessive condition has also been mapped using DBCRs, namely, Alstrom syndrome (OMIM 203800). This autosomal recessive disorder was caused by a paternally inherited frameshift mutation of the *ALMS1* gene, inherited along with a maternally inherited translocation that disrupted the other copy of the gene, producing a clinically affected compound heterozygote (54, 55).

Structural variation, both from CNVs and from balanced chromosomal rearrangements, can have an effect on the expression of genes near to, and some distance from, the breakpoints. This can be caused by the physical dissociation of the gene and its regulatory *cis* elements (termed "Position Effect" (56) or, more specifically, "*cis*-ruption" (57)). This type of disruption has been shown to be the basis of a number of different disorders in both human and mouse.

Perhaps, the most studied disorder caused by "*cis*-ruption" is aniridia (OMIM 106210), a congenital eye malformation characterized by the absence of the iris and caused by haploinsufficiency of the *PAX6* gene on human chromosome 11p13. Mapping of human reciprocal translocations found a number of chromosome 11 breakpoints that mapped downstream (3′) of the *PAX6* coding region, the furthest being 125 kb away and situated in the last intron of the *ELP4* gene. This initially led to the assignation of another aniridia locus, but heterozygous loss of *ELP4* was shown not to cause the aniridia phenotype. Further studies utilizing mouse–human somatic cell hybrids showed that, in fact, the translocation breakpoints disrupted long-range *cis*-regulatory elements of *PAX6* (58).

Position effects caused by disruption of *cis*-regulatory elements have been reported for distances of up to 1.5 Mb both telomeric and centromeric to the gene in question (59). The dissociation of *cis*-regulatory elements in human disorders is most readily recognized when the phenotype is the same as that caused by point mutations or other loss of function mutations of the causative gene. However, disruption of these elements may only produce developmental stage-specific or tissue-specific effects and can have varying phenotypic results. An example of this can be seen with the *SOX9* gene and its involvement in three different human disorders, namely: campomelic dysplasia (CD, OMIM 114290), acampomelic campomelic dysplasia (ACD), and Pierre-Robin sequence (PRS, OMIM 261800). Coding sequence mutations resulting in haploinsufficiency of *SOX9* are thought to be the basis of CD and translocation breakpoints that map upstream of *SOX9* have also been reported in these patients, suggesting disruption of *cis*-regulatory elements. However, the severity of the phenotype in these translocation patients is variable and there appears to be some correlation between severity and the distance between the breakpoint and *SOX9* (60). Breakpoints have been classified into distinct clusters: a proximal cluster 50–375 kb upstream of *SOX9* and a distal cluster 789–932 kb away (61). Most breakpoints that fall within the proximal cluster result in mild to severe campomelia (bowing of the long bones) while distal breakpoints result in campomelic dysplasia without campomelia (ACD). The existence of a further cluster, located even further upstream of the *SOX9* transcription start site (more than 1 Mb away) and associated solely with the craniofacial abnormality PRS, has also been suggested (59). The distance at which a breakpoint occurs from the relevant gene can, therefore, have a significant impact on the phenotype observed. Similarly, a breakpoint has been found approximately 1.3 Mb downstream of *SOX9* in a patient with ACD and male-to-female sex reversal (62), suggesting that *cis*-regulatory elements are not only situated some distance from the gene itself, but can also be either up- or downstream.

Structural rearrangements can also have an effect on gene expression by placing the coding region of a gene under the influence of a foreign promoter. An example of this can be seen in Aromatase excess syndrome (OMIM 139300), an autosomal dominant disorder characterized by high systemic oestrogen levels, short stature, and prepubertal gynaecomastia/premature breast development. Aromatase is involved in oestrogen biosynthesis and is encoded by the *CYP19* gene, which is under the control of tissue-specific promoters. Inversions have been reported which place the coding region of the *CYP19* gene under the control of the promoter of a ubiquitously expressed gene that is normally situated some distance away, on the opposite strand. This rearrangement results in a gain of function mutation of *CYP19* and overexpression/aberrant expression of aromatase (63, 64). The formation of cryptic promoters is, therefore, another way in which chromosomal rearrangements can locally disrupt *cis* interactions and have an effect on gene expression.

The presence of a structural chromosomal rearrangement may also disrupt *cis* interactions over a larger scale. Intrachromosomal interactions have been shown to occur over distances of up to 90 Mb in lymphoblastoid cells (65) and co-associations between parts of the same chromosome situated 25 Mb apart have also been observed (33, 66, 67). Structural rearrangements, such as reciprocal translocations, have the power to disrupt these interactions by repositioning the resultant derivative chromosomes within the nucleus (see below).

4. Aberrant Nuclear Positioning and Expression

Chromosomes are arranged within the nucleus in a non-random fashion and, in lymphoblastoid cells, occupy a radial position with respect to their gene density, with gene-rich chromosomes being situated at the centre of the nucleus and gene poor towards the periphery (68, 69). The organization of chromosomes within the nucleus impacts upon the position of translocation chromosomes (68, 70, 71), and the artificial tethering of a chromosome to the nuclear periphery can have a direct and reversible effect on gene expression (72–74). Similarly, movement of genes towards the centre or periphery of the nucleus has been correlated with increases and decreases in gene expression levels, respectively (72, 73, 75, 76).

The occurrence of a structural rearrangement, such as a reciprocal translocation, can result in the derived chromosomes being shifted from their usual positions within the nucleus and the genes contained therein being placed into an anomalous chromatin environment (70, 77). As well as the disruption of *cis* interactions, *trans* interactions may also be disrupted by this reorganization. Transcribed alleles have been shown to localize at "transcription factories" (see above) or at aggregations of splicing factors, termed "nuclear speckles" or 'SC35 defined domains' (78). This co-occurrence of genes had led to the hypothesis that gene positioning within the nucleus is not random and that preferential associations occur (79, 80) and may have a functional basis. Thus, relocation of chromosomes within the nucleus may disrupt these interactions and affect the expression of not only those genes situated on the chromosomes involved in the translocation, but also on others in a genome-wide fashion.

There is also evidence that the relocation of derivative chromosomes can have a knock-on effect on the organization and placement of other chromosomes within the nucleus. The presence of a reciprocal translocation between human chromosomes 11 and 22, t(11;22)(q23;q11) within a cell causes reorganization of chromosome territories within the nucleus. The derivative chromosomes that result from this translocation are shifted into an anomalous position in the centre of the nucleus, to the exclusion of other, non-translocated (normal) chromosomes which are shifted more peripherally (77). Therefore, not only are any local *cis* interactions disrupted by the breakpoints of the chromosome rearrangement, but *trans* interactions are also potentially disrupted throughout the genome. This could explain why the large-scale gene expression changes seen in the cells of these translocation individuals involve genes on nearly every chromosome, not just those directly involved in the translocation.

Similarly, while every chromosome occupies a specific territory in a non-random radial position within the nucleus, the boundaries between these territories are not impenetrable and intermingling between chromosomes can occur (81, 82). This mixing of chromosome territories is thought to be transcription dependent as transcriptional inhibition affects the amount of interweaving that occurs between sets of chromosomes (81). If transcription-dependent associations between sets of chromosomes are frequent enough to influence the organization of chromosomes within the nucleus, it is not unreasonable to hypothesize that an alteration in the organization of chromosomes, caused by a structural rearrangement, could have an effect on transcription and subsequent gene expression levels.

5. Conclusion

Structural genomic variation, whether it results in the gain or loss of genetic material or is present in a balanced form, such as reciprocal translocations and inversions, can have a profound and dramatic effect on normal gene expression. This can occur from changes in the copy number of a gene or genes by affecting the copy number of regulatory elements or physically separating regulatory elements from their associated gene, all resulting in the perturbation of normal expression. Similarly, large-scale structural rearrangements can result in reorganization of genetic material within the nucleus, potentially disrupting both local *cis* interactions and also genome-wide *trans* interactions. Many of the mechanisms underlying the gene expression changes observed in association with structural variation remain to be elucidated, but this type of genomic alteration has already been shown to contribute significantly to normal phenotypic variation and evolution and to form the basis of many clinical phenotypes and traits. Future work should provide more insight into the precise extent of this contribution.

References

1. Adams, D.J., Dermitzakis, E.T., Cox, T., Smith, J., Davies, R., Banerjee, R., Bonfield, J., Mullikin, J.C., Chung, Y.J., Rogers, J., and Bradley, A. (2005) Complex haplotypes, copy number polymorphisms and coding variation in two recently divergent mouse strains *Nat. Genet.* **37,** 532–536
2. Egan, C.M., Sridhar, S., Wigler, M., and Hall, I.M. (2007) Recurrent DNA copy number variation in the laboratory mouse *Nat.Genet.* **39,** 1384–1389
3. Henrichsen, C.N., Vinckenbosch, N., Zollner, S., Chaignat, E., Pradervand, S., Schutz, F., Ruedi, M., Kaessmann, H., and Reymond, A. (2009) Segmental copy number variation shapes tissue transcriptomes *Nat.Genet.* **41,** 424–429
4. Henrichsen, C.N., Chaignat, E., and Reymond, A. (2009) Copy number variants, diseases and gene expression *Hum.Mol.Genet.* **18,** R1–R8
5. Guryev, V., Saar, K., Adamovic, T., Verheul, M., van Heesch, S.A., Cook, S., Pravenec, M., Aitman, T., Jacob, H., Shull, J.D., Hubner, N., and Cuppen, E. (2008) Distribution and functional impact of DNA copy number variation in the rat *Nat.Genet.* **40,** 538–545
6. Dopman, E.B. and Hartl, D.L. (2007) A portrait of copy-number polymorphism in Drosophila melanogaster *Proc.Natl.Acad.Sci. USA* **104,** 19920–19925
7. Emerson, J.J., Cardoso-Moreira, M., Borevitz, J.O., and Long, M. (2008) Natural selection shapes genome-wide patterns of copy-number polymorphism in Drosophila melanogaster *Science* **320,** 1629–1631
8. Redon, R., Ishikawa, S., Fitch, K.R., Feuk, L., Perry, G.H., Andrews, T.D., Fiegler, H., Shapero, M.H., Carson, A.R., Chen, W., Cho, E.K., Dallaire, S., Freeman, J.L., Gonzalez, J.R., Gratacos, M., Huang, J., Kalaitzopoulos, D., Komura, D., MacDonald, J.R., Marshall, C.R., Mei, R., Montgomery, L., Nishimura, K., Okamura, K., Shen, F., Somerville, M.J., Tchinda, J., Valsesia, A., Woodwark, C., Yang, F., Zhang, J., Zerjal, T., Zhang, J., Armengol, L., Conrad, D.F., Estivill, X., Tyler-Smith, C., Carter, N.P., Aburatani, H., Lee, C., Jones, K.W., Scherer, S.W., and Hurles, M.E. (2006) Global variation in copy number in the human genome *Nature* **444,** 444–454
9. Tuzun, E., Sharp, A.J., Bailey, J.A., Kaul, R., Morrison, V.A., Pertz, L.M., Haugen, E., Hayden, H., Albertson, D., Pinkel, D., Olson, M.V., and Eichler, E.E. (2005) Fine-scale structural variation of the human genome *Nat. Genet.* **37,** 727–732
10. Sharp, A.J., Locke, D.P., McGrath, S.D., Cheng, Z., Bailey, J.A., Vallente, R.U., Pertz, L.M., Clark, R.A., Schwartz, S., Segraves, R., Oseroff, V.V., Albertson, D.G., Pinkel, D., and Eichler, E.E. (2005) Segmental duplications and copy-number variation in the human genome *Am.J.Hum.Genet.* **77,** 78–88
11. Perry, G.H., Tchinda, J., McGrath, S.D., Zhang, J., Picker, S.R., Caceres, A.M., Iafrate, A.J., Tyler-Smith, C., Scherer, S.W., Eichler, E.E., Stone, A.C., and Lee, C. (2006) Hotspots for copy number variation in chimpanzees and humans *Proc.Natl.Acad.Sci. USA* **103,** 8006–8011
12. Kehrer-Sawatzki, H. and Cooper, D.N. (2007) Structural divergence between the human and chimpanzee genomes *Hum.Genet.* **120,** 759–778
13. Lee, A.S., Gutierrez-Arcelus, M., Perry, G.H., Vallender, E.J., Johnson, W.E., Miller, G.M., Korbel, J.O., and Lee, C. (2008) Analysis of copy number variation in the rhesus macaque genome identifies candidate loci for evolutionary and human disease studies *Hum.Mol.Genet.* **17,** 1127–1136
14. Chen, W.K., Swartz, J.D., Rush, L.J., and Alvarez, C.E. (2009) Mapping DNA structural variation in dogs *Genome Res.* **19,** 500–509
15. Nicholas, T.J., Cheng, Z., Ventura, M., Mealey, K., Eichler, E.E., and Akey, J.M. (2009) The genomic architecture of segmental duplications and associated copy number variants in dogs *Genome Res.* **19,** 491–499
16. Liu, G.E., Hou, Y., Zhu, B., Cardone, M.F., Jiang, L., Cellamare, A., Mitra, A., Alexander, L.J., Coutinho, L.L., Dell'aquila, M.E., Gasbarre, L.C., Lacalandra, G., Li, R.W., Matukumalli, L.K., Nonneman, D., Regitano, L.C., Smith, T.P., Song, J., Sonstegard, T.S., Van Tassell, C.P., Ventura, M., Eichler, E.E., McDaneld, T.G., and Keele, J.W. (2010) Analysis of copy number variations among diverse cattle breeds *Genome Res.*
17. Conrad, D.F., Pinto, D., Redon, R., Feuk, L., Gokcumen, O., Zhang, Y., Aerts, J., Andrews, T.D., Barnes, C., Campbell, P., Fitzgerald, T., Hu, M., Ihm, C.H., Kristiansson, K., Macarthur, D.G., MacDonald, J.R., Onyiah, I., Pang, A.W., Robson, S., Stirrups, K., Valsesia, A., Walter, K., Wei, J., Tyler-Smith, C., Carter, N.P., Lee, C., Scherer, S.W., and Hurles, M.E. (2010) Origins and functional impact of copy number variation in the human genome *Nature* **464,** 704–712
18. Ogilvie, C.M., Braude, P., and Scriven, P.N. (2001) Successful pregnancy outcomes after

preimplantation genetic diagnosis (PGD) for carriers of chromosome translocations *Hum. Fertil.(Camb.)* **4,** 168–171

19. Warburton, D. (1991) De novo balanced chromosome rearrangements and extra marker chromosomes identified at prenatal diagnosis: clinical significance and distribution of breakpoints *Am.J.Hum.Genet* **49,** 995–1013
20. Lupski, J.R., Oca-Luna, R.M., Slaugenhaupt, S., Pentao, L., Guzzetta, V., Trask, B.J., Saucedo-Cardenas, O., Barker, D.F., Killian, J.M., Garcia, C.A., Chakravarti, A., and Patel, P.I. (1991) DNA duplication associated with Charcot-Marie-Tooth disease type 1A *Cell* **66,** 219–232
21. Balikova, I., Lehesjoki, A.E., de Ravel, T.J., Thienpont, B., Chandler, K.E., Clayton-Smith, J., Traskelin, A.L., Fryns, J.P., and Vermeesch, J.R. (2009) Deletions in the VPS13B (COH1) gene as a cause of Cohen syndrome *Hum. Mutat.* **30,** E845–E854
22. Breckpot, J., Takiyama, Y., Thienpont, B., Van Vooren, S., Vermeesch, J.R., Ortibus, E., and Devriendt, K. (2008) A novel genomic disorder: a deletion of the SACS gene leading to spastic ataxia of Charlevoix-Saguenay *Eur.J.Hum. Genet.* **16,** 1050–1054
23. Lesnik Oberstein, S.A., Kriek, M., White, S.J., Kalf, M.E., Szuhai, K., den Dunnen, J.T., Breuning, M.H., and Hennekam, R.C. (2006) Peters Plus syndrome is caused by mutations in B3GALTL, a putative glycosyltransferase *Am.J.Hum.Genet.* **79,** 562–566
24. Watkins-Chow, D.E. and Pavan, W.J. (2008) Genomic copy number and expression variation within the C57BL/6J inbred mouse strain *Genome Res.* **18,** 60–66
25. Dathe, K., Kjaer, K.W., Brehm, A., Meinecke, P., Nurnberg, P., Neto, J.C., Brunoni, D., Tommerup, N., Ott, C.E., Klopocki, E., Seemann, P., and Mundlos, S. (2009) Duplications involving a conserved regulatory element downstream of BMP2 are associated with brachydactyly type A2 *Am.J.Hum.Genet.* **84,** 483–492
26. Weterman, M.A., van Ruissen, F., de Wissel, M., Bordewijk, L., Samijn, J.P., van der Pol, W.L., Meggouh, F., and Baas, F. (2010) Copy number variation upstream of PMP22 in Charcot-Marie-Tooth disease *Eur.J.Hum.Genet.* **18,** 421–428
27. Lee, J.A., Madrid, R.E., Sperle, K., Ritterson, C.M., Hobson, G.M., Garbern, J., Lupski, J.R., and Inoue, K. (2006) Spastic paraplegia type 2 associated with axonal neuropathy and apparent PLP1 position effect *Ann.Neurol.* **59,** 398–403
28. Brown, K.K., Reiss, J.A., Crow, K., Ferguson, H.L., Kelly, C., Fritzsch, B., and Morton, C.C. (2010) Deletion of an enhancer near DLX5 and DLX6 in a family with hearing loss, craniofacial defects, and an inv(7)(q21.3q35) *Hum.Genet.* **127,** 19–31
29. Hogart, A., Leung, K.N., Wang, N.J., Wu, D.J., Driscoll, J., Vallero, R.O., Schanen, N.C., and LaSalle, J.M. (2009) Chromosome 15q11-13 duplication syndrome brain reveals epigenetic alterations in gene expression not predicted from copy number *J.Med.Genet.* **46,** 86–93
30. Carelle-Calmels, N., Saugier-Veber, P., Girard-Lemaire, F., Rudolf, G., Doray, B., Guerin, E., Kuhn, P., Arrive, M., Gilch, C., Schmitt, E., Fehrenbach, S., Schnebelen, A., Frebourg, T., and Flori, E. (2009) Genetic compensation in a human genomic disorder *N.Engl.J.Med.* **360,** 1211–1216
31. Orozco, L.D., Cokus, S.J., Ghazalpour, A., Ingram-Drake, L., Wang, S., van Nas, A., Che, N., Araujo, J.A., Pellegrini, M., and Lusis, A.J. (2009) Copy number variation influences gene expression and metabolic traits in mice *Hum. Mol.Genet.* **18,** 4118–4129
32. Iborra, F.J., Pombo, A., Jackson, D.A., and Cook, P.R. (1996) Active RNA polymerases are localized within discrete transcription "factories" in human nuclei *J.Cell Sci.* **109 (Pt 6),** 1427–1436
33. Osborne, C.S., Chakalova, L., Brown, K.E., Carter, D., Horton, A., Debrand, E., Goyenechea, B., Mitchell, J.A., Lopes, S., Reik, W., and Fraser, P. (2004) Active genes dynamically colocalize to shared sites of ongoing transcription *Nat.Genet.* **36,** 1065–1071
34. Sexton, T., Umlauf, D., Kurukuti, S., and Fraser, P. (2007) The role of transcription factories in large-scale structure and dynamics of interphase chromatin *Semin.Cell Dev.Biol.* **18,** 691–697
35. Merla, G., Howald, C., Henrichsen, C.N., Lyle, R., Wyss, C., Zabot, M.T., Antonarakis, S.E., and Reymond, A. (2006) Submicroscopic deletion in patients with Williams-Beuren syndrome influences expression levels of the nonhemizygous flanking genes *Am.J.Hum.Genet.* **79,** 332–341
36. Stranger, B.E., Forrest, M.S., Dunning, M., Ingle, C.E., Beazley, C., Thorne, N., Redon, R., Bird, C.P., de Grassi, A., Lee, C., Tyler-Smith, C., Carter, N., Scherer, S.W., Tavare, S., Deloukas, P., Hurles, M.E., and Dermitzakis, E.T. (2007) Relative impact of nucleotide and copy number variation on gene expression phenotypes *Science* **315,** 848–853
37. Molina, J., Carmona-Mora, P., Chrast, J., Krall, P.M., Canales, C.P., Lupski, J.R., Reymond, A., and Walz, K. (2008) Abnormal social behaviors and altered gene expression rates in a mouse model for Potocki-Lupski syndrome *Hum.Mol. Genet.* **17,** 2486–2495

38. Gabellini, D., Green, M.R., and Tupler, R. (2002) Inappropriate gene activation in FSHD: a repressor complex binds a chromosomal repeat deleted in dystrophic muscle *Cell* **110,** 339–348
39. Gabellini, D., D'Antona, G., Moggio, M., Prelle, A., Zecca, C., Adami, R., Angeletti, B., Ciscato, P., Pellegrino, M.A., Bottinelli, R., Green, M.R., and Tupler, R. (2006) Facioscapulohumeral muscular dystrophy in mice overexpressing FRG1 *Nature* **439,** 973–977
40. Muncke, N., Wogatzky, B.S., Breuning, M., Sistermans, E.A., Endris, V., Ross, M., Vetrie, D., Catsman-Berrevoets, C.E., and Rappold, G. (2004) Position effect on PLP1 may cause a subset of Pelizaeus-Merzbacher disease symptoms *J.Med.Genet* **41,** e121
41. Lee, J.A. and Lupski, J.R. (2006) Genomic rearrangements and gene copy-number alterations as a cause of nervous system disorders *Neuron* **52,** 103–121
42. Fraser, P. and Bickmore, W. (2007) Nuclear organization of the genome and the potential for gene regulation *Nature* **447,** 413–417
43. Heard, E. and Bickmore, W. (2007) The ins and outs of gene regulation and chromosome territory organisation *Curr.Opin.Cell Biol.* **19,** 311–316
44. Denoeud, F., Kapranov, P., Ucla, C., Frankish, A., Castelo, R., Drenkow, J., Lagarde, J., Alioto, T., Manzano, C., Chrast, J., Dike, S., Wyss, C., Henrichsen, C.N., Holroyd, N., Dickson, M.C., Taylor, R., Hance, Z., Foissac, S., Myers, R.M., Rogers, J., Hubbard, T., Harrow, J., Guigo, R., Gingeras, T.R., Antonarakis, S.E., and Reymond, A. (2007) Prominent use of distal 5′ transcription start sites and discovery of a large number of additional exons in ENCODE regions *Genome Res.* **17,** 746–759
45. Birney, E., Stamatoyannopoulos, J.A., Dutta, A., Guigo, R., Gingeras, T.R., Margulies, E.H., Weng, Z., Snyder, M., Dermitzakis, E.T., Thurman, R.E., Kuehn, M.S., Taylor, C.M., Neph, S., Koch, C.M., Asthana, S., Malhotra, A., Adzhubei, I., Greenbaum, J.A., Andrews, R.M., Flicek, P., Boyle, P.J., Cao, H., Carter, N.P., Clelland, G.K., Davis, S., Day, N., Dhami, P., Dillon, S.C., Dorschner, M.O., Fiegler, H., Giresi, P.G., Goldy, J., Hawrylycz, M., Haydock, A., Humbert, R., James, K.D., Johnson, B.E., Johnson, E.M., Frum, T.T., Rosenzweig, E.R., Karnani, N., Lee, K., Lefebvre, G.C., Navas, P.A., Neri, F., Parker, S.C., Sabo, P.J., Sandstrom, R., Shafer, A., Vetrie, D., Weaver, M., Wilcox, S., Yu, M., Collins, F.S., Dekker, J., Lieb, J.D., Tullius, T.D., Crawford, G.E., Sunyaev, S., Noble, W.S., Dunham, I., Denoeud, F., Reymond, A., Kapranov, P., Rozowsky, J., Zheng, D., Castelo, R., Frankish, A., Harrow, J., Ghosh, S., Sandelin, A., Hofacker, I.L., Baertsch, R., Keefe, D., Dike, S., Cheng, J., Hirsch, H.A., Sekinger, E.A., Lagarde, J., Abril, J.F., Shahab, A., Flamm, C., Fried, C., Hackermuller, J., Hertel, J., Lindemeyer, M., Missal, K., Tanzer, A., Washietl, S., Korbel, J., Emanuelsson, O., Pedersen, J.S., Holroyd, N., Taylor, R., Swarbreck, D., Matthews, N., Dickson, M.C., Thomas, D.J., Weirauch, M.T., Gilbert, J., Drenkow, J., Bell, I., Zhao, X., Srinivasan, K.G., Sung, W.K., Ooi, H.S., Chiu, K.P., Foissac, S., Alioto, T., Brent, M., Pachter, L., Tress, M.L., Valencia, A., Choo, S.W., Choo, C.Y., Ucla, C., Manzano, C., Wyss, C., Cheung, E., Clark, T.G., Brown, J.B., Ganesh, M., Patel, S., Tammana, H., Chrast, J., Henrichsen, C.N., Kai, C., Kawai, J., Nagalakshmi, U., Wu, J., Lian, Z., Lian, J., Newburger, P., Zhang, X., Bickel, P., Mattick, J.S., Carninci, P., Hayashizaki, Y., Weissman, S., Hubbard, T., Myers, R.M., Rogers, J., Stadler, P.F., Lowe, T.M., Wei, C.L., Ruan, Y., Struhl, K., Gerstein, M., Antonarakis, S.E., Fu, Y., Green, E.D., Karaoz, U., Siepel, A., Taylor, J., Liefer, L.A., Wetterstrand, K.A., Good, P.J., Feingold, E.A., Guyer, M.S., Cooper, G.M., Asimenos, G., Dewey, C.N., Hou, M., Nikolaev, S., Montoya-Burgos, J.I., Loytynoja, A., Whelan, S., Pardi, F., Massingham, T., Huang, H., Zhang, N.R., Holmes, I., Mullikin, J.C., Ureta-Vidal, A., Paten, B., Seringhaus, M., Church, D., Rosenbloom, K., Kent, W.J., Stone, E.A., Batzoglou, S., Goldman, N., Hardison, R.C., Haussler, D., Miller, W., Sidow, A., Trinklein, N.D., Zhang, Z.D., Barrera, L., Stuart, R., King, D.C., Ameur, A., Enroth, S., Bieda, M.C., Kim, J., Bhinge, A.A., Jiang, N., Liu, J., Yao, F., Vega, V.B., Lee, C.W., Ng, P., Shahab, A., Yang, A., Moqtaderi, Z., Zhu, Z., Xu, X., Squazzo, S., Oberley, M.J., Inman, D., Singer, M.A., Richmond, T.A., Munn, K.J., Rada-Iglesias, A., Wallerman, O., Komorowski, J., Fowler, J.C., Couttet, P., Bruce, A.W., Dovey, O.M., Ellis, P.D., Langford, C.F., Nix, D.A., Euskirchen, G., Hartman, S., Urban, A.E., Kraus, P., Van Calcar, S., Heintzman, N., Kim, T.H., Wang, K., Qu, C., Hon, G., Luna, R., Glass, C.K., Rosenfeld, M.G., Aldred, S.F., Cooper, S.J., Halees, A., Lin, J.M., Shulha, H.P., Zhang, X., Xu, M., Haidar, J.N., Yu, Y., Ruan, Y., Iyer, V.R., Green, R.D., Wadelius, C., Farnham, P.J., Ren, B., Harte, R.A., Hinrichs, A.S., Trumbower, H., and Clawson, H. (2007) Identification and analysis of functional elements in 1% of the human genome by the ENCODE pilot project *Nature* **447,** 799–816
46. Reymond, A., Henrichsen, C.N., Harewood, L., and Merla, G. (2007) Side effects of genome structural changes *Curr.Opin.Genet.Dev.* **17,** 381–386
47. Cahan, P., Li, Y., Izumi, M., and Graubert, T.A. (2009) The impact of copy number variation

on local gene expression in mouse hematopoietic stem and progenitor cells *Nat.Genet.* **41,** 430–437

48. Lindenbaum, R.H., Clarke, G., Patel, C., Moncrieff, M., and Hughes, J.T. (1979) Muscular dystrophy in an X; 1 translocation female suggests that Duchenne locus is on X chromosome short arm *J.Med.Genet* **16,** 389–392
49. Verellen-Dumoulin, C., Freund, M., De Meyer, R., Laterre, C., Frederic, J., Thompson, M.W., Markovic, V.D., and Worton, R.G. (1984) Expression of an X-linked muscular dystrophy in a female due to translocation involving Xp21 and non-random inactivation of the normal X chromosome *Hum.Genet* **67,** 115–119
50. Ray, P.N., Belfall, B., Duff, C., Logan, C., Kean, V., Thompson, M.W., Sylvester, J.E., Gorski, J.L., Schmickel, R.D., and Worton, R.G. (1985) Cloning of the breakpoint of an X;21 translocation associated with Duchenne muscular dystrophy *Nature* **318,** 672–675
51. Cremers, F.P., Brunsmann, F., Berger, W., van Kerkhoff, E.P., van de Pol, T.J., Wieringa, B., Pawlowitzki, I.H., and Ropers, H.H. (1990) Cloning of the breakpoints of a deletion associated with choroidermia *Hum.Genet* **86,** 61–64
52. Attree, O., Olivos, I.M., Okabe, I., Bailey, L.C., Nelson, D.L., Lewis, R.A., McInnes, R.R., and Nussbaum, R.L. (1992) The Lowe's oculocerebrorenal syndrome gene encodes a protein highly homologous to inositol polyphosphate-5-phosphatase *Nature* **358,** 239–242
53. Gleeson, J.G., Allen, K.M., Fox, J.W., Lamperti, E.D., Berkovic, S., Scheffer, I., Cooper, E.C., Dobyns, W.B., Minnerath, S.R., Ross, M.E., and Walsh, C.A. (1998) Doublecortin, a brain-specific gene mutated in human X-linked lissencephaly and double cortex syndrome, encodes a putative signaling protein *Cell* **92,** 63–72
54. Collin, G.B., Marshall, J.D., Ikeda, A., So, W.V., Russell-Eggitt, I., Maffei, P., Beck, S., Boerkoel, C.F., Sicolo, N., Martin, M., Nishina, P.M., and Naggert, J.K. (2002) Mutations in ALMS1 cause obesity, type 2 diabetes and neurosensory degeneration in Alstrom syndrome *Nat.Genet* **31,** 74–78
55. Hearn, T., Renforth, G.L., Spalluto, C., Hanley, N.A., Piper, K., Brickwood, S., White, C., Connolly, V., Taylor, J.F., Russell-Eggitt, I., Bonneau, D., Walker, M., and Wilson, D.I. (2002) Mutation of ALMS1, a large gene with a tandem repeat encoding 47 amino acids, causes Alstrom syndrome *Nat.Genet* **31,** 79–83
56. Kleinjan, D.J. and van Heyningen, V. (1998) Position effect in human genetic disease *Hum. Mol.Genet* **7,** 1611–1618
57. Kleinjan, D.J. and Coutinho, P. (2009) Cis-ruption mechanisms: disruption of cis-regulatory control as a cause of human genetic disease *Brief.Funct.Genomic.Proteomic.* **8,** 317–332
58. Lauderdale, J.D., Wilensky, J.S., Oliver, E.R., Walton, D.S., and Glaser, T. (2000) 3′ deletions cause aniridia by preventing PAX6 gene expression *Proc.Natl.Acad.Sci. USA* **97,** 13755–13759
59. Benko, S., Fantes, J.A., Amiel, J., Kleinjan, D.J., Thomas, S., Ramsay, J., Jamshidi, N., Essafi, A., Heaney, S., Gordon, C.T., McBride, D., Golzio, C., Fisher, M., Perry, P., Abadie, V., Ayuso, C., Holder-Espinasse, M., Kilpatrick, N., Lees, M.M., Picard, A., Temple, I.K., Thomas, P., Vazquez, M.P., Vekemans, M., Crollius, H.R., Hastie, N.D., Munnich, A., Etchevers, H.C., Pelet, A., Farlie, P.G., FitzPatrick, D.R., and Lyonnet, S. (2009) Highly conserved non-coding elements on either side of SOX9 associated with Pierre Robin sequence *Nat.Genet.* **41,** 359–364
60. Gordon, C.T., Tan, T.Y., Benko, S., FitzPatrick, D., Lyonnet, S., and Farlie, P.G. (2009) Long-range regulation at the SOX9 locus in development and disease *J.Med.Genet.* **46,** 649–656
61. Leipoldt, M., Erdel, M., Bien-Willner, G.A., Smyk, M., Theurl, M., Yatsenko, S.A., Lupski, J.R., Lane, A.H., Shanske, A.L., Stankiewicz, P., and Scherer, G. (2007) Two novel translocation breakpoints upstream of SOX9 define borders of the proximal and distal breakpoint cluster region in campomelic dysplasia *Clin. Genet.* **71,** 67–75
62. Velagaleti, G.V., Bien-Willner, G.A., Northup, J.K., Lockhart, L.H., Hawkins, J.C., Jalal, S.M., Withers, M., Lupski, J.R., and Stankiewicz, P. (2005) Position effects due to chromosome breakpoints that map approximately 900 Kb upstream and approximately 1.3 Mb downstream of SOX9 in two patients with campomelic dysplasia *Am.J.Hum.Genet* **76,** 652–662
63. Demura, M., Martin, R.M., Shozu, M., Sebastian, S., Takayama, K., Hsu, W.T., Schultz, R.A., Neely, K., Bryant, M., Mendonca, B.B., Hanaki, K., Kanzaki, S., Rhoads, D.B., Misra, M., and Bulun, S.E. (2007) Regional rearrangements in chromosome 15q21 cause formation of cryptic promoters for the CYP19 (aromatase) gene *Hum.Mol.Genet.* **16,** 2529–2541
64. Shozu, M., Sebastian, S., Takayama, K., Hsu, W.T., Schultz, R.A., Neely, K., Bryant, M., and Bulun, S.E. (2003) Estrogen excess associated with novel gain-of-function mutations affecting the aromatase gene *N.Engl.J.Med.* **348,** 1855–1865
65. Lieberman-Aiden, E., van Berkum, N.L., Williams, L., Imakaev, M., Ragoczy, T., Telling, A., Amit, I., Lajoie, B.R., Sabo, P.J., Dorschner, M.O., Sandstrom, R., Bernstein, B., Bender, M.A.,

Groudine, M., Gnirke, A., Stamatoyannopoulos, J., Mirny, L.A., Lander, E.S., and Dekker, J. (2009) Comprehensive mapping of long-range interactions reveals folding principles of the human genome *Science* **326,** 289–293
66. Mitchell, J.A. and Fraser, P. (2008) Transcription factories are nuclear subcompartments that remain in the absence of transcription *Genes Dev.* **22,** 20–25
67. Simonis, M., Klous, P., Splinter, E., Moshkin, Y., Willemsen, R., de Wit, E., van Steensel, B., and de Laat, W. (2006) Nuclear organization of active and inactive chromatin domains uncovered by chromosome conformation capture-on-chip (4 C) *Nat.Genet.* **38,** 1348–1354
68. Croft, J.A., Bridger, J.M., Boyle, S., Perry, P., Teague, P., and Bickmore, W.A. (1999) Differences in the localization and morphology of chromosomes in the human nucleus *J.Cell Biol.* **145,** 1119–1131
69. Boyle, S., Gilchrist, S., Bridger, J.M., Mahy, N.L., Ellis, J.A., and Bickmore, W.A. (2001) The spatial organization of human chromosomes within the nuclei of normal and emerin-mutant cells *Hum.Mol.Genet* **10,** 211–219
70. Taslerova, R., Kozubek, S., Lukasova, E., Jirsova, P., Bartova, E., and Kozubek, M. (2003) Arrangement of chromosome 11 and 22 territories, EWSR1 and FLI1 genes, and other genetic elements of these chromosomes in human lymphocytes and Ewing sarcoma cells *Hum.Genet.* **112,** 143–155
71. Taslerova, R., Kozubek, S., Bartova, E., Gajduskova, P., Kodet, R., and Kozubek, M. (2006) Localization of genetic elements of intact and derivative chromosome 11 and 22 territories in nuclei of Ewing sarcoma cells *J.Struct.Biol.* **155,** 493–504
72. Finlan, L.E., Sproul, D., Thomson, I., Boyle, S., Kerr, E., Perry, P., Ylstra, B., Chubb, J.R., and Bickmore, W.A. (2008) Recruitment to the nuclear periphery can alter expression of genes in human cells *PLoS.Genet.* **4,** e1000039
73. Reddy, K.S., Rajangam, S., and Thomas, I.M. (1999) Structural chromosomal anomaly in mental retardation *Indian J.Pediatr.* **66,** 937–940
74. Deniaud, E. and Bickmore, W.A. (2009) Transcription and the nuclear periphery: edge of darkness? *Curr.Opin.Genet.Dev.* **19,** 187–191
75. Williams, R.R., Azuara, V., Perry, P., Sauer, S., Dvorkina, M., Jorgensen, H., Roix, J., McQueen, P., Misteli, T., Merkenschlager, M., and Fisher, A.G. (2006) Neural induction promotes large-scale chromatin reorganisation of the Mash1 locus *J.Cell Sci.* **119,** 132–140
76. Szczerbal, I., Foster, H.A., and Bridger, J.M. (2009) The spatial repositioning of adipogenesis genes is correlated with their expression status in a porcine mesenchymal stem cell adipogenesis model system *Chromosoma* **118,** 647–663
77. Harewood, L., Schutz, F., Boyle, S., Perry, P., Delorenzi, M., Bickmore, W.A., and Reymond, A. (2010) The effect of translocation-induced nuclear reorganization on gene expression *Genome Res.*
78. Brown, J.M., Green, J., das Neves, R.P., Wallace, H.A., Smith, A.J., Hughes, J., Gray, N., Taylor, S., Wood, W.G., Higgs, D.R., Iborra, F.J., and Buckle, V.J. (2008) Association between active genes occurs at nuclear speckles and is modulated by chromatin environment *J.Cell Biol.* **182,** 1083–1097
79. Misteli, T. (2004) Spatial positioning; a new dimension in genome function *Cell* **119,** 153–156
80. Osborne, C.S., Chakalova, L., Mitchell, J.A., Horton, A., Wood, A.L., Bolland, D.J., Corcoran, A.E., and Fraser, P. (2007) Myc dynamically and preferentially relocates to a transcription factory occupied by Igh *PLoS.Biol.* **5,** e192
81. Branco, M.R. and Pombo, A. (2006) Intermingling of chromosome territories in interphase suggests role in translocations and transcription-dependent associations *PLoS.Biol.* **4,** e138
82. Goetze, S., Mateos-Langerak, J., Gierman, H.J., de Leeuw, W., Giromus, O., Indemans, M.H., Koster, J., Ondrej, V., Versteeg, R., and van Driel, R. (2007) The three-dimensional structure of human interphase chromosomes is related to the transcriptome map *Mol.Cell Biol.* **27,** 4475–4487

Chapter 9

The Challenges of Studying Complex and Dynamic Regions of the Human Genome

Edward J. Hollox

Abstract

Recent work has emphasised that the human genome is not simple and static, but complex and dynamic. This review focuses on the regions that are particularly hard to dissect and analyse, yet hold clues to how the genome changes during evolution and disease. I begin by summarising recent key advances in the understanding of the variable structure of our genome, and then I discuss a medley of methods that may allow us to analyse this structure in fine detail. In the final part, I describe potential future developments in this field, and make an argument that, just as we routinely genotype single-nucleotide polymorphisms now and will routinely re-sequence genomes in the near future, we should be aiming to physically re-map the individual human genome for each individual we study.

Key words: Copy number variation, CNV, Defensin, Linkage, Next-generation sequencing, Sequence read depth, Haplotype, Human genome, Reference, Recombination, Duplication, Subtelomeric

1. Introduction

With limited time and budget, the research community often determines priorities by focusing on research questions, where data are straightforward to get: what is sometimes called the "low-hanging fruit". However, those of us who have spent late summers gathering fruit know that the fruits at the top of the tree, requiring precarious balancing on ladders, are usually the ripest, largest, and sweetest. So, to extend the cliché, this review focuses on how to reach the "high-hanging fruit" of the genome: genomic regions that are structurally variable but are particularly challenging to characterise accurately, and to determine their medical significance. I describe the methods and concepts needed to understand how we approach such regions, and suggest how research in this direction may progress in the future. I also illustrate these theoretical issues with data both from the beta-defensin region at 8p23.1,

Lars Feuk (ed.), *Genomic Structural Variants: Methods and Protocols*, Methods in Molecular Biology, vol. 838,
DOI 10.1007/978-1-61779-507-7_9, © Springer Science+Business Media, LLC 2012

which has been a focus of my research, and other examples from the published literature.

Of course, the fact that data are difficult to get does not necessarily make that data more important or interesting. Why should we be interested in complex and dynamic regions of the genome? Firstly, genes are present in these regions, and it appears that genes involved in direct interaction with the environment, such as immunity and cell surface receptor genes, are particularly enriched in these regions (1, 2). Secondly, for some dynamic regions, there is rapid change in structure between species and this is sometimes accompanied by evidence of rapid evolution at the amino acid level, suggesting that these regions are often crucibles of gene evolution (3–6). Thirdly, structural differences between chromosomes prevent homologous recombination and exchange of genetic information, potentially facilitating sympatric speciation. Finally, such regions are hotspots for clinically relevant rearrangements, both in the germ line and soma, with consequences for chromosomal genomic disorders and cancer (7–11).

After years of controversy, in the mid-1950s, it became clear that DNA was the hereditary material and the human karyotype had 46 chromosomes. Since then, the two different scales of analysing human variation, namely, at the molecular level through DNA and protein polymorphism and at the karyotype level by analysing chromosomal abnormalities, have progressed almost independently. With improving technology, molecular variation has been measured at a larger and larger scale, and cytogenetic variation at a smaller and smaller scale until recently, when the two approaches have coincided. This has led to more formal definitions of structural variation, which includes copy number variation (CNV) and other types of variation, such as translocations and inversions.

The development of array comparative genomic hybridisation (array-CGH) technology allowed DNA copy number to be assayed at a submicroscopic scale on the whole genome (12). It soon became clear that there was considerable variation in copy number of certain genomic regions between different healthy individuals. A flurry of research papers combined with improvements in array-CGH technology allowed more reliable estimates of the copy number range and region boundaries (1, 13, 14), resulting in an oligonucleotide tiling-path array-CGH study which, at the time of writing, represents the zenith of this area of research (15). This study comprehensively searched for CNV of regions larger than 500 bp in 40 individuals of European and Sub-Saharan African descent. Two genomes compared at random show differences in copy number of about 24 Mb (0.78% of the genome). Importantly, the diploid copy number could be called on only 61% of the CNVs validated, and of these 77% were deletions (diploid copy numbers of 0, 1, 2), 16% were duplications (diploid copy numbers of 2, 3, 4), and less than 1% were multiallelic (diploid copy numbers of 3 or more). This bias probably reflects the fact that the quantitative difference between deletion copy numbers is greater than duplications, which are in turn greater than multiallelic loci. It is likely that a large proportion of the validated but untypable CNVs are multiallelic, and these validated but untypable CNVs are very enriched for RefSeq genes ($p=1.16\times10^{-13}$, two tailed fishers exact test, when comparing CNV counts for two genomes compared at random). These regions, which are technically challenging to define and type, are the focus of this review.

2. Limitations of a Reference Genome

Genome-wide analysis of structural variation would have not been possible without the human genome sequence. Yet we are beginning to take the genome sequence assembly for granted, and perhaps make assumptions about our data, and the human reference sequence, that at best may not always be justified and at worst may lead us astray. An example is the presentation of data from the array-CGH method. The genomic DNA hybridises to target DNA sequences, whether they are BAC clones or oligonucleotides, which have themselves been designed using the human reference genome sequence. Any gain of signal at a particular target DNA sequence, representing a gain of copy number of the test genomic DNA, is represented as an increase in signal at that DNA sequence in the human reference genome. But we have no evidence that the extra copy (or copies) of DNA is actually at that locus in our test DNA. Often, it is a good assumption based on the likely mutational mechanisms of CNV: ectopic recombination between misaligned repeat sequences can cause duplications, and these duplications can occur in *cis*, at the same locus. However, this may not always be the case, most notably for retrotransposed genes, where an increase in exon copy number is marked as an increase at that exon, where in fact the extra copy of that exon is on another chromosome. For example, the *PRKRA* gene, which encodes an activator of an interferon-induced protein kinase, is polymorphically duplicated, with one copy on chromosome 2 and the polymorphically retrotransposed copy on chromosome 6 (15).

The struggle in reconciling the concept of a single reference genome with structurally variable and complex regions can be illustrated by the beta-defensin copy number variable region on human chromosome band 8p23.1. A detailed review of this region is beyond the scope of this review, but see ref. 16. At either end of 8p23.1 are two regions, called REPP and REPD, rich in olfactory receptor repeats and endogenous retroviral elements (17, 18). A polymorphic inversion exists between REPP and REPD which carries single-copy sequence encoding several genes, including *GATA4* (18, 19). Embedded within the repeat-rich region is a large segmental duplication that carries several beta-defensin genes and the *SPAG11* gene, called the beta-defensin repeat region. This region varies in copy number as a whole unit of at least 260 kb, and probably larger, with a modal copy number of 4 and a range of copy numbers from 1 to 12 (16, 20). The history of the reference assembly of this region across several builds of the human genome is shown in Fig. 1. It is important to note the persistence of gaps even in the latest genome assembly, presumably due to the polymorphic nature of this region. Both physical and genetic mapping approaches, described below, show that the current assembly is only an approximation of the actual situation. Using these approaches, it has been shown that beta-defensin repeats can be at either REPD, REPP, or both, and can be present at multiple copies at each locus ("Likely arrangement" in Fig. 1) (21).

Other dedicated studies hint at the diversity lurking in regions rich in segmental duplications, which are often poorly assembled. A landmark study involving manual assembly, curation, and experimental verification of the subtelomeric regions showed their extraordinary structural diversity (22). In particular, subtelomeric duplications have been translocated across the subtelomeric regions

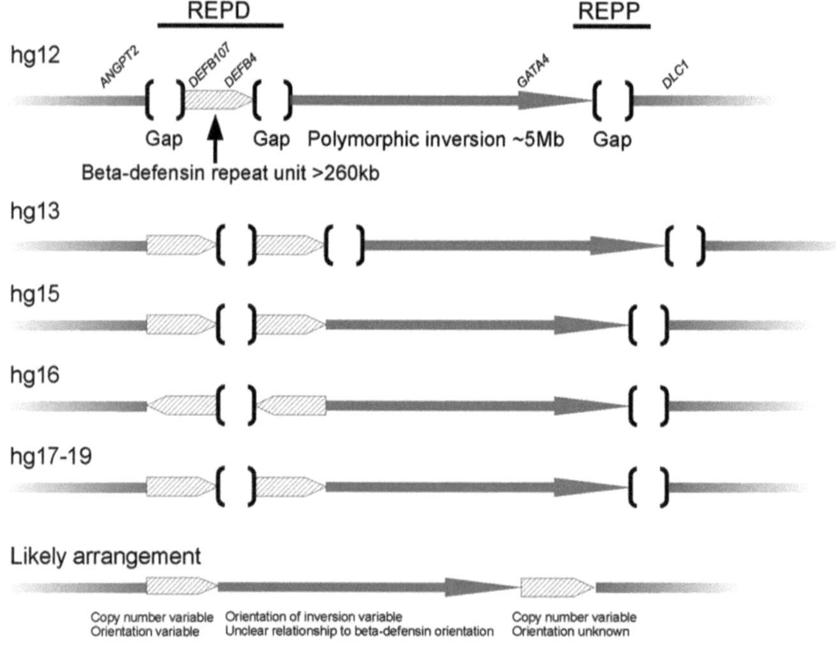

Fig. 1. A history of the reference assembly of the beta-defensin region at 8p23.1. Each assembly release is shown from hg12 (released June 2002), with the likely arrangement deduced from both physical and genetic mapping approaches shown at the bottom of the diagram. Not drawn to scale. Reference assemblies from the University of California Santa Cruz Genome Browser (http://genome.ucsc.edu).

of different chromosome arms. These duplications sponsor further deletions, duplications, and other rearrangements which may, in turn, generate more translocations, some of which are clinically important. This genomic maelstrom can carry genes with it; for example, the WASH gene, which is involved in reorganising the actin cytoskeleton in response to extracellular stimuli, is distributed across several chromosomes and is copy number variable (23, 24). The situation at the beta-defensin repeat region echoes these subtelomeric regions: initial duplication events are likely to have sponsored the large polymorphic inversion and large clinically relevant rearrangements (16, 25, 26). This seems to be a common theme; segmental duplication and structurally variable regions sponsor large-scale chromosomal rearrangements with clinical consequences (8). It is likely that purifying selection against rearrangements with clinical effect, both *in utero* and in adulthood, act as a brake on an ever-increasing cascade of genome rearrangements.

The latest genome assembly build (hg19) contains 357 gaps (Fig. 2), and many of these are likely to be due to polymorphic structural variation because, like the beta-defensin region, these gaps are flanked by segmental duplication-rich regions. Although the reference human genome assembles a contiguous sequence for other repeat-rich regions (such as the cadherin region on chromosome 16), this may be due to assembly from BACs taken from one structural variant allele, and does not necessarily rule out large-scale variation in chromosome structure. One approach to tackling the gaps, and appraising the level of structural variation in the genome, is to develop alternative assemblies in certain regions of the genome. The current build has alternative assemblies for three regions: the MHC region on chromosome 6 (27), the MAPT region on chromosome 17 (28), and the UGT2B17 region on chromosome 4 (29, 30). The number of regions, and the number of alternative assemblies for each region, should increase with gap regions targeted.

The reference human genome is constructed from physical and genetic maps of genomic DNA from a handful of individuals of western European descent, and we have very little knowledge about the structural variation in non-European populations. Second-generation high-throughput sequencing of genomes from Korean, Chinese, and Yoruba individuals allows us to gain a more complete picture of global DNA sequence variation, a picture that becomes clearer with the completion of the 1,000 Genomes Project (http://www.1000genomes.org/). Second-generation sequencing methods provide short sequence reads from random regions of the genome, preventing *de novo* assembly of the genome. These short sequence reads are mapped back onto the reference genome, effectively obscuring the vast majority of structural variation. Sequence read depth (SRD, i.e. the number of sequence traces mapping to a given locus) can be a useful indicator of variation in DNA copy

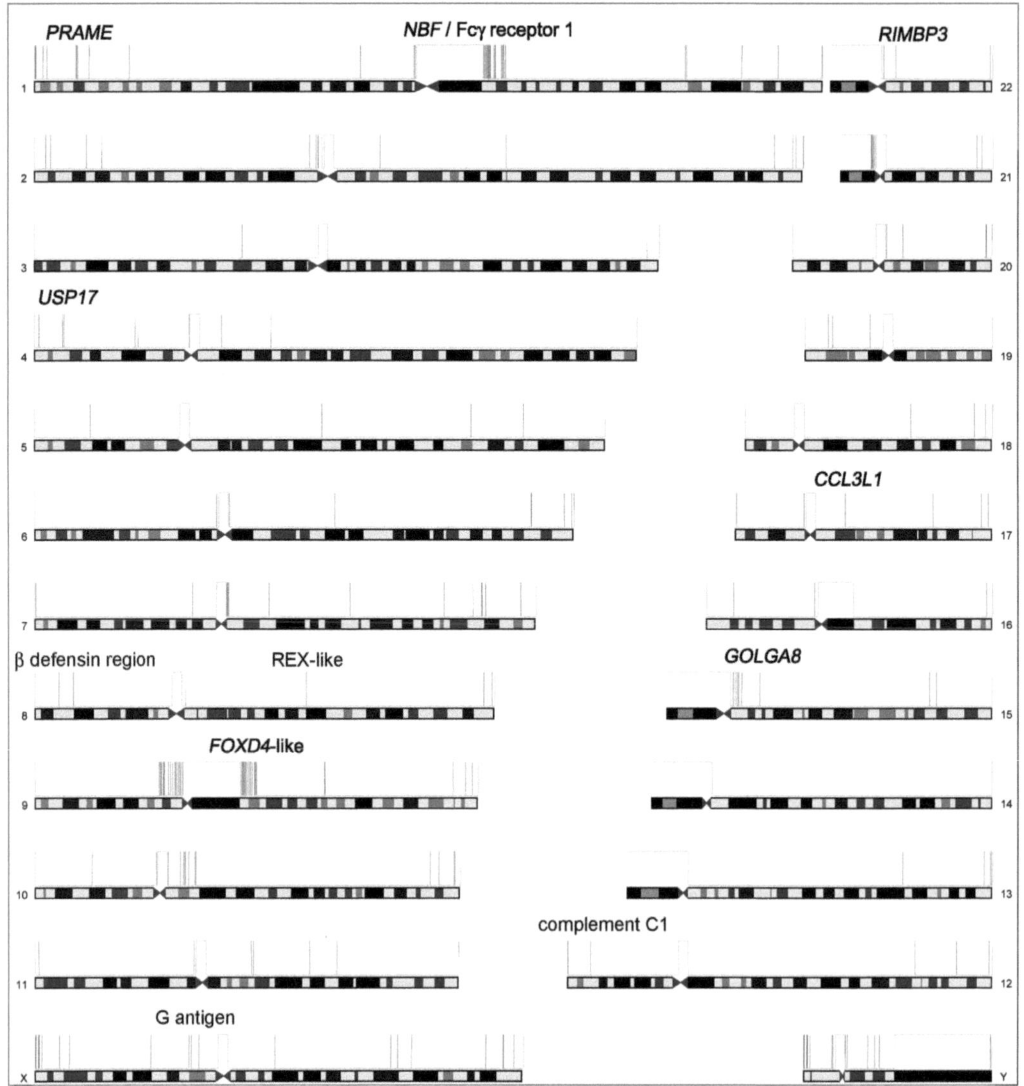

Fig. 2. Gaps in the human genome assembly. The current human genome assembly (hg19) is shown with *vertical grey bars* showing locations of gaps. Selected genes or gene families within segmental duplications flanking these gaps are highlighted. Drawn using the Genome Graphs function of the UCSC Genome Browser (60).

number; for example, in a heterozygous deletion, we would expect half the number of sequence reads to map to a locus compared to a reference sample (30–32). This has the potential to be a very powerful tool for detecting and typing copy number variation, but, like array-CGH, provides no information on the physical structure of the copy number variable locus.

Finally, it is worth remembering that the reference genome is the euchromatic part of the genome and the heterochromatic region is uncharacterised, including the p-arms of acrocentric chromosomes (33). This is mainly because it is repeat rich and unclonable, but it is functionally important. Heterochromatin includes rDNA genes which, in addition to coding for rRNA, also affect chromatin structure and gene regulation in other genomic regions. rDNA and other heterochromatic regions show a large amount of structural variation which is not well-characterised at present (34–36).

3. Diplotype, Genotype, and Haplotype

The different concepts produced by understanding structural variation reinforce the importance of using consistent language to describe our observations. For structural variation where there is no change in copy number, such as a polymorphic inversion, these are straightforward and are analogous to a single-nucleotide polymorphism; the locus is diploid and must have one of three genotypes, homozygous normal, heterozygous, and homozygous inverted (37). Ideally, the "normal" and "inverted" alleles should be defined with respect to the ancestral orientation of the region, but often these alleles are defined arbitrarily.

For regions that show CNV, in most cases, we know or assume that the locus maps to one chromosome and therefore, in a diploid cell, there are two chromosomes that carry that locus but with variable number of copies of the region. Importantly, we often do not know which homologous chromosome carries which number of copies because most direct measurements, including quantitative PCR approaches, array CGH, and SRD, give the total copy number summed over both chromosomes. A copy number genotype is where the copy number of the individual chromosome is known. For example, a beta-defensin copy number of 4, which is the modal copy number for nearly all populations, can represent copy number genotypes of 0, 4, or 1, 3, or 2, 2.

The term "copy number haplotype" is often used interchangeably with genotype. However, using the analogy of SNP haplotypes, where a haplotype conveys knowledge of the physical arrangement of genotypes on a chromosome, a copy number haplotype could be when the physical position of the chromosome of different copies is known. These copies may or may not be distinguished by sequence differences within the copies; for example, a copy number haplotype of the beta-defensin region would define how many copies were at the distal and the proximal site on a chromosome or the position of copies with a given nucleotide change (Fig. 3). This principle can be extended for more complex multiallelic loci, where there are more copies in the genome that are spread between different chromosomes. For example, the WASH gene has seven copies assembled in the human genome distributed across 6 chromosomes (38).

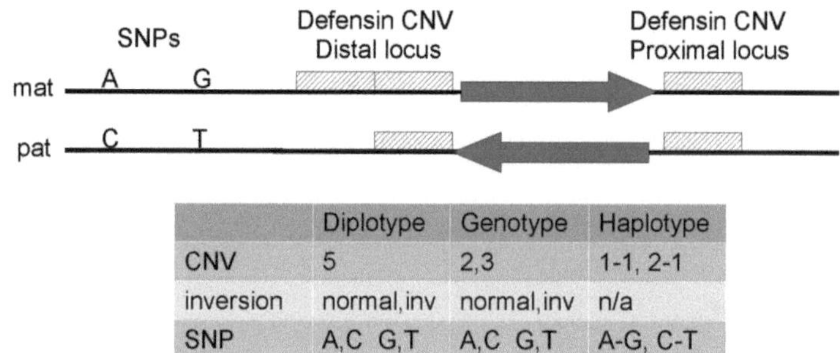

	Diplotype	Genotype	Haplotype
CNV	5	2,3	1-1, 2-1
inversion	normal, inv	normal, inv	n/a
SNP	A,C G,T	A,C G,T	A-G, C-T

Fig. 3. Defining diplotype, genotype, and haplotype for CNV. An example from the beta-defensin region is shown, representing SNP variation, inversion polymorphism, and CNV and the diplotype, genotype, and haplotype for each type of variation in that region.

4. Generating Diversity by Meiotic Recombination and Segregation

Structural variation between generations can be generated by Mendelian segregation of structurally variable alleles. This is due to the fact that, as discussed above, most copy number measurements are of diplotype rather than of genotype, and Mendelian segregation analysis in families shows the segregation of alleles forming genotypes. Indeed, segregation analysis in families is a powerful method of determining copy number genotype from copy number diplotype. An example of this is shown in Fig. 4, where two parents have copy number diplotypes of four each, but analysis of the children shows diplotypes of 3 and 5, because the parental genotypes are 2, 2 and 1, 3. Analysis of this region by array-CGH shows no apparent copy number change for the parents because four copies is modal and the copy number of the reference DNA often used in array-CGH. However, analysis of the children would show four children with gain of copy (copy number 5) and two with a loss of copy number (copy number 3). Because the parents seemed to show no copy number change, these would be often incorrectly interpreted as *de novo* changes.

The situation can become more complicated if the different copies of copy number variable loci are on different chromosomes. If we take the example of the WASH gene mentioned above, then simple Mendelian segregation could generate a wide range of different copy numbers from two parents that have the same copy number (Fig. 5). Again, such a situation would be viewed as *de novo* mutation unless there was an awareness of the underlying complexity of the locus.

If a region with the same sequence is at two different loci on the same chromosome and both loci are copy number variable, then simple allelic recombination in the intervening sequence

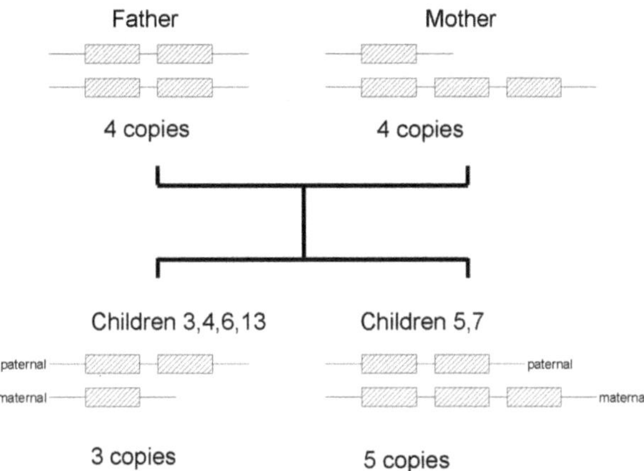

Fig. 4. Diplotype to genotype using segregation of CNV in families. Copy number and deduced genotypes for the beta-defensin region in CEPH family 1375. The parents have the same diplotype but different genotypes, and segregation of the alleles, not de novo mutation, causes apparent loss or gain of copy number in the children. Data from Abu Bakar et al. (21).

between these loci can generate new haplotypes, which may carry a different number of copies of the region. This is illustrated again by variation at the beta-defensin region on chromosome 8p23.1. Here, the CNV region is at either end of 8p23.1 (Fig. 1), and recombination within the 5 Mb single-copy region between the two CNV regions can change copy number (Fig. 6) (21). Other structural variation can play an important role in these cases. For the beta-defensin region, REPP and REPD (Fig. 1) have sponsored a polymorphic inversion of the single-copy region between them (18). Recombination between inversion alleles of opposite orientation forms dicentric and acentric chromosomes that are not mitotically stable, which effectively suppresses viable products of recombination in inversion heterozygotes (16, 39, 40). Recombination is permissive in inversion homozygotes (Fig. 7). This means that the frequency of copy number change caused by allelic recombination depends critically on the inversion genotypes of the parents. On a population level, this translates to the frequency of copy number change caused by allelic recombination depending on the frequency of the inverted allele in the population. This illustrates the potential interdependence and crosstalk of structural variation between different loci.

The processes described above may produce what could be regarded as "cryptic" mutational events. Nevertheless, these events do alter copy number between generations and can alter the copy number on individual chromosomes. Genuine mutational events generating *de novo* copy number change are where

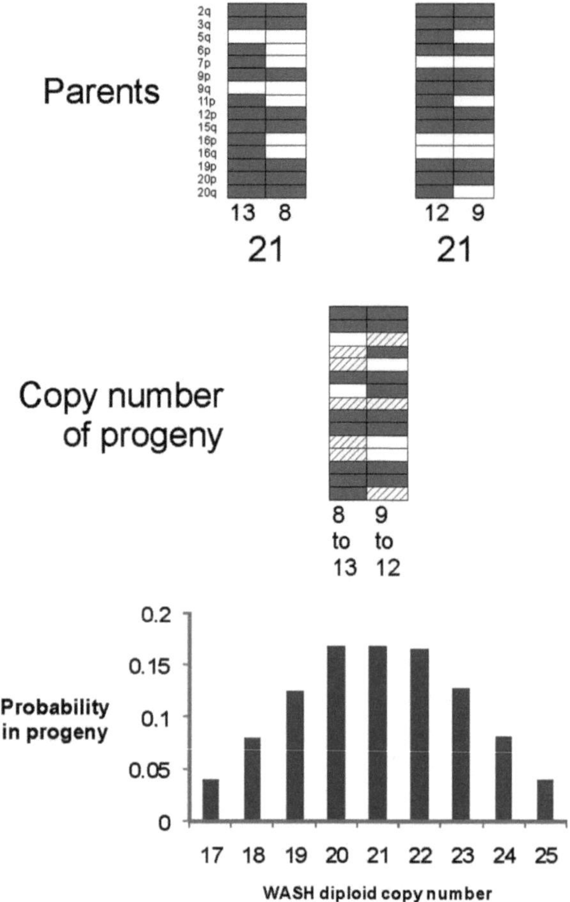

Fig. 5. Mendelian segregation of a gene CNV on multiple chromosomes. Location and copy number of WASH genes in each parent are shown, with each box representing a WASH gene polymorphic occupancy site either occupied (*grey*) or absent (*white*). The occupancy plot of all potential progeny for the WASH genes is shown below the parental plots; each haploid genome can either be occupied with a WASH gene at each site (*grey*), or be absent (*white*), or either (hatched) depending on which of the two homologous chromosomes was inherited from the parent. Total haploid and diploid copy numbers are shown below the occupancy plot. A probability/frequency distribution of total diploid WASH copy number of progeny from this mating is shown at the bottom. Data are from reference (38); the two individuals chosen as hypothetical parents for this figure are unrelated individuals H1 and H5 in that paper.

the molecular mechanism involves the locus in question. These involve processes, such as non-allelic homologous recombination and non-homologous end joining, which are discussed in depth in another chapter by Lupski and colleagues and in a recent review (41). Methods for the dissection of the processes involving copy number change are in their infancy, and determining the relative roles of each process at each locus is essential for understanding the evolution of these complex genomic regions.

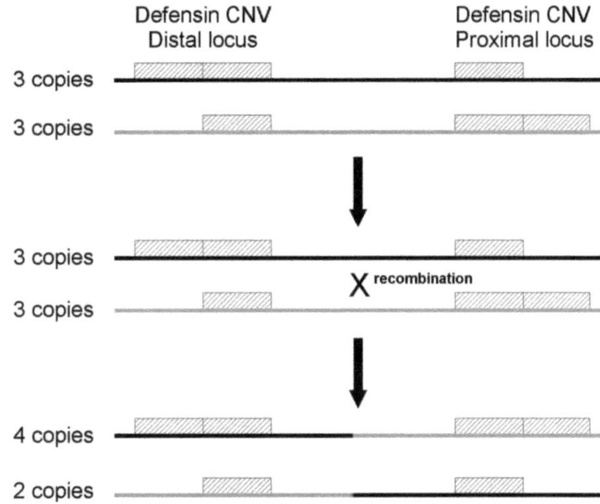

Fig. 6. Allelic recombination between CNV at two positions. An example of how allelic recombination in single-copy sequence between a CNV mapping to two locations can generate chromosomes with new copy numbers of the CNV region. A single copy of the CNV region is shown as a hatched box, with the two homologous chromosomes shown as *dark grey* and *pale grey*.

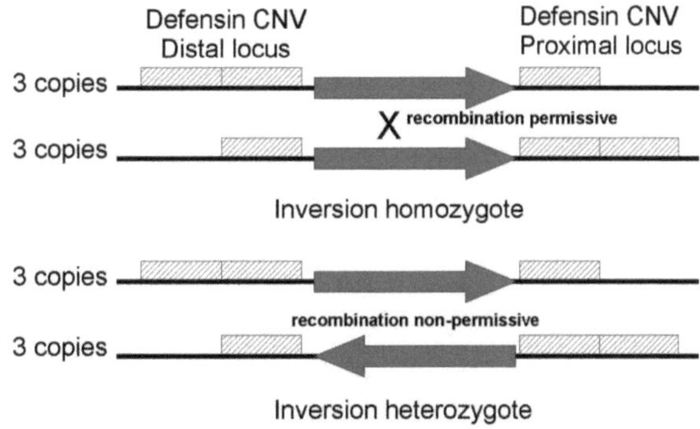

Fig. 7. Inversions can prevent allelic recombination between CNV at two positions. A single copy of the CNV region is shown as a hatched box, and the single-copy sequence between the defensin CNVs shown as an inversion polymorphism by the *grey arrow*.

5. Genetic Re-mapping Approaches to Investigating Structural Variation

New structural variants, and clues to their nature, can be determined by genetic mapping (19). Linkage and association analyses allow the investigation of the co-segregation of structural variants and chromosomal markers in families and in populations, respectively. By analysing the pattern of co-segregation and the location of the co-segregating markers, we can make inferences about the existence and location of structural variation.

Inversion polymorphisms can be discovered by analysing segregation of markers in families. This relies on individual cases of a single-recombination event which, given a genetic map placing the markers in an inverted order, results in an apparent triple-recombination event. Of the three apparent recombination events, the first and the third represent the boundaries of the inversion, with the second representing the real crossover event (16, 19). Such characteristic inversion signatures can be identified using haplotype reconstruction from pedigrees and software, such as chrompic from the CRI-MAP (42) suite, and can identify the long-inversion polymorphism at chromsomal region 8p23.1. These inversion signatures may also result in differences between genetic maps, depending on the relative frequencies of the inversion alleles. For example, the genetic map based on microsatellite markers typed on large Icelandic pedigrees places the markers in the 8p23.1 inversion in the inverted orientation, in contrast to the genetic map produced by CEPH (43). However, this approach is dependent on high-density genotyping; at lower densities or for small inversions, only two or three adjacent markers show this signature and are likely to be interpreted as genotyping errors.

Once a reliable assay for typing any structural variant is developed, this assay can be applied to type the DNA from the Centre d'Etude du Polymorphisme Humain (CEPH) families. This allows confirmation of the overall linkage of the structural variant to its expected position in the genome using standard linkage mapping approaches. Recombination and segregation in individual families can also allow important inferences to be made about genomic location, including if there is any heterogeneity. Segregation analysis can confirm that all variants of the structural variation (such as copies of a multicopy CNV) are on the same chromosome, and recombination can confirm whether all variants are at the same position on that chromosome.

The segregation analysis approach is best illustrated by an example: in this case, again, using the CNV of the beta-defensin region on chromosome 8. The repeat unit that is copy number variable is large (at least 260 kb) and contains several short tandem repeats (STRs, also known as microsatellites). These STRs may vary in length between defensin repeats, and because of the multi-copy nature of the region the maximum expected number of STR variants themselves vary; for an individual with four copies of the defensin region per diploid genome, we could have between one and four STR length variants. These STR loci are particularly valuable for two reasons: firstly, they provide a "digital" method for estimating the number of copies of the defensin repeat region per diploid genome and secondly, each STR variant can tag a particular defensin repeat. By tagging individual defensin repeats using STR length variants, we can follow segregation of the defensin repeats through families. Using this approach, it was found that some defensin repeats segregated with the alternative chromosomal region following a recombination event within 8p23.1. This was observed several times, leading to the conclusion that the defensin repeat region could either be at the locus annotated on the human reference genome or at a locus 5 Mb proximal (21). This genetic re-mapping pointed to the defensin repeats being at both the REPP and REPD olfactory repeat regions, in contrast to the current genome assemblies (Fig. 1).

6. Physical Re-mapping Approaches to Investigating Structural Variation

We can use physical re-mapping approaches to identify different forms of structural variation. Perhaps, the most thorough approach is the comparison of two human genomes that have been assembled independently. Ideally, these would be two randomly sampled haploid genomes, but the two most completely assembled genomes are mosaics of several diploid genomes from several individuals: Celera Genomic's R27c and NCBI's Build 35 assembly. The analysis provided a comprehensive analysis of structural variation in these two genomes, which covered the full spectrum of variant sizes, resulting in a large expansion in the catalogue of structural variants (44). The limitation is obvious; it requires a complete independent assembly of a human genome and is not scalable to large numbers of samples.

The most successful approach so far for analysing large numbers of samples has been paired-end mapping (PEM) (45–47). This relies on physically isolating DNA fragments within a known size range, and sequencing the ends of each fragment. These ends can be mapped on to the human reference genome, and discrepancies in size and orientation indicate that the test genome is structurally different from the human reference genome in that region. The main challenge of these methods is the physical isolation of DNA fragments of a given size because any errors or artefacts at this stage result in false-positive identification of structural variation. PEM was pioneered using fosmid clones to physically isolate DNA fragments of known size (45). Fosmids are vectors that accept a very restricted range of DNA fragment size (around 40 kb) because of the lambda phage packaging step used in producing fosmid libraries. However, this approach requires that a fosmid library is made for each sample, which is time consuming and expensive. There is a lower limit of 8 kb for the size of deletions and duplications detected because of the variation around 40 kb of the cloned fragment sizes. In order to overcome these disadvantages, an alternative approach for physical isolation of DNA fragments has been used. Smaller fragments can be isolated by shearing genomic DNA to generate fragments of the desired size (47, 48);

this is potentially a more flexible approach but depends critically on the size range that shearing produces. In addition, the process involves circularisation and self-ligation of sheared fragments and this process may produce false positives due to ligation between fragments.

A general approach to physical mapping is to randomly shear genomic DNA into fragments, and test how frequently marker A is on the same DNA fragment as marker B. The closer they are, the more likely they are to be on the same fragment, and by testing multiple independent sampling of the genome we can measure how frequently they are on the same fragment and therefore estimate statistically their physical distance. This approach is taken in radiation hybrid (RH) mapping, where cells from an individual are irradiated by X-rays to fragment the chromosomes (49). This cell is then fused with a rodent cell, and hybrid cells, which include fragments of human chromosomes, selected and propagated. The result is a series of cell lines, where random fragments of the human genome have integrated into the rodent genome. These cell lines can be tested for co-localisation of marker A with marker B, often using PCR. Controlling the dose of radiation can control the size of DNA fragment produced, but establishing and maintaining these radiation hybrid cell lines is a large amount of work even for cells from one human, let alone a panel of different individuals needed to physically re-map the genome.

An approach that uses the same principle of fragmentation and sampling is called Happy mapping (50). This approach fragments genomic DNA and aliquots limiting dilutions of this fragmented genomic DNA so that any single marker is present only in around half of these aliquots. Using PCR to screen for marker A and marker B, we can then calculate whether the number of co-amplifications of both from the same aliquot is higher than expected by random assortment of independent fragments, and therefore the probability that they occur on the same fragment. Such an approach has been successful for physical mapping of a number of organisms (51–53), and it is attractive for physical re-mapping because it uses genomic DNA, not cells, and has the potential for automation. Parallel typing of thousands of markers may be possible with second-generation sequencing technologies. However, because the method relies on PCR amplification of markers from single molecules of genomic DNA, PCR contamination can be a problem.

7. Future Developments in Structural Variation Analysis

The drive to reduce the cost of DNA sequencing opens up many new approaches that use DNA sequencing as a tool rather than as an end in itself. Commercial second-generation sequencing platforms, which are currently the Roche 454 Pyrosequencing method, Applied Biosystem's Solid ligation-based approach, and Illumina's Genome Analyzer, produce DNA sequence at much lower cost than dideoxy-terminator Sanger sequencing. New approaches will appear, including Helicos single-molecule sequencing and Pacific Biosciences real-time sequencing, which have the potential for long sequencing reads (54). With this in mind, it is clear that SRD has plenty of potential for detecting CNV (5, 31, 32). Indeed, attempts have been made but there are still key impediments to it being used routinely, such as the current cost of re-sequencing an entire genome. This is falling rapidly, but routine genomic re-sequencing is still outside the budget of most labs. Within the foreseeable future, sequencing hundreds of samples is still cost-efficient only in large sequencing centres, physically separating data generation from analysis and discouraging experimentation. Perhaps in a few years, the ability to re-sequence a genome will really come to the experimenter's bench, and be as routine as agarose gel electrophoresis. It is also not yet clear how different technologies vary in their coverage of particular regions. This can be viewed optimistically, in that by using several platforms with different biases an unbiased coverage of the genome can be obtained. Pessimistically, it suggests that the genome may have to be re-sequenced several times using several platforms for that unbiased view of structural variation using SRD.

In general, a deeper SRD is required to accurately type CNVs of higher copy number or of smaller genomic size. As an example, Fig. 8 shows the SRD needed to distinguish four from five copies of a CNV of a given size at a confidence level comparable with SNP genotyping chips, based on data from two second-generation sequencing platforms. This challenge is a key test of the typing ability of a technology, and would be requirement for many multiallelic CNVs, such as *FCGR3A/B* (55, 56) and the beta-defensin region. It can be seen that there is a reciprocal relationship between SRD and size of event distinguished, which predicts that a linear increase in SRD will greatly improve its power to detect CNV initially, but after a certain point, as the curve begins to flatten, ever greater increases in SRD are required to continue improving CNV measurement and resolution. The difference between the Roche 454 and Illumina sequencing methods is due to the variance in the SRD of a constant diploid region of both methods (57). This variance is much lower for the Roche 454 method compared to the Illumina method, perhaps due to the longer reads and increased uniqueness of a Roche 454 read when mapping to a repeat-rich human genome. Longer sequence reads, therefore, increase the power to detect differences in SRD which would be interpreted as CNV.

Sequencing approaches require significant amounts (several micrograms) of DNA starting material. This is usually generated by various amplification procedures, including whole-genome

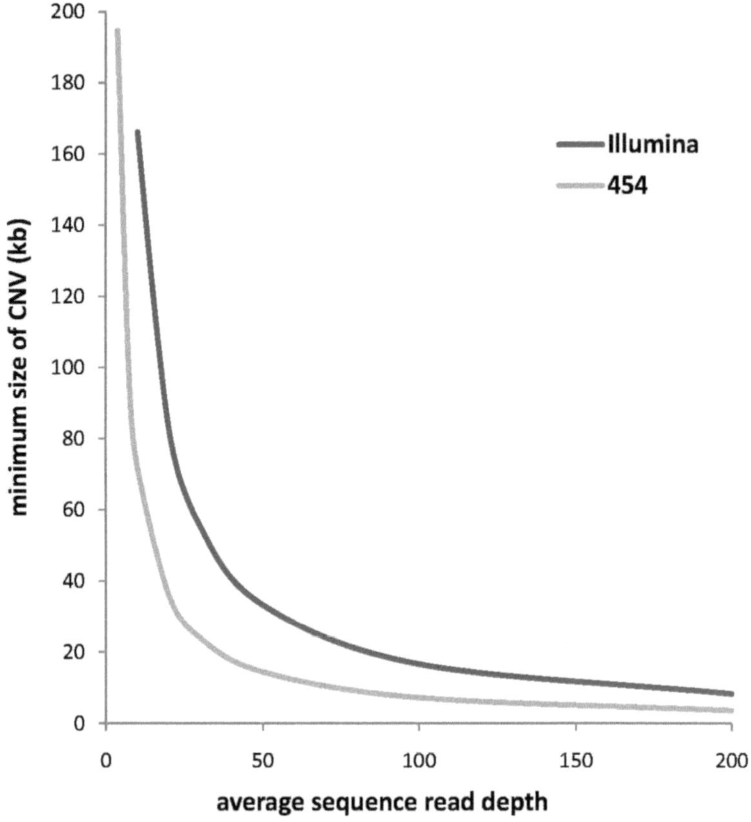

Fig. 8. Sequence read depth and size of CNV typed. Using published data from a genome sequenced using 454 technology (Watson) and a genome sequenced using Illumina Solexa technology (na18507) (32, 60, 61), the average read depth required to distinguish four copies from five copies of a CNV of a given size was calculated. Distinguishing four from five copies is stringently defined as the SRD $(\mu + 3\sigma)^{4copies} <$ the SRD $(\mu + 3\sigma)^{5copies}$, which predicts a type II error rate of 1×10^{-9}.

amplification, but there are biases in these methods because small differences in amplification efficiency are magnified during exponential amplification resulting in artefactual variation in copy number in the amplified fragment pool. Such procedures may not work at all with particularly low-quality and damaged DNA, which is frequently the condition of clinically important samples. Molecular copy number counting (MCC), which is related to Happy mapping, relies on multiple aliquots of genomic DNA at limiting dilution so that a given target molecule is either present once or absent in a given aliquot (58). Amplification of the target locus using essentially standard PCR amplification of a small amplicon gives a number of positive aliquots, which, for a heterozygous deletion, is expected to be half the number of positive aliquots for a reference locus. Although not yet adapted for high throughput, this method has potential for automation and, importantly, the accurate typing of diplotype (4 from 5 copies, for example) is dependent (in theory) only on the number of aliquots. Therefore, it may be unique in the ability for typing higher copy numbers accurately and precisely.

For large-scale structural variation, including balanced translocations and inversions where there is no net change in copy number, the future seems to lie in combining new sequencing approaches with physical approaches to mapping chromosomes. These physical approaches could include pulsed-field gel electrophoresis, which separates DNA fragments up to several Mb, fluorescence in situ hybridisation that gives direct information on the localisation of particular DNA sequences within the chromosome, or Happy mapping. Such approaches are still time intensive, labour intensive, and of low throughput, but a dedicated focusing of technology development in these areas leads to an understanding of structural variation between individuals and species in the context of evolution and the architecture of chromosomes in the nucleus. New sequencing methods give us high-throughput sequence re-mapping of the genome; we should now aim for high-throughput physical re-mapping of the genome.

Acknowledgments

My work is currently supported by a Medical Research Council New Investigator Award (GO801123) and a Wellcome Trust project grant (no.087663). Thanks to Dr Lee Machado for comments on an earlier draft of the manuscript.

References

1. Redon, R., Ishikawa, S., Fitch, K. R., Feuk, L., Perry, G. H., Andrews, T. D., Fiegler, H., Shapero, M. H., Carson, A. R., Chen, W., Cho, E. K., Dallaire, S., Freeman, J. L., Gonzalez, J. R., Gratacos, M., Huang, J., Kalaitzopoulos, D., Komura, D., MacDonald, J. R., Marshall, C. R., Mei, R., Montgomery, L., Nishimura, K., Okamura, K., Shen, F., Somerville, M. J., Tchinda, J., Valsesia, A., Woodwark, C., Yang, F., Zhang, J., Zerjal, T., Zhang, J., Armengol, L., Conrad, D. F., Estivill, X., Tyler-Smith, C., Carter, N. P., Aburatani, H., Lee, C., Jones, K. W., Scherer, S. W., and Hurles, M. E. (2006) Global Variation in Copy Number in the Human Genome. *Nature. 444*, 444–454.
2. Traherne, J. A., Martin, M., Ward, R., Ohashi, M., Pellett, F., Gladman, D., Middleton, D., Carrington, M., and Trowsdale, J. (2010) Mechanisms of Copy Number Variation and Hybrid Gene Formation in the KIR Immune Gene Complex. *Hum. Mol. Genet. 19*, 737–751.
3. Johnson, M. E., Viggiano, L., Bailey, J. A., Abdul-Rauf, M., Goodwin, G., Rocchi, M., and Eichler, E. E. (2001) Positive Selection of a Gene Family during the Emergence of Humans and African Apes. *Nature. 413*, 514–519.
4. Johnson, M. E., Cheng, Z., Morrison, V. A., Scherer, S., Ventura, M., Gibbs, R. A., Green, E. D., and Eichler, E. E. (2006) Recurrent Duplication-Driven Transposition of DNA during Hominoid Evolution. *Proceedings of the National Academy of Sciences. 103*, 17626–17631.
5. Marques-Bonet, T., Kidd, J. M., Ventura, M., Graves, T. A., Cheng, Z., Hillier, L. D. W., Jiang, Z., Baker, C., Malfavon-Borja, R., and Fulton, L. A. (2009) A Burst of Segmental Duplications in the African Great Ape Ancestor. *Nature. 457*, 877–881.
6. Perry, G. H., Yang, F, Marques-Bonet, T., Murphy, C., Fitzgerald, T., Lee, A. S., Hyland, C., Stone, A. C., Hurles, M. E., Tyler-Smith,

C., Eichler, E.E., Carter, N. P., Lee, C., and Redon, R (2008) Copy number variation and evolution in humans and chimpanzees. *Genome Res. 18*, 1698–1710.

7. Mefford, H. C., Sharp, A. J., Baker, C., Itsara, A., Jiang, Z., Buysse, K., Huang, S., Maloney, V. K., Crolla, J. A., Baralle, D., Collins, A., Mercer, C., Norga, K., de Ravel, T., Devriendt, K., Bongers, E. M., de Leeuw, N., Reardon, W., Gimelli, S., Bena, F., Hennekam, R. C., Male, A., Gaunt, L., Clayton-Smith, J., Simonic, I., Park, S. M., Mehta, S. G., Nik-Zainal, S., Woods, C. G., Firth, H. V., Parkin, G., Fichera, M., Reitano, S., Lo Giudice, M., Li, K. E., Casuga, I., Broomer, A., Conrad, B., Schwerzmann, M., Räber, L., Gallati, S., Striano, P., Coppola, A., Tolmie, J. L., Tobias, E. S., Lilley, C., Armengol, L., Spysschaert, Y., Verloo, P., De Coene, A., Goossens, L., Mortier, G., Speleman, F., van Binsbergen, E., Nelen, M. R., Hochstenbach, R., Poot, M., Gallagher, L., Gill, M., McClellan, J., King, M. C., Regan, R., Skinner, C., Stevenson, R. E., Antonarakis, S. E., Chen, C., Estivill, X., Menten, B., Gimelli, G., Gribble, S., Schwartz, S., Sutcliffe, J. S., Walsh, T., Knight, S. J., Sebat, J., Romano, C., Schwartz, C. E., Veltman, J. A., de Vries, B. B., Vermeesch, J. R., Barber, J. C., Willatt, L., Tassabehji, M., and Eichler, E. E. (2008) Recurrent rearrangements of chromosome 1q21.1 and variable pediatric phenotypes. *N. Engl. J. Med. 359*, 1685–1699.
8. Mefford, H. C., and Eichler, E. E. (2009) Duplication Hotspots, Rare Genomic Disorders, and Common Disease. *Curr. Opin. Genet. Dev. 19*, 196–204.
9. Sharp, A. J., Mefford, H. C., Li, K., Baker, C., Skinner, C., Stevenson, R. E., Schroer, R. J., Novara, F., De Gregori, M., Ciccone, R., Broomer, A., Casuga, I., Wang, Y., Xiao, C., Barbacioru, C., Gimelli, G., Bernardina, B. D., Torniero, C., Giorda, R., Regan, R., Murday, V., Mansour, S., Fichera, M., Castiglia, L., Failla, P., Ventura, M., Jiang, Z., Cooper, G. M., Knight, S. J., Romano, C., Zuffardi, O., Chen, C., Schwartz, C. E., and Eichler, E. E. (2008) A recurrent 15q13.3 microdeletion syndrome associated with mental retardation and seizures. *Nat. Genet. 40*, 322–328.
10. Camps, J., Grade, M., Nguyen, Q. T., Hörmann, P., Becker, S., Hummon, A. B., Rodriguez, V., Chandrasekharappa, S., Chen, Y., Difilippantonio, M. J., Becker, H., Ghadimi, B. M., and Ried, T (2008) Chromosomal breakpoints in primary colon cancer cluster at sites of structural variants in the genome. *Cancer Res. 68*, 1284–1295.
11. Lupski, J. R. (2007) Genomic Rearrangements and Sporadic Disease. *Nat. Genet. 39*, S43–47.
12. Pinkel, D., Segraves, R., Sudar, D., Clark, S., Poole, I., Kowbel, D., Collins, C., Kuo, W. L., Chen, C., Zhai, Y., Dairkee, S. H., Ljung, B. M., Gray, J. W., and Albertson, D. G. (1998) High Resolution Analysis of DNA Copy Number Variation using Comparative Genomic Hybridization to Microarrays. *Nat. Genet. 20*, 207–211.
13. Iafrate, A. J., Feuk, L., Rivera, M. N., Listewnik, M. L., Donahoe, P. K., Qi, Y., Scherer, S. W., and Lee, C. (2004) Detection of Large-Scale Variation in the Human Genome. *Nat. Genet. 36*, 949–951.
14. Sharp, A. J., Locke, D. P., McGrath, S. D., Cheng, Z., Bailey, J. A., Vallente, R. U., Pertz, L. M., Clark, R. A., Schwartz, S., Segraves, R., Oseroff, V. V., Albertson, D. G., Pinkel, D., and Eichler, E. E. (2005) Segmental Duplications and Copy-Number Variation in the Human Genome. *Am. J. Hum. Genet. 77*, 78–88.
15. Conrad, D. F., Pinto, D., Redon, R., Feuk, L., Gokcumen, O., Zhang, Y., Aerts, J., Andrews, T. D., Barnes, C., Campbell, P., Fitzgerald, T., Hu, M., Ihm, C. H., Kristiansson, K., Macarthur, D. G., Macdonald, J. R., Onyiah, I., Pang, A. W., Robson, S., Stirrups, K., Valsesia, A., Walter, K., and Wei, J. Wellcome Trust Case Control Consortium, Tyler-Smith C, Carter NP, Lee C, Scherer SW, Hurles ME. (2010) Origins and functional impact of copy number variation in the human genome. *Nature. 464*, 704–712.
16. Hollox, E. J., Barber, J. C. K., Brookes, A. J., and Armour, J. A. L. (2008) Defensins and the Dynamic Genome: What we can Learn from Structural Variation at Human Chromosome Band 8p23. 1. *Genome Res. 18*, 1686–1697.
17. Sugawara, H., Harada, N., Ida, T., Ishida, T., Ledbetter, D. H., Yoshiura, K., Ohta, T., Kishino, T., Niikawa, N., and Matsumoto, N. (2003) Complex Low-Copy Repeats Associated with a Common Polymorphic Inversion at Human Chromosome 8p23. *Genomics. 82*, 238–244.
18. Giglio, S., Broman, K. W., Matsumoto, N., Calvari, V., Gimelli, G., Neumann, T., Ohashi, H., Voullaire, L., Larizza, D., Giorda, R., Weber, J. L., Ledbetter, D. H., and Zuffardi, O. (2001) Olfactory receptor-gene clusters, genomic-inversion polymorphisms, and common chromosome rearrangements. *Am J Hum Genet. 68*, 874–883.
19. Broman, K. W., Matsumoto, N., Giglio, S., Martin, C. L., Roseberry, J. A., Zuffardi, O., Ledbetter, D. H., and Weber, J. L. (2003) Common Long Human Inversion Polymorphism on Chromosome 8p. *Statistics and Science: A Festschrift for Terry Speed.*

20. Hollox, E. J., Armour, J. A., and Barber, J. C. (2003) Extensive Normal Copy Number Variation of a Beta-Defensin Antimicrobial-Gene Cluster. *Am. J. Hum. Genet.* *73*, 853–858.
21. Abu Bakar, S., Hollox, E. J., and Armour, J. A. L. (2009) Allelic Recombination between Distinct Genomic Locations Generates Copy Number Diversity in Human β-Defensins. *Proceedings of the National Academy of Sciences.* *106*, 853–858.
22. Linardopoulou, E. V., Williams, E. M., Fan, Y., Friedman, C., Young, J. M., and Trask, B. J. (2005) Human Subtelomeres are Hot Spots of Interchromosomal Recombination and Segmental Duplication. *Nature.* *437*, 94–100.
23. Linardopoulou, E. V., Parghi, S. S., Friedman, C., Osborn, G. E., Parkhurst, S. M., and Trask, B. J. (2007) Human Subtelomeric WASH Genes Encode a New Subclass of the WASP Family. *PLoS Genet.* *3*, e237.
24. Liu, R., Abreu-Blanco, M. T., Barry, K. C., Linardopoulou, E. V., Osborn, G. E., and Parkhurst, S. M. (2009) Wash Functions Downstream of Rho and Links Linear and Branched Actin Nucleation Factors. *Development.* *136*, 2849–2860.
25. Barber, J. C., Maloney, V. K., Huang, S., Bunyan, D. J., Cresswell, L., Kinning, E., Benson, A., Cheetham, T., Wyllie, J., Lynch, S. A., Zwolinski, S., Prescott, L., Crow, Y., Morgan, R., and Hobson, E. (2008) 8p23.1 Duplication Syndrome; a Novel Genomic Condition with Unexpected Complexity Revealed by Array CGH. *Eur. J. Hum. Genet.* *16*, 18–27.
26. Flint, J., and Knight, S. (2003) The use of Telomere Probes to Investigate Submicroscopic Rearrangements Associated with Mental Retardation. *Curr. Opin. Genet. Dev.* *13*, 310–316.
27. Horton, R., Gibson, R., Coggill, P., Miretti, M., Allcock, R. J., Almeida, J., Forbes, S., Gilbert, J. G., Halls, K., Harrow, J. L., Hart, E., Howe, K., Jackson, D. K., Palmer, S., Roberts, A. N., Sims, S., Stewart, C. A., Traherne, J. A., Trevanion, S., Wilming, L., Rogers, J., de Jong, P. J., Elliott, J. F., Sawcer, S., Todd, J. A., Trowsdale, J., and Beck, S. (2008) Variation analysis and gene annotation of eight MHC haplotypes: the MHC Haplotype Project. *Immunogenetics.* *60*, 1–18.
28. Stefansson, H., Helgason, A., Thorleifsson, G., Steinthorsdottir, V., Masson, G., Barnard, J., Baker, A., Jonasdottir, A., Ingason, A., Gudnadottir, V. G., Desnica, N., Hicks, A., Gylfason, A., Gudbjartsson, D. F., Jonsdottir, G. M., Sainz, J., Agnarsson, K., Birgisdottir, B., Ghosh, S., Olafsdottir, A., Cazier, J. B., Kristjansson, K., Frigge, M. L., Thorgeirsson, T. E., Gulcher, J. R., Kong, A., and Stefansson, K. (2005) A common inversion under selection in Europeans. *Nat. Genet.* *37*, 129–137.
29. Zody, M. C., Jiang, Z., Fung, H. C., Antonacci, F., Hillier, L. W., Cardone, M. F., Graves, T. A., Kidd, J. M., Cheng, Z., Abouelleil, A., Chen, L., Wallis, J., Glasscock, J., Wilson, R. K., Reily, A. D., Duckworth, J., Ventura, M., Hardy, J., Warren, W. C., and Eichler, E. E. (2008) Evolutionary toggling of the MAPT 17q21.31 inversion region. *Nat. Genet.* *40*, 1076–1083.
30. Bailey, J. A., Gu, Z., Clark, R. A., Reinert, K., Samonte, R. V., Schwartz, S., Adams, M. D., Myers, E. W., Li, P. W., and Eichler, E. E. (2002) Recent Segmental Duplications in the Human Genome. *Science.* *297*, 1003–1007.
31. Yoon, S., Xuan, Z., Makarov, V., Ye, K., and Sebat, J. (2009) Sensitive and Accurate Detection of Copy Number Variants using Read Depth of Coverage. *Genome Res.* *19*, 1586–1592.
32. Alkan, C., Kidd, J. M., Marques-Bonet, T., Aksay, G., Antonacci, F., Hormozdiari, F., Kitzman, J. O., Baker, C., Malig, M., and Mutlu, O. (2009) Personalized copy number and segmental duplication maps using next-generation sequencing. *Nat. Genet.* *41*, 1061–1067.
33. Eichler, E. E., Clark, R. A., and She, X. (2004) An Assessment of the Sequence Gaps: Unfinished Business in a Finished Human Genome. *Nature Reviews Genetics.* *5*, 345–354.
34. Codina-Pascual, M., Navarro, J., Oliver-Bonet, M., Kraus, J., Speicher, M., Arango, O., Egozcue, J., and Benet, J. (2006) Behaviour of Human Heterochromatic Regions during the Synapsis of Homologous Chromosomes. *Human Reproduction.* *21*, 1490–1497.
35. Craig-Holmes, A., Moore, F., and Shaw, M. (1973) Polymorphism of Human C-Band Heterochromatin. I. Frequency of Variants. *Am. J. Hum. Genet.* *25*, 181–192.
36. Craig-Holmes, A., Moore, F., and Shaw, M. (1975) Polymporphism of Human C-Band Heterochromatin. II. Family Studies with Suggestive Evidence for Somatic Crossing Over. *Am. J. Hum. Genet.* *27*, 178–189.
37. de la Chapelle, A., Schröder, J., Stenstrand, K., Fellman, J., Herva, R., Saarni, M., Anttolainen, I., Tallila, I., Tervilä, L., Husa, L., Tallqvist, G., Robson, E. B., Cook, P. J., and Sanger, R. (1974) Pericentric inversions of human chromosomes 9 and 10. *Am. J. Hum. Genet.* *26*(6), 746–766.
38. Linardopoulou, E. V., Parghi, S. S., Friedman, C., Osborn, G. E., Parkhurst, S. M., and Trask,

B. J. (2007) Human Subtelomeric WASH Genes Encode a New Subclass of the WASP Family. *PLoS Genet. 3*, e237.

39. Giglio, S., Calvari, V., Gregato, G., Gimelli, G., Camanini, S., Giorda, R., Ragusa, A., Guerneri, S., Selicorni, A., Stumm, M., Tonnies, H., Ventura, M., Zollino, M., Neri, G., Barber, J., Wieczorek, D., Rocchi, M., and Zuffardi, O. (2002) Heterozygous Submicroscopic Inversions Involving Olfactory Receptor-Gene Clusters Mediate the Recurrent t(4;8)(p16;p23) Translocation. *Am. J. Hum. Genet. 71*, 276–285.
40. Hoffmann, A. A., and Rieseberg, L. H. (2008) Revisiting the Impact of Inversions in Evolution: From Population Genetic Markers to Drivers of Adaptive Shifts and Speciation? *Annual review of ecology, evolution, and systematics. 39*, 21–42.
41. Hastings, P., Lupski, J. R., Rosenberg, S. M., and Ira, G. (2009) Mechanisms of Change in Gene Copy Number. *Nature Reviews Genetics 10*, 551–564.
42. Lander, E. S., and Green, P. (1987) Construction of Multilocus Genetic Linkage Maps in Humans. *Proc. Natl. Acad. Sci. USA. 84*, 2363–2367.
43. Kim, S. H., Ma, X., Weremowicz, S., Ercolino, T., Powers, C., Mlynarski, W., Bashan, K. A., Warram, J. H., Mychaleckyj, J., and Rich, S. S. (2004) Identification of a Locus for Maturity-Onset Diabetes of the Young on Chromosome 8p23. *Diabetes. 53*, 1375–1384.
44. Khaja, R., Zhang, J., MacDonald, J. R., He, Y., Joseph-George, A. M., Wei, J., Rafiq, M. A., Qian, C., Shago, M., Pantano, L., Aburatani, H., Jones, K., Redon, R., Hurles, M., Armengol, L., Estivill, X., Mural, R. J., Lee, C., Scherer, S. W., and Feuk, L. (2006) Genome Assembly Comparison Identifies Structural Variants in the Human Genome. *Nat. Genet. 38*, 1413–1418.
45. Tuzun, E., Saini, S. S., Yang, H., Alagappan, D., Higgs, S., and Christadoss, P. (2006) Genetic Evidence for the Involvement of Fcgamma Receptor III in Experimental Autoimmune Myasthenia Gravis Pathogenesis. *J. Neuroimmunol. 174*, 157–167.
46. Kidd, J. M., Cooper, G. M., Donahue, W. F., Hayden, H. S., Sampas, N., Graves, T., Hansen, N., Teague, B., Alkan, C., Antonacci, F., Haugen, E., Zerr, T., Yamada, N. A., Tsang, P., Newman, T. L., Tuzun, E., Cheng, Z., Ebling, H. M., Tusneem, N., David, R., Gillett, W., Phelps, K. A., Weaver, M., Saranga, D., Brand, A., Tao, W., Gustafson, E., McKernan, K., Chen, L., Malig, M., Smith, J. D., Korn, J. M., McCarroll, S. A., Altshuler, D. A., Peiffer, D. A., Dorschner, M., Stamatoyannopoulos, J., Schwartz, D., Nickerson, D. A., Mullikin, J. C., Wilson, R. K., Bruhn, L., Olson, M. V., Kaul, R., Smith, D. R., and Eichler, E. E. (2008) Mapping and Sequencing of Structural Variation from Eight Human Genomes. *Nature. 453*, 56–64.
47. Korbel, J. O., Urban, A. E., Affourtit, J. P., Godwin, B., Grubert, F., Simons, J. F., Kim, P. M., Palejev, D., Carriero, N. J., Du, L., Taillon, B. E., Chen, Z., Tanzer, A., Saunders, A. C., Chi, J., Yang, F., Carter, N. P., Hurles, M. E., Weissman, S. M., Harkins, T. T., Gerstein, M. B., Egholm, M., and Snyder, M. (2007) Paired-End Mapping Reveals Extensive Structural Variation in the Human Genome. *Science. 318*, 420–426.
48. Bashir, A., Volik, S., Collins, C., Bafna, V., and Raphael, B. J. (2008) Evaluation of Paired-End Sequencing Strategies for Detection of Genome Rearrangements in Cancer. *PLoS Computational Biology. 4.*
49. Walter, M. A., Spillett, D. J., Thomas, P., Weissenbach, J., and Goodfellow, P. N. (1994) A Method for Constructing Radiation Hybrid Maps of Whole Genomes. *Nat. Genet. 7*, 22–28.
50. Dear, P. H., and Cook, P. R. (1993) Happy Mapping: Linkage Mapping using a Physical Analogue of Meiosis. *Nucleic Acids Res. 21*, 13–20.
51. Piper, M. B., Bankier, A. T., and Dear, P. H. (1998) A HAPPY Map of Cryptosporidium Parvum. *Genome Res. 8*, 1299–1307.
52. Dear, P. H., Bankier, A. T., and Piper, M. B. (1998) A High-Resolution Metric HAPPY Map of Human Chromosome 14. *Genomics. 48*, 232–241.
53. Eichinger, L., Pachebat, J. A., Glöckner, G., Rajandream, M. A., Sucgang, R., Berriman, M., Song, J., Olsen, R., Szafranski, K., Xu, Q., Tunggal, B., Kummerfeld, S., Madera, M., Konfortov, B. A., Rivero, F., Bankier, A. T., Lehmann, R., Hamlin, N., Davies, R., Gaudet, P., Fey, P., Pilcher, K., Chen, G., Saunders, D., Sodergren, E., Davis, P., Kerhornou, A., Nie, X., Hall, N., Anjard, C., Hemphill, L., Bason, N., Farbrother, P., Desany, B., Just, E., Morio, T., Rost, R., Churcher, C., Cooper, J., Haydock, S., van Driessche, N., Cronin, A., Goodhead, I., Muzny, D., Mourier, T., Pain, A., Lu, M., Harper, D., Lindsay, R., Hauser, H., James, K., Quiles, M., Madan Babu, M., Saito, T., Buchrieser, C., Wardroper, A., Felder, M., Thangavelu, M., Johnson, D., Knights, A., Loulseged, H., Mungall, K., Oliver, K., Price, C., Quail ,M. A., Urushihara, H., Hernandez, J., Rabbinowitsch, E., Steffen, D., Sanders, M., Ma, J., Kohara, Y., Sharp, S., Simmonds, M., Spiegler, S., Tivey, A., Sugano, S., White, B., Walker, D., Woodward, J., Winckler, T., Tanaka, Y., Shaulsky, G., Schleicher, M.,

Weinstock, G., Rosenthal, A., Cox, E. C., Chisholm, R. L., Gibbs, R., Loomis, W. F., Platzer, M., Kay, R. R., Williams, J., Dear, P. H., Noegel, A. A., Barrell, B., and Kuspa, A. (2005) *Nature. 435*, 43–57.

54. Mardis, E. R. (2008) Next-Generation DNA Sequencing Methods. *Annu Rev Genomics Hum Genet. 9*, 387–402.

55. Hollox, E. J., Detering, J. C., Dehnugara, T. (2009) An integrated approach for measuring copy number variation at the FCGR3 (CD16) locus. Hum Mutat. *30*, 477–484.

56. Fanciulli, M., Vyse, T., and Aitman, T. (2008) Copy Number Variation of Fc Gamma Receptor Genes and Disease Predisposition. *Cytogenet Genome Res. 123*, 161–168.

57. Alkan C, Kidd JM, Marques-Bonet T, Aksay G, Antonacci F, Hormozdiari F, Kitzman JO, Baker C, Malig M, Mutlu O. (2009) Personalized Copy Number and Segmental Duplication Maps using Next-Generation Sequencing. *Nat. Genet. 41*, 1061–1067.

58. Daser A, Thangavelu M, Pannell R, Forster A, Sparrow L, Chung G, Dear P H, Rabbitts TH. (2006) Interrogation of Genomes by Molecular Copy-Number Counting (MCC). *Nat. Methods. 3*, 447–453.

59. Kent, W. J., Sugnet, C. W., Furey, T. S., Roskin, K. M., Pringle, T. H., and Zahler, A. M. (2002) The Human Genome Browser at UCSC. *Genome Res. 12*, 996–1006.

60. Bentley, D. R., Balasubramanian, S., Swerdlow, H. P., Smith, G. P., Milton, J., Brown, C. G., Hall, K. P., Evers, D. J., Barnes, C. L., Bignell, H. R., Boutell, J. M., Bryant, J., Carter, R. J., Keira Cheetham, R., Cox, A. J., Ellis, D. J., Flatbush, M. R., Gormley, N. A., Humphray, S. J., Irving, L. J., Karbelashvili, M. S., Kirk, S. M., Li, H., Liu, X., Maisinger, K. S., Murray, L. J., Obradovic, B., Ost, T., Parkinson, M. L., Pratt, M. R., Rasolonjatovo, I. M., Reed, M. T., Rigatti, R., Rodighiero, C., Ross, M. T., Sabot, A., Sankar, S. V., Scally, A., Schroth, G. P., Smith, M. E., Smith, V. P., Spiridou, A., Torrance, P. E., Tzonev, S. S., Vermaas, E. H., Walter, K., Wu, X., Zhang, L., Alam, M. D., Anastasi, C., Aniebo, I. C., Bailey, D. M., Bancarz, I. R., Banerjee, S., Barbour, S. G., Baybayan, P. A., Benoit, V. A., Benson, K. F., Bevis, C., Black, P. J., Boodhun, A., Brennan, J. S., Bridgham, J. A., Brown, R. C., Brown, A. A., Buermann, D. H., Bundu, A. A., Burrows, J. C., Carter, N. P., Castillo, N., Chiara, E., Catenazzi, M., Chang, S., Neil Cooley, R., Crake, N. R., Dada, O. O., Diakoumakos, K. D., Dominguez-Fernandez, B., Earnshaw, D. J., Egbujor, U. C., Elmore, D. W., Etchin, S. S., Ewan, M. R., Fedurco, M., Fraser, L. J., Fuentes Fajardo, K. V., Scott Furey, W., George, D., Gietzen, K. J., Goddard, C. P., Golda, G. S., Granieri, P. A., Green, D. E., Gustafson, D. L., Hansen, N. F., Harnish, K., Haudenschild, C. D., Heyer, N. I., Hims, M. M., Ho, J. T., Horgan, A. M., Hoschler, K., Hurwitz, S., Ivanov, D. V., Johnson, M. Q., James, T., Huw Jones, T. A., Kang, G. D., Kerelska, T. H., Kersey, A. D., Khrebtukova, I., Kindwall, A. P., Kingsbury, Z., Kokko-Gonzales, P. I., Kumar, A., Laurent, M. A., Lawley, C. T., Lee, S. E., Lee, X., Liao, A. K., Loch, J. A., Lok, M., Luo, S., Mammen, R. M., Martin, J. W., McCauley, P. G., McNitt, P., Mehta, P., Moon, K. W., Mullens, J. W., Newington, T., Ning, Z., Ling Ng, B., Novo, S. M., O'Neill, M. J., Osborne, M. A., Osnowski, A., Ostadan, O., Paraschos, L. L., Pickering, L., Pike, A. C., Pike, A. C., Chris Pinkard, D., Pliskin, D. P., Podhasky, J., Quijano, V. J., Raczy, C., Rae, V. H., Rawlings, S. R., Chiva Rodriguez, A., Roe, P. M., Rogers, J., Rogert Bacigalupo, M. C., Romanov, N., Romieu, A., Roth, R. K., Rourke, N. J., Ruediger, S. T., Rusman, E., Sanches-Kuiper, R. M., Schenker, M. R., Seoane, J. M., Shaw, R. J., Shiver, M. K., Short, S. W., Sizto, N. L., Sluis, J. P., Smith, M. A., Ernest Sohna Sohna, J., Spence, E. J., Stevens, K., Sutton, N., Szajkowski, L., Tregidgo, C. L., Turcatti, G., Vandevondele, S., Verhovsky, Y., Virk, S. M., Wakelin, S., Walcott, G. C., Wang, J., Worsley, G. J., Yan, J., Yau, L., Zuerlein, M., Rogers, J., Mullikin, J. C., Hurles, M. E., McCooke, N. J., West, J. S., Oaks, F. L., Lundberg, P. L., Klenerman, D., Durbin, R., and Smith, A. J. (2008) Accurate whole human genome sequencing using reversible terminator chemistry. *Nature. 456*(7218), 53–59.

61. Wheeler, D. A., Srinivasan, M., Egholm, M., Shen, Y., Chen, L., McGuire, A., He, W., Chen, Y. J., Makhijani, V., Roth, G. T., Gomes, X., Tartaro, K., Niazi, F., Turcotte, C. L., Irzyk, G. P., Lupski, J. R., Chinault, C., Song, X. Z., Liu, Y., Yuan, Y., Nazareth, L., Qin, X., Muzny, D. M., Margulies, M., Weinstock, G. M., Gibbs, R. A., and Rothberg, J. M. (2008) The complete genome of an individual by massively parallel DNA sequencing. *Nature. 452*, 872–876.

Chapter 10

Population Genetic Nature of Copy Number Variation

Per Sjödin and Mattias Jakobsson

Abstract

Copy number variation has recently received considerable attention, and copy number variants (CNVs) have been shown to be both common in mammalian genomes and important for understanding genetic and phenotypic variation. As empirical knowledge and detection methods are quickly advancing, evolutionary theories about CNVs are rapidly updated and often revised. Here, we review recent progress on understanding CNVs, and we discuss some key issues for future research. In essence, we discuss four major forces in population genetics, recombination, mutation, selection, and demography, in relation to CNVs.

Key words: Copy number variation, Recombination, Mutation, Selection, Demography

1. Introduction

Large numbers of duplicated gene regions have been identified in genome sequences of several different organisms (1–5) and large numbers of differences have been detected in gene content between related species (6–10). This observation indicates that beneficial copy number variants (CNVs, see Table 1 for definitions) sometimes arise and increase in frequency, potentially via positive selection (12–17). However, CNVs also have the potential to disrupt genes, which suggests that many CNVs may be deleterious, preventing them to rise to high frequency (18–21). Recent studies have found that CNVs are polymorphic within the human population and that CNVs are widespread in the human genome (22–25). CNVs have also been associated with genetic diseases (24, 26–31) suggesting that CNVs may be causal for some common human diseases. Furthermore, it has been postulated that CNVs may represent a major genetic component underlying phenotypic variation (29), in addition to being a source of genetic variation among individuals and among human groups of different ethnic origin.

Lars Feuk (ed.), *Genomic Structural Variants: Methods and Protocols*, Methods in Molecular Biology, vol. 838,
DOI 10.1007/978-1-61779-507-7_10, © Springer Science+Business Media, LLC 2012

Table 1
Definitions adapted from Scherer et al. (11)

Structural variant	"... the umbrella term to encompass a group of genomic alterations involving segments of DNA typically larger than 1 kb..." "... may be quantitative (copy number variants comprising deletions, insertions, and duplications) and/or positional (translocations) or orientational (inversions)"
SD	"A segment of DNA >1 kb in size that occurs in two or more copies per haploid genome, with the different copies sharing >90% sequence identity"
Indel	"... collective abbreviation to describe relative gain or loss of a segment of one or more nucleotides ..."
CNV	"... at least 1 kb in size ..." "... genomic copy number gains (insertions or duplications) or losses (deletions or null genotypes) relative to a designated reference genome sequence"
CNP	"... a CNV that occurs in more than 1% of the population"
CNVR or a CNV locus	A CNV but also used to refer to "... a multiplex arrangement of variant units in close proximity, forming a CNV region"

Since CNVs are likely to be functionally important, they are also likely to be of evolutionary importance. In order to understand evolutionary changes, or the contribution of genetic variation to phenotypic variation, we need to consider CNVs in the light of population genetics. Here, we focus on four major population genetic forces: recombination, mutation, selection, and demography. It is well-known that these forces are not mutually independent. As an example, consider a new mutation in a particular region of a genome. This mutation would, according to some empirical studies, lower the probability of recombination occurring in heterozygote individuals. The recombination rate in the region decreases and selection is less efficient in this part of the genome due to the Hill–Robertson effect. These correlations are probably quite weak for mutations involving single nucleotides, but might be much stronger for mutations involving large stretches of DNA, such as CNVs.

2. Recombination

Since recombination events occur in pairs of chromosomes, it is particularly important to understand how a CNV that is heterozygote in an individual affects recombination. A heterozygote CNV will create a considerable length difference between the chromosomes, which in itself may have a large impact on the recombination process. Large sequence differences inhibit the homologous recombination process (32, 33) and it is therefore not surprising that homologous recombination is inhibited by a heterozygous length difference (34–38). A specific case of homologous recombination, nonallelic homologous recombination (NAHR), may however be *more* frequent in heterozygote individuals than in homozygote individuals (39). A likely explanation for this effect is that the unpaired DNA loop in heterozygote individuals is free to pair with nonallelic loci during meiosis.

The presence of a CNV may also affect the relative probability that a recombination event results in a crossing-over event or a gene conversion event. In fact, a recent study (40) suggests that gene conversion may be less inhibited by heterozygous sequence differences than recombination resulting in crossing-over events. If this result extrapolates to heterozygous length differences, the effect of gene conversion on CNVs may be appreciable. In particular, similar to how the bias for G/C nucleotides in gene conversion tracts mimic positive selection for increased GC content (for a recent review, see ref. 41), a bias to retain either the short or the long allele would mimic selection for a change in genome size. The effect would be double: if the short allele is preferred, changes that make the genome smaller (deletions) would be promoted, while changes that enlarge the genome (insertions) would be disfavored (and vice versa if the bias is for the long allele). However, the support for gene conversion favoring the long, or short, allele in a heterozygote has not been unanimous: an early study showed a preference for the long allele (42), whereas a later study implied a more complicated scenario where length and structure of the indel also matter (43).

3. Mutation Process

Based on a rough estimate that each CNV affects at most 0.008% of the genome (1,447 CNVRs covering 12% of the human genome (24)), the infinite sites model (where every mutation hits a new place in the genome) could be a reasonable mutation model for CNVs. However, CNVs are frequently clustered into hotspots with high mutation rates and these CNV hotspots can potentially be hit by recurrent mutations (44, 45). As an illustration, by comparing CNVs in chimpanzees to known CNVs in humans, it was found that only 24 out of 355 CNVs (6.8%) were specific to chimpanzees (46, 47). In other words, 93.2% of CNVs found in chimpanzee were also polymorphic in humans—a much higher fraction of shared polymorphisms than expected between these species. This observation is best explained by recurrent mutations hitting specific regions in the primate genome, i.e., CNV hotspots. If CNVs are characterized by being confined to hypermutable hotspots, population genetic modeling of CNVs will need to incorporate these properties. For instance, it may lead to an inflation of derived allele frequencies, which in turn can cause false inference of positive selection. In general, both the infinite allele model and infinite site model may be inappropriate for CNVs that may be better modeled by finite allele models such as the K-allele model (48) or the stepwise mutation model (49).

Several studies have found a striking enrichment of segmental duplications (SDs, Table 1) in CNV regions (24). The same genomic region may sometimes be identified as both a SD and a CNV, but because these two types of structural variation are commonly detected using different technologies and methods, it is often difficult to sort out their relationship, and the apparent enrichment of SDs in CNVs can be an artifact (24, 50). However, CNVs are also overrepresented in older SDs (51), indicating a more causal role for SDs in creating CNVs. It has been suggested that SDs predispose to NAHR events (24, 45, 47). NAHR events typically generate novel structural variation, and this would account for both CNV hotspots and the commonly observed genomic overlap between CNVs and SDs. Although SDs and CNVs are closely related concepts—SDs are basically fixed CNVs—SDs have been a major research focus in their own right and are known to be associated with interspecific and recurrent synteny breaks (see ref. 52 for a recent review). Interestingly, the chromosomal distribution of SDs in primates (including humans) seems to be the outlier among mammals: while SDs are typically organized in tandem (<1 Mb apart) in all investigated mammals (dog, cow, and mouse), the distance between SD copies in primates is much longer and more varied (53). The burst of Alu activity in the primate ancestor some 40 million years ago (54) is a strong candidate explanation as the genomic distribution of old SDs is strongly correlated with Alus while young SDs and CNVs are not (55) implying that SD/CNV formation is a dynamic process that changes over time.

If a distinction is made between the CNVs that overlap with SDs and those that do not, two distinct classes of CNVs emerge. The CNVs that do not overlap SDs typically do not belong to the copy number polymorphism (CNP) class (i.e., they segregate at frequencies below 1%, see Table 1), and they seem to be much less affected by recurrent mutations than CNVs that overlap SDs (Table 2). It appears that one class represents CNVs in hotspots, while the other class represents CNVs with low frequencies that behave like normal biallelic markers (51). As mentioned above, an individual that is heterozygote for a CNV may have an increased NAHR rate (39). This connection could provide a unifying mechanism for the two classes of CNVs since a CNV destined to become fixed, thereby creating a SD, is expected to be in a heterozygote state half of the time until fixation (56). As a consequence, a positive feedback loop will be initiated since the increased rate of NAHR will lead to an increase in the mutation rate for novel structural variation, which in turn acts as a stimulant for the NAHR

Table 2
Contrast between CNVs that do not overlap SDs and those that do

	CNVs outside SDs	CNVs overlapping SDs	References
Population frequency	–	+	(51)
Length	–	+	(21)
Gene-rich regions	(+)	+	(51)
Gene-poor regions	+	–	(51)
Synteny breaks	–	+	(21)
dN/dS	–	+	(21)
Environmental response genes	(+)	+	(51)
Disease genes	+	–	(21)
Alu	–	+	(55)

rate. A prediction, admittedly not easily tested, is that most hotspot CNVs should be initiated by the fixation of a duplication—not a deletion—since duplications necessarily provide the substrate for NAHR events which deletions do not.

The relative rate of NAHR-mediated duplications and deletions is also an important factor for understanding these hotspots. Relevant empirical data is however scarce as most studies report the number of duplications and deletions observed in the genome, which depends on both mutation rate and selection (see below). Although limited to male-specific NAHR and to a few CNV hotspots, Turner et al. (45) studied the de novo mutation rate directly and found that NAHR more often generated deletions than insertions.

Besides NAHR, nonhomologous end joining (NHEJ) has been suggested as a recombination-based mechanism for CNV formation. However, while recurrent CNV events are strongly associated with NAHR, CNVs due to NHEJ are rarely recurrent (50). Consequently, while the de novo mutation rate of CNVs due to NAHR has been estimated to be almost on the same order of magnitude as for microsatellites, CNV formations due to NHEJ probably have a much lower mutation rate similar to the rate estimated for SNPs (50). Retrotransposition is another major mechanism for CNV formation but in contrast to NAHR and NHEJ, it does not give rise to deletions, only to insertions of transcribed sequence. Finally, a novel replication-based mechanism, fork stalling and template switching (FoSTeS), has been proposed to account for CNVs that are difficult to explain by NAHR, NHEJ, or retrotranspositions. Together, these four mechanisms are believed to be responsible for the majority of CNVs (see ref. 31 for a review).

4. Selection

While the detrimental effect of an indel is likely to increase with its length (57), there is also accumulating evidence that deletions are more deleterious than insertions (58–62). For instance, CNV duplications are more common among CNVs with high frequency and CNV deletions are more common in CNVs with low frequency (19). Furthermore, a significantly lower proportion of CNV deletions than CNV duplications overlap with genes (24). This difference in selective constraint between deletions and insertions may be explained by the fact that deletions have two cut-points while insertions only have one: the probability that an insertion disrupts an important sequence motif does not depend on the length of the insertion, but the probability of disruption increases with the length of a deletion (58).

Genes involved in environmental response have been found to be overrepresented in CNVs in several organisms, including humans (23, 24, 63), mice (5), dogs (64), cows (53), and possibly also in fruit flies (16). A more detailed picture emerges when we separate CNVs that overlap SDs and those that do not. CNVs that do not overlap SDs show no (21), or considerably weaker (51), enrichment of environmental genes. These CNVs are instead overrepresented in gene-poor regions in contrast to CNVs that overlap SDs (51). This trend has been explained by positive selection for CNV changes in genes involved in environmental response (65, 66). Indeed, many of the differences listed in Table 10.2 between CNVs outside of SDs and those that overlap SDs are potentially explained by positive selection for CNVs/SDs in regions with a high density of genes involved in environmental response. However, some of these differences, such as the difference between frequency and length, could also be explained if CNVs that overlap SDs are more affected by recurrent events (this would inflate the frequency and also the detected length if several overlapping events are counted as one). Other differences could be secondary effects. CNVs may, for instance, be common in gene-rich regions as an effect of certain sequence motifs (e.g., non-B_DNA forming sequence), which are enriched in gene regions at the same time as they increase the rate of CNV formations (67). More conservative explanations are now being considered for the connection between genes involved in environmental response and CNVs: these genes may be enriched in CNVs not as a result of positive selection, but instead due to relaxation of selective constraint (21, 68).

Positive selection for CNVs is nonetheless likely to have played a significant role in at least some cases and one such example is described in a study by Perry et al. (15). In this study, the authors showed that positive selection, driven by new starch-rich diets, increased the number of copies of the salivary amylase gene (*AMY1*) in some human populations. Finally, many arguments for why gene duplications may be favorable are also relevant for CNV duplications (69).

5. Demography

On the one hand, demographic history has a strong effect on patterns of genetic variation and CNVs are not likely to be an exception. On the other hand, the influence of demography on polymorphic CNVs (>1%, Table 1) is potentially overshadowed if CNVs are typically hypermutable and often affected by recurrent mutations (Table 2). However, comparisons of inferences of human population structure based on microsatellites and SNPs—two types of polymorphisms with very different mutational models and mutation rates—show great similarities between types of polymorphisms (70, 71). Despite an observed skew of CNVs toward rare alleles, potentially caused by purifying selection (19, 72, 73), it has been of interest to assess whether patterns of copy number variation across populations match the corresponding patterns for other types of loci, such as SNPs (24, 73, 74). In the extreme case of strong purifying selection, CNVs would have such low frequencies that they would not be shared even among closely related populations. Alternatively, if CNVs are generally unaffected by selection, they would display similar distributions across populations to the alleles of other types of neutral polymorphic loci. Discordance in inferences of human population structure based on CNVs and those based on SNPs would therefore suggest that most CNVs are sufficiently transient that they have not followed the same pattern of events (e.g., divergences, migrations, and admixture) in human history that typical neutral genetic variants have experienced. In contrast, agreement in CNV-based and SNP-based inferences would suggest that some CNVs are old enough to have migrated out of Africa with human founding populations, and that many other CNVs are old enough to be affected by recent human history of various divergences and migration events.

The greatest amount of genetic variation (based on mtDNA, microsatellites, and SNPs) has consistently been identified within African populations and variation outside of Africa has been shown to be a subset of the African diversity (18, 70, 75, 76). Most genome-wide population level investigations of CNVs have been conducted in the four HapMap populations, Yoruba from West Africa, European Americans, Han Chinese, and Japanese (18, 19,

24, 67, 68, 77–83), or in populations with similar ethnicities (84). Although the results for the HapMap populations are not always consistent, CNVs show patterns of variation similar to SNPs: higher diversity and more unique CNVs in African populations than in the European or East Asian populations (18, 79, 82, 84). Redon et al. (24) also showed that individuals can be correctly assigned to populations based on CNVs, demonstrating that CNV patterns of variation are at least partly shaped by human demographic history. However, some studies of CNVs that investigate samples from multiple non-HapMap populations failed to detect population stratification (72, 85) and conclude that there is limited evidence for population stratification of CNVs in geographically distinct human populations.

The CNV patterns across populations in a different and much larger data set, the HGDP panel of individuals (more than a thousand individuals representing more than fifty populations across the globe (86)) was recently investigated (73, 74, 87). By examining patterns of variation in CNVs and SNPs in the same individuals from the same populations, Jakobsson et al. (73) found that inferences of population structure based on CNVs largely accorded with those based on SNPs, but it was unclear how much ascertainment bias—exclusion of troublesome SNPs (potentially due to being located in a CNV region) from commercial SNP-typing arrays—affect the results by reducing CNV calls in the ascertained populations (73, 88). Examining the same populations, Itsara et al. (87) suggested that variation in genotyping intensity (due to variation in the experimental material) across the genome can produce false CNV calls, which may affect inferences of population stratification, and argue that there is limited evidence for stratification of CNVs in geographically distinct populations. However, in a recent study, Wang et al. (74) reanalyzed the CNV data of Jakobsson et al. (73) after excluding high-variance (in genotyping intensity) individuals from the analysis, and found that the similarity of SNP-based and CNV-based inference of population structure increases, especially when accounting for the smaller number of CNV loci studied compared to the number of SNPs. By using the data from the stringent CNV calls in Wang et al. (74), we compute the percentage of CNVs that are private to geographic regions. After correcting for sample size differences across geographic regions (89), we find a greater percentage of private CNVs in Africa, followed by Eurasia, and similar levels for East Asia, Oceania, and the Americas (Fig. 1). In summary, several studies (including the computation above) support the view that neither the rate of (recurrent) mutations nor the amount of purifying selection against CNVs has been great enough to erase the underlying signature of past human migrations from patterns of copy number variation across populations.

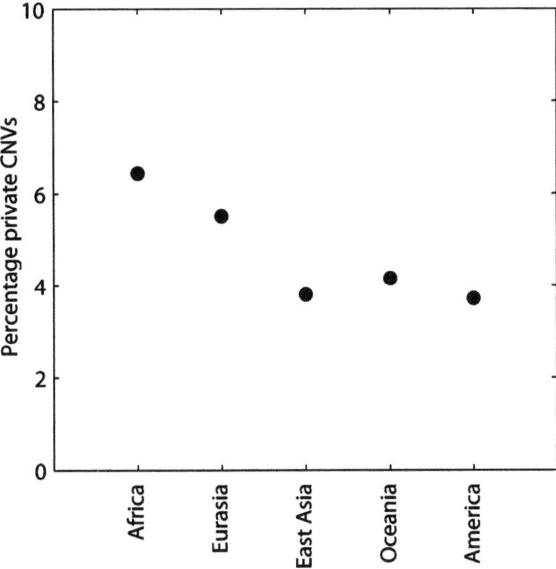

Fig. 1. Percentage of autosomal nonsingleton CNVs that is private to a particular geographic region. The remaining 76.36% of the CNVs are found in more than one region. The percentages correspond to a sample of ten individuals from each region and the CNV calls were performed using the PennCNV algorithm (79), and only including individuals with "standard deviation of the log *R* ratio" less than 0.22 (see ref. 74 for details).

6. Linkage disequilibrium and CNVs

Several studies report a severe lack of LD between CNVs and flanking SNPs (19, 24, 83), implying that searching for CNV-associated diseases may be a waste of effort (unless the disease causing CNV is included as a marker in the study). Redon et al. (24) considered three possible explanations for this lack of LD. First, some CNV duplications might represent transposition events that would generate linkage disequilibrium around the (unknown) acceptor locus but not the donor locus. Second, some CNVs might undergo recurrent mutations or reversions. Third, CNVs might occur preferentially in genomic regions with lower densities of SNP markers. These three hypotheses have all received subsequent support although it is clear that at least part of the explanation is the lack of SNPs in regions associated with CNVs and SDs (88). Redon et al. (24) argued that under their second explanation, CNV duplications should be in lower linkage disequilibrium with flanking SNPs than CNV deletions. Interestingly, although Redon et al. failed to detect a difference in LD between CNV duplications and CNV deletions, two subsequent studies showed that LD is indeed lower for duplications than for deletions (67, 90). In contrast, Schrider and Hahn (90) do not interpret this finding as evidence for the lack of LD being caused by recurrent mutations, but instead propose that because the location of the duplicated copy is (sometimes) unknown, the relevant flanking SNPs are also unknown.

7. Future Directions

Many issues concerning CNVs will be resolved with upcoming advances in sequencing technology. At present, chromosomes are reconstructed from millions of short sequence reads but the upcoming "third generation" sequencing platforms will allow reading large parts of chromosomes in a single read (91, 92). This will resolve the problem of telling individual CNVs and SDs apart as well as determining the exact breakpoints for a CNV. It will also lead to a new perspective where the study of general structural variation will be more straightforward since inversions and translocations will be much easier to characterize. Finally, as small-scale "indels" and larger "structural variation" apparently share many features—such as the excess of de novo deletions compared to insertions, and deletions being more deleterious than insertions—both small-scale and large-scale structural mutations might be governed by the same population genetic properties. As hinted at in Conrad et al. (67), the different categories of structural variation should perhaps be studied as variation of a common theme, and for this purpose the upcoming sequencing technology is particularly promising.

Acknowledgments

We thank M. Lascoux for helpful comment on the manuscript. This work was supported by grants from Carl Trygger's foundation and by the Swedish Research Council Formas.

References

1. Lynch, M., and Conery, J.S. (2000) The evolutionary fate and consequences of duplicated genes *Science* **290**, 1151–5.
2. Rubin, G.M., Yandell, M.D., Wortman, J.R., Gabor Miklos, G.L., Nelson, C.R., Hariharan, I.K., Fortini, M.E., Li, P.W., Apweiler, R., Fleischmann, W., Cherry, J.M., Henikoff, S., Skupski, M.P., Misra, S., Ashburner, M., Birney, E., Boguski, M.S., Brody, T., Brokstein, P., Celniker, S.E., Chervitz, S.A., Coates, D., Cravchik, A., Gabrielian, A., Galle, R.F., Gelbart, W.M., George, R.A., Goldstein, L.S., Gong, F., Guan, P., Harris, N.L., Hay, B.A., Hoskins, R.A., Li, J., Li, Z., Hynes, R.O., Jones, S.J., Kuehl, P.M., Lemaitre, B., Littleton, J.T., Morrison, D.K., Mungall, C., O'Farrell, P.H., Pickeral, O.K., Shue, C., Vosshall, L.B., Zhang, J., Zhao, Q., Zheng, X.H., and Lewis, S. (2000) Comparative genomics of the eukaryotes *Science* **287**, 2204–15.
3. International Human Genome Sequencing Consortium (2001) Initial sequencing and analysis of the human genome *Nature* **409**, 860–921.

4. Zhang, J. (2003) Evolution by gene duplication: an update *Trends Ecol Evol* **18**, 292–8.
5. She, X., Cheng, Z., Zöllner, S., Church, D.M., and Eichler, E.E. (2008) Mouse segmental duplication and copy number variation *Nat Genet* **40**, 909–14.
6. Dujon, B., Sherman, D., Fischer, G., Durrens, P., Casaregola, S., Lafontaine, I., De Montigny, J., Marck, C., Neuvéglise, C., Talla, E., Goffard, N., Frangeul, L., Aigle, M., Anthouard, V., Babour, A., Barbe, V., Barnay, S., Blanchin, S., Beckerich, J.M., Beyne, E., Bleykasten, C., Boisramé, A., Boyer, J., Cattolico, L., Confanioleri, F., De Daruvar, A., Despons, L., Fabre, E., Fairhead, C., Ferry-Dumazet, H., Groppi, A., Hantraye, F., Hennequin, C., Jauniaux, N., Joyet, P., Kachouri, R., Kerrest, A., Koszul, R., Lemaire, M., Lesur, I., Ma, L., Muller, H., Nicaud, J.M., Nikolski, M., Oztas, S., Ozier-Kalogeropoulos, O., Pellenz, S., Potier, S., Richard, G.F., Straub, M.L., Suleau, A., Swennen, D., Tekaia, F., Wésolowski-Louvel, M., Westhof, E., Wirth, B., Zeniou-Meyer, M., Zivanovic, I., Bolotin-Fukuhara, M., Thierry, A., Bouchier, C., Caudron, B., Scarpelli, C., Gaillardin, C., Weissenbach, J., Wincker, P., and Souciet, J.L. (2004) Genome evolution in yeasts *Nature* **430**, 35–44.
7. Parkinson, J., Mitreva, M., Whitton, C., Thomson, M., Daub, J., Martin, J., Schmid, R., Hall, N., Barrell, B., Waterston, R.H., McCarter, J.P., and Blaxter, M.L. (2004) A transcriptomic analysis of the phylum Nematoda *Nat Genet* **36**, 1259–67.
8. Demuth, J.P., De Bie, T., Stajich, J.E., Cristianini, N., and Hahn, M.W. (2006) The evolution of mammalian gene families *PLoS One* **1**, e85.
9. Hahn, M.W., Hanm M.V., and Han, S.G. (2007) Gene family evolution across 12 Drosophila genomes *PLoS Genet* **3**, 2135–46.
10. Opazo, J.C., Hoffmann, F.G., and Storz, J.F. (2008) Differential loss of embryonic globin genes during the radiation of placental mammals *Proc Natl Acad Sci USA* **105**, 12950–5.
11. Scherer, S.W., Lee, C., Birney, E., Altshuler, D.M., Eichler, E.E., Carter, N.P., Hurles, M.E., and Feuk, L. (2007) Challenges and standards in integrating surveys of structural variation *Nat Genet* **39**, S7–15.
12. Johnson, M.E., Viggiano, L., Bailey, J.A., Abdul-Rauf, M., Goodwin, G., Rocchi, M., and Eichler, E.E. (2001) Positive selection of a gene family during the emergence of humans and African apes *Nature* **413**, 514–19.
13. Nguyen, D.Q., Webber, C., and Ponting, C.P. (2006) Bias of selection on human copy-number variants *PLoS Genet* **2**, e20.
14. Heger, A., and Ponting, C. P. (2007) Evolutionary rate analyses of orthologs and paralogs from 12 Drosophila genomes *Genome Res* **17**, 1837–49.
15. Perry, G.H., Dominy, N.J., Claw, K.G., Lee, A.S., Fiegler, H., Redon, R., Werner, J., Villanea, F.A., Mountain, J.L., Misra, R., Carter, N.P., Lee, C., and Stone, A.C. (2007) Diet and the evolution of human amylase gene copy number variation *Nat Genet* **39**, 1256–60.
16. Emerson, J.J., Cardoso-Moreira, M., Borevitz, J.O., and Long, M. (2008) Natural selection shapes genome-wide patterns of copy-number polymorphism in Drosophila melanogaster *Science* **320**, 1629–31.
17. Xue, Y., Sun, D., Daly, A., Yang, F., Zhou, X., Zhao, M., Huang, N., Zerjal, T., Lee, C., Carter, N.P., Hurles, M.E., and Tyler-Smith, C. (2008) Adaptive evolution of UGT2B17 copy-number variation *Am J Hum Genet* **83**, 337–46.
18. Conrad, D.F., Andrews, T.D., Carter, N.P., Hurles, M.E., and Pritchard, J.K. (2006) A high-resolution survey of deletion polymorphism in the human genome *Nat Genet* **38**, 75–81.
19. Locke, D.P., Sharp, A.J., McCarroll, S.A., McGrath, S.D., Newman, T.L., Cheng, Z., Schwartz, S., Albertson, D.G., Pinkel, D., Altshuler, D.M, and Eichler, E.E. (2006) Linkage disequilibrium and heritability of copy-number polymorphisms within duplicated regions of the human genome *Am J Hum Genet* **79**, 275–90.
20. Dopman, E.B., and Hartl, D.L. (2007) A portrait of copy-number polymorphism in Drosophila melanogaster *Proc Natl Acad Sci USA* **104**, 19920–5.
21. Nguyen, D.Q., Webber, C., Hehir-Kwa, J., Pfundt, R., Veltman, J., and Ponting, C.P. (2008) Reduced purifying selection prevails over positive selection in human copy number variant evolution *Genome Res* **18**, 1711–23.
22. Iafrate, A.J., Feuk, L., Rivera, M.N., Listewnik, M.L., Donahoe, P.K., Qi, Y., Scherer, S.W., and Lee, C. (2004) Detection of large-scale variation in the human genome *Nat Genet* **36**, 949–51.
23. Sebat, J., Lakshmi, B., Troge, J., Alexander, J., Young, J., Lundin, P., Månér, S., Massa, H., Walker, M., Chi, M., Navin, N., Lucito, R., Healy, J., Hicks, J., Ye, K., Reiner, A., Gilliam, T.C., Trask, B., Patterson, N., Zetterberg, A., and Wigler, M. (2004) Large-scale copy number polymorphism in the human genome *Science* **305**, 525–8.
24. Redon, R., Ishikawa, S., Fitch, K.R., Feuk, L., Perry, G.H., Andrews, T.D., Fiegler, H.,

Shapero, M.H., Carson, A.R., Chen, W., Cho, E.K., Dallaire, S., Freeman, J.L., González, J.R., Gratacòs, M., Huang, J., Kalaitzopoulos, D., Komura, D., MacDonald, J.R., Marshall, C.R., Mei, R., Montgomery, L., Nishimura, K., Okamura, K., Shen, F., Somerville, M.J., Tchinda, J., Valsesia, A., Woodwark, C., Yang, F., Zhang, J., Zerjal, T., Zhang, J., Armengol, L., Conrad, D.F., Estivill, X., Tyler-Smith, C., Carter, N.P., Aburatani, H., Lee, C., Jones, K.W., Scherer, S.W., and Hurles, M.E. (2006) Global variation in copy number in the human genome *Nature* **444**, 444–54.

25. Wong, K.K., deLeeuw, R.J., Dosanjh, N.S., Kimm, L.R., Cheng, Z., Horsman, D.E., MacAulay, C., Ng, R.T., Brown, C.J., Eichler, E.E., and Lam, W.L. (2007) A comprehensive analysis of common copy-number variations in the human genome *Am J Hum Genet* **80**, 91–104.
26. Lupski, J.R., de Oca-Luna, R.M., Slaugenhaupt, S., Pentao, L., Guzzetta, V., Trask, B.J., Saucedo-Cardenas, O., Barker, D.F., Killian, J.M., Garcia, C.A., Chakravarti, A., and Patel, P.I. (1991) DNA duplication associated with Charcot-Marie-Tooth disease type 1A *Cell* **66**, 219–32.
27. Singleton, A.B., Farrer, M., Johnson, J., Singleton, A., Hague, S., Kachergus, J., Hulihan, M., Peuralinna, T., Dutra, A., Nussbaum, R., Lincoln, S., Crawley, A., Hanson, M., Maraganore, D., Adler, C., Cookson, M.R., Muenter, M., Baptista, M., Miller, D., Blancato, J., Hardy, J., and Gwinn-Hardy, K. (2003) α-synuclein locus triplication causes Parkinson's disease *Science* **302**, 841.
28. Gonzalez, E., Kulkarni, H., Bolivar, H., Mangano, A., Sanchez, R., Catano, G., Nibbs, R.J., Freedman, B.I., Quinones, M.P., Bamshad, M.J., Murthy, K.K., Rovin, B.H., Bradley, W., Clark, R.A., Anderson, S.A., O'connell, R.J., Agan, B.K., Ahuja, S.S., Bologna, R., Sen, L., Dolan, M.J., and Ahuja, S.K. (2005) The influence of CCL3L1 gene-containing segmental duplications on HIV-1/AIDS susceptibility *Science* **307**, 1434–40.
29. Beckmann, J.S., Estivill, X., and Antonarakis, S.E. (2007) Copy number variants and genetic traits: closer to the resolution of phenotypic to genotypic variability *Nat Rev Genet* **8**, 639–46.
30. Mefford, H.C., Sharp, A.J., Baker, C., Itsara, A., Jiang, Z., Buysse, K., Huang, S., Maloney, V.K., Crolla, J.A., Baralle, D., Collins, A., Mercer, C., Norga, K., de Ravel, T., Devriendt, K., Bongers, E.M., de Leeuw, N., Reardon, W., Gimelli, S., Bena, F., Hennekam, R.C., Male, A., Gaunt, L., Clayton-Smith, J., Simonic, I., Park, S.M., Mehta, S.G., Nik-Zainal, S., Woods, C.G., Firth, H.V., Parkin, G., Fichera, M., Reitano, S., Lo Giudice, M., Li, K.E., Casuga, I., Broomer, A., Conrad, B., Schwerzmann, M., Räber, L., Gallati, S., Striano, P., Coppola, A., Tolmie, J.L., Tobias, E.S., Lilley, C., Armengol, L., Spysschaert, Y., Verloo, P., De Coene, A., Goossens, L., Mortier, G., Speleman, F., van Binsbergen, E., Nelen, M.R., Hochstenbach, R., Poot, M., Gallagher, L., Gill, M., McClellan, J., King, M.C., Regan, R., Skinner, C., Stevenson, R.E., Antonarakis, S.E., Chen, C., Estivill, X., Menten, B., Gimelli, G., Gribble, S., Schwartz, S., Sutcliffe, J.S., Walsh, T., Knight, S.J., Sebat, J., Romano, C., Schwartz, C.E., Veltman, J.A., de Vries, B.B., Vermeesch, J.R., Barber, J.C., Willatt, L., Tassabehji, M., and Eichler, E.E. (2008) Recurrent rearrangements of chromosome 1q21.1 and variable pediatric phenotypes *N Engl J Med* **359**, 1685–99.
31. Zhang, F., Gu, W., Hurles, M.E., and Lupski, J.R. (2009) Copy number variation in human health, disease, and evolution *Annu Rev Genomics Hum Genet* **10**, 451–81.
32. McKim, K.S., Peters, K., and Rose, A.M. (1993) Two types of sites required for meiotic chromosome pairing in Caenorhabditis elegans *Genetics* **134**, 749–68.
33. Hammarlund, M., Davis, M.W., Nguyen, H., Dayton, D., and Jorgensen, E.M. (2005) Heterozygous insertions alter crossover distribution but allow crossover interference in Caenorhabditis elegans *Genetics* **171**, 1047–56.
34. Navarro, A., Betrán, E., Barbadilla, A., and Ruiz, A. (1997) Recombination and gene flux caused by gene conversion and crossing over in inversion heterokaryotypes *Genetics* **146**, 695–709.
35. Shaw, C.J., and Lupski, J.R. (2004) Implications of human genome architecture for rearrangement-based disorders: the genomic basis of disease *Hum Mol Genet* **13**, R57–64.
36. Lupski, J.R., and Stankiewicz, P., (2005) Genomic disorders: molecular mechanisms for rearrangements and conveyed phenotypes *PLoS Genet* **1**, e49.
37. Erdogan, F., Chen, W., Kirchhoff, M., Kalscheuer, V.M., Hultschig, C., Müller, I., Schulz, R., Menzel, C., Bryndorf, T., Ropers, H.H., and Ullmann, R. (2006) Impact of low copy repeats on the generation of balanced and unbalanced chromosomal aberrations in mental retardation *Cytogenet Genome Res* **115**, 247–53.
38. Lindsay, S.J., Khajavi, M., Lupski, J.R., and Hurles, M.E. (2006) A chromosomal rearrangement hotspot can be identified from

population genetic variation and is coincident with a hotspot for allelic recombination *Am J Hum Genet* **79**, 890–902.
39. Sun, X., Zhang, Y., Yang, S., Chen, J.Q., Hohn, B., and Tian, D. (2008) Insertion DNA Promotes Ectopic Recombination during Meiosis in Arabidopsis *Mol Biol Evol* **25**, 2079–83.
40. Welz-Voegele, C., and Jinks-Robertson, S. (2008) Sequence divergence impedes crossover more than noncrossover events during mitotic gap repair in yeast *Genetics* **179**, 1251–62.
41. Duret, L., and Galtier, N. (2009) Biased gene conversion and the evolution of mammalian genomic landscapes *Annu Rev Genomics Hum Genet* **10**, 285–311.
42. Lamb, B.C. (1985) The effects of mispair and nonpair correction in hybrid DNA on base ratios (G + C content) and total amounts of DNA *Mol Biol Evol* **2**, 175–88.
43. Bill, C.A., Taghian, D.G., Duran, W.A., and Nickoloff, J.A. (2001) Repair bias of large loop mismatches during recombination in mammalian cells depends on loop length and structure *Mutat Res* **485**, 255–65.
44. White, S.J., Vissers, L.E., Geurts van Kessel, A., de Menezes, R.X., Kalay, E., Lehesjoki, A.E., Giordano, P.C., van de Vosse, E., Breuning, M.H., Brunner, H.G., den Dunnen, J.T., and Veltman, J.A.(2007) Variation of CNV distribution in five different ethnic populations *Cytogenet Genome Res* **118**, 19–30.
45. Turner, D.J., Miretti, M., Rajan, D., Fiegler, H., Carter, N.P., Blayney, M.L., Beck, S., and Hurles, M.E. (2008) Germline rates of de novo meiotic deletions and duplications causing several genomic disorders *Nat Genet* **40**, 90–5.
46. Kehrer-Sawatski, H., and Cooper, D.N. (2008) Comparative analysis of copy number variation in primate genomes *Cytogenet Genome Res* **123**, 288–96.
47. Perry, G.H., Tchinda, J., McGrath, S.D., Zhang, J., Picker, S.R., Cáceres, A.M., Iafrate, A.J., Tyler-Smith, C., Scherer, S.W., Eichler, E.E., Stone, A.C., and Lee, C. (2006) Hotspots for copy number variation in chimpanzees and humans *Proc Natl Acad Sci USA* **103**, 8006–11.
48. Ewens, W. J. (2004) Mathematical Population Genetics. Second Revised Edition. Springer-Verlag, New York.
49. Ohta, T., and Kimura, M. (1973) A model of mutation appropriate to estimate the number of electrophoretically detectable alleles in a finite population *Genet Res* **22**, 201–4.
50. Conrad, D.F., and Hurles, M.E. (2007) The population genetics of structural variation *Nat Genet* **39**, S30–6.
51. Cooper, G.M., Nickerson, D.A., and Eichler, E.E. (2007) Mutational and selective effects on copy-number variants in the human genome *Nat Genet* **39**, S22–9.
52. Marques-Bonet, T., Girirajan, S., and Eichler, E.E. (2009) The origins and impact of primate segmental duplications *Trends Genet* **25**, 443–54.
53. Liu, G.E., Ventura, M., Cellamare, A., Chen, L., Cheng, Z., Zhu, B., Li, C., Song, J., and Eichler, E.E. (2009) Analysis of recent segmental duplications in the bovine genome *BMC Genomics* **10**, 571.
54. Bailey, J.A., Liu, G.E., and Eichler, E.E. (2003) An Alu Transposition Model for the Origin and Expansion of Human Segmental Duplications *Am J Hum Genet* **73**, 823–34.
55. Kim, P.M., Lam, H.Y., Urban, A.E., Korbel, J.O., Affourtit, J., Grubert, F., Chen, X., Weissman, S., Snyder, M., and Gerstein, M.B. (2008) Analysis of copy number variants and segmental duplications in the human genome: Evidence for a change in the process of formation in recent evolutionary history *Genome Res* **18**, 1865–74.
56. Tian, D., Wang, Q., Zhang, P., Araki, H., Yang, S., Kreitman, M., Nagylaki, T., Hudson, R., Bergelson, J., and Chen, J.Q. (2008) Single-nucleotide mutation rate increases close to insertions/deletions in eukaryotes *Nature* **455**, 105–8.
57. Petrov, D.A., and Hartl, D.L. (2000) Pseudogene evolution and natural selection for a compact genome *J Heredity* **91**, 221–7.
58. Petrov, D.A. (2002) Mutational Equilibrium Model of Genome Size Evolution *Theor Pop Biol* **61**, 533–46.
59. Taylor, M.S., Ponting, C.P., and Copley, R.R. (2004) Occurrence and Consequences of Coding Sequence Insertions and Deletions in Mammalian Genomes *Genome Res* **14**, 555–66.
60. Taylor, M.S., Kai, C., Kawai, J., Carninci, P., Hayashizaki Y, and Semple, C.A. (2006) Heterotachy in mammalian promoter evolution *PLoS Genet* **2**, e30.
61. Kim, J., He, X., and Sinha, S. (2009) Evolution of Regulatory Sequences in 12 Drosophila Species *PLoS Genet* **5**, e1000330.
62. Sjödin, P., Bataillon, T., and Schierup, M.H. (2010) Insertion and deletion processes in recent human history *PLoS One* **5**, e8650.
63. Tuzun, E., Sharp, A.J., Bailey, J.A., Kaul, R., Morrison, V.A., Pertz, L.M., Haugen, E., Hayden, H., Albertson, D., Pinkel, D.,

Olson, M.V., and Eichler, E.E. (2005) Fine-scale structural variation of the human genome *Nat Genet* **37**, 727–32.

64. Nicholas, T.J., Cheng, Z., Ventura, M., Mealey, K., Eichler, E.E., and Akey, J.M. (2009) The genomic architecture of segmental duplications and associated copy number variants in dogs *Genome Res* **19**, 491–9.
65. Mouse Genome Sequencing Consortium (2002) Initial sequencing and comparative analysis of the mouse genome *Nature* **420**, 520–62.
66. Rat Genome Sequencing Project Consortium (2004) Genome sequence of the Brown Norway rat yields insights into mammalian evolution *Nature* **428**, 493–521.
67. Conrad, D.F., Pinto, D., Redon, R., Feuk, L., Gokcumen, O., Zhang, Y., Aerts, J., Andrews, T.D., Barnes, C., Campbell, P., Fitzgerald, T., Hu, M., Ihm, C.H., Kristiansson, K., Macarthur, D.G., Macdonald, J.R., Onyiah, I., Pang, A.W., Robson, S., Stirrups, K., Valsesia, A., Walter, K., Wei, J.; The Wellcome Trust Case Control Consortium, Tyler-Smith, C., Carter, N.P., Lee, C., Scherer, S.W., and Hurles, M.E. (2010) Origins and functional impact of copy number variation in the human genome *Nature* **464**, 704–12.
68. Perry, G.H., Yang, F., Marques-Bonet, T., Murphy, C., Fitzgerald, T., Lee, A.S., Hyland, C., Stone, A.C., Hurles, M.E., Tyler-Smith, C., Eichler, E.E., Carter, N.P., Lee, C., and Redon, R. (2008) Copy number variation and evolution in humans and chimpanzees *Genome Res* **18**, 1698–710.
69. Korbel, J.O., Kim, P.M., Chen, X., Urban, A.E., Weissman, S., Snyder, M., and Gerstein, M.B. (2008) The current excitement about copy-number variation: how it relates to gene duplications and protein families *Curr Opin Struct Biol* **18**, 366–74.
70. Rosenberg, N.A., Mahajan, S., Ramachandran, S., Zhao, C., Pritchard, J.K., and Feldman, M.W. (2005) Clines, clusters, and the effect of study design on the inference of human population structure *PLoS Genet* **1**, e70.
71. Li, J.Z., Absher, D.M., Tang, H., Southwick, A.M., Casto, A.M., Ramachandran, S., Cann, H.M., Barsh, G.S., Feldman, M., Cavalli-Sforza, L.L., and Myers, R.M. (2008) Worldwide human relationships inferred from genome-wide patterns of variation *Science* **319**, 1100–4.
72. Sharp, A.J., Locke, D.P., McGrath, S.D., Cheng, Z., Bailey, J.A., Vallente, R.U., Pertz, L.M., Clark, R.A., Schwartz, S., Segraves, R., Oseroff, V.V., Albertson, D.G., Pinkel, D., and Eichler, E.E. (2005) Segmental duplications and copy-number variation in the human genome *Am J Hum Genet* 77, 78–88.
73. Jakobsson, M., Scholz, S.W., Scheet, P., Gibbs, J.R., VanLiere, J.M., Fung, H.C., Szpiech, Z.A., Degnan, J.H., Wang, K., Guerreiro, R., Bras, J.M., Schymick, J.C., Hernandez, D.G., Traynor, B.J., Simon-Sanchez, J., Matarin, M., Britton, A., van de Leemput, J., Rafferty, I., Bucan, M., Cann, H.M., Hardy, J.A., Rosenberg, N.A., and Singleton, A.B. (2008) Genotype, haplotype, and copy-number variation in worldwide human populations *Nature* **451**, 998–1003.
74. Wang, C., Szpiech, Z.A., Degnan, J.H., Jakobsson, M., Pemberton, T.J., Hardy, J.A., Singleton, A. B., and Rosenberg, N.A. (2010) Comparing Spatial Maps of Human Population-Genetic Variation Using Procrustes Analysis *Stat Appl in Genet and Mol Biol* **9**, Article 13.
75. Cann, H.M., Cohen, D., and Dausset, J. (1987) Diagnosis of genetic disease by linkage analysis *Birth Defects Orig Artic Ser* **23**, 33–60.
76. Garrigan, D., and Hammer, M.F. (2006) Reconstructing human origins in the genomic era *Nat Rev Genet* 7, 669–80.
77. The International HapMap Consortium (2005) A haplotype map of the human genome *Nature* **437**, 1299–320.
78. McCarroll, S.A., Hadnott, T.N., Perry, G.H., Sabeti, P.C., Zody, M.C., Barrett, J.C., Dallaire, S., Gabriel, S.B., Lee, C., Daly, M.J., Altshuler, D.M., and International HapMap Consortium (2006) Common deletion polymorphisms in the human genome *Nat Genet* **38**, 86–92.
79. Wang, K., Li, M., Hadley, D., Liu, R., Glessner, J., Grant, S.F., Hakonarson, H., and Bucan, M. (2007) PennCNV: an integrated hidden Markov model designed for high-resolution copy number variation detection in whole-genome SNP genotyping data *Genome Res* **17**, 1665–74.
80. Kidd, J.M., Cooper, G.M., Donahue, W.F., Hayden, H.S., Sampas, N., Graves, T., Hansen, N., Teague, B., Alkan, C., Antonacci, F., Haugen, E., Zerr, T., Yamada, N.A., Tsang, P., Newman, T.L., Tüzün, E., Cheng, Z., Ebling, H.M., Tusneem, N., David, R., Gillett, W., Phelps, K.A., Weaver, M., Saranga, D., Brand, A., Tao, W., Gustafson, E., McKernan, K., Chen, L., Malig, M., Smith, J.D., Korn, J.M., McCarroll, S.A., Altshuler, D.A., Peiffer, D.A., Dorschner, M., Stamatoyannopoulos, J., Schwartz, D., Nickerson, D.A., Mullikin, J.C., Wilson, R.K., Bruhn, L., Olson, M.V., Kaul, R., Smith, D.R., and Eichler, E.E. (2008) Mapping and sequencing of structural variation from eight human genomes *Nature* **453**, 56–64.

81. Armengol, L., Villatoro, S., González, J.R., Pantano, L., García-Aragonés, M., Rabionet, R., Cáceres, M., and Estivill, X. (2009) Identification of copy number variants defining genomic differences among major human groups *PLoS One* **4**, e7230.
82. Takahashi, N., Satoh, Y., Kodaira, M., and Katayama, H. (2008) Large-scale copy number variants (CNVs) detected in different ethnic human populations *Cytogenet Genome Res* **123**, 224–33.
83. Kato, M., Kawaguchi, T., Ishikawa, S., Umeda, T., Nakamichi, R., Shapero, M.H., Jones, K.W., Nakamura, Y., Aburatani, H., and Tsunoda, T. (2010) Population-genetic nature of copy number variations in the human genome *Hum Mol Genet* **19**, 761–73.
84. Hinds, D.A., Kloek, A.P., Jen, M., Chen, X., and Frazer, K.A. (2006) Common deletions and SNPs are in linkage disequilibrium in the human genome *Nat Genet* **38**, 82–5.
85. de Ståhl, T.D., Sandgren, J., Piotrowski, A., Nord, H., Andersson, R., Menzel, U., Bogdan, A., Thuresson, A.C., Poplawski, A., von Tell, D., Hansson, C.M., Elshafie, A.I., Elghazali, G., Imreh, S., Nordenskjöld, M., Upadhyaya, M., Komorowski, J., Bruder, C.E., and Dumanski, J.P. (2008) Profiling of copy number variations (CNVs) in healthy individuals from three ethnic groups using a human genome 32 K BAC-clone-based array *Hum Mutat* **29**, 398–408.
86. Cann, H.M., de Toma, C., Cazes, L., Legrand, M.F., Morel, V., Piouffre, L., Bodmer, J., Bodmer, W.F., Bonne-Tamir, B., Cambon-Thomsen, A., Chen, Z., Chu, J., Carcassi, C., Contu, L., Du, R., Excoffier, L., Ferrara, G.B., Friedlaender, J.S., Groot, H., Gurwitz, D., Jenkins, T., Herrera, R.J., Huang, X., Kidd, J., Kidd, K.K., Langaney, A., Lin, A.A., Mehdi, S.Q., Parham, P., Piazza, A., Pistillo, M.P., Qian, Y., Shu, Q., Xu, J., Zhu, S., Weber, J.L., Greely, H.T., Feldman, M.W., Thomas, G., Dausset, J., and Cavalli-Sforza LL. (2002) A human genome diversity cell line panel *Science* **296**, 261–2.
87. Itsara, A., Cooper, G.M., Baker, C., Girirajan, S., Li, J., Absher, D., Krauss, R.M., Myers, R.M., Ridker, P.M., Chasman, D.I., Mefford, H., Ying, P., Nickerson, D.A., and Eichler, E.E. (2009) Population analysis of large copy number variants and hotspots of human genetic disease *Am J Hum Genet* **84**, 148–61.
88. McCarroll, S.A., Kuruvilla, F.G., Korn, J.M., Cawley, S., Nemesh, J., Wysoker, A., Shapero, M.H., de Bakker, P.I., Maller, J.B., Kirby, A., Elliott, A.L., Parkin, M., Hubbell, E., Webster, T., Mei, R., Veitch, J., Collins, P.J., Handsaker, R., Lincoln, S., Nizzari, M., Blume, J., Jones, K.W., Rava, R., Daly, M.J., Gabriel, S.B., and Altshuler, D. (2008) Integrated detection and population-genetic analysis of SNPs and copy number variation *Nat Genet* **40**, 1166–74.
89. Kalinowski, S.T. (2004) Counting alleles with rarefaction: private alleles and hierarchical sampling designs *Conserv Genet* **5**, 539–43.
90. Schrider, D.R., and Hahn, M.W. (2010) Lower linkage disequilibrium at CNVs is due to both recurrent mutation and transposing duplications *Mol Biol Evol* **27**, 103–11.
91. Branton, D., Deamer, D.W., Marziali, A., Bayley, H., Benner, S.A., Butler, T., Di Ventra, M., Garaj, S., Hibbs, A., Huang, X., Jovanovich, S.B., Krstic, P.S., Lindsay, S., Ling, X.S., Mastrangelo, C.H., Meller, A., Oliver, J.S., Pershin, Y.V., Ramsey, J.M., Riehn, R., Soni, G.V., Tabard-Cossa, V., Wanunu, M., Wiggin, M., and Schloss, J.A. (2008) The potential and challenges of nanopore sequencing *Nat Biotechnol* **26**, 1146–53.
92. Eid, J., Fehr, A., Gray, J., Luong, K., Lyle, J., Otto, G., Peluso, P., Rank, D., Baybayan, P., Bettman, B., Bibillo, A., Bjornson, K., Chaudhuri, B., Christians, F., Cicero, R., Clark, S., Dalal, R., Dewinter, A., Dixon, J., Foquet, M., Gaertner, A., Hardenbol, P., Heiner, C., Hester, K., Holden, D., Kearns, G., Kong, X., Kuse, R., Lacroix, Y., Lin, S., Lundquist, P., Ma, C., Marks, P., Maxham, M., Murphy, D., Park, I., Pham, T., Phillips, M., Roy, J., Sebra, R., Shen, G., Sorenson, J., Tomaney, A., Travers, K., Trulson, M., Vieceli, J., Wegener, J., Wu, D., Yang, A., Zaccarin, D., Zhao, P., Zhong, F., Korlach, J., and Turner, S. (2009) Real-time DNA sequencing from single polymerase molecules *Science* **323**, 133–8.

Chapter 11

Detection and Interpretation of Genomic Structural Variation in Mammals

Ira M. Hall and Aaron R. Quinlan

Abstract

Structural variation (SV) encompasses diverse types of genomic variants including deletions, duplications, inversions, transpositions, translocations, and complex rearrangements, and is now recognized to be an abundant class of genetic variation in mammals. Different individuals, or strains, of a given species can differ by thousands of variants. However, despite a large number of studies over the past decade and impressive progress on many fronts, there remain significant gaps in our knowledge, particularly in species other than human. Arguably the most relevant among these are genetically tractable models such as mouse, rat, and dog. The emergence of efficient and affordable DNA sequencing technologies presents an opportunity to make rapid progress toward understanding the nature, origin, and function of SV in these, and other, domesticated species. Here, we summarize the current state of knowledge of SV in mammals, with a focus on the similarities and differences between domesticated species and human. We then present methods to identify SV breakpoints from next-generation sequence (NGS) data by paired-end mapping, split-read mapping, and local assembly, and discuss challenges that arise when interpreting these data in lineages with complex breeding histories and incomplete reference genomes. We further describe technical modifications that allow for identification of variants involving repetitive DNA elements such as transposons and segmental duplications. Finally, we explore a few of the key biological insights that can be gained by applying NGS methods to model organisms.

Key words: Structural variation, Copy number variation, Mammals, Model systems, Paired-end mapping, Split-read mapping, Breakpoint assembly, Mutation mechanism, Next-generation sequencing, Genomic rearrangements

1. Introduction

Genomic differences underlie the vast majority of heritable phenotypic differences and provide the raw material for evolution. They come in a broad range of shapes and sizes, from single-nucleotide polymorphisms (SNPs) to chromosomal rearrangements involving many megabases of DNA. As a rule, our appreciation for the different

Lars Feuk (ed.), *Genomic Structural Variants: Methods and Protocols*, Methods in Molecular Biology, vol. 838, DOI 10.1007/978-1-61779-507-7_11, © Springer Science+Business Media, LLC 2012

classes of genetic variation has been directly linked to our ability to detect them, and most conceptual advances have had their roots in technological advances. For most of the twentieth century geneticists focused on large-scale genomic rearrangements because these were the only variants that were visible by early cytogenetic methods. However, given their large size (>3 Mb) and frequent association with either cancer, sporadic disease, or long periods of genome evolution, genomic rearrangements were generally thought to be rare within species. The discovery of the structure of DNA and the elaboration of the genetic code spawned great interest in the role of small-scale variants such as SNPs, and the development of molecular cloning, DNA sequencing, and PCR during the 1970s and 1980s allowed for researchers to directly ascertain these variants in an increasingly high-throughput manner. By the turn of the century, as thousands of automated DNA sequencers were churning out data for the Human Genome Project, conventional wisdom held that the overall structure of a genome was relatively static, and that most natural variation could be explained by small-scale sequence differences. This is best exemplified by an oft-repeated statement of this era that the phenotypic differences between two humans could be explained by the presence of 1 SNP in every 1,000 bp, and the differences between a human and chimpanzee by 1 in 100.

The first convincing evidence that genome architecture was more dynamic came from the work of E. Eichler and colleagues who, upon aligning the draft human genome to itself, discovered that roughly 5% of DNA is contained within recent segmental duplications (SDs) (1). SDs are defined as multicopy segments larger than 1 kb that share greater than 90% pairwise nucleotide identity. The prevalence of highly similar duplications indicated that the human genome had undergone substantial large-scale structural mutation over recent evolutionary time. Similar levels of segmental duplication have been reported in all mammalian genomes sequenced thus far including various primates, mouse, rat, dog, and cow (2–6).

The availability of a reference genome sequence enabled the construction of genome-wide microarrays containing bacterial artificial chromosomes (BACs) or oligonucleotide probes, and comparative genomic hybridization to these arrays [array comparative genome hybridization (aCGH)] reveals DNA copy number variation (CNV) between two genomes. The first two studies to apply this technology to "normal" human individuals discovered surprising levels of CNV (7, 8), and this observation led to a profound conceptual shift in our view of genome variation (9). While the first studies detected merely a handful of CNVs in pairwise comparisons, over the next 6 years a large number of genome-wide mapping studies using progressively higher resolution oligonucleotide microarrays, fosmid end-sequencing, and massively parallel paired-end sequencing have reported several hundred to several thousand variants in pairwise comparisons between humans, depending on the resolution and scope of the methods employed. Thus, structural variation (SV) is an abundant class of genetic variation in mammals. We operationally define SV as differences in the copy number, orientation, or location of genomic segments exceeding 100 bp in size. This definition encompasses diverse types of genomic variants including deletions, duplications, inversions, transpositions, translocations, and complex rearrangements. SVs affect many genes and by some measures a larger fraction of the genome than SNPs, and the extent to which these differences underlie phenotypic variation and disease is currently an active line of investigation in many laboratories around the world.

There have been numerous studies and many important findings in this field over the past decade. There are also many unresolved questions, particularly in species other than human. Arguably the greatest challenge in answering these questions is methodological; accurate and unbiased interpretations depend upon accurate and unbiased SV detection, and this technical goal has not yet been achieved. Now, next-generation sequencing technologies offer the unprecedented opportunity to, at least in theory, map virtually all classes of SV at extremely high resolution and at reasonable cost. It should be possible to make rapid progress toward a better understanding of SV in all mammals. Unfortunately, interpretation of next-generation sequence (NGS) data in complex and repetitive genomes presents a number of nontrivial computational difficulties, and these difficulties can be exacerbated in experiments involving organisms that are not as well characterized as human.

In this review, we focus on detection and interpretation of SV in model mammalian species such as mouse, rat, and dog. To provide context, we first summarize the current state of knowledge of SV in mammals, with a focus on the similarities and differences between domesticated species and human. We then present methods to identify SV breakpoints from NGS data, and discuss specific challenges that can arise when interpreting these data in lineages with complex breeding histories and incomplete reference genomes. Finally, we explore a few of the biological insights that can be gained by applying NGS methods to model systems.

1.1. Abundance of Genomic SV in Domesticated Mammals

Over the past 6 years, there have been 11 SV mapping studies in mouse (3, 10–19), one in rat (20), two in dog (6, 21) and one in cow (22). All of these except two mouse studies (17, 18) have used aCGH, and thus were only able to detect relatively large CNVs. One common theme emerging from these studies is that all mammalian species examined thus far display abundant SV in their genomes, and appear to have roughly similar overall levels. For example, the first aCGH studies in mouse were published shortly after the first human studies (10, 11), and these studies discovered roughly similar numbers of CNVs. Subsequent studies using progressively higher resolution arrays and genetically diverse panels of inbred mouse strains (3, 12–16) identified many thousands of CNVs. In one study alone, Cutler et al. identified 2,094 CNVs among 41 inbred strains (13). The single highest resolution aCGH study to date (15) identified ~300 CNVs between two "classical" inbred strains, which is consistent with the most comprehensive aCGH study in human (23) if resolution differences are taken into account. Genome-wide experiments using a common platform (~385,000-probe NimbleGen arrays) in mouse (14), rat (20), dog (21), and cow (22) discovered an average of 10–20 CNVs per sample. Similar to previous findings in human (24, 25), aCGH experiments in mouse (3) and dog (6) using microarrays targeting segmental duplications discovered remarkable levels of CNV in these dynamic regions of genome.

The only comprehensive NGS-based study in mouse (our own) discovered 7,196 SVs in a single strain comparison (18), which is a remarkable level of variation and substantially more than has been reported in human studies if resolution differences are taken into account (26–30). However, this high level of SV is mostly accounted for by transposable element variants (TEVs), which comprise ~70% of all SV. This result is not entirely surprising given that retroelements are known to be very active in mouse (17, 31). Independent of the abundance of transposons, a direct comparison of SV levels measured by NGS is complicated by the fact that vastly different SV detection algorithms have been employed by different studies; however, at a first approximation the levels of non-TE variation appear to be similar.

While these experiments demonstrate that overall levels of SV are roughly similar, there are a few important caveats. First, direct comparisons between species are complicated by methodological differences. It remains a possibility that important differences in SV levels will be found once reference genome assemblies are completed to an equal level of accuracy and all species are examined on a common SV discovery platform, such as genome-wide sequencing using a single analysis pipeline. Second, in contrast to humans, domesticated species have been subjected to artificial selection and directed breeding, and as a consequence different strains, or breeds (hereafter referred to as strains), within a species may exhibit vastly different levels of genetic relatedness. For example, comparison of two mouse strains that are closely related by their breeding history will yield far fewer SVs than a comparison between strains that were generated by different founding stock. In addition, experiments involving wild populations or geographically isolated subspecies will detect substantially more variation than those involving inbred lines. Finally, the relative abundance of different SV classes may vary between different species, and thus the relative levels of SV that are detected may differ depending on the experimental platform that is used. For example, as mentioned above, retrotransposons are especially active in rodents and thus genome-wide sequencing experiments that are able to detects variable TE insertions are likely to find substantially more SV in rodents than similar experiments in human or dog.

2. Genomic Distribution of SV

A true comparison of the genomic distribution of SV is confounded by the fact that studies have rarely mapped the physical location of variable genomic segments, and thus we must rely upon the location of the affected segment in the reference genome and upon analyses of recent segmental duplications (which are often either SVs themselves or hotspots for SV). Nevertheless, these are useful proxies and much has been learned.

2.1. Correlation with Segmental Duplications

The first notable observation is that in all species examined thus far structural variants are highly enriched at sites of segmental duplication in the reference genome. Typical estimates range from four- to tenfold enrichment relative to a random model (14, 32–34). The breakpoints of large-scale rearrangements that occurred during mammalian evolution are also highly correlated with SDs (35–37). This enrichment has generally been explained by the propensity of duplicated sequences to promote nonallelic homologous recombination (NAHR), and there are many clear examples of this among the de novo SVs that cause human genomic disorders (38). However, there are suggestions from the literature (39–45) and direct evidence from our own work (18) that homology-independent mechanisms also preferentially occur in SDs. The cause of this is not presently clear and there are various possibilities, but regardless of mechanism the preferential localization of SVs in segmentally duplicated genomic regions is a common theme. Since segmental duplications have a markedly nonuniform genomic distribution (46), patterns of structural variation are also nonuniform and differ substantially from other classes of variation such as SNPs.

2.2. Distribution of Segmental Duplications and Associated SV

An important consideration is that patterns of SD are very different between primates and other mammalian species. In all species, most SDs are intrachromosomal, but in primates the majority of SDs are interspersed, with only ~30% being present in a "tandem" configuration (<1 mb apart) (2). In contrast, SDs in other mammalian genomes are primarily tandem. For example, in the highest-quality nonprimate reference genome, the mouse, ~90% of SDs are present in local clusters of tandem duplications (3, 47). The observation that a tandem configuration predominates in the genomes of diverse mammals including mouse (3), rat (5), dog (6), and cow (4) indicates that this is the ancestral state and that the primate lineage underwent a marked shift (2). It is not known whether the different patterns are due to differences in mutational mechanism or selection, but these differences inevitably affect the distribution of SV. Moreover, SDs present in a tandem configuration may have higher rates of spontaneous mutation, since the frequency of NAHR depends upon the proximity of participating duplicons. As suggested by She et al. (3), this effect might partially explain the extremely high mutation rates documented at certain SD loci in our own previous study of spontaneous SV among closely related mouse strains (19).

2.3. Transposable Elements

Another important difference is the relative activity of different transposon classes. Overall transposon activity is much greater in rodents relative to human and dog (48) and thus TEV comprises a greater fraction of overall SV. For example, while aCGH experiments suggest similar levels of large-scale CNV (see above), DNA sequence-based studies suggest that there are at least 5,000 TEVs between two inbred mouse strains (such as DBA/2J and C57BL/6J) (18), and less than 1,000 between two humans (49). Moreover, there are differences in the relative activity of different element classes. Whereas LINE and LTR transposons comprise the vast majority of TEV in mouse, human TEV is dominated by Alu-SINE elements (49). Differences in transposon activity do not merely affect TEV, but can also lead to other forms of SV. For example, LINE element machinery can cause retrotransposition of host transcripts leading to retrogenes, and indeed there are hundreds of variable retrogenes between inbred mouse strains (18). LINE elements can also cause double-stranded breaks even at sites where insertion does not occur (50), and inaccurate repair of DNA breaks can contribute to new SV. Because LINE elements are much more active in rodents than human or dog, these mechanisms may potentially lead to distinct patterns of SV among species. Moreover, while all TE classes can generate new SV through NAHR, the genomic distribution of various TE classes may differ. For example, Alu-SINE elements are enriched at boundaries of segmental duplications in the human genome (51), suggesting that NAHR between these elements has been a major force driving duplications. In contrast, there is an enrichment of LTR and LINE elements, but not SINEs, at mouse SDs (3). Given that SINEs preferentially insert into GC-rich regions (which are gene-rich) and LINEs and LTRs preferentially insert into AT-rich genomic regions (which are gene-poor), the relative frequency of NAHR between different TE classes can lead to significant differences in the genomic distribution of SV (3).

2.4. Complex Variants

An unresolved question concerns the prevalence of complex variants, defined as SVs that appear to have arisen from a single mutational event yet contain multiple adjacent breakpoints in proximity (e.g., <1 kb apart). Studies of complex disease-causing rearrangements at several loci in the human genome led J. Lupski, P. Hastings, and colleagues to propose a new SV formation mechanism, termed FoSTeS (52) or MMBIR (53), that involves template switching during DNA replication and/or repair. We hereafter refer to this mechanism by the more general term "template switching." We have shown that complex rearrangements involving multiple adjacent breakpoints are common in the mouse genome and account for ~16% of all SVs. It remains to be seen whether genome-wide studies in human (or other species) will find a similar level of complex variation, or whether the mouse genome is particularly prone to this mechanism. We expect that ongoing analysis of 1000 Genomes Project data (http://www.1000genomes.org/) will help resolve this question. Complex variants can have a strong and somewhat misleading effect on the genomic distribution of SV since complex variants manifest as local clusters of SV calls, and these can only be disentangled into a single variant call in datasets with subkilobase resolution.

3. Breeding Effects

Patterns of variation are strongly influenced by breeding history and selection, and in this respect model genetic systems differ substantially from human. Artificial selection has fixed genetic variants underlying desired traits (and linked variants via hitchhiking), and inbreeding has "flushed" strongly deleterious variants from inbred lineages. The extent to which these effects may be apparent is dependent on the genetic composition of the founding stock, the precise breeding history, and the nature and strength of artificial selection. These variables may differ considerably among different species and strains. For example, laboratory mouse strains have their origin in "fancy" mice, which were derived at least several hundred years ago by interbreeding of divergent subspecies. After hundreds of years of selective breeding, scientists at the turn of the century derived a relatively large number of inbred strains from a relatively small pool of genetically diverse progenitor fancy mice (54). This breeding history gives the mouse genome a unique composition, such that each genome is a mosaic of segments with different subspecific origins. Thus, in pairwise comparisons between strains whose genomes have different mosaic patterns, regions of the genome with a different subspecific origin show very high levels of variation and regions of the genome that are identical by descent (IBD) show very low levels of variation (55). Similar patterns exist in rat (56) and to a lesser extent in dog (48). As we discuss later, these patterns can complicate SV detection. In addition to these effects, regions of the genome harboring genes that have been selected for during domestication may show little or no SV among all strains of a given species.

4. SV Discovery by Sequencing

While the studies conducted to date illustrate that, in general, the landscape of SV in domesticated mammalian species is similar to that in primates, fundamental questions regarding SV frequency, size, genomic distribution, mechanistic causes, and phenotypic impact remain unanswered. In large part, these questions persist because of the inherent limitations of aCGH for studying this class of variation. The resolution of aCGH methods is typically limited to 10–100 kb, and aCGH is blind to balanced rearrangements such as inversions and reciprocal translocations. Moreover, aCGH has limited sensitivity to detect lesions arising from repetitive sequences such as segmental duplications and transposable elements. Given the relatively high activity of TEs in rodent genomes (17, 18, 31), and the fact that segmental duplications are hotspots for SV, the inability to screen for mutations in duplicated sequence precludes the detection of a substantial and functionally relevant portion of structural variation (18, 29).

The recent proliferation of accurate, high-throughput DNA sequencing techniques eliminates many of the biases inherent to microarrays and provides a powerful and economical approach for genome-wide SV characterization. Current sequencing techniques can localize SV breakpoints much more precisely and, given sufficient sequence depth and/or read length, allow one to infer the causal mechanism of SV formation by characterizing the nucleotide sequences flanking SV breakpoints (discussed in more detail below). The specific sequencing methods used by the different available technologies are diverse, yet with respect to the detection of SV they can be broadly classified into two categories: shorter, paired-end sequences and longer, contiguous sequences. In the following two sections, we discuss the merits and weaknesses of the two molecular approaches in the context of SV discovery and characterization.

4.1. SV Discovery with Paired-End Mapping

Recently, substantial focus has been placed on the development of computational methods to exploit so-called paired-end sequences for the discovery of diverse classes of SV. The fundamental principles of this sequencing approach have been described in detail elsewhere (57–59); the basic premise is that the two respective ends (hereafter referred to as "matepairs," or "pairs") of millions of larger DNA molecules are sequenced from an experimental (or "test") genome and compared to a reference genome. Prior to sequencing, DNA fragments from the test genome are carefully restricted to a predictable size range (e.g., 500 bp). Paired-end mapping (PEM) approaches proceed by aligning the sequenced matepairs to a reference genome, and use the expected size distribution and orientation of the pairs to infer whether the structure of the test genome agrees with that of the reference. So-called concordant matepairs align with the expected distance and orientation and indicate that the structure of the test genome agrees with that of the reference genome. The corollary is that "discordant" matepairs, which align with an unexpected distance and/or orientation, suggest possible structural variation between the test and reference genomes.

Each class of structural variation (e.g., deletion, insertion, inversion, etc.) has a characteristic discordant mapping "signature" (Fig. 1). PEM approaches must carefully detect and exclude concordant matepairs so that putative SV can be confidently identified from the remaining discordant pairs. In order to rule out remaining alignment artifacts and chimeric molecules, most SV discovery algorithms screen for multiple discordant matepairs (typically two or more) that have the same signature and support

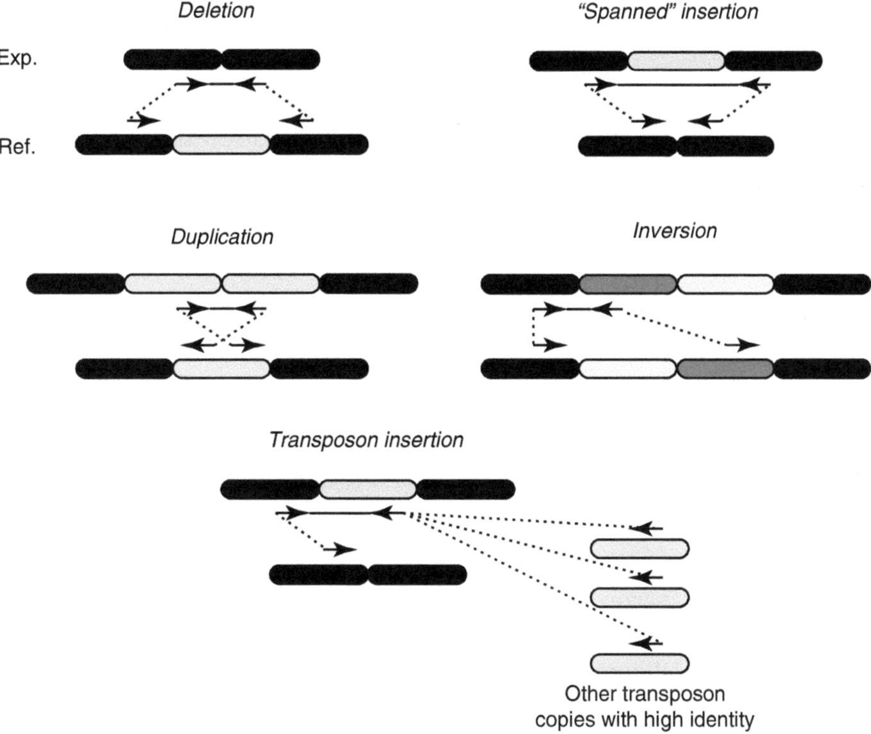

Fig. 1. Paired-end mapping signatures. Shown are five different classes of structural variation and the matepair mapping patterns that result. In each case, the experimental (Exp.) genome is shown on top and the reference genome (Ref.) below. The genomic segment affected by the SV is shown in the reference genome in *gray* and unaffected segments in *black*. Each matepair derived from the experimental genome has a known size (e.g., 500 bp) and the two respective reads are in opposing orientation (*black arrows*), as shown beneath the experimental genome. Structural variation is apparent when matepairs map to the reference genome with an unpredicted size and/or orientation. *Dotted lines* connect the actual matepair sequences, obtained from the experimental genome, to the reference genome. The orientation of the alignment in the reference genome is indicated by the direction of the *arrows*. Note that each SV class gives a distinctive pattern. Note also that the read mapping to the transposon insertion in the bottom-most example will also map to all other similar copies of that transposon class (three copies are shown).

the same SV breakpoint. A complementary approach to PEM is to use the depth of sequence coverage (DOC) from concordant matepairs to detect duplications and deletions in the test genome (18, 58, 60–62). DOC approaches are analogous to aCGH and are simplified in species such as mouse where the genome of a given strain is largely homozygous.

4.2. Detection of Repeats

Mammalian genomes are highly repetitive and a significant fraction of SV involves multicopy elements such as transposons and segmental duplications. Accurate detection of SV at repetitive loci requires certain technical modifications. For example, when a rearrangement occurs within a segmental duplication, the discordant matepairs that indicate the mutation will align to the locus where the mutation occurred, as well as to all other similar copies of the repeated sequence. Similarly, when a mobile element transposes to an otherwise nonrepetitive location in a test genome, one end of the discordant matepairs will align uniquely to the region flanking the insertion site while the end sequenced from the mobile element insertion will align to all other similar mobile elements in the reference genome (Fig. 1). The identification of such variants necessitates sensitive read alignment such that many mappings (i.e., hundreds or thousands) are reported for matepairs derived from repetitive elements. In addition, SV discovery algorithms must be able to cluster discordant mappings such that only a single mapping for a given matepair is included in a single variant call. At the time of writing two published algorithms are capable of this (18, 29); however, we expect that other extant algorithms will be extended accordingly.

4.3. Technical Challenges Affecting Accurate SV Discovery by PEM

In our research, we have found that false-positive variant calls primarily arise from three sources: (1) insufficiently sensitive sequence alignment; (2) incomplete removal of sequencing artifacts; and (3) matepairs originating from poorly assembled genomic regions.

4.3.1. Sequence Alignment

PEM using NGS data generally relies upon relatively short reads (<100 bp), and the error rate increases steadily as longer sequences are obtained. As a result, the already difficult task of accurately aligning millions of short sequences to a repetitive genome is exacerbated by the need to account for errors and polymorphism in the aligned sequences. The most typical misalignment artifact arises when the paired sequences have sufficient polymorphisms and/or errors to prevent the aligner from detecting the proper alignment(s) which would indicate that the matepair is concordant with the reference genome. Instead, the only alignment(s) detected will erroneously suggest that the matepair is discordant. When this error occurs in a systematic fashion at specific genomic loci (as in our experience it often does), multiple erroneously aligned matepairs will cluster at each of these loci and be identified by the PEM-detection algorithm, resulting in a substantially elevated false-positive SV discovery rate.

This problem is especially pernicious given the breeding history domesticated mammals. Because of this history, both SNPs and SVs are nonrandomly distributed throughout the genome, and the variation observed between any two lines will largely mimic pairwise differences in haplotype structure. In the laboratory mouse, for example, haplotypes of different subspecific origin can harbor extremely high rates of polymorphism (e.g., ~1 per 200 bp), while haplotypes with the same subspecific origin generally have extremely low rates (e.g., ~1 per 10 kb) (55, 63). Between two classical inbred strains divergent haplotypes comprise roughly one-third of the genome (63). Thus, genomic regions where the test genome and the reference genome have different subspecific origins will be greatly enriched in problematic alignments. These regional effects can confound data interpretation because they lead to very different false discovery rates in different parts of the genome, and this effect varies depending on the strains that are compared. Since the rates of polymorphism may be as much as fivefold higher between divergent haplotypes in domesticated species than among human individuals, more sensitive sequence alignment is required in these species to achieve similar levels of accuracy.

We have found that an effective approach to mitigating alignment artifacts is to use a tiered, increasingly sensitive alignment scheme (18). Such an approach begins with a fast, yet less sensitive aligner such as BWA (64) to quickly identify the majority of the easily identified concordant matepairs. The remaining discordant matepairs are iteratively scrutinized with successively more sensitive aligners (e.g., Novoalign (65) or Mosaik (66)) and settings until one is confident that only truly discordant matepairs remain.

4.3.2. Sequencing Artifacts

Additional complications to PEM approaches are caused by experimental artifacts that arise during DNA library construction and sequencing. The most common and problematic of these artifacts are the "duplicate" matepairs; that is, a single matepair that is sequenced multiple times solely owing to artifacts in the library construction and/or sequencing processes. Duplicate molecules lead to spurious positive SV calls by falsely creating clusters of seemingly independent matepairs that suggest the same SV breakpoint. Duplicate matepairs arise either by PCR amplification of insufficiently complex DNA libraries, or in the case of the Illumina/Solexa platform, when the base-calling software incorrectly calls multiple sequences from a single cluster. Such duplicates can be identified after sequence alignment by screening for matepairs that have identical alignment coordinates. Software packages such as SAMTOOLS (67) and PICARD (68) provide utilities for removing duplicates sequences; however they do not allow duplicates to be detected from matepairs that have approximately the same alignment coordinates. We find this to be a necessary consideration as sequencing errors at the beginning or end of reads can cause bona fide duplicates to have alignment coordinates that differ by 1 or 2 bp. Moreover, neither SAMTOOLS nor PICARD have utilities for removing duplicates from datasets that contain multiple mappings for matepairs that align to nonunique sequence. As discussed above, including these mappings is a necessary requirement for identifying variants that involve segmental duplications and transposons. Removing duplicates in such datasets is complicated because, depending on alignment sensitivity and the number of mappings recorded, only a subset of mappings may be shared between duplicate reads. It is therefore necessary to examine *all* mappings for any that might be duplicates, and then to remove *all* of the mappings for all but one of the duplicate matepairs.

4.3.3. Reference Genome Effects

An insidious source of false positives arises from the incomplete nature of current reference genome assemblies. Repetitive genomic loci are notoriously difficult to assemble accurately. Such loci may be assembled improperly or incompletely, or may even be entirely missing from the reference genome. This is a major technical issue for SV detection for two reasons. First, improper or incomplete assembly yields a PEM signature that is indistinguishable from true SV. Second, matepairs arising from sequences that are not present in the reference genome—namely, centromeres, telomeres, and assembly gaps—can often be aligned to other genomic locations. Insofar as these erroneous alignments occur in a systematic fashion (as we are convinced they often do) false-positive SV calls will result. Artifacts caused by the reference genome can be difficult to identify and disregard. For example, in humans such effects might only be apparent after analyzing numerous genomes and noticing that certain variants were called in all samples (so-called monomorphic variants). Indeed, two sequencing-based studies that analyzed multiple humans reported an abnormal number of monomorphic variants (24, 61). Highly inbred species such as mouse and rat offer an important advantage in this respect since it is possible to re-sequence an individual from the reference strain as a control, and there should be very few new genetic differences between closely related inbred individuals (19). Our recent study represents the first to use this control, and we identified 405 high-confidence “variants” between our C57BL/6J individual and the reference (18). Only 10–20 of these appear to be real variants, and most of the remainder represent artifacts caused by the reference genome itself. While these confounding effects may be relatively mild in human and mouse, they will present significant obstacles for applying PEM in domesticated species with less complete reference genomes including rat, dog, cow, cat, pig, and others. This argues for a continued effort to improve the quality of reference genome assemblies in diverse organisms.

In some respects, accurate PEM can also depend on genome annotations. For example, some strategies (such as our own) remove reads that map to simple sequence repeats (SSRs), and others may attempt to identify transposon insertions by aligning matepairs directly to annotated TE sequences. Moreover, even if SV detection is entirely independent of genome annotations, the manner in which variants are classified and interpreted is inherently dependent on them. Genome annotations are not as comprehensive nor as accurate in genomes other than human, and in this respect SV discovery and/or interpretation can be more difficult.

4.4. SV Discovery with Split-Read Mapping

A more powerful approach to SV discovery is the use of longer (e.g., >200 bp) DNA sequences to characterize SV breakpoints at single base-pair resolution. DNA sequences from a test genome that span the site of an SV breakpoint will align to the reference genome in "split" fashion (69). That is, distinct segments of the DNA sequence will align to different loci in the reference genome, and the distance and orientation of these alignments indicate the type of rearrangement that occurred in the test genome (Fig. 2). The fundamental advantage of this approach is that a single "split" read can identify the exact nucleotide at which the breakpoint occurred. Multiple "split" reads corroborating the same breakpoint can be assembled with programs such as PHRAP (70) to generate a consensus sequence describing the breakpoint locus. Importantly, by aligning the consensus sequence to the reference genome, one can infer the causal mechanism based on the sequence homology in the regions flanking the breakpoint (Fig. 2). For example, extensive (e.g., >50 bp) sequence homology is a hallmark of NAHR, while breakpoints exhibiting microhomology or no homology suggest nonhomologous end joining (NHEJ) or template switching (52, 53). Complex variants that contain multiple breakpoints in close proximity or that have accumulated insertions of DNA directly into the breakpoint itself, most likely arose via template switching (71). These studies are crucial because there is substantial uncertainty about the relative role of different SV formation mechanisms (18, 24, 72), and obtaining a more coherent understanding of these molecular forces is a necessary prerequisite for understanding the etiology of human diseases that are caused by de novo structural variation, namely, genomic disorders and cancer. Thus, future research should be focused on characterizing as many SV breakpoints as possible from diverse germline and somatic genomes. In this regard, the use of long DNA sequences for split-read detection in model organisms is a very attractive approach.

4.4.1. Technical Caveats

A technical consideration for this approach is that longer sequences often overlap or contain repetitive DNA such as transposons, segmental duplications, or SSRs. Thus, sequence reads whose alignment(s) do not meet the criteria for "concordance" with the reference genome (e.g., >90% identity and >90% length) could

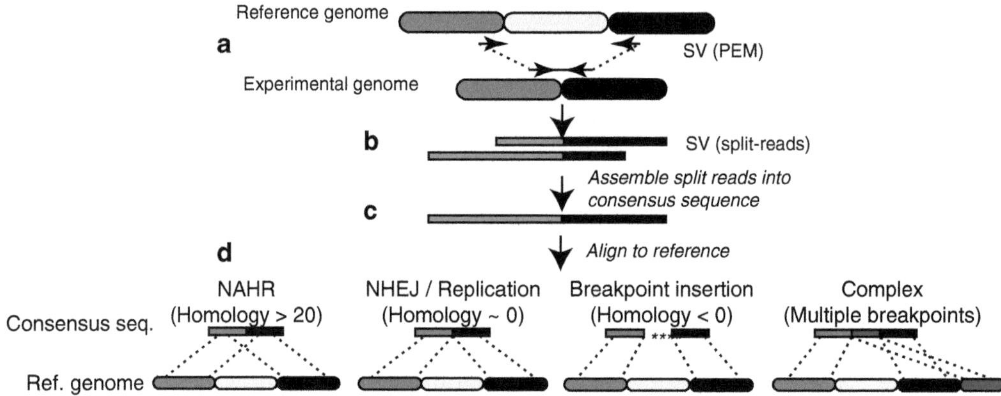

Fig. 2. Breakpoint isolation. At top is shown the reference genome and beneath that the experimental genome. (**a**) Breakpoints are localized to a genomic interval by paired-end mapping, where the reads are shown as *black arrows*, with the direction of the *arrow* indicating read orientation, and the alignments to the reference are shown as *dotted lines*. Note that the matepair maps the reference genome with a larger span than predicted, indicating a deletion in the experimental genome corresponding to the *light-gray* segment in the reference. (**b**) Long-reads that map to the predicted breakpoint region in split fashion (as shown in panel **d**) identify the breakpoint at single-nucleotide resolution. (**c**) Assembly of the long-reads produces a consensus sequence that describes the SV breakpoint. (**d**) Alignment of the consensus breakpoint sequence to the reference genome can reveal the molecular mechanism that generated the SV. From left, when the breakpoint contains significant homology to the regions flanking the deletion, the likely cause is NAHR. When the breakpoint contains little or no homology, the likely cause is NHEJ or replication-based template switching. When DNA has been inserted directly into the breakpoint, this manifests as a gap in the alignment to the reference. If the inserted DNA segment was generated by template-independent DNA synthesis, the likely cause is NHEJ. If the inserted DNA originated from elsewhere in the genome, the likely cause is template switching. At right, complex variants are apparent by the presence of multiple adjacent, often intertwined breakpoints. Most of these are likely due to template switching.

have smaller subsequences that align to hundreds or thousands of locations with similar identity. For this reason, one must define strict criteria for determining which alignments are retained for each distinct segment of a read when screening for putative split-read alignments. We also note that many of the technical challenges presented by short-read PEM, as discussed in the section above, are true for split-read mapping (SRM) as well. Indeed, while accurate alignments are much easier to obtain for long-reads, the difficulties presented by mapping SVs in repetitive elements are very similar, and the artifacts caused by duplicate reads and reference genome quality are likewise an issue.

4.4.2. Current Limitations of SRM

The current limitation to this approach is the cost of obtaining a sufficient number of long DNA sequences from a test genome. At the time of writing, traditional capillary sequencing and Roche/454 pyrosequencing are the primary means for generating longer reads. However, the low throughput and high cost of these technologies prohibit genome-wide SV breakpoint discovery in large genomes. Sequencing technologies from Pacific Biosciences, Life Technologies, and Ion Torrent promise to reduce the cost of generating sufficiently long DNA sequences, but whether the throughput of these systems will be on the scale necessary for genome-wide SV discovery in mammalian genomes is not clear. In the interim, a hybrid approach seems to be the most economical. In one possible hybrid strategy, SV breakpoints would localized to small genomic intervals using genome-wide PEM. Then, DNA capture techniques (73) would be used to isolate breakpoint-containing genomic regions, and breakpoints would be sequenced with a long-read technology. Breakpoints could then be characterized and/or genotyped on a large scale among many individuals or strains using the aforementioned "split-read" approach (see Fig. 2).

5. Application of NGS Methods to Model Mammals

Given the decreasing costs of DNA sequencing, it seems likely that all major lines of mouse, rat, and dog will be fully sequenced in the next few years using next-generation technologies. Indeed, the Mouse Genomes Project is already sequencing 17 inbred strains (http://www.sanger.ac.uk/resources/mouse/genomes/). These NGS data will not be sufficient for de novo whole-genome assembly, but they will be sufficient to generate genome-wide variation maps. To the extent that these maps are comprehensive and accurate they should resolve most outstanding questions regarding the prevalence and genomic distribution of SV in model mammals. What else might we learn about structural variation in the coming years?

5.1. Mechanisms of SV Formation

One unresolved question is the mechanistic origins of SV. This question has not yet been fully addressed in any species because mapping breakpoints to single-nucleotide resolution has historically been a laborious process. The large number of breakpoints that will be characterized over the next few years should allow for a direct comparison of the relative role of NAHR, NHEJ, and template switching among different species. Given their close evolutionary relationship one might expect that the contribution of different mechanisms would be conserved among all mammals; however, differences in the genomic distribution of segmental duplications indicates that there may be mechanistic differences. This important question should be resolved in the next few years.

A far more difficult question to address is how genetic factors or environmental conditions affect SV genesis. This subject has profound implications for our understanding of evolution and disease. It is possible, perhaps even likely, that genetic variation in genes that affect genome stability causes certain individuals to be more or less susceptible to genomic rearrangements. This could cause an increased risk of sporadic disease in these individuals and their progeny. Similar effects during evolution might lead to very different rates and/or patterns of genome evolution in specific lineages, as has been well-documented for gibbon (2). Furthermore, it was recently proposed that DNA replication-based mutations may be promoted by cellular stress (53, 71), and this raises the intriguing question of whether certain environmental conditions may affect rates of SV formation. While there is scant evidence for either of the above hypotheses, there is clearly strong historical precedent from cancer research that both genes and environment affect susceptibility to a disease marked by genomic rearrangements.

In our view, the only way to adequately address these questions is through development of a high-throughput screening method that is capable of measuring the effects of many different genetic backgrounds and environmental compounds in an unbiased manner. This is best accomplished in the laboratory mouse. While the screening methods for such studies do not yet exist and are difficult to envision with current technologies, we are hopeful that ongoing development of long-read single molecule sequencing technologies will increase the sensitivity of SV detection to a sufficient extent that the frequency of *rare* SVs can be measured *within* somatic and germline cell populations by SRM. This or a similar method could serve as a quantitative genome-wide measure of structural mutation rates, which would allow for unbiased identification of specific factors that modulate genome stability.

5.2. Somatic Variation

Another interesting question that could readily be addressed in mice is the prevalence of somatically acquired SV. There are intriguing suggestions from studies in human and mouse that individuals are composed of genetically variable somatic cell populations (74–78), and the extent to which this is true has important implications for diverse fields of biology including sporadic disease, cancer, aging, and stem cell therapy. Mice offer the obvious advantage of allowing many different tissues to be examined for SV in an inbred background. However, the major obstacle for studying somatic variation is obtaining pure samples of a given lineage. Crude tissue samples are generally composed of many different cell-types with diverse developmental histories, and current genome-wide methods cannot detect variants that are rare within a population of cells (as most somatic mutations are expected to be). Recent advances in stem cell technology provide a means to copy the genome of individual somatic cells through induced expression of 3–4 genes (79). By generating transgenic mice in which all somatic cells contain drug-inducible versions of these genes (80), it is possible to clone single somatic cells from diverse lineages (81). Application of sequence-based SV discovery methods to stem cell lines generated by this technology could answer a number of unresolved questions. For example, how prevalent is somatic variation? Are genomic patterns of somatic SV different from germline SV? Are somatic variants generated by the same mechanisms as germline variants, or do they more closely resemble the aberrations found in cancers? Are different developmental lineages more or less susceptible to new mutations?

5.3. The Genetics of Gene Expression

There is great interest in how genetic variation affects heritable variability in gene expression. The typical approach is to treat gene expression as a quantitative trait and to use conventional mapping methods to identify associated genetic variation (so-called eQTLs). Notably, aCGH-based experiments in human (82) and mouse (15, 16) have shown that CNVs make a significant contribution to gene expression and underlie as much as 20% of heritable differences. However, this line of investigation is limited in human due to the difficulty of obtaining a sufficient number of samples from interesting tissues and/or cell-types. In rodents, gene expression can be assessed in virtually any tissue in a stable and renewable inbred background. The availability of recombinant inbred lines and segregating populations make these organisms ideal for investigating this topic. Some interesting questions that could be addressed include the following. What types of variants have the greatest contribution to gene expression? Do these operate in *cis* or *trans*? What sorts of genetic factors are involved? At a systems level, what is the genetic architecture of gene expression control? One interesting line of investigation is the role of transposons in epigenetic gene control. TEs, in particular, LTRs, are known to silence genes and to serve as alternative promoters through epigenetic processes such as DNA methylation and RNA-interference. There are roughly ~5,000 variable TE insertions between two classical inbred mouse strains, many of which lie within or near genes (18), and likely many more TEVs among wild-derived strains. Therefore, assessing the functional effects of TE-mediated gene control is a tractable problem that can be addressed with current genomic methods.

5.4. Mapping Phenotypic Variation

The relationship between genetic and phenotypic variation is a fundamental question in biology, and at present the contribution of SV is unclear. Model mammals offer significant practical advantages for addressing this question. Most domesticated species display extremely high levels of phenotypic diversity between strains yet little within strains, and due to breeding history and artificial selection their genomes generally have simpler patterns of variation. These features allow trait mapping to be accomplished with a much smaller number of individuals. Moreover, artificial selection favors penetrant alleles with large phenotypic effects, which are the easiest to map, and due to inbreeding model mammals suffer from many of the same genetic diseases and susceptibilities that plague humans.

Comprehensive SV maps will be immediately useful for identifying functional variants in genomic regions identified by genome-wide association studies (GWAS). GWAS has proven to be a powerful mapping strategy in humans and model species, and using this approach many genomic regions have been identified that show a significant association with traits. There has recently been impressive success using GWAS in dogs (83), and current efforts to generate more diverse mapping populations in mouse through the "Collaborative Cross" project (84) promise to greatly increase the power of this approach. One limitation of GWAS is that pinning down the causal variant(s) embedded within a large associated region can be difficult. This is especially true in domesticated species since, due to breeding bottlenecks, LD can extend over large genomic distances, often several megabases (48, 56, 63). Whole-genome variation maps will make it far easier to identify and test candidate causal variants, and this will help to reveal the role of common SVs in phenotypic variation.

However, it is increasingly clear that for many (if not most) complex traits common genetic variation explains a minor fraction of the phenotypic variance. What, then, is the genetic basis for this missing heritability? Most current hypotheses revolve around a role for rare variants (i.e., recent mutations) with large phenotypic effects (85), and in this respect SVs are prime candidates: they are often large, can affect gene dosage and/or structure, and rates of SV formation are relatively high compared to SNPs, particularly at hotspots (38). How might model species help to resolve this issue?

The only reliable way to assess the role of recent mutations is by direct detection. Ideally, this would be done by whole-genome sequencing of all the individuals in a mapping population, but this will remain prohibitively expensive for at least 5 years. In the interim, a potential alternative is to perform direct genotyping on the most informative set of the genetic differences obtained by sequencing widely used strains. This approach is powerful in domesticated species because most relevant genetic variation can be captured by sequencing a small number of individuals. This includes recent mutations that are not well-tagged by SNPs. In this context, the high-resolution SV detection methods that we describe above are perfectly suited for assessing the functional impact of both ancient and recent SV. One potential approach is to develop high-throughput PCR assays to directly genotype SV breakpoints. Coupled with conventional SNP genotyping, this method could be useful as a more direct form of trait mapping. Alternatively, one could imagine a more powerful genotyping system based upon sequence capture of all SV breakpoints, all putative functional SNPs, as well as an adequate number of haplotype-tagging SNPs, followed by DNA sequence-based genotyping with NGS technologies.

Using such methods we expect that significant breakthroughs will be obtained in coming years and that a coherent understanding of the causes and consequences of genomic structural variation will emerge.

References

1. Bailey, J. A., Yavor, A. M., Massa, H. F., Trask, B. J., and Eichler, E. E. (2001) Segmental duplications: organization and impact within the current human genome project assembly, *Genome Res 11*, 1005–1017.
2. Marques-Bonet, T., Girirajan, S., and Eichler, E. E. (2009) The origins and impact of primate segmental duplications, *Trends Genet 25*, 443–454.
3. She, X., Cheng, Z., Zollner, S., Church, D. M., and Eichler, E. E. (2008) Mouse segmental duplication and copy number variation, *Nat Genet 40*, 909–914.
4. Liu, G. E., Ventura, M., Cellamare, A., Chen, L., Cheng, Z., Zhu, B., Li, C., Song, J., and Eichler, E. E. (2009) Analysis of recent segmental duplications in the bovine genome, *BMC Genomics 10*, 571.
5. Tuzun, E., Bailey, J. A., and Eichler, E. E. (2004) Recent segmental duplications in the working draft assembly of the brown Norway rat, *Genome Res 14*, 493–506.
6. Nicholas, T. J., Cheng, Z., Ventura, M., Mealey, K., Eichler, E. E., and Akey, J. M. (2009) The genomic architecture of segmental duplications and associated copy number variants in dogs, *Genome Res 19*, 491–499.
7. Iafrate, A. J., Feuk, L., Rivera, M. N., Listewnik, M. L., Donahoe, P. K., Qi, Y., Scherer, S. W., and Lee, C. (2004) Detection of large-scale variation in the human genome, *Nat Genet 36*, 949–951.
8. Sebat, J., Lakshmi, B., Troge, J., Alexander, J., Young, J., Lundin, P., Maner, S., Massa, H., Walker, M., Chi, M., Navin, N., Lucito, R., Healy, J., Hicks, J., Ye, K., Reiner, A., Gilliam, T. C., Trask, B., Patterson, N., Zetterberg, A., and Wigler, M. (2004) Large-scale copy number polymorphism in the human genome, *Science 305*, 525–528.
9. Feuk, L., Carson, A. R., and Scherer, S. W. (2006) Structural variation in the human genome, *Nature Reviews Genetics 7*, 85–97.
10. Li, J., Jiang, T., Mao, J. H., Balmain, A., Peterson, L., Harris, C., Rao, P. H., Havlak, P., Gibbs, R., and Cai, W. W. (2004) Genomic segmental polymorphisms in inbred mouse strains, *Nat Genet 36*, 952–954.
11. Snijders, A. M., Nowak, N. J., Huey, B., Fridlyand, J., Law, S., Conroy, J., Tokuyasu, T., Demir, K., Chiu, R., Mao, J. H., Jain, A. N., Jones, S. J., Balmain, A., Pinkel, D., and Albertson, D. G. (2005) Mapping segmental and sequence variations among laboratory mice using BAC array CGH, *Genome Res 15*, 302–311.
12. Adams, D. J., Dermitzakis, E. T., Cox, T., Smith, J., Davies, R., Banerjee, R., Bonfield, J., Mullikin, J. C., Chung, Y. J., Rogers, J., and Bradley, A. (2005) Complex haplotypes, copy number polymorphisms and coding variation in two recently divergent mouse strains, *Nat Genet 37*, 532–536.
13. Cutler, G., Marshall, L. A., Chin, N., Baribault, H., and Kassner, P. D. (2007) Significant gene content variation characterizes the genomes of inbred mouse strains, *Genome Res 17*, 1743–1754.
14. Graubert, T. A., Cahan, P., Edwin, D., Selzer, R. R., Richmond, T. A., Eis, P. S., Shannon, W. D., Li, X., McLeod, H. L., Cheverud, J. M., and Ley, T. J. (2007) A high-resolution map of segmental DNA copy number variation in the mouse genome, *PLoS Genet 3*, e3.
15. Cahan, P., Li, Y., Izumi, M., and Graubert, T. A. (2009) The impact of copy number variation on local gene expression in mouse hematopoietic stem and progenitor cells, *Nat Genet 41*, 430–437.
16. Henrichsen, C. N., Vinckenbosch, N., Zollner, S., Chaignat, E., Pradervand, S., Schutz, F., Ruedi, M., Kaessmann, H., and Reymond, A. (2009) Segmental copy number variation shapes tissue transcriptomes, *Nat Genet 41*, 424–429.
17. Akagi, K., Li, J., Stephens, R. M., Volfovsky, N., and Symer, D. E. (2008) Extensive variation between inbred mouse strains due to endogenous L1 retrotransposition, *Genome Res 18*, 869–880.
18. Quinlan, A. R., Clark, R. A., Sokolova, S., Leibowitz, M. L., Zhang, Y., Hurles, M. E., and Hall, I. M. (2010) Genome-wide mapping and assembly of structural variant breakpoints in the mouse genome, *Genome Research In Press.*
19. Egan, C. M., Sridhar, S., Wigler, M., and Hall, I. M. (2007) Recurrent DNA copy number variation in the laboratory mouse, *Nat Genet 39*, 1384–1389.
20. Guryev, V., Saar, K., Adamovic, T., Verheul, M., van Heesch, S. A., Cook, S., Pravenec, M., Aitman, T., Jacob, H., Shull, J. D., Hubner, N., and Cuppen, E. (2008) Distribution and functional impact of DNA copy number variation in the rat, *Nat Genet 40*, 538–545.
21. Chen, W. K., Swartz, J. D., Rush, L. J., and Alvarez, C. E. (2009) Mapping DNA structural variation in dogs, *Genome Res 19*, 500–509.
22. Liu, G. E., Hou, Y., Zhu, B., Cardone, M. F., Jiang, L., Cellamare, A., Mitra, A., Alexander, L. J., Coutinho, L. L., Dell'aquila, M. E.,

Gasbarre, L. C., Lacalandra, G., Li, R. W., Matukumalli, L. K., Nonneman, D., Regitano, L. C., Smith, T. P., Song, J., Sonstegard, T. S., Van Tassell, C. P., Ventura, M., Eichler, E. E., McDaneld, T. G., and Keele, J. W. Analysis of copy number variations among diverse cattle breeds, *Genome Res.*

23. Conrad, D. F., Pinto, D., Redon, R., Feuk, L., Gokcumen, O., Zhang, Y., Aerts, J., Andrews, T. D., Barnes, C., Campbell, P., Fitzgerald, T., Hu, M., Ihm, C. H., Kristiansson, K., Macarthur, D. G., Macdonald, J. R., Onyiah, I., Pang, A. W., Robson, S., Stirrups, K., Valsesia, A., Walter, K., Wei, J., Tyler-Smith, C., Carter, N. P., Lee, C., Scherer, S. W., and Hurles, M. E. (2009) Origins and functional impact of copy number variation in the human genome, *Nature.*

24. Kidd, J. M., Cooper, G. M., Donahue, W. F., Hayden, H. S., Sampas, N., Graves, T., Hansen, N., Teague, B., Alkan, C., Antonacci, F., Haugen, E., Zerr, T., Yamada, N. A., Tsang, P., Newman, T. L., Tuzun, E., Cheng, Z., Ebling, H. M., Tusneem, N., David, R., Gillett, W., Phelps, K. A., Weaver, M., Saranga, D., Brand, A., Tao, W., Gustafson, E., McKernan, K., Chen, L., Malig, M., Smith, J. D., Korn, J. M., McCarroll, S. A., Altshuler, D. A., Peiffer, D. A., Dorschner, M., Stama-toyannopoulos, J., Schwartz, D., Nickerson, D. A., Mullikin, J. C., Wilson, R. K., Bruhn, L., Olson, M. V., Kaul, R., Smith, D. R., and Eichler, E. E. (2008) Mapping and sequencing of structural variation from eight human genomes, *Nature 453*, 56–64.

25. Redon, R., Ishikawa, S., Fitch, K. R., Feuk, L., Perry, G. H., Andrews, T. D., Fiegler, H., Shapero, M. H., Carson, A. R., Chen, W., Cho, E. K., Dallaire, S., Freeman, J. L., Gonzalez, J. R., Gratacos, M., Huang, J., Kalaitzopoulos, D., Komura, D., Macdonald, J. R., Marshall, C. R., Mei, R., Montgomery, L., Nishimura, K., Okamura, K., Shen, F., Somerville, M. J., Tchinda, J., Valsesia, A., Woodwark, C., Yang, F., Zhang, J., Zerjal, T., Armengol, L., Conrad, D. F., Estivill, X., Tyler-Smith, C., Carter, N. P., Aburatani, H., Lee, C., Jones, K. W., Scherer, S. W., and Hurles, M. E. (2006) Global variation in copy number in the human genome, *Nature 444*, 444–454.

26. Bentley, D. R., Balasubramanian, S., Swerdlow, H. P., Smith, G. P., Milton, J., Brown, C. G., Hall, K. P., Evers, D. J., Barnes, C. L., Bignell, H. R., Boutell, J. M., Bryant, J., Carter, R. J., Keira Cheetham, R., Cox, A. J., Ellis, D. J., Flatbush, M. R., Gormley, N. A., Humphray, S. J., Irving, L. J., Karbelashvili, M. S., Kirk, S. M., Li, H., Liu, X., Maisinger, K. S., Murray, L. J., Obradovic, B., Ost, T., Parkinson, M. L., Pratt, M. R., Rasolonjatovo, I. M., Reed, M. T., Rigatti, R., Rodighiero, C., Ross, M. T., Sabot, A., Sankar, S. V., Scally, A., Schroth, G. P., Smith, M. E., Smith, V. P., Spiridou, A., Torrance, P. E., Tzonev, S. S., Vermaas, E. H., Walter, K., Wu, X., Zhang, L., Alam, M. D., Anastasi, C., Aniebo, I. C., Bailey, D. M., Bancarz, I. R., Banerjee, S., Barbour, S. G., Baybayan, P. A., Benoit, V. A., Benson, K. F., Bevis, C., Black, P. J., Boodhun, A., Brennan, J. S., Bridgham, J. A., Brown, R. C., Brown, A. A., Buermann, D. H., Bundu, A. A., Burrows, J. C., Carter, N. P., Castillo, N., Chiara, E. C. M., Chang, S., Neil Cooley, R., Crake, N. R., Dada, O. O., Diakoumakos, K. D., Dominguez-Fernandez, B., Earnshaw, D. J., Egbujor, U. C., Elmore, D. W., Etchin, S. S., Ewan, M. R., Fedurco, M., Fraser, L. J., Fuentes Fajardo, K. V., Scott Furey, W., George, D., Gietzen, K. J., Goddard, C. P., Golda, G. S., Granieri, P. A., Green, D. E., Gustafson, D. L., Hansen, N. F., Harnish, K., Haudenschild, C. D., Heyer, N. I., Hims, M. M., Ho, J. T., Horgan, A. M., Hoschler, K., Hurwitz, S., Ivanov, D. V., Johnson, M. Q., James, T., Huw Jones, T. A., Kang, G. D., Kerelska, T. H., Kersey, A. D., Khrebtukova, I., Kindwall, A. P., Kingsbury, Z., Kokko-Gonzales, P. I., Kumar, A., Laurent, M. A., Lawley, C. T., Lee, S. E., Lee, X., Liao, A. K., Loch, J. A., Lok, M., Luo, S., Mammen, R. M., Martin, J. W., McCauley, P. G., McNitt, P., Mehta, P., Moon, K. W., Mullens, J. W., Newington, T., Ning, Z., Ling Ng, B., Novo, S. M., O'Neill, M. J., Osborne, M. A., Osnowski, A., Ostadan, O., Paraschos, L. L., Pickering, L., Pike, A. C., Chris Pinkard, D., Pliskin, D. P., Podhasky, J., Quijano, V. J., Raczy, C., Rae, V. H., Rawlings, S. R., Chiva Rodriguez, A., Roe, P. M., Rogers, J., Rogert Bacigalupo, M. C., Romanov, N., Romieu, A., Roth, R. K., Rourke, N. J., Ruediger, S. T., Rusman, E., Sanches-Kuiper, R. M., Schenker, M. R., Seoane, J. M., Shaw, R. J., Shiver, M. K., Short, S. W., Sizto, N. L., Sluis, J. P., Smith, M. A., Ernest Sohna Sohna, J., Spence, E. J., Stevens, K., Sutton, N., Szajkowski, L., Tregidgo, C. L., Turcatti, G., Vandevondele, S., Verhovsky, Y., Virk, S. M., Wakelin, S., Walcott, G. C., Wang, J., Worsley, G. J., Yan, J., Yau, L., Zuerlein, M., Mullikin, J. C., Hurles, M. E., McCooke, N. J., West, J. S., Oaks, F. L., Lundberg, P. L., Klenerman, D., Durbin, R., and Smith, A. J. (2008) Accurate whole human genome sequencing using reversible terminator chemistry, *Nature 456*, 53–59.

27. Wang, J., Wang, W., Li, R., Li, Y., Tian, G., Goodman, L., Fan, W., Zhang, J., Li, J., Guo, Y., Feng, B., Li, H., Lu, Y., Fang, X., Liang, H., Du, Z., Li, D., Zhao, Y., Hu, Y., Yang, Z., Zheng, H., Hellmann, I., Inouye, M., Pool, J.,

Yi, X., Zhao, J., Duan, J., Zhou, Y., Qin, J., Ma, L., Li, G., Zhang, G., Yang, B., Yu, C., Liang, F., Li, W., Li, S., Ni, P., Ruan, J., Li, Q., Zhu, H., Liu, D., Lu, Z., Li, N., Guo, G., Ye, J., Fang, L., Hao, Q., Chen, Q., Liang, Y., Su, Y., San, A., Ping, C., Yang, S., Chen, F., Li, L., Zhou, K., Ren, Y., Yang, L., Gao, Y., Yang, G., Li, Z., Feng, X., Kristiansen, K., Wong, G. K., Nielsen, R., Durbin, R., Bolund, L., Zhang, X., and Yang, H. (2008) The diploid genome sequence of an Asian individual, *Nature 456*, 60–65.

28. Ahn, S. M., Kim, T. H., Lee, S., Kim, D., Ghang, H., Kim, B. C., Kim, S. Y., Kim, W. Y., Kim, C., Park, D., Lee, Y. S., Kim, S., Reja, R., Jho, S., Kim, C. G., Cha, J. Y., Kim, K. H., Lee, B., Bhak, J., and Kim, S. J. (2009) The first Korean genome sequence and analysis: Full genome sequencing for a socio-ethnic group, *Genome Res.*
29. Hormozdiari, F., Alkan, C., Eichler, E. E., and Sahinalp, S. C. (2009) Combinatorial algorithms for structural variation detection in high-throughput sequenced genomes, *Genome Res 19*, 1270–1278.
30. McKernan, K. J., Peckham, H. E., Costa, G. L., McLaughlin, S. F., Fu, Y., Tsung, E. F., Clouser, C. R., Duncan, C., Ichikawa, J. K., Lee, C. C., Zhang, Z., Ranade, S. S., Dimalanta, E. T., Hyland, F. C., Sokolsky, T. D., Zhang, L., Sheridan, A., Fu, H., Hendrickson, C. L., Li, B., Kotler, L., Stuart, J. R., Malek, J. A., Manning, J. M., Antipova, A. A., Perez, D. S., Moore, M. P., Hayashibara, K. C., Lyons, M. R., Beaudoin, R. E., Coleman, B. E., Laptewicz, M. W., Sannicandro, A. E., Rhodes, M. D., Gottimukkala, R. K., Yang, S., Bafna, V., Bashir, A., Macbride, A., Alkan, C., Kidd, J. M., Eichler, E. E., Reese, M. G., De La Vega, F. M., and Blanchard, A. P. (2009) Sequence and structural variation in a human genome uncovered by short-read, massively parallel ligation sequencing using two-base encoding, *Genome Res.*
31. Kazazian, H. H., Jr. (2004) Mobile elements: drivers of genome evolution, *Science 303*, 1626–1632.
32. Sharp, A. J., Locke, D. P., McGrath, S. D., Cheng, Z., Bailey, J. A., Vallente, R. U., Pertz, L. M., Clark, R. A., Schwartz, S., Segraves, R., Oseroff, V. V., Albertson, D. G., Pinkel, D., and Eichler, E. E. (2005) Segmental duplications and copy-number variation in the human genome, *American Journal of Human Genetics 77*, 78–88.
33. Tuzun, E., Sharp, A. J., Bailey, J. A., Kaul, R., Morrison, V. A., Pertz, L. M., Haugen, E., Hayden, H., Albertson, D., Pinkel, D., Olson, M. V., and Eichler, E. E. (2005) Fine-scale structural variation of the human genome, *Nature Genetics 37*, 727–732.
34. Perry, G. H., Tchinda, J., McGrath, S. D., Zhang, J., Picker, S. R., Caceres, A. M., Iafrate, A. J., Tyler-Smith, C., Scherer, S. W., Eichler, E. E., Stone, A. C., and Lee, C. (2006) Hotspots for copy number variation in chimpanzees and humans, *Proc Natl Acad Sci U S A 103*, 8006–8011.
35. Bourque, G., Pevzner, P. A., and Tesler, G. (2004) Reconstructing the genomic architecture of ancestral mammals: lessons from human, mouse, and rat genomes, *Genome Res 14*, 507–516.
36. Bailey, J. A., Baertsch, R., Kent, W. J., Haussler, D., and Eichler, E. E. (2004) Hotspots of mammalian chromosomal evolution, *Genome Biol 5*, R23.
37. Murphy, W. J., Larkin, D. M., Everts-van der Wind, A., Bourque, G., Tesler, G., Auvil, L., Beever, J. E., Chowdhary, B. P., Galibert, F., Gatzke, L., Hitte, C., Meyers, S. N., Milan, D., Ostrander, E. A., Pape, G., Parker, H. G., Raudsepp, T., Rogatcheva, M. B., Schook, L. B., Skow, L. C., Welge, M., Womack, J. E., O'Brien S, J., Pevzner, P. A., and Lewin, H. A. (2005) Dynamics of mammalian chromosome evolution inferred from multispecies comparative maps, *Science 309*, 613–617.
38. Lupski, J. R. (2007) Genomic rearrangements and sporadic disease, *Nat Genet 39*, S43-47.
39. Stankiewicz, P., Shaw, C. J., Dapper, J. D., Wakui, K., Shaffer, L. G., Withers, M., Elizondo, L., Park, S. S., and Lupski, J. R. (2003) Genome architecture catalyzes nonrecurrent chromosomal rearrangements, *Am J Hum Genet 72*, 1101–1116.
40. Lee, J. A., Inoue, K., Cheung, S. W., Shaw, C. A., Stankiewicz, P., and Lupski, J. R. (2006) Role of genomic architecture in PLP1 duplication causing Pelizaeus-Merzbacher disease, *Hum Mol Genet 15*, 2250–2265.
41. Bauters, M., Van Esch, H., Friez, M. J., Boespflug-Tanguy, O., Zenker, M., Vianna-Morgante, A. M., Rosenberg, C., Ignatius, J., Raynaud, M., Hollanders, K., Govaerts, K., Vandenreijt, K., Niel, F., Blanc, P., Stevenson, R. E., Fryns, J. P., Marynen, P., Schwartz, C. E., and Froyen, G. (2008) Nonrecurrent MECP2 duplications mediated by genomic architecture-driven DNA breaks and break-induced replication repair, *Genome Res 18*, 847–858.
42. Carvalho, C. M., Zhang, F., Liu, P., Patel, A., Sahoo, T., Bacino, C. A., Shaw, C., Peacock, S., Pursley, A., Tavyev, Y. J., Ramocki, M. B., Nawara, M., Obersztyn, E., Vianna-Morgante,

A. M., Stankiewicz, P., Zoghbi, H. Y., Cheung, S. W., and Lupski, J. R. (2009) Complex rearrangements in patients with duplications of MECP2 can occur by fork stalling and template switching, *Hum Mol Genet 18*, 2188–2203.

43. Kim, P. M., Lam, H. Y., Urban, A. E., Korbel, J. O., Affourtit, J., Grubert, F., Chen, X., Weissman, S., Snyder, M., and Gerstein, M. B. (2008) Analysis of copy number variants and segmental duplications in the human genome: Evidence for a change in the process of formation in recent evolutionary history, *Genome Res 18*, 1865–1874.
44. Bailey, J. A., Church, D. M., Ventura, M., Rocchi, M., and Eichler, E. E. (2004) Analysis of segmental duplications and genome assembly in the mouse, *Genome Res 14*, 789–801.
45. Hampton, O. A., Den Hollander, P., Miller, C. A., Delgado, D. A., Li, J., Coarfa, C., Harris, R. A., Richards, S., Scherer, S. E., Muzny, D. M., Gibbs, R. A., Lee, A. V., and Milosavljevic, A. (2009) A sequence-level map of chromosomal breakpoints in the MCF-7 breast cancer cell line yields insights into the evolution of a cancer genome, *Genome Res 19*, 167–177.
46. Bailey, J. A., Gu, Z., Clark, R. A., Reinert, K., Samonte, R. V., Schwartz, S., Adams, M. D., Myers, E. W., Li, P. W., and Eichler, E. E. (2002) Recent segmental duplications in the human genome, *Science 297*, 1003–1007.
47. Church, D. M., Goodstadt, L., Hillier, L. W., Zody, M. C., Goldstein, S., She, X., Bult, C. J., Agarwala, R., Cherry, J. L., DiCuccio, M., Hlavina, W., Kapustin, Y., Meric, P., Maglott, D., Birtle, Z., Marques, A. C., Graves, T., Zhou, S., Teague, B., Potamousis, K., Churas, C., Place, M., Herschleb, J., Runnheim, R., Forrest, D., Amos-Landgraf, J., Schwartz, D. C., Cheng, Z., Lindblad-Toh, K., Eichler, E. E., and Ponting, C. P. (2009) Lineage-specific biology revealed by a finished genome assembly of the mouse, *PLoS Biol 7*, e1000112.
48. Lindblad-Toh, K., Wade, C. M., Mikkelsen, T. S., Karlsson, E. K., Jaffe, D. B., Kamal, M., Clamp, M., Chang, J. L., Kulbokas, E. J., 3rd, Zody, M. C., Mauceli, E., Xie, X., Breen, M., Wayne, R. K., Ostrander, E. A., Ponting, C. P., Galibert, F., Smith, D. R., DeJong, P. J., Kirkness, E., Alvarez, P., Biagi, T., Brockman, W., Butler, J., Chin, C. W., Cook, A., Cuff, J., Daly, M. J., DeCaprio, D., Gnerre, S., Grabherr, M., Kellis, M., Kleber, M., Bardeleben, C., Goodstadt, L., Heger, A., Hitte, C., Kim, L., Koepfli, K. P., Parker, H. G., Pollinger, J. P., Searle, S. M., Sutter, N. B., Thomas, R., Webber, C., Baldwin, J., Abebe, A., Abouelleil, A., Aftuck, L., Ait-Zahra, M., Aldredge, T., Allen, N., An, P., Anderson, S., Antoine, C., Arachchi, H., Aslam, A., Ayotte, L., Bachantsang, P., Barry, A., Bayul, T., Benamara, M., Berlin, A., Bessette, D., Blitshteyn, B., Bloom, T., Blye, J., Boguslavskiy, L., Bonnet, C., Boukhgalter, B., Brown, A., Cahill, P., Calixte, N., Camarata, J., Cheshatsang, Y., Chu, J., Citroen, M., Collymore, A., Cooke, P., Dawoe, T., Daza, R., Decktor, K., DeGray, S., Dhargay, N., Dooley, K., Dorje, P., Dorjee, K., Dorris, L., Duffey, N., Dupes, A., Egbiremolen, O., Elong, R., Falk, J., Farina, A., Faro, S., Ferguson, D., Ferreira, P., Fisher, S., FitzGerald, M., Foley, K., Foley, C., Franke, A., Friedrich, D., Gage, D., Garber, M., Gearin, G., Giannoukos, G., Goode, T., Goyette, A., Graham, J., Grandbois, E., Gyaltsen, K., Hafez, N., Hagopian, D., Hagos, B., Hall, J., Healy, C., Hegarty, R., Honan, T., Horn, A., Houde, N., Hughes, L., Hunnicutt, L., Husby, M., Jester, B., Jones, C., Kamat, A., Kanga, B., Kells, C., Khazanovich, D., Kieu, A. C., Kisner, P., Kumar, M., Lance, K., Landers, T., Lara, M., Lee, W., Leger, J. P., Lennon, N., Leuper, L., LeVine, S., Liu, J., Liu, X., Lokyitsang, Y., Lokyitsang, T., Lui, A., Macdonald, J., Major, J., Marabella, R., Maru, K., Matthews, C., McDonough, S., Mehta, T., Meldrim, J., Melnikov, A., Meneus, L., Mihalev, A., Mihova, T., Miller, K., Mittelman, R., Mlenga, V., Mulrain, L., Munson, G., Navidi, A., Naylor, J., Nguyen, T., Nguyen, N., Nguyen, C., Nicol, R., Norbu, N., Norbu, C., Novod, N., Nyima, T., Olandt, P., O'Neill, B., O'Neill, K., Osman, S., Oyono, L., Patti, C., Perrin, D., Phunkhang, P., Pierre, F., Priest, M., Rachupka, A., Raghuraman, S., Rameau, R., Ray, V., Raymond, C., Rege, F., Rise, C., Rogers, J., Rogov, P., Sahalie, J., Settipalli, S., Sharpe, T., Shea, T., Sheehan, M., Sherpa, N., Shi, J., Shih, D., Sloan, J., Smith, C., Sparrow, T., Stalker, J., Stange-Thomann, N., Stavropoulos, S., Stone, C., Stone, S., Sykes, S., Tchuinga, P., Tenzing, P., Tesfaye, S., Thoulutsang, D., Thoulutsang, Y., Topham, K., Topping, I., Tsamla, T., Vassiliev, H., Venkataraman, V., Vo, A., Wangchuk, T., Wangdi, T., Weiand, M., Wilkinson, J., Wilson, A., Yadav, S., Yang, S., Yang, X., Young, G., Yu, Q., Zainoun, J., Zembek, L., Zimmer, A., and Lander, E. S. (2005) Genome sequence, comparative analysis and haplotype structure of the domestic dog, *Nature 438*, 803–819.
49. Xing, J., Zhang, Y., Han, K., Salem, A. H., Sen, S. K., Huff, C. D., Zhou, Q., Kirkness, E. F., Levy, S., Batzer, M. A., and Jorde, L. B. (2009) Mobile elements create structural variation: analysis of a complete human genome, *Genome Res 19*, 1516–1526.

50. Cordaux, R., and Batzer, M. A. (2009) The impact of retrotransposons on human genome evolution, *Nat Rev Genet 10*, 691–703.

51. Bailey, J. A., Liu, G., and Eichler, E. E. (2003) An Alu transposition model for the origin and expansion of human segmental duplications, *Am J Hum Genet 73*, 823–834.

52. Lee, J. A., Carvalho, C. M., and Lupski, J. R. (2007) A DNA replication mechanism for generating nonrecurrent rearrangements associated with genomic disorders, *Cell 131*, 1235–1247.

53. Hastings, P. J., Ira, G., and Lupski, J. R. (2009) A microhomology-mediated break-induced replication model for the origin of human copy number variation, *PLoS Genet 5*, e1000327.

54. Wade, C. M., and Daly, M. J. (2005) Genetic variation in laboratory mice, *Nat Genet 37*, 1175–1180.

55. Wade, C. M., Kulbokas, E. J., 3rd, Kirby, A. W., Zody, M. C., Mullikin, J. C., Lander, E. S., Lindblad-Toh, K., and Daly, M. J. (2002) The mosaic structure of variation in the laboratory mouse genome, *Nature 420*, 574–578.

56. Saar, K., Beck, A., Bihoreau, M. T., Birney, E., Brocklebank, D., Chen, Y., Cuppen, E., Demonchy, S., Dopazo, J., Flicek, P., Foglio, M., Fujiyama, A., Gut, I. G., Gauguier, D., Guigo, R., Guryev, V., Heinig, M., Hummel, O., Jahn, N., Klages, S., Kren, V., Kube, M., Kuhl, H., Kuramoto, T., Kuroki, Y., Lechner, D., Lee, Y. A., Lopez-Bigas, N., Lathrop, G. M., Mashimo, T., Medina, I., Mott, R., Patone, G., Perrier-Cornet, J. A., Platzer, M., Pravenec, M., Reinhardt, R., Sakaki, Y., Schilhabel, M., Schulz, H., Serikawa, T., Shikhagaie, M., Tatsumoto, S., Taudien, S., Toyoda, A., Voigt, B., Zelenika, D., Zimdahl, H., and Hubner, N. (2008) SNP and haplotype mapping for genetic analysis in the rat, *Nat Genet 40*, 560–566.

57. Medvedev, P., Stanciu, M., and Brudno, M. (2009) Computational methods for discovering structural variation with next-generation sequencing, *Nat Methods 6*, S13-S20.

58. Du, J., Bjornson, R. D., Zhang, Z. D., Kong, Y., Snyder, M., and Gerstein, M. B. (2009) Integrating sequencing technologies in personal genomics: optimal low cost reconstruction of structural variants, *PLoS Comput Biol 5*, e1000432.

59. Bashir, A., Volik, S., Collins, C., Bafna, V., and Raphael, B. J. (2008) Evaluation of paired-end sequencing strategies for detection of genome rearrangements in cancer, *PLoS Comput Biol 4*, e1000051.

60. Alkan, C., Kidd, J. M., Marques-Bonet, T., Aksay, G., Antonacci, F., Hormozdiari, F., Kitzman, J. O., Baker, C., Malig, M., Mutlu, O., Sahinalp, S. C., Gibbs, R. A., and Eichler, E. E. (2009) Personalized copy number and segmental duplication maps using next-generation sequencing, *Nat Genet 41*, 1061–1067.

61. Yoon, S., Xuan, Z., Makarov, V., Ye, K., and Sebat, J. (2009) Sensitive and accurate detection of copy number variants using read depth of coverage, *Genome Res.*

62. Chiang, D. Y., Getz, G., Jaffe, D. B., O'Kelly, M. J., Zhao, X., Carter, S. L., Russ, C., Nusbaum, C., Meyerson, M., and Lander, E. S. (2009) High-resolution mapping of copy-number alterations with massively parallel sequencing, *Nat Methods 6*, 99–103.

63. Frazer, K. A., Eskin, E., Kang, H. M., Bogue, M. A., Hinds, D. A., Beilharz, E. J., Gupta, R. V., Montgomery, J., Morenzoni, M. M., Nilsen, G. B., Pethiyagoda, C. L., Stuve, L. L., Johnson, F. M., Daly, M. J., Wade, C. M., and Cox, D. R. (2007) A sequence-based variation map of 8.27 million SNPs in inbred mouse strains, *Nature 448*, 1050–1053.

64. Li, H., and Durbin, R. (2009) Fast and Accurate Short Read Alignment with Burrows-Wheeler Transform, *Bioinformatics.*

65. Novoalign. (www. novocraft.com).

66. Mosiak. (http://code.google.com/p/mosaik-aligner/).

67. Li, H., Handsaker, B., Wysoker, A., Fennell, T., Ruan, J., Homer, N., Marth, G., Abecasis, G., and Durbin, R. (2009) The Sequence Alignment/Map format and SAMtools, *Bioinformatics 25*, 2078–2079.

68. Picard. (http://picard.sourceforge.net/).

69. Mills, R. E., Luttig, C. T., Larkins, C. E., Beauchamp, A., Tsui, C., Pittard, W. S., and Devine, S. E. (2006) An initial map of insertion and deletion (INDEL) variation in the human genome, *Genome Res 16*, 1182–1190.

70. Green, P. (unpublished) http://www.phrap.org/phredphrapconsed.html.

71. Hastings, P. J., Lupski, J. R., Rosenberg, S. M., and Ira, G. (2009) Mechanisms of change in gene copy number, *Nat Rev Genet 10*, 551–564.

72. Korbel, J. O., Urban, A. E., Affourtit, J. P., Godwin, B., Grubert, F., Simons, J. F., Kim, P. M., Palejev, D., Carriero, N. J., Du, L., Taillon, B. E., Chen, Z., Tanzer, A., Saunders, A. C., Chi, J., Yang, F., Carter, N. P., Hurles, M. E., Weissman, S. M., Harkins, T. T., Gerstein, M. B., Egholm, M., and Snyder, M. (2007) Paired-end mapping reveals extensive structural variation in the human genome, *Science 318*, 420–426.

73. Mamanova, L., Coffey, A. J., Scott, C. E., Kozarewa, I., Turner, E. H., Kumar, A., Howard, E., Shendure, J., and Turner, D. J. Target-

enrichment strategies for next-generation sequencing, *Nat Methods 7*, 111–118.

74. Liang, Q., Conte, N., Skarnes, W. C., and Bradley, A. (2008) Extensive genomic copy number variation in embryonic stem cells, *Proc Natl Acad Sci U S A 105*, 17453–17456.
75. Bruder, C. E., Piotrowski, A., Gijsbers, A. A., Andersson, R., Erickson, S., de Stahl, T. D., Menzel, U., Sandgren, J., von Tell, D., Poplawski, A., Crowley, M., Crasto, C., Partridge, E. C., Tiwari, H., Allison, D. B., Komorowski, J., van Ommen, G. J., Boomsma, D. I., Pedersen, N. L., den Dunnen, J. T., Wirdefeldt, K., and Dumanski, J. P. (2008) Phenotypically concordant and discordant monozygotic twins display different DNA copy-number-variation profiles, *Am J Hum Genet 82*, 763–771.
76. Piotrowski, A., Bruder, C. E., Andersson, R., de Stahl, T. D., Menzel, U., Sandgren, J., Poplawski, A., von Tell, D., Crasto, C., Bogdan, A., Bartoszewski, R., Bebok, Z., Krzyzanowski, M., Jankowski, Z., Partridge, E. C., Komorowski, J., and Dumanski, J. P. (2008) Somatic mosaicism for copy number variation in differentiated human tissues, *Hum Mutat 29*, 1118–1124.
77. Lam, K. W., and Jeffreys, A. J. (2007) Processes of de novo duplication of human alpha-globin genes, *Proc Natl Acad Sci U S A 104*, 10950–10955.
78. Flores, M., Morales, L., Gonzaga-Jauregui, C., Dominguez-Vidana, R., Zepeda, C., Yanez, O., Gutierrez, M., Lemus, T., Valle, D., Avila, M. C., Blanco, D., Medina-Ruiz, S., Meza, K., Ayala, E., Garcia, D., Bustos, P., Gonzalez, V., Girard, L., Tusie-Luna, T., Davila, G., and Palacios, R. (2007) Recurrent DNA inversion rearrangements in the human genome, *Proc Natl Acad Sci U S A 104*, 6099–6106.
79. Takahashi, K., and Yamanaka, S. (2006) Induction of pluripotent stem cells from mouse embryonic and adult fibroblast cultures by defined factors, *Cell 126*, 663–676.
80. Boland, M. J., Hazen, J. L., Nazor, K. L., Rodriguez, A. R., Gifford, W., Martin, G., Kupriyanov, S., and Baldwin, K. K. (2009) Adult mice generated from induced pluripotent stem cells, *Nature 461*, 91–94.
81. Wernig, M., Lengner, C. J., Hanna, J., Lodato, M. A., Steine, E., Foreman, R., Staerk, J., Markoulaki, S., and Jaenisch, R. (2008) A drug-inducible transgenic system for direct reprogramming of multiple somatic cell types, *Nat Biotechnol 26*, 916–924.
82. Stranger, B. E., Forrest, M. S., Dunning, M., Ingle, C. E., Beazley, C., Thorne, N., Redon, R., Bird, C. P., de Grassi, A., Lee, C., Tyler-Smith, C., Carter, N., Scherer, S. W., Tavare, S., Deloukas, P., Hurles, M. E., and Dermitzakis, E. T. (2007) Relative impact of nucleotide and copy number variation on gene expression phenotypes, *Science 315*, 848–853.
83. Shearin, A. L., and Ostrander, E. A. Leading the way: canine models of genomics and disease, *Dis Model Mech 3*, 27–34.
84. Iraqi, F. A., Churchill, G., and Mott, R. (2008) The Collaborative Cross, developing a resource for mammalian systems genetics: a status report of the Wellcome Trust cohort, *Mamm Genome 19*, 379–381.
85. Manolio, T. A., Collins, F. S., Cox, N. J., Goldstein, D. B., Hindorff, L. A., Hunter, D. J., McCarthy, M. I., Ramos, E. M., Cardon, L. R., Chakravarti, A., Cho, J. H., Guttmacher, A. E., Kong, A., Kruglyak, L., Mardis, E., Rotimi, C. N., Slatkin, M., Valle, D., Whittemore, A. S., Boehnke, M., Clark, A. G., Eichler, E. E., Gibson, G., Haines, J. L., Mackay, T. F., McCarroll, S. A., and Visscher, P. M. (2009) Finding the missing heritability of complex diseases, *Nature 461*, 747–753.

Chapter 12

Structural Genetic Variation in the Context of Somatic Mosaicism

Jan P. Dumanski and Arkadiusz Piotrowski

Abstract

Somatic mosaicism is the result of postzygotic de novo mutation occurring in a portion of the cells making up an organism. Structural genetic variation is a very heterogeneous group of changes, in terms of numerous types of aberrations that are included in this category, involvement of many mechanisms behind the generation of structural variants, and because structural variation can encompass genomic regions highly variable in size. Structural variation rapidly evolved as the dominating type of changes behind human genetic diversity, and the importance of this variation in biology and medicine is continuously increasing.

In this review, we combine the evidence of structural variation in the context of somatic cells. We discuss the normal and disease-related somatic structural variation. We review the recent advances in the field of monozygotic twins and other models that have been studied for somatic mutations, including other vertebrates. We also discuss chromosomal mosaicism in a few prime examples of disease genes that contributed to understanding of the importance of somatic heterogeneity. We further highlight challenges and opportunities related to this field, including methodological and practical aspects of detection of somatic mosaicism. The literature devoted to interindividual variation versus papers reporting on somatic variation suggests that the latter is understudied and underestimated. It is important to increase our awareness about somatic mosaicism, in particular, related to structural variation. We believe that further research of somatic mosaicism will prove beneficial for better understanding of common sporadic disorders.

Key words: Copy number variation, Heterogeneity, Mosaicism, Chimerism, Aneuploidy, Somatic cell, Monozygotic twins, Mitochondrial genome, Nuclear genome, Uniparental disomy

1. Introduction, Definitions, and Scope

The focus of this review is based on normal and disease-related structural genetic variation in the context of somatic cells. Structural variation has emerged over the past 5 years as the dominating type of human interindividual variation. It is therefore reasonable to assume that this variation will also be prevalent between different populations of normal somatic cells within the same organism. The extent of somatic genetic variation and its normal phenotypic and

Lars Feuk (ed.), *Genomic Structural Variants: Methods and Protocols*, Methods in Molecular Biology, vol. 838,
DOI 10.1007/978-1-61779-507-7_12,

disease-related functional consequences are not as well studied as the interindividual genetic differences. Our scope is also to highlight challenges as well as opportunities related to the analyses of genetic variation in somatic cells. Structural genetic variation accounts for a heterogeneous group of alterations, both in terms of the many types of aberrations that are included in this category and their variable genomic size. This structural variation includes deletions, duplications, insertions, inversions, translocations, retrotranspositions, and other complex rearrangements that are not easily classified (1, 2). The common denominator underlying these aberrations is that they change the chromosomal architecture, and several mechanisms behind the development of structural variation have been proposed (3). The best explored subtype of structural variation involves changes affecting copy number of DNA segments (CNV), for fragments ranging from <1 kb to many megabases, including entire chromosomes (aneuploidy). Uniparental disomy (UPD) will also be discussed, as it can be considered a special case of structural variation. UPD is a result of structural changes; the most common of which is related to meiotic or mitotic nondisjunction. UPD is a mechanism for disturbance of gene expression through either loss of imprinting or reduction to homozygosity and has been associated with many diseases (4–7).

Somatic mosaicism is defined by the presence of genetically distinct lineages of somatic cells in a single organism that are derived from the same zygote. The term mosaicism can be applied to all types of aberrations, from point mutations to aneuploidies. The latter group of copy number alterations involved in mosaicism is actually one of the best studied, due to recent application of molecular cytogenetic analyses in the routine diagnostics of patients referred to clinical geneticists (8–10). Chimerism is a closely related but distinct term, which should not be confused with mosaicism. Chimerism is also defined by the presence of two or more genetically distinct cell lines within the same organism. However, in the case of chimerism, there is a fusion of cells from two different zygotes within a single embryo. Human chimerism is commonly recognized due to the presence of both 46,XX and 46,XY cell lines, which can manifest clinically and are detectable in cytogenetic analyses. In animal studies, creation of chimeras is a routine laboratory procedure aimed at developing knock-out models for specific genes. Various types of human chimeras have been described (11–16), but chimerism is outside the scope of this review. Finally, the cancer field is a prime example of numerous distinct disorders caused predominantly by somatic mutations, which frequently involve structural rearrangements. However, tumor-specific mutations are not our focus and we refer to reviews on this topic (17–21).

1.1. Physiological Somatic Mosaicism; Shortening of Telomeres, Rearrangements of Immunoglobulin/T-Cell Receptor Genes, and Mitochondrial DNA Deletions in Aging

The variation of telomere length represents a very special case of structural genetic variation, which is relevant both in the context of interindividual and somatic variation. The length of telomeres functions as a clock for the number of cell divisions, limiting the replicative capacity of cells, which is important for cell senescence, aging, and cancer. For a detailed discussion on this subject, we refer to reports summarizing the state of this field (22–26). The somatic rearrangement of immunoglobulin (Ig) and T-cell receptor (TCR) genes in B and T lymphocytes is another example of physiological somatic mosaicism. The Ig and TCR genes are inactive in most cells, but undergo remarkable, tightly controlled reshuffling in order to activate them, which leads to individual B- or T-lymphocyte producing a monospecific antibody or TCR, respectively. This process involves recombination of VDJ (or VJ) distinct gene regions and class-switch recombination, both processes are induced by specific enzymes; RAG1 and RAG2 recombinases (27). In contrast to the tightly controlled processes of telomere shortening and B-/T-lymphocyte recombination, other known examples of somatic mosaicism for structural rearrangements are apparently a result of stochastic random processes. Evidence suggesting aberrant rearrangements of immunoglobulin genes in somatic cells other than B lymphocytes has been reported. In one of our projects, aiming at characterization of candidate genes from a region of human chromosome 22 that is located in the vicinity of immunoglobulin lambda chain genes (IGL), we observed rearrangements of IGL locus that were apparently not restricted to B lymphocytes (28). One sample was peripheral blood DNA from a patient with schwannomatosis, a rare disease which is related but clinically distinct to neurofibromatosis type-2 and presents with multiple peripheral schwannomas. The level of IGL locus deletion detected in this sample suggested that the majority of nucleated cells in the peripheral blood of this patient contained rearranged IGL locus, highly reminiscent of rearrangements which are normally restricted to B cells. The other sample suggesting a similar IGL rearrangement was a benign tumor (schwannoma) from a patient with schwannomatosis. Thus, although the activation of RAG1 and RAG2 recombinases is a strictly controlled cellular process, rare mistakes are possible.

Although the focus of this review is on CNVs in the nuclear genome, rearrangements of mitochondrial DNA (mtDNA) should be discussed as this topic is quite relevant to the somatic mosaicism for structural DNA variation. A somatic mammalian cell contains 1,000–10,000 copies of the mitochondrial genome and mutations of the mtDNA very frequently involve deletions. There is a random distribution of mtDNA molecules during cell division into daughter cells and this can result in different number of mutant mtDNA molecules in two daughter cells. In other words, a cell carrying a low number of mutated mtDNA molecules can give rise to a daughter cell with high level of mutations (29, 30). The first pathogenic mtDNA mutations in humans were reported in 1988 with affected patients either having homoplasmy (presence of only mutated DNA in cells) or heteroplasmy, i.e., a mixture of wild-type and mutated mtDNA (31, 32). More than 100 different types of mtDNA mutations have so far been associated with inherited mitochondrial diseases (33). Normal aging is also a process frequently associated with mtDNA mutations. There has to be a minimal threshold of heteroplasmic mtDNA damage in a cell to cause respiratory chain deficiency. It has been shown that different types of heteroplasmic mtDNA mutations have different thresholds for induction of respiratory chain dysfunction, ranging from 90% for some tRNA point mutations to 60% for mtDNA deletions (reviewed in ref. 34). It is generally accepted that mtDNA deletions accumulate in different tissues during the human aging process (35, 36). However, there are only a few reports on accumulation of mtDNA point mutations, which may seem surprising. One suggested explanation for the paucity of reports on acquired mtDNA point mutations in aging could be the lack of reliable methods for accurately quantifying low levels of point mutations in aging tissues (30). An alternative explanation might be related to the different mutation rates for point mutations and structural rearrangements in the mtDNA, which is not yet well defined. In the nuclear genome, the locus-specific mutation rate for CNVs has been suggested to be 2–4 orders of magnitude higher than the corresponding number for single-nucleotide polymorphisms (SNPs) (3, 37, 38).

1.2. Monozygotic Twins Are an Excellent Model for Studies of Somatic Mosaicism

MonoZygotic twins (MZ, also called identical) and DiZygotic twins (DZ, also called fraternal) represent an extraordinary resource for the understanding of many disorders and phenotypes, including complex genetic diseases. Approximately 3% of deliveries are twin births, and ~10% of these are monozygotic (39–41). There are three types of monozygotic pregnancies with regard to placenta and membranes; dichorionic diamniotic pregnancy; monochorionic diamniotic pregnancy; and monochorionic monoamniotic pregnancy. The second type is by far the most common (39). Monozygotic twins span the boundary between two individuals. They are two different persons, but can also be treated as a single subject, genetically matched at conception and present in two copies. MZ co-twins provide the most closely matched genetic controls possible. They are age matched; they essentially share the same environment throughout gestation and usually similar environments thereafter. Most importantly, their nuclear and mitochondrial genomes are identical at conception. As somatic mosaicism is defined by the presence of genetically different populations of somatic cells in a single organism, monozygotic twins are therefore a suitable model for the analysis of mosaicism, since any genetic differences in twins derived from the same zygote represent an indisputable example of somatic variation. DZ twins on the other hand arise when two oocytes are fertilized by two different sperm cells and are independently implanted in the uterus. Two DZ twins therefore share only as much of their genomes as any two siblings would.

Twin biology has been studied for decades and numerous twin repositories exist (39, 42). Twins are a powerful tool for studying various diseases, evaluating quantitative-trait loci, estimating heritability, studying differences in gene expression, and testing hypotheses regarding gene–environment interactions. Traditionally, phenotypic discordance in MZ twins has been attributed to the environment and variable penetrance. The "classical twin method" has been used since the 1920s as a way of determining whether a trait is influenced mainly by environmental or genetic components. This method compares the concordance rates of MZ twins that are reared together with the same measures as same-sex DZ twins that are also reared together. A heritable phenotype will be more concordant in MZ twins than in DZ twins (43, 44). This type of analysis has been applied to a large number of phenotypes, for instance, asthma, ADHD, autism, psoriasis, cancer, and obesity (45–54).

The classical twin method relies on several general assumptions:

1. That MZ twins are 100% genetically identical, while DZ twins share as much of their genomes as any two siblings.
2. That investigators can reliably distinguish between MZ and DZ twins.
3. That the results from twins can be generalized to fit the single-born population.
4. That the environments of MZ and DZ twins are controlled and are approximately equal.

A criticism of the conclusions from some of the classical twin studies concern the fact that differing environments have been considered as equal (55). It is generally accepted that the results from phenotypically discordant MZ twin studies can be generalized to the single-born population. An illustrative example supporting this notion is provided by phenotypically discordant MZ twins in identification of mutations in the *IRF6* gene implicated in Van der Woude syndrome (VWS, OMIM 119300) (56). In this study, mutation analysis of genes from the candidate VWS region was done in a MZ twins discordant for the VWS. A nonsense mutation was identified in the *IRF6* gene, which was further supported by the identification of additional mutations in other VWS families.

1.3. Identification of Frequent CNVs in Monozygotic Twins Has Broad Implications

Prior to our report (57), it has been generally assumed that MZ twins are genetically identical and that phenotypic differences between MZ co-twins from the same pair are mainly due to environmental factors. Rare examples of genetic (41, 58) and, more recently, epigenetic differences between MZ twins (59–62) have, however, been described. The former are mainly related to aneuploidies. In collaboration with the Swedish Twin Registry in Stockholm and Dutch Twin Registry in Amsterdam, we performed the analysis of nine MZ twins discordant for Parkinson's disease (or closely related neurodegenerative phenotypes) and ten pairs of healthy phenotypically unselected MZ twins. We applied 32K BAC clone array and Illumina 300K SNP bead array. Both platforms were robust, sensitive, and complementary in the detection and validation of CNVs between MZ twins (57). From the beginning of CNV analysis in MZ twins, we observed results indicative of frequent CNV differences between co-twins. Essentially every pair of monozygotic twins showed evidence for somatic CNVs.

We also showed that the identification of CNVs can explain disease-related discordance between two co-twins and one example is illustrative (57). In one twin pair, we detected subtle (affecting a minority of cells) but genomically large aberrations on chromosome 4 (affecting 85 Mb in 10–15% of cells) and 11q, affecting 22 Mb in 20–25% of cells (for details, see Fig. 1 in ref. 57). The very large size of the deletions on chromosomes 4 and 11 implicated the involvement of some tumor-specific rearrangements. We therefore searched the literature and discovered that deleted regions on chromosomes 4 and 11 were previously described as involved in the development of chronic lymphocytic leukemia (CLL, OMIM 151400) (63, 64). At the time of our CNV analysis, however, we were unaware whether the relevant twin had been diagnosed with CLL. When reassessing the clinical history of the co-twin with deletions on chromosomes 4 and 11, we confirmed that CLL was actually a part of his clinical description. In essence, we uncovered a minimal residual CLL disease in this twin. These results thus show that the analysis of CNV differences in discordant MZs can be used to identify pathogenic changes and it illustrates the importance of performing the analysis in the affected tissue. Another noteworthy aspect of the CNV analysis in MZ twins is detection of overlapping patterns of CNVs in nine studied pairs that were discordant for neurodegenerative phenotypes. Thirty-one loci deviated in three phenotypically discordant MZ pairs and four loci deviated in four MZ pairs (ref. 57, Supplementary data, Tables S3 and S4). Thus, although we studied only nine discordant pairs, these chromosomal segments emerged as candidates for the development of Parkinson's disease. There is, however, one important drawback. We only studied one tissue (blood), which is distant (in the perspective of ontological development) from dopaminergic neurons that cause the Parkinson's disease. Therefore, we do not know whether the critical cells causing the disorder are affected by these CNVs. Nevertheless, these results are suggesting that overlapping patterns of CNVs can be detected, even from a limited number of twins.

Our results challenge the first assumption of the above-mentioned classical twin method by showing that essentially all of MZ twin pairs analyzed had one or more within-pair differences for CNVs. Thus, they are not 100% genetically identical. This tenet is valid for MZ twins that are both healthy concordant and discordant for a disease (57). This opens up the field for analyses of cohorts of phenotypically discordant MZ twins as a way of characterizing genetic de novo predisposition factors by uncovering CNV differences between affected and unaffected MZ twins of the same pair. The advantage offered by a MZ twin-model is that any within-pair difference must be occurring de novo; i.e., at anytime during the twins' lifetime, but likely early, during gestation, which therefore affects many cell lineages. Furthermore, the analyses of CNV differences between MZ twins are not confounded by hundreds or thousands of CNVs that normally expected to occur between two unrelated subjects (1, 2, 65). Thus, the use of MZ twins should allow us to define genotype–phenotype associations from a smaller number of studied subjects. Because there exist numerous cohorts of MZ twins discordant for many diseases (42), this approach could be implemented for evaluating the impact of CNVs on many conditions. It has also been recently suggested that epigenetic differences in DNA methylation between two MZ co-twins could be responsible for phenotypic discordance, e.g., in aging and schizophrenia (59–62, 66). This is another level of analysis that can be applied in the context of phenotypically discordant MZ twins. The epigenetic analyses and studies of structural genetic differences should be seen as two complementary fields, which clearly provide an opportunity for synergy.

Our CNV findings influence the understanding of genotypic somatic cell diversity in general and in monozygotic twins in particular (57). The use of term "identical" in the context of genetic makeup of monozygotic twins should be revised. Monozygotic twins are apparently genetically identical only at conception (67). We would also like to stress that, although our published data seems to suggest a difference (57), it is not currently justified to conclude that MZ twins phenotypically discordant for Parkinson disease (or related phenotypes) are more frequently affected by CNV than are concordant phenotypically unselected MZ twins. To assess this question, a more thorough analysis, to determine the normal frequency and genomic distribution of de novo CNV (or "baseline CNV") between MZ twins, performed in a comprehensive set of phenotypically unselected MZ twins, is needed.

Our continued work aims to delineate the "baseline CNV" in MZ twins by studying a considerably larger cohort of twins (68), and the Illumina SNP beadchips are used as the primary analysis platform. The results obtained from one pair of twins should be mentioned here, as they highlight the issue of considerably increased sensitivity for the detection of somatic CNVs. In one pair of healthy female MZ twins, we uncovered a very low level of somatic mosaicism for aneuploidy of chromosomes X and Y. Genotyping of blood DNA from both twins using Illumina 610 SNP array indicated copy number imbalance for chromosome X in one twin. Comparison of B-allele frequency (BAF) values between twins revealed loss of heterozygozity for the entire chromosome X in twin 1 affecting a minority of cells, since Log *R* ratio values did not deviate. Fluorescence in situ hybridization (FISH) analysis confirmed monosomy X (45,X) in 7% of proband nucleated blood cells. Unexpectedly, FISH analysis also showed that the other twin contains 45,X and 46,XY lineages, both present in 1% of cells (68). The mechanism behind formation of these aneuploidies suggests aberrant chromosome segregation events in meiosis and mitoses following conception. A probable mechanism is an error in the meiotic division of either the father or the mother; 24,XY gamete meeting 23,X gamete, alternatively 24,XX and 23,Y gametes joined together. This has been followed by trisomy rescue in the mitotic divisions of the embryo, giving rise to one normal predominant 46,XX cell lineage and two less prevalent abnormal lineages (45,X and 46,XY). The latter lines were distributed with different frequencies to the twins. The SNP genotyping shows 5,056 heterozygous chromosome X SNPs in both twins. This indicates that the mechanism behind our results is incompatible with a normal 46,XY zygote affected by mitotic nondisjunction events leading to mosaic aneuploidies. The explanation behind the normal phenotype with no expression of Ullrich-Turner syndrome and no discordance for phenotypic sex of the twins studied here is likely the low number of aneuploid cells. This report (68) contributes to our understanding of the frequency of structural genomic variation in normal monozygotic twins. Furthermore, we show excellent sensitivity of the Illumina platform for detection of aberrations that are present in a minority of the studied cells, using BAF values as the primary tool for somatic CNV discovery. Our conclusions regarding the sensitivity of detection of mosaicism that is present in a small proportion of studied cells are in agreement with a very recent study reporting the detection of aneuploidies in the range of 5% of cells using Illumina SNP beadchips (10) (see also below).

1.4. Other Examples of Mosaicism for Structural Variation with No Obvious Relation to Disease

Examples of aneuploidy and other large subchromosomal alterations in the context of somatic mosaicism, which were not coupled with an aberrant phenotype, represent the bulk of available information. The methods for the discovery of CNVs are continuously improving, which helps in more efficient identification of cases with somatic mosaicism. This in turn may facilitate the discovery of somatic chromosomal aberrations in normal subjects, which were previously detected in the constitutional nonmosaic state and were then usually associated with an abnormal phenotype. Monosomy

X, trisomy 13, tetrasomy 5p, trisomy and tetrasomy 9p, and marker chromosome 10 are some examples of somatic mosaicism in normal subjects that support this concept (68–73). The lower level of cells affected by the abnormal allele is the probable explanation behind the normal phenotype. Another intriguing aspect of somatic mosaicism comes from reports suggesting the frequent presence of aneuploid cells in developing as well as adult neurons in the human brain (74–76). The high degree of observed aneuploid cells (10–40%) suggests that this process may have significance in the brain development or aging.

There are also reports of considerably smaller genomic rearrangements in normal tissues. In our recent study, we investigated genome-wide in vivo somatic mosaicism using array-CGH with microarray of tiling path of BAC clones (77). We analyzed 34 normal tissue samples from three adult subjects and identified at least six somatic CNVs, which size ranged from 82 to 176 kb, often encompassing known genes, potentially affecting gene function. The majority of described CNVs were previously shown to be polymorphic between unrelated subjects, suggesting that some CNVs reported as germline might represent somatic events, since in most studies of this kind, only one tissue is typically examined and analysis of parents of the studied subjects is not routinely performed. For example, one somatic CNV on chromosome 8 (chr8:86.5–86.9 Mb, hg17) was reported as highly variable (224 out of 270 subjects) in the study of Redon et al. (1). Another striking result was that five out of six reported somatic CNVs were flanking the gaps in the human genome sequence. The gaps in the assembly of human DNA sequence are challenging to bridge, correspond to considerable fraction of the human genome and have been classified into three groups (78–80). It is accepted that class III "structural" gaps are due to sequence features that are refractory to cloning in bacterial vectors. Indeed, we have been involved in work with such a region, around the constitutional t(11;22) translocation breakpoint. We have shown that palindromes around translocation breakpoints are unstable and cause usually reliable vectors (such as BACs) to rearrange their inserts during culturing in *Escherichia coli* (81). The sequencing by synthesis approach has been very recently shown as a way to bridge some of the existing type III gaps (80). One possible explanation behind the lack of consensus DNA sequence across class I and especially class II gaps might be the extreme somatic variability of DNA sequences from these loci in individual cells that are used for construction of libraries, which are a substrate for sequencing.

Retro-transposition is another example of a genomically confined structural rearrangement which occurs as a somatic event. Coufal et al. (82) detected an increase in the copy number of endogenous L1 elements in the hippocampus, and in several other regions of adult human brain, when compared to the copy number of endogenous L1s in heart or liver samples from the same donor. These data suggest that de novo L1 retrotransposition events may contribute to individual somatic mosaicism. In addition, there are examples of gene-specific effects of somatic CNVs. Rearrangements in *RHD* and *RHCE* gene loci from chromosome 1 are considered as one of the reasons for Rh blood antigen mosaicism, which may lead to hematologic complications (83). Furthermore, the processes of copy number change behind human α-globin gene deletions were studied directly in genomic DNA by using single-DNA-molecule methods. These deletions proved to be common in both blood and sperm and may possibly have advantageous influence on the phenotype, especially in regions with high risk of malaria (84). It should also be noted that a low-level subchromosomal mosaicism might often escape detection. For instance, by using sensitive PCR-based approach, Flores et al. (85) demonstrated low-level mosaicism for inversion polymorphisms (down to 10^{-5} inverted versus wild-type structures) in blood, which would not have been detected by other methods.

2. Somatic Mosaicism for Structural Rearrangements and Genetic Disorders

2.1. Mosaicism and Genes Causing Monogenic Disorders

Mutations in specific disease genes causing Mendelian disorders that are coupled with somatic mosaicism are being reported at an ever increasing rate. Mosaicism can result in a milder disease phenotype or can unmask an expression of a mutation that would otherwise be lethal to the embryo. Somatic changes have also been shown to cause reversion of disease phenotype. It is likely that many instances of somatic mosaicism are not clinically recognized since the patient may show a borderline, mild clinical phenotype due to a low level of cells carrying a mutation. Another reason underlying this ascertainment bias is that mosaicism is primarily relevant for sporadic cases (de novo mutations) with no previous family history of a disease. The steadily growing body of data indicates that somatic mosaicism for pathogenic mutations affecting known disease genes should be seen as a rule applicable to the vast majority of disease-related genes, rather than as an exception. The spectrum of inherited mutations affecting known disease genes is highly variable and point mutations are usually more common than structural rearrangements. This clinically oriented field has already delivered important lessons regarding genotype and phenotype relationships, mutation frequencies, and disease expressivity as well as insights into genetic fitness of patients affected with different mutations. As there are several comprehensive reviews on this subject (86–93), we will therefore discuss in depth only selected aspects of a few well-known disease genes affected by pathogenic structural rearrangements and which also display somatic mosaicism.

Duchenne muscular dystrophy (DMD) is an X-chromosome linked, lethal neuromuscular disorder, affecting one in 3,500 live born males. The *DMD* gene is the largest known gene in humans spanning 2.4 Mbp and containing 79 exons (94–96). This gene has been extensively studied and shows exciting findings with regard to somatic mosaicism for mutations that cause the disease. The mutation spectrum of the *DMD* gene is atypical as up to 75% of DMD cases are due to structural rearrangements; i.e., a deletion or duplication of one or more exons. The high incidence of DMD and early lethality of the disorder is coupled with high de novo mutation frequency. The gene contains two mutational hot spots; the most common distal (exons 45–52) as well as the proximal across exons 2–7 (97). There is a difference in distribution of rearrangements within the gene in patients showing mosaicism versus nonmosaic cases, which is intriguing. Deletions in patients showing somatic mosaicism are preferentially clustered around exon 2 (98, 99). This suggests that the mechanism behind generation of these structural rearrangements is different in mitosis versus meiosis. Another noteworthy aspect of *DMD* gene research is the estimation of mutation rates for structural rearrangements and point mutations, which indicate that the former is several orders of magnitude higher than the corresponding number for point mutations (37). It should be mentioned in this context that *DMD* gene is not rich in segmental duplications (also called low copy repeats, LCRs), which are known to confer a high risk of deletions/duplications due to nonallelic homologous recombination between highly similar copies of sequences present in the segmental duplications. The third interesting aspect of DMD gene is a reversion of disease phenotype in muscle fibers of DMD patients, via mitotic rearrangements restoring the reading frame and allowing some dystrophin expression to occur. In several cases, the revertant mutation appeared to be in the distal deletion hotspot, supporting the suggestion that this region is inherently unstable. Somatic reversions have also been described for many other diseases (100–102) (reviewed in refs. 88, 90).

Neurofibromatosis Type 1 (NF1) is an autosomal dominant tumor syndrome caused by mutations in the *NF1* gene on chromosome 17 (103–105). Approximately, 5% of NF1 patients are affected by large (1.2–1.4 Mb) disease causing deletions, which remove the entire *NF1* gene, along with several other genes located in its vicinity (106, 107). Most of these large deletions are the result of nonallelic homologous recombination between large segmental duplications, flanking the *NF1* gene. Prior to the important study by Kehrer-Sawatzki et al. (106), mosaicism for the *NF1* gene deletions had been detected in a few cases (108, 109). However, the unexpectedly high frequency of mosaicism reaching 40% was detected when sporadic NF1 patients were specifically targeted for detailed analysis of deletions using DNA derived from several tissues. Mosaic patients also lacked the cognitive defects and facial dysmorphology typically associated with *NF1* microdeletions, suggesting a clear genotype–phenotype correlation in NF1. In patients with mosaicism, the proportion of cells with the deletion was 91–100% in peripheral leukocytes but was much lower (51–80%) in buccal smears or peripheral skin fibroblasts. Therefore, the analysis of several tissues is recommended for all sporadic NF1 patients, and especially patients with NF1 microdeletions. Detailed analysis of the deletion breakpoints revealed other surprising results. In contrast to the typical *NF1* deletion of 1.4 Mb (occurring between the major segmental duplications flaking the *NF1* gene, also known as Type 1 deletions), seven of the eight mosaic deletions were 1.2 Mb in size (known as type 2 deletions) and were the product of recombination between the *SUZ12* gene and a highly similar pseudogene (106, 109). Thus, type I *NF1* microdeletions occur by intrachromosomal recombination during meiosis, while the type II deletions are mediated by intrachromosomal recombination during mitosis. This scenario is reminiscent of the above described findings for the *DMD* gene, pointing again to a different mechanism behind generation of some structural rearrangements in meiosis and mitosis. Intriguingly, another study (110) revealed that 12 out of 13 mosaic type 2 deletions in the *NF1* gene were found in females. The marked female preponderance among mosaic type 2 deletions contrasts with the equal sex distribution noted for type 1 deletion and the other *NF1* microdeletions, with breakpoints located elsewhere within the *NF1* locus.

2.2. Chromosomal and Subchromosomal Mosaicism

Chromosomal mosaicism is a common phenomenon. For instance, data from analyses of early human embryos after in vitro fertilization show that approximately 50% of these are mosaic for a chromosomal anomaly (aneuploidy) and the aberrations are mainly due to nondisjunction (111–114). Furthermore, aneuploidy is a common cause of developmental disorders with a frequency ~50% in spontaneous abortions (115, 116). In addition, based on single cell analyses, a surprisingly high rate (70%) of mitotic chromosomal abnormalities has recently been reported in the blastomeres of in vitro fertilized preimplantation embryos. This high rate of chromosomal aberrations is not reflected in newborns, which implies that there is a negative selection against embryonic cells with chromosomal abnormalities (117). Only a few aneuploidies that are present in all studied cells (nonmosaics) have been seen in live born individuals, whereas mosaic aneuploidy is better tolerated. Consequently, mosaicism has been reported for a large number of chromosomal abnormalities and is associated with a variety of phenotypes (115, 116, 118). Somatic aneuploidy can arise from meiotic events with abnormal zygote or mitotically, with a normal zygote and a subsequent nondisjunction or anaphase lag during a somatic cell division.

The application of array-CGH in analyses of large collections of routine samples referred to clinical genetic laboratories has considerably improved the resolution of detection of small subchromosomal changes, which are beyond the detection level of cytogenetic analyses (8, 9). Furthermore, this technology also improved detection of mosaicism down to 10–20% of cells containing a variant genotype. BAC-clone-based array-CGH allows a discovery rate for pathogenic mosaic aneuploidy of about 0.4% patients referred to clinical geneticists (8, 9). The recent improvement has been application of Illumina SNP-based arrays in the routine diagnostics. The frequency of mosaicism for aneuploidies (or large subchromosomal aberrations) in live born individuals has been raised to 1% (10). The 21 patients with mosaicism for large chromosomal aberrations reported by Conlin et al. (10) corresponded to 10% of all abnormalities diagnosed in the laboratory during the same period, from more than 2,000 analyzed patients. Another important improvement has been detection of cells with a variant genotype in the range of less than 5%. This high sensitivity was possible by analysis of BAF output from Illumina beadchips, which is a measure of the heterozygozity level for informative SNP probes. The values reflecting fluorescent intensities of probes on the array (Log *R* ratio, LRR) were not sensitive enough at this low level of mosaicism. The results reported by Conlin et al. regarding the sensitivity for detection of very low level of cells containing an abnormality are in agreement with our analysis from a pair of healthy MZ twins with a low level of somatic mosaicism for aneuploidy of chromosomes X and Y. Using essentially the same analysis platform, we were able to readily detect monosomy X in 7% of cells from one twin, but were unable to observe 1% of cells containing monosomy X, which was present in the other co-twin (68).

One of the advantages related to array-CGH or SNP-based arrays in the analysis of mosaicism is the use of DNA derived from blood cells in vivo, rather than cells that have been manipulated by in vitro culturing. Routine chromosome analysis usually involves phytohemagglutinin (PHA) stimulated T cells followed by in vitro culturing, arrest in mitotic division, and standard analyses of 20 metaphase cells. FISH analysis, another widely applied technique, is also typically performed on T-cells caught in metaphase and assessed for gain or loss of a specific locus. A single abnormal cell, among only 20 that have been counted, might be interpreted as an artifact of cell culture. Furthermore, the aneuploid cells (or cells containing other aberrations) might be under-represented in the T-cell population that is studied, if they are less competitive for growth than normal cells in vivo or do not respond well to the PHA growth stimuli in vitro. It has been shown that analysis on T-cell stimulated cultures may not detect somatic chromosomal mosaicism in Pallister–Killian syndrome patients and an additional chromosome analysis on skin fibroblasts is routinely performed in these subjects (119, 120). Thus, it is possible that many cases of mosaicism go undetected because aneuploid cells are under-represented in the cells used for analyses. Another noteworthy aspect related to the choice of analyzed somatic cells is that several studies have shown varying levels of chromosomal mosaicism in different tissues and at different locations within the same tissue (121–123).

It seems that brain might be the organ that is particularly prone to aneuploidies, some of which may be associated with aberrant phenotypes. A recent study by Iourov et al. (124) showed a higher incidence of aneuploid cells in ataxia telangiectasia brains (20–50%) compared to that of normal individuals (~10%). The genomic distribution of aneuploidies appeared stochastic. However, in the same study, it was shown that the cerebral cortex of Alzheimer's disease patients is tenfold enriched in cells that harbor chromosome 21 hypo- or hyperploidies. In another study from the same group, a low-level aneuploidy was found associated with autism (125). Furthermore, autism has been recently linked to mosaicism for chromosome 4p12-p16 duplication (126). In the interesting study of familial hypophosphatemic rickets, three family members shared a 52 kb causal mosaic deletion in the *PHEX* gene. The mother showed, however, no disease associated phenotype, while her two children were affected. This observation was due to predominant presence of the wild-type allele in the mother in contrast to that of the children (127). This example indicates that a high-level mosaicism for the aberrant allele was required to cause an abnormal phenotype. On the other hand, the level of cells containing an aberrant allele is not always positively correlated to the severity of phenotype. Certain disorders, such as developmental delay, can apparently be caused by mosaicism in the small fraction of cells, as shown in an 11p13 deletion patient (9).

2.3. Somatic Mosaicism for UPD

UPD can be considered a special case of structural genetic variation. UPD usually does not change the copy number of affected chromosome(s) or chromosomal segment. UPD is, however, a result of a structural rearrangement, most commonly due to meiotic or mitotic nondisjunction/anaphase lag, alternatively mitotic recombination. UPD can affect the entire chromosome, large parts of a single chromosome (usually via formation of a marker chromosome) or smaller chromosomal segments (segmental UPD). UPD can also be genome-wide maternal or paternal, but this form is very rare (5–7, 128). The simplest definition of UPD in the context of a single affected chromosome is "inheritance of both homologues of a pair of chromosomes from one parent only" (7). There are two main consequences of UPD from the disease point of view; an imprinting disorder via loss of imprinting or inheritance of a recessive trait in a non-Mendelian fashion. The latter is mediated by "reduction to homozygosity" causing a recessive phenotype to appear, which is inherited from a heterozygous parent. UPD has been shown to cause cancers and many developmental disorders. The list of conditions that have been associated with UPD is continuously growing (4–7) and this trend is likely to continue due to application of SNP-based arrays with ultra-high resolution analysis of normal and disease-related samples. UPD cannot be detected by cytogenetic analyses or by standard array-CGH. However, BAF track values from SNP-based arrays, such as Illumina bead chips are sensitive tools for the detection of constitutional (nonmosaic) and mosaic form of UPD (10). Beckwith–Wiedemann syndrome is an example of a disorder which is frequently associated mosaicism for UPD. About 20% of Beckwith–Wiedemann syndrome patients are mosaic for paternal UPD of 11p15 and the severity of the phenotype has been linked to the level of mosaicism (129).

3. Somatic Structural Genetic Variation in Other Vertebrates

To date, there is surprisingly few published data sets concerning mosaicism for structural rearrangements in other vertebrates. Aneuploidy has been the most commonly reported alteration and neurons seem to be particularly prone to this variation, which is similar to humans. Rehen et al. (130) during analysis of mouse neuroblasts found that up to 33% cells were aneuploid as a result of mitotic nondisjunction. Further investigation showed that the aneuploidies were also preserved in mouse adult neurons. Even higher rate of aneuploidy (78%, with losses of whole chromosomes being more prevalent than gains) was reported in brain cells of teleost fish, which might be associated with exceptional ability of the teleost fish to generate new cells in the brain (131). However, functional implications of such high heterogeneity of neuronal cells for chromosomal imbalances remain unknown. Aneuploidy and chromosome loss followed by reduplication of homologous chromosome that result in UPD were described in murine embryonic stem cells (132). Mouse embryonic stem cells have also recently been reported as polymorphic with regard to smaller CNVs (133). However, the two latter studies addressed in vitro cultured cells and the in vivo significance of these studies remains to be elucidated. Another interesting result comes from the studies of L1 element activity in transgenic mouse and rat models during embryogenesis. Mouse and human L1 elements were transcriptionally active both in germ and embryonic cells. However, the integration of L1 elements was more frequent in embryonic cells. The presence of somatic retrotransposition was further corroborated in wild-type, nontransgenic organisms. Another interesting aspect of this study was the observed somatic mosaicism on DNA level that was generated via an RNA intermediate. The L1 element which is transcribed already in the germ cells is transferred to the zygote during fertilization and is subsequently integrated into the genome of the embryo (134). In summary, there are some similarities in the pattern of somatic rearrangements between humans and other vertebrates. The reported changes in other species appear to be especially common in neurons and the early stage embryos. The amount of available data is, however, limited.

4. Concluding Remarks; Challenges, Opportunities, and Open Questions

There are a number of challenges in the field of somatic genetic variation. The primary one is perhaps that the accepted dogma sets out that the genome of normal somatic cells is static and essentially identical. As reports suggesting otherwise are so far relatively rare, they are considered as exceptions confirming the general rule. A comparison of the number of papers devoted to the study of interindividual genetic variation versus papers reporting on somatic variation clearly suggest that the latter aspect is under studied and consequently under estimated. It is important to increase our awareness and understanding of somatic mosaicism for various genetic changes.

Sporadic disorders, defined as a lack of similar cases among closest relatives of an affected patient, are by far the most common in medicine. Studies of differences in the genetic or epigenetic makeup of appropriate target cells, responsible for generation of a disease phenotype, in comparison with other normal cells of the same patient, might therefore be informative. However, current biobanking procedures for sample collection do not address the issue of somatic mosaicism, as only one normal control tissue is usually collected from each patient.

There is one crucial methodological challenge in scoring and subsequent validation of somatic CNVs (or other types of structural variants) in MZ twins and other samples. These CNVs will typically occur in a proportion of cells, which makes the analysis more demanding than scoring "germline CNVs" that are expected to be present in 100% of studied cells. Except for investigations of solid tumors, the studies of genetic variation between different populations of somatic cells are challenging because the cells that might present with a variant genotype, possibly causing a phenotype, will usually be mixed with other types of surrounding normal cells in any tissue. The enrichment of target cells with a differing genotype is therefore necessary and this is coupled with methodological obstacles, requiring microdissection or sorting of cells. However, methodology is continuously improving. As briefly mentioned above, standard array-CGH is capable of detecting CNVs affecting approximately 10–20% of cells (8, 9, 57). The application of SNP arrays, for instance, Illumina beadchips, via BAF as a tool for detection of somatic mosaicism on the genome-wide scale, introduced a considerable improvement, allowing detection of mosaicism in the range of 5% of affected cells (10, 68). Furthermore, the same platform and also using BAF allows an efficient assessment of UPD, which represents an understudied aspect of genome biology in the context of somatic mosaicism, especially when shorter UPD segments are considered. Another aspect of recent methodological development is related to reliable whole genome profiling of as few cells as possible. Clinical samples are often limited on the amount of DNA that is available for analysis; relevant examples are preimplantation diagnosis, analysis of circulating tumor cells or micrometastases. The protocols for the analysis of a single or a few cells are improving and it has recently been shown that 5–10 cells generate reliable genomic profiles with an effective resolution of 500 kb (135). Considering the above, it is likely to assume that the rate of discovery for short somatic CNVs that are present in a small proportion of studied cells will continue to accelerate. The desirable goal would be discovery of aberrations down to a few 100 bp in genomic extent that are present in less than 5% of cells, using global platforms for genome analysis. Reaching this goal in a comprehensive number of samples/patients will require improvement of methods for experimental validation that allow a considerably higher degree of multiplexing than it is possible today.

Regions in the human genome that are rich in segmental duplications are apparently the most variable somatically (77, 85) and also between different normal individuals (1, 2). Variation in these regions occurs via the mechanism of nonallelic homologous recombination between highly similar copies of redundant sequences present in the segmental duplications. Considering this, it is unfortunate that there is an increasing lack of genome-wide and high-resolution platforms that would allow studies of variation in these complex and very variable segments of the human genome. BAC-clone-based genomic arrays are less commonly used, as they have a limited resolution of analysis, which is determined by the size of genomic insert DNA in the clones spotted on the array. Nevertheless, these tools are efficient in picking up structural variation in regions that are rich in segmental duplications. The commercially available oligonucleotide-based platforms for CNV analyses are primarily designed using repeat-free and nonredundant fraction of the genome.

A concern can be raised in the context of studies of genetic variation in general and somatic variation in particular, related the use of lymphoblastoid cell lines (LCLs). These cultured cells might acquire a new genotype, which was not present in the original B lymphocytes that gave rise to the LCL. For instance, the HapMap samples have been widely used for analysis of various types of genetic variation, including structural variation. These samples are based on LCLs, which are Epstein–Barr virus (EBV) transformed B lymphocytes and have been cultured extensively in vitro. The in vitro induced CNVs are a realistic possibility to consider, especially for CNVs that have been characterized in one cell line only. The latter category represents roughly 50% of all characterized CNVs (2) and some of these might therefore represent in vitro culturing artifacts. An alternative scenario may be that the extensive in vitro culturing of LCLs conceals a somatic CNV that might be present in the studied subject, from which the cell line is derived. LCLs are polyclonal in the beginning, then become gradually oligoclonal and monoclonal after prolonged culturing (136, 137). Furthermore, it has been shown that cells affected by some chromosomal rearrangements are less efficiently cultured in vitro, when compared to normal euploid cells (119, 120). This combination might lead to a selective removal of cells with a variant genotype.

Finally, there is a number of important questions that should be addressed in the context of somatic structural genetic variation.

1. Are there significant tissue-specific differences in the frequency and/or genomic distribution of structural genetic variants or other mutations? If so, is there a process of tissue-specific selection for a certain variant, which might be responsible for generation of a phenotype in the relevant tissue/organ? A concept of selection from a pool of preexisting somatic mutations in disease development has recently been proposed (138).
2. Is the extent of somatic structural variation in commonly used experimental animals similar to what has been observed in human samples?
3. How frequent is large-scale chromosomal mosaicism in generally healthy subjects? We refer to variation that is similar to that described above for healthy monozygotic twins with aneuploidies of chromosomes X and Y and to an increasing number of other studies reporting somatic mosaicism for various types of large chromosomal aberrations in apparently normal subjects (68–73, 139–146).
4. The results from mutational analyses of well-characterized disease genes, such as *DMD* and *NF1*, suggest that there is a difference in the mechanisms behind structural rearrangements that occur during mitosis and meiosis. The extent of these differences requires further analysis.
5. Cohorts of monozygotic twins that are discordant for many common diseases are possible to collect. The interesting question is how helpful will these twin samples be for delineation of genetic and epigenetic factors predisposing for common complex disorders?
6. Yet another question relates to the normal process of aging and the frequency/distribution of structural variants in the nuclear

genome. As mentioned above, the mitochondrial genome accumulates deletions with increased age. It is tempting to speculate that a similar scenario is also frequently occurring in the nuclear genome. Two recent papers point to this as a possibility (85, 147).

Acknowledgments

We thank Drs. Nils-Göran Larsson, Lars Forsberg, Patrick Buckley, Teresita Diaz de Ståhl, and Kenneth Nilsson for review of the manuscript. This work was supported by the Ellison Medical Foundation, the Swedish Cancer Society, the Swedish Research Council to JPD; and by the Foundation for Polish Science, the Foundation for the Development of Polish Pharmacy and Medicine to A.P.

References

1. Redon R, Ishikawa S, Fitch KR, et al. (2006) Global variation in copy number in the human genome. Nature;444:444–54.
2. Conrad DF, Pinto D, Redon R, et al. (2010) Origins and functional impact of copy number variation in the human genome. Nature; 464:704–712.
3. Hastings PJ, Lupski JR, Rosenberg SM, Ira G. (2009) Mechanisms of change in gene copy number. Nat Rev Genet;10:551–64.
4. Feinberg AP. (2002) Genomic imprinting and cancer. In: Vogelstein B, Kinzler KW, eds. The genetic basis of human cancer. Second ed. New York: McGraw-Hill.
5. Engel E. (2006) A fascination with chromosome rescue in uniparental disomy: Mendelian recessive outlaws and imprinting copyrights infringements. Eur J Hum Genet;14:1158–69.
6. Kotzot D. (2008) Prenatal testing for uniparental disomy: indications and clinical relevance. Ultrasound Obstet Gynecol;31:100–5.
7. Kotzot D. (2008) Complex and segmental uniparental disomy updated. J Med Genet;45: 545–56.
8. Ballif BC, Rorem EA, Sundin K, et al. (2006) Detection of low-level mosaicism by array CGH in routine diagnostic specimens. Am J Med Genet A;140:2757–67.
9. Cheung SW, Shaw CA, Scott DA, et al. (2007) Microarray-based CGH detects chromosomal mosaicism not revealed by conventional cytogenetics. Am J Med Genet A;143A:1679–86.
10. Conlin LK, Thiel BD, Bonnemann CG, et al. (2010) Mechanisms of mosaicism, chimerism and uniparental disomy identified by SNP array analysis. Human Molecular Genetics;19 1263–75.
11. Strain L, Warner JP, Johnston T, Bonthron DT. (1995) A human parthenogenetic chimaera. Nat Genet;11:164–9.
12. Hall JG. (1996) Twinning: mechanisms and genetic implications. Curr Opin Genet Dev;6:343–7.
13. van Dijk BA, Boomsma DI, de Man AJ. (1996) Blood group chimerism in human multiple births is not rare. Am J Med Genet;61:264–8.
14. Bianchi DW, Lo YM. (2001) Fetomaternal cellular and plasma DNA trafficking: the Yin and the Yang. Ann N Y Acad Sci;945:119–31.
15. Quaini F, Urbanek K, Beltrami AP, et al. (2002) Chimerism of the transplanted heart. N Engl J Med;346:5–15.
16. Malan V, Vekemans M, Turleau C. (2006) Chimera and other fertilization errors. Clin Genet;70:363–73.
17. Vogelstein B, Kinzler KW. (2002) The genetic basis of human cancer. Second Edition ed. New York: McGrawh-Hill.
18. Vogelstein B, Kinzler KW. (2004) Cancer genes and the pathways they control. Nature medicine;10:789–99.
19. Stratton MR, Campbell PJ, Futreal PA. (2009) The cancer genome. Nature;458:719–24.

20. Salk JJ, Fox EJ, Loeb LA. (2010) Mutational heterogeneity in human cancers: origin and consequences. Annu Rev Pathol;5:51–75.
21. Frank S. (2010) Somatic evolutionary genomics: Mutations during development cause highly variable genetic mosaicism with risk of cancer and neurodegeneration. PNAS USA;107:1725–30.
22. Rubin H. (2002) The disparity between human cell senescence in vitro and lifelong replication in vivo. Nat Biotechnol;20:675–81.
23. Takubo K, Izumiyama-Shimomura N, Honma N, et al. (2002) Telomere lengths are characteristic in each human individual. Exp Gerontol;37:523–31.
24. Nakamura K, Izumiyama-Shimomura N, Sawabe M, et al. (2002) Comparative analysis of telomere lengths and erosion with age in human epidermis and lingual epithelium. J Invest Dermatol;119:1014–9.
25. Baird DM. (2008) Telomere dynamics in human cells. Biochimie;90:116–21.
26. Baird DM, Britt-Compton B, Rowson J, Amso NN, Gregory L, Kipling D. (2006) Telomere instability in the male germline. Hum Mol Genet;15:45–51.
27. Strachan T, Read A. (2004) Human Molecular Genetics 3. Third edition ed. New York: Garland Publishing.
28. Buckley P, Mantripragada K, Díaz de Ståhl T, et al. (2005) Identification of genetic aberrations on chromosome 22 outside the NF2 locus in schwannomatosis and neurofibromatosis type 2. Hum Mut;26:540–9.
29. Terzioglu M, Larsson NG. (2007) Mitochondrial dysfunction in mammalian ageing. Novartis Found Symp;287:197–208; discussion 13.
30. Trifunovic A, Larsson NG. (2008) Mitochondrial dysfunction as a cause of ageing. J Intern Med;263:167–78.
31. Holt IJ, Harding AE, Morgan-Hughes JA. (1988) Deletions of muscle mitochondrial DNA in patients with mitochondrial myopathies. Nature;331:717–9.
32. Wallace DC, Singh G, Lott MT, et al. (1988) Mitochondrial DNA mutation associated with Leber's hereditary optic neuropathy. Science;242:1427–30.
33. DiMauro S, Hirano M. (2005) Mitochondrial encephalomyopathies: an update. Neuromuscul Disord;15:276–86.
34. Larsson NG, Clayton DA. (1995) Molecular genetic aspects of human mitochondrial disorders. Annu Rev Genet;29:151–78.
35. Mohamed SA, Hanke T, Erasmi AW, et al. (2006) Mitochondrial DNA deletions and the aging heart. Exp Gerontol;41:508–17.
36. Lee HC, Pang CY, Hsu HS, Wei YH. (1994) Differential accumulations of 4,977 bp deletion in mitochondrial DNA of various tissues in human ageing. Biochim Biophys Acta;1226:37–43.
37. van Ommen GJ. (2005) Frequency of new copy number variation in humans. Nat Genet;37:333–4.
38. Lupski JR. (2007) Genomic rearrangements and sporadic disease. Nat Genet;39:S43–7.
39. Hall JG. (2003) Twinning. Lancet;362:735–43.
40. Martin JA, Hamilton BE, Sutton PD, et al. (2007) Births: Final Data for 2005. National Vital Statistics Reports;56.
41. Gringras P, Chen W. (2001) Mechanisms for differences in monozygous twins. Early Hum Dev;64:105–17.
42. Busjahn A, Hur YM. (2006) Twin registries: an ongoing success story. Twin Res Hum Genet;9:705.
43. Merriman C. (1924) The intellectual resemblance of twins. Psycological Monographs; 33:1–58.
44. Siemens H. (1924) Zwillingspathologie: Ihre Bedeutung; ihre Methodik; ihre bisherigen Ergebnisse. Berlin: Springer Verlag.
45. Nystad W, Roysamb E, Magnus P, Tambs K, Harris JR. (2005) A comparison of genetic and environmental variance structures for asthma, hay fever and eczema with symptoms of the same diseases: a study of Norwegian twins. International journal of epidemiology; 34:1302–9.
46. Harris JR, Magnus P, Samuelsen SO, Tambs K. (1997) No evidence for effects of family environment on asthma. A retrospective study of Norwegian twins. American journal of respiratory and critical care medicine; 156:43–9.
47. Ahmadi KR, Lanchbury JS, Reed P, et al. (2003) Novel association suggests multiple independent QTLs within chromosome 5q21–33 region control variation in total humans IgE levels. Genes and immunity;4:289–97.
48. Faraone SV, Perlis RH, Doyle AE, et al. (2005) Molecular genetics of attention-deficit/hyperactivity disorder. Biological psychiatry;57:1313–23.
49. Folstein S, Rutter M. (1977) Genetic influences and infantile autism. Nature;265:726–8.
50. Grjibovski AM, Olsen AO, Magnus P, Harris JR. (2007) Psoriasis in Norwegian twins: contribution of genetic and environmental effects. J Eur Acad Dermatol Venereol;21:1337–43.
51. Heflin LH, Meyerowitz BE, Hall P, et al. (2005) Cancer as a risk factor for long-term cognitive deficits and dementia. Journal of the National Cancer Institute;97:854–6.

52. Lichtenstein P, Holm NV, Verkasalo PK, et al. (2000) Environmental and heritable factors in the causation of cancer–analyses of cohorts of twins from Sweden, Denmark, and Finland. N Engl J Med;343:78–85.
53. Strachan DP, Wong HJ, Spector TD. (2001) Concordance and interrelationship of atopic diseases and markers of allergic sensitization among adult female twins. The Journal of allergy and clinical immunology;108:901–7.
54. Stunkard AJ, Foch TT, Hrubec Z. (1986) A twin study of human obesity. Jama;256:51–4.
55. Joseph J. (2002) Twin studies in psychiatry and psychology: science or pseudoscience? The Psychiatric quarterly;73:71–82.
56. Kondo S, Schutte BC, Richardson RJ, et al. (2002) Mutations in IRF6 cause Van der Woude and popliteal pterygium syndromes. Nat Genet;32:285–9.
57. Bruder C, Piotrowski A, Gijsbers A, et al. (2008) Phenotypically Concordant and Discordant Monozygotic Twins Display Different DNA Copy-Number-Variation Profiles. Am J Hum Genet;82:763–71.
58. Machin GA. (1996) Some causes of genotypic and phenotypic discordance in monozygotic twin pairs. Am J Med Genet;61:216–28.
59. Fraga MF, Ballestar E, Paz MF, et al. (2005) Epigenetic differences arise during the lifetime of monozygotic twins. Proc Natl Acad Sci USA;102:10604–9.
60. Kaminsky ZA, Tang T, Wang SC, et al. (2009) DNA methylation profiles in monozygotic and dizygotic twins. Nat Genet;41:240–5.
61. Petronis A, Gottesman, II, Kan P, et al. (2003) Monozygotic twins exhibit numerous epigenetic differences: clues to twin discordance? Schizophrenia bulletin;29:169–78.
62. Petronis A. (2006) Epigenetics and twins: three variations on the theme. Trends Genet;22:347–50.
63. Summersgill B, Thornton P, Atkinson S, et al. (2002) Chromosomal imbalances in familial chronic lymphocytic leukaemia: a comparative genomic hybridisation analysis. Leukemia;16:1229–32.
64. Ripolles L, Ortega M, Ortuno F, et al. (2006) Genetic abnormalities and clinical outcome in chronic lymphocytic leukemia. Cancer Genet Cytogenet;171:57–64.
65. Korbel JO, Urban AE, Affourtit JP, et al. (2007) Paired-End Mapping Reveals Extensive Structural Variation in the Human Genome. Science;318:420–6.
66. Tsujita T, Niikawa N, Yamashita H, et al. (1998) Genomic discordance between monozygotic twins discordant for schizophrenia. The American journal of psychiatry;155:422–4.
67. Machin G. (2009) Non-identical monozygotic twins, intermediate twin types, zygosity testing, and the non-random nature of monozygotic twinning: a review. Am J Med Genet C Semin Med Genet;151 C:110–27.
68. Razzaghian HR, Shahi MH, Forsberg LA, et al. (2010) Somatic mosaicism for chromosome X and Y aneuploidies in monozygotic twins heterozygous for sickle cell disease mutation. Am J Med Genet A;152A:2595–2598.
69. Stumm M, Musebeck J, Tonnies H, et al. (2002) Partial trisomy 9p12p21.3 with a normal phenotype. J Med Genet; 39:141–4.
70. McAuliffe F, Winsor EJ, Chitayat D. (2005) Tetrasomy 9p mosaicism associated with a normal phenotype. Fetal Diagn Ther;20:219–22.
71. Di Giacomo MC, Susca FC, Resta N, Bukvic N, Vimercati A, Guanti G. (2007) Trisomy 13 mosaicism in a phenotypically normal child: description of cytogenetic and clinical findings from early pregnancy beyond 2 years of age. Am J Med Genet A;143:518–20.
72. Sung PL, Chang SP, Wen KC, et al. (2009) Small supernumerary marker chromosome originating from chromosome 10 associated with an apparently normal phenotype. Am J Med Genet A;149A:2768–74.
73. Venci A, Bettio D. (2009) Tetrasomy 5p mosaicism due to an additional isochromosome 5p in a man with normal phenotype. Am J Med Genet A;149A:2889–91.
74. Pack SD, Weil RJ, Vortmeyer AO, et al. (2005) Individual adult human neurons display aneuploidy: detection by fluorescence in situ hybridization and single neuron PCR. Cell cycle;4:1758–60.
75. Iourov IY, Liehr T, Vorsanova SG, Kolotii AD, Yurov YB. (2006) Visualization of interphase chromosomes in postmitotic cells of the human brain by multicolour banding (MCB). Chromosome Res;14:223–9.
76. Yurov YB, Iourov IY, Vorsanova SG, et al. (2007) Aneuploidy and confined chromosomal mosaicism in the developing human brain. PLoS ONE;2:e558.
77. Piotrowski A, Bruder C, Andersson R, et al. (2008) Somatic mosaicism for copy number variation in differentiated human tissues. Human Mutation;29:1118–24.
78. Eichler EE, Clark RA, She X. (2004) An assessment of the sequence gaps: unfinished business in a finished human genome. Nat Rev Genet;5:345–54.
79. Bovee D, Zhou Y, Haugen E, et al. (2008) Closing gaps in the human genome with fosmid resources generated from multiple individuals. Nat Genet;40:96–101.

80. Garber M, Zody MC, Arachchi HM, et al. (2009) Closing gaps in the human genome using sequencing by synthesis. Genome Biol;10:R60.
81. Tapia-Paez I, Kost-Alimova M, Hu P, et al. (2001) The position of t(11;22)(q23;q11) constitutional translocation breakpoint is conserved among its carriers. Hum Genet;109:167–77.
82. Coufal NG, Garcia-Perez JL, Peng GE, et al. (2009) L1 retrotransposition in human neural progenitor cells. Nature;460:1127–31.
83. Kormoczi GF, Dauber EM, Haas OA, et al. (2007) Mosaicism due to myeloid lineage restricted loss of heterozygosity as cause of spontaneous Rh phenotype splitting. Blood;110:2148–57.
84. Lam KW, Jeffreys AJ. (2006) Processes of copy-number change in human DNA: the dynamics of {alpha}-globin gene deletion. Proc Natl Acad Sci USA;103:8921–7.
85. Flores M, Morales L, Gonzaga-Jauregui C, et al. (2007) Recurrent DNA inversion rearrangements in the human genome. Proc Natl Acad Sci USA;104:6099–106.
86. Hall JG. (1988) Review and hypotheses: somatic mosaicism: observations related to clinical genetics. Am J Hum Genet;43:355–63.
87. Gottlieb B, Beitel LK, Trifiro MA. (2001) Somatic mosaicism and variable expressivity. Trends Genet;17:79–82.
88. Youssoufian H, Pyeritz RE. (2002) Mechanisms and consequences of somatic mosaicism in humans. Nat Rev Genet;3: 748–58.
89. Erickson RP. (2003) Somatic gene mutation and human disease other than cancer. Mutat Res;543:125–36.
90. Hirschhorn R. (2003) In vivo reversion to normal of inherited mutations in humans. J Med Genet;40:721–8.
91. Notini AJ, Craig JM, White SJ. (2008) Copy number variation and mosaicism. Cytogenet Genome Res;123:270–7.
92. Lutskiy MI, Park JY, Remold SK, Remold-O'Donnell E. (2008) Evolution of highly polymorphic T cell populations in siblings with the Wiskott-Aldrich Syndrome. PLoS One;3:e3444.
93. Gottlieb B, Chalifour LE, Mitmaker B, et al. (2009) BAK1 gene variation and abdominal aortic aneurysms. Hum Mutat;30:1043–7.
94. Mandel JL. (1989) Dystrophin. The gene and its product. Nature;339:584–6.
95. Den Dunnen JT, Grootscholten PM, Dauwerse JG, et al. (1992) Reconstruction of the 2.4 Mb human DMD-gene by homologous YAC recombination. Hum Mol Genet;1:19–28.
96. Roberts RG, Coffey AJ, Bobrow M, Bentley DR. (1993) Exon structure of the human dystrophin gene. Genomics;16:536–8.
97. White SJ, den Dunnen JT. (2006) Copy number variation in the genome; the human DMD gene as an example. Cytogenet Genome Res;115:240–6.
98. Passos-Bueno MR, Bakker E, Kneppers AL, et al. (1992) Different mosaicism frequencies for proximal and distal Duchenne muscular dystrophy (DMD) mutations indicate difference in etiology and recurrence risk. Am J Hum Genet;51:1150–5.
99. White SJ, Aartsma-Rus A, Flanigan KM, et al. (2006) Duplications in the DMD gene. Hum Mutat;27:938–45.
100. Kvittingen EA, Rootwelt H, Berger R, Brandtzaeg P. (1994) Self-induced correction of the genetic defect in tyrosinemia type I. The Journal of clinical investigation;94: 1657–61.
101. Ellis NA, Lennon DJ, Proytcheva M, Alhadeff B, Henderson EE, German J. (1995) Somatic intragenic recombination within the mutated locus BLM can correct the high sister-chromatid exchange phenotype of Bloom syndrome cells. Am J Hum Genet;57:1019–27.
102. Gregory JJ, Jr., Wagner JE, Verlander PC, et al. (2001) Somatic mosaicism in Fanconi anemia: evidence of genotypic reversion in lymphohematopoietic stem cells. Proc Natl Acad Sci USA;98:2532–7.
103. Cawthon RM, Weiss R, Xu GF, et al. (1990) A major segment of the neurofibromatosis type 1 gene: cDNA sequence, genomic structure, and point mutations. Cell;62:193–201.
104. Viskochil D, Buchberg AM, Xu G, et al. (1990) Deletions and translocation interrupt a cloned gene at the neurofibromatosis type 1 locus. Cell;62:187–92.
105. Wallace MR, Marchuk DA, Andersen LB, et al. (1990) Type 1 neurofibromatosis gene: Identification of a large transcript disrupted in three NF1 patients. Science;249:181–6.
106. Kehrer-Sawatzki H, Kluwe L, Sandig C, et al. (2004) High frequency of mosaicism among patients with neurofibromatosis type 1 (NF1) with microdeletions caused by somatic recombination of the JJAZ1 gene. Am J Hum Genet;75:410–23.
107. Mantripragada KK, Thuresson AC, Piotrowski A, et al. (2006) Identification of novel deletion breakpoints bordered by segmental duplications in the NF1 locus using high resolution array-CGH. J Med Genet;43:28–38.

108. Rasmussen SA, Colman SD, Ho VT, et al. (1998) Constitutional and mosaic large NF1 gene deletions in neurofibromatosis type 1. J Med Genet;35:468–71.
109. Petek E, Jenne DE, Smolle J, et al. (2003) Mitotic recombination mediated by the JJAZF1 (KIAA0160) gene causing somatic mosaicism and a new type of constitutional NF1 microdeletion in two children of a mosaic female with only few manifestations. J Med Genet;40:520–5.
110. Steinmann K, Cooper DN, Kluwe L, et al. (2007) Type 2 NF1 deletions are highly unusual by virtue of the absence of nonallelic homologous recombination hotspots and an apparent preference for female mitotic recombination. Am J Hum Genet;81:1201–20.
111. Bielanska M, Tan SL, Ao A. (2002) Chromosomal mosaicism throughout human preimplantation development in vitro: incidence, type, and relevance to embryo outcome. Human reproduction (Oxford, England);17:413–9.
112. Munne S, Bahce M, Sandalinas M, et al. (2004) Differences in chromosome susceptibility to aneuploidy and survival to first trimester. Reprod Biomed Online;8:81–90.
113. Delhanty JD. (2005) Mechanisms of aneuploidy induction in human oogenesis and early embryogenesis. Cytogenet Genome Res;111:237–44.
114. Munne S. (2006) Chromosome abnormalities and their relationship to morphology and development of human embryos. Reprod Biomed Online;12:234–53.
115. Kalousek DK. (2000) Pathogenesis of chromosomal mosaicism and its effect on early human development. Am J Med Genet;91: 39–45.
116. Hassold T, Hall H, Hunt P. (2007) The origin of human aneuploidy: where we have been, where we are going. Hum Mol Genet;16 Spec No. 2:R203–8.
117. Vanneste E, Voet T, Le Caignec C, et al. (2009) Chromosome instability is common in human cleavage-stage embryos. Nature medicine;15:577–83.
118. Hassold TJ, Jacobs PA. (1984) Trisomy in man. Annu Rev Genet;18:69–97.
119. Reeser SL, Wenger SL. (1992) Failure of PHA-stimulated i(12p) lymphocytes to divide in Pallister-Killian syndrome. Am J Med Genet;42:815–9.
120. Priest JH, Rust JM, Fernhoff PM. (1992) Tissue specificity and stability of mosaicism in Pallister-Killian+i(12p) syndrome: relevance for prenatal diagnosis. Am J Med Genet;42:820–4.
121. Kingston HM, Nicolini U, Haslam J, Andrews T. (1993) 46,XY/47,XY, + 17p+mosaicism in amniocytes associated with fetal abnormalities despite normal fetal blood karyotype. Prenat Diagn;13:637–42.
122. Magenis E, Webb MJ, Spears B, Opitz JM. (1999) Blaschkolinear malformation syndrome in complex trisomy-7 mosaicism. Am J Med Genet;87:375–83.
123. Kayser M, Henderson LB, Kreutzman J, Schreck R, Graham JM, Jr. (2000) Blaschkolinear skin pigmentary variation due to trisomy 7 mosaicism. Am J Med Genet;95: 281–4.
124. Iourov IY, Vorsanova SG, Liehr T, Yurov YB. (2009) Aneuploidy in the normal, Alzheimer's disease and ataxia-telangiectasia brain: differential expression and pathological meaning. Neurobiology of disease;34:212–20.
125. Yurov YB, Vorsanova SG, Iourov IY, et al. (2007) Unexplained autism is frequently associated with low-level mosaic aneuploidy. J Med Genet;44:521–5.
126. Kakinuma H, Ozaki M, Sato H, Takahashi H. (2008) Variation in GABA-A subunit gene copy number in an autistic patient with mosaic 4 p duplication (p12p16). Am J Med Genet B Neuropsychiatr Genet;147B:973–5.
127. Saito T, Nishii Y, Yasuda T, et al. (2009) Familial hypophosphatemic rickets caused by a large deletion in PHEX gene. European journal of endocrinology / European Federation of Endocrine Societies;161: 647–51.
128. Wilson M, Peters G, Bennetts B, et al. (2008) The clinical phenotype of mosaicism for genome-wide paternal uniparental disomy: two new reports. Am J Med Genet A;146A: 137–48.
129. Smith AC, Shuman C, Chitayat D, et al. (2007) Severe presentation of Beckwith-Wiedemann syndrome associated with high levels of constitutional paternal uniparental disomy for chromosome 11p15. Am J Med Genet A;143A:3010–5.
130. Rehen SK, McConnell MJ, Kaushal D, Kingsbury MA, Yang AH, Chun J. (2001) Chromosomal variation in neurons of the developing and adult mammalian nervous system. Proc Natl Acad Sci USA;98:13361–6.
131. Rajendran RS, Zupanc MM, Losche A, Westra J, Chun J, Zupanc GK. (2007) Numerical chromosome variation and mitotic segregation defects in the adult brain of teleost fish. Developmental neurobiology;67:1334–47.
132. Cervantes RB, Stringer JR, Shao C, Tischfield JA, Stambrook PJ. (2002) Embryonic stem cells and somatic cells differ in mutation

frequency and type. Proc Natl Acad Sci USA;99:3586–90.

133. Liang Q, Conte N, Skarnes WC, Bradley A. (2008) Extensive genomic copy number variation in embryonic stem cells. Proc Natl Acad Sci USA;105:17453–6.

134. Kano H, Godoy I, Courtney C, et al. (2009) L1 retrotransposition occurs mainly in embryogenesis and creates somatic mosaicism. Genes & development;23:1303–12.

135. Geigl JB, Obenauf AC, Waldispuehl-Geigl J, et al. (2009) Identification of small gains and losses in single cells after whole genome amplification on tiling oligo arrays. Nucleic Acids Res;37:e105.

136. Nilsson K, Ponten J. (1975) Classification and biological nature of established human hematopoietic cell lines. Int J Cancer;15: 321–41.

137. Giovanella B, Nilsson K, Zech L, Yim O, Klein G, Stehlin JS. (1979) Growth of diploid, Epstein-Barr virus-carrying human lymphoblastoid cell lines heterotransplanted into nude mice under immunologically privileged conditions. Int J Cancer;24:103–13.

138. Gottlieb B, Beitel LK, Alvarado C, et al. (2010) Selection and mutation in the "new" genetics: an emerging hypothesis. Hum Genet;127:491–501.

139. Blouin JL, Avramopoulos D, Pangalos C, Antonarakis SE. (1993) Normal phenotype with paternal uniparental isodisomy for chromosome 21. Am J Hum Genet;53:1074–8.

140. Bernardini L, Sinibaldi L, Ceccarini C, Novelli A, Dallapiccola B. (2005) Reproductive history of a healthy woman with mosaic duplication of chromosome 4p. Prenat Diagn;25:283–5.

141. Loitzsch A, Bartsch O. (2006) Healthy 12-year-old boy with mosaic inv dup(15) (q13). Am J Med Genet A;140:640–3.

142. Santos M, Mrasek K, Rigola MA, Starke H, Liehr T, Fuster C. (2007) Identification of a "cryptic mosaicism" involving at least four different small supernumerary marker chromosomes derived from chromosome 9 in a woman without reproductive success. Fertility and sterility;88:969 e11–7.

143. Frey NV, Leid CE, Nowell PC, et al. (2008) Trisomy 8 in an allogeneic stem cell transplant recipient representative of a donor-derived constitutional abnormality. Am J Hematol;83: 846–9.

144. Hockner M, Utermann B, Erdel M, Fauth C, Utermann G, Kotzot D. (2008) Molecular characterization of a de novo ring chromosome 6 in a growth retarded but otherwise healthy woman. Am J Med Genet A;146: 925–9.

145. Liehr T, Ewers E, Kosyakova N, et al. (2009) Handling small supernumerary marker chromosomes in prenatal diagnostics. Expert Rev Mol Diagn;9:317–24.

146. Iwarsson E, Sahlen S, Nordgren A. (2009) Jumping translocation in a phenotypically normal male: A study of mosaicism in spermatozoa, lymphocytes, and fibroblasts. Am J Med Genet A;149A:1706–11.

147. Yang D, McCrann DJ, Nguyen H, et al. (2007) Increased polyploidy in aortic vascular smooth muscle cells during aging is marked by cellular senescence. Aging cell;6:257–60.

Chapter 13

Online Resources for Genomic Structural Variation

Tam P. Sneddon and Deanna M. Church

Abstract

Genomic structural variation (SV) can be thought of on a continuum from a single base pair insertion/deletion (INDEL) to large megabase-scale rearrangements involving insertions, deletions, duplications, inversions, or translocations of whole chromosomes or chromosome arms. These variants can occur in coding or noncoding DNA, they can be inherited or arise sporadically in the germline or somatic cells. Many of these events are segregating in the population and can be considered common alleles while others are new alleles and thus rare events. All species studied to date harbor structural variants and these may be benign, contributing to phenotypes such as sensory perception and immunity, or pathogenic resulting in genomic disorders including DiGeorge/velocardiofacial, Smith-Margenis, Williams-Beuren, and Prader-Willi syndromes. As structural variants are identified, validated, and their significance, origin, and prevalence are elucidated, it is of critical importance that these data be collected and collated in a way that can be easily accessed and analyzed.

This chapter describes current structural variation online resources (see Fig. 1 and Table 1), highlights the challenges in capturing, storing, and displaying SV data, and discusses how dbVar and DGVa, the genomic structural variation databases developed at NCBI and EBI, respectively, were designed to address these issues.

Key words: Copy number variant, Insertion/deletion, Structural variation, dbVar, DGVa, Database of Genomic Variants

1. Introduction

In 1991, Charcot-Marie-Tooth (CMT) disease was the first autosomal dominant disease associated with a gene dosage effect due to an inherited DNA rearrangement (1). It is now widely accepted that copy number variants (CNVs) account for a number of genomic disorders including DiGeorge/velocardiofacial, Smith-Margenis, Williams-Beuren, and Prader-Willi syndromes and, with increased genotype–phenotype correlations, an increasing number of new genomic disorders such as a learning disability phenotype associated

Lars Feuk (ed.), *Genomic Structural Variants: Methods and Protocols*, Methods in Molecular Biology, vol. 838,
DOI 10.1007/978-1-61779-507-7_13, © Springer Science+Business Media, LLC 2012

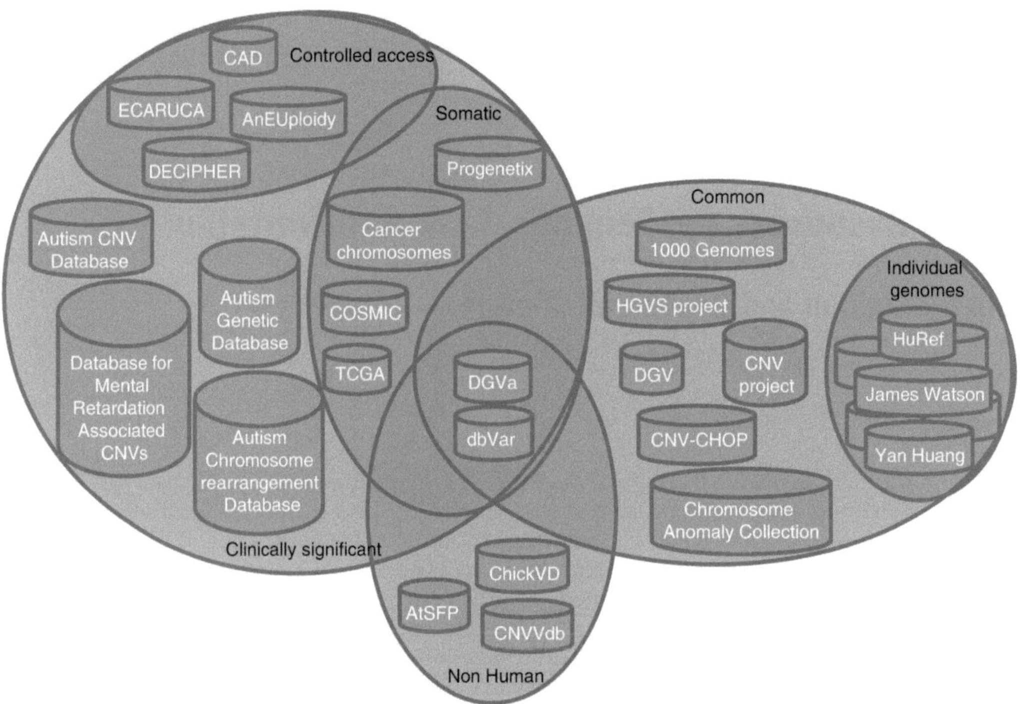

Fig. 1. Illustration of online structural variation resources discussed in this chapter and how they overlap in content, with dbVar and DGVa containing variants from all three categories: common variants (including individual genomes), clinically significant variants (including somatic/cancer and controlled access), and nonhuman variation. See Table 1 for a list of resource URLs.

with a 17q21.31 microdeletion (2, 3) and most recently with autism spectrum disorder associated with a 16p11.2 microdeletion (4).

In addition, evidence for copy number variation in disease resistance and susceptibility in humans is accumulating with publications on *CCL3L1* and susceptibility to HIV/AIDS (5), *FCGR3B* and risk of systemic lupus erythematosus (6) and several independent studies correlating copy number of the beta defensin genes with predisposition to Crohn's disease (7), risk of psoriasis (8), and sporadic prostate cancer (9).

Copy number variation among genes is not restricted to a disease phenotype. Many genes that are found to be CNV (both in humans and in mouse) are involved in environmental response, for example, sensory perception (olfactory receptors) and immunity (defensins) (10–15).

Although single-nucleotide polymorphisms (SNPs) were initially thought to contribute the majority of human genomic variation (16, 17), it is now recognized that structural variation represents a significant, and at present poorly understood, contribution to an individual's genetic makeup. It is only within the past 6 years, aided by the development of technologies such as high-throughput sequencing, paired-end mapping (PEM), and array comparative

Table 1
Online resources for structural genomic variation

Resource	URL
AnEUploidy	http://www.aneuploidy.eu/
AtSFP	http://signal.salk.edu/cgi-bin/AtSFP
Autism Chromosome Rearrangement Database	http://projects.tcag.ca/autism/
Autism CNV Database	http://projects.tcag.ca/autism_500k/
Autism Genetic Database	http://wren.bcf.ku.edu
Cancer Chromosomes	http://www.ncbi.nlm.nih.gov/cancerchromosomes
The Cancer Genome Atlas (TCGA)	http://cancergenome.nih.gov/
ChickVD	http://chicken.genomics.org.cn/
Chromosome Abnormality Database (CAD)	http://www.ukcad.org.uk./cocoon/ukcad/
Chromosome Anomaly Collection	http://www.ngrl.org.uk/wessex/collection/
CNVVdb	http://cnvvdb.genomics.sinica.edu.tw/
Copy Number Variation (CNV) Project	http://www.sanger.ac.uk/humgen/cnv/
Copy Number Variation Project at the Children's Hospital of Philadelphia (CHOP)	http://cnv.chop.edu/
COSMIC	http://www.sanger.ac.uk/genetics/CGP/cosmic/
Database for Mental Retardation Associated CNVs	http://bioinfo.ut.ee/dbcard/
Database of Genomic Variants (DGV)	http://projects.tcag.ca/variation/
dbVar	http://www.ncbi.nlm.nih.gov/dbvar/
DECIPHER	https://decipher.sanger.ac.uk
DGVa	http://www.ebi.ac.uk/dgva/
ECARUCA	http://www.ecaruca.net/
Human Genome Structural Variation (HGSV) Project	http://hgsv.washington.edu/
HuRef	http://huref.jcvi.org/
James Watson	http://jimwatsonsequence.cshl.edu
Progenetix	http://www.progenetix.net
1000 Genomes	http://www.1000genomes.org
Yan Huang	http://yh.genomics.org.cn/

genome hybridization (aCGH), that the extent of structural variation in phenotypically normal individuals has been investigated.

As advances in technology are making it easier, faster, and cheaper to sequence and analyze the genomes both within and between individuals of many species, numerous SV resources are being made available to access and analyze the data (Table 1). Although not an exhaustive list, a number of these current resources are described below and illustrated in Fig. 1.

2. Resources

2.1. Common Structural Variation

There are many variants commonly found in individuals that are healthy, or who have not been phenotyped. This is not to say that they are not of phenotypic consequence, merely that they have no known association with any disease phenotype. Several resources provide access to common structural variation data online and are described below.

2.1.1. Individual Genomes

At present there are three "complete" human genome assemblies, the GRCh37 reference sequence (18), Celera, and HuRef (Venter) (19). The alignment of these assemblies to each other provides structural variants and can be viewed in NCBI MapViewer (20). In addition, an increasing number of individual genomes, including James Watson (21) and Yan Huang (22), have been sequenced using next-generation technology (21–25). Both James Watson's and Yan Huang's genome sequences were aligned to NCBI36 (the previous version of GRCh37). The resulting SVs are available for download and can be viewed via the James Watson Genome Browser (26) and the Yan Huang GBrowser (27).

The Copy Number Variation Project

The Genome Structural Variation Consortium Copy Number Variation (CNV) Project (26) is a collaboration between groups at the Wellcome Trust Sanger Institute (Cambridge, UK), Harvard Medical School (Boston, USA), and the Hospital for Sick Children (Toronto, Canada). The CNV Project provides data from the analysis of copy number variation in the 270 Phase I and II HapMap samples using aCGH with a genome-wide Whole Genome TilePath (WGTP) array consisting of ~27,000 bacterial artificial chromosome (BAC) clones using a single male reference, Coriell individual NA10851 (27). The project also conducted a CNV discovery project to identify common CNVs greater than 500 bp in size using a set of NimbleGen arrays consisting of ~42 million probes. They analyzed 40 females with European or African ancestry, against the same single male reference sample, Coriell individual NA10851. The SV data are available for download from the CNV Project Web site (26) which provides links to view the data as tracks in the UCSC (28) or Ensembl (29) genome browsers.

2.1.2. Human Genome Structural Variation Project

The Human Genome Structural Variation (HGVS) Project (30) based at the University of Washington (Seattle, USA) aims to characterize structural variation at the sequence level. The project involves sequencing the ends of fosmids and BACs from multiple individuals and aligning them to NCBI35. The database currently contains results from an initial analysis of eight individuals (31, 32). The data, including validated SV sites and novel sequence not in NCBI35, are available for download and a link provides access to this data as tracks in the UCSC genome browser (30).

2.1.3. 1000 Genomes

The 1000 Genomes Project (33) is an international research consortium formed to create the most detailed and medically useful picture to date of human genetic variation from the sequencing and analysis of 1,000 individuals. The pilot paper describes 1 million short insertions and deletions, and 20,000 structural variants (34). The project subsequently validated 22,025 deletions and 6,000 SVs, including insertions and tandem duplications (35). These data are available via FTP from the Web site and a link provides access to the pilot data tracks in the Ensembl and UCSC Genome Browsers (33).

2.1.4. The Copy Number Variation Project at the Children's Hospital of Philadelphia

The CNV Project at the Children's Hospital of Philadelphia (CHOP) (36) represents an effort to identify all frequent copy number variations that exist in the human genome. Ongoing research uses the Illumina HumanHap 550 BeadChip to generate genotype data which is analyzed for CNVs using Illumina's BeadStudio software in combination with in-house CNV detection methodologies. The database currently contains CNVs identified in 2,026 healthy children (37). These data are available for download in NCB35 coordinates (and NCBI36 using the UCSC liftover tool (52)) and can be viewed in a genome browser (36).

2.1.5. Chromosome Anomaly Collection

The Chromosome Anomaly Collection (38) contains cases of unbalanced chromosome abnormalities (UBCAs) without phenotypic effect that have been directly transmitted from parents to children. The Collection also includes the cytogenetically visible euchromatic variants that can now be regarded as part of the continuum of copy number variation in the human genome. Cytogenetic data are represented on ideograms and provided in tables on the Web site.

2.1.6. Nonhuman Structural Variation

A number of studies have used aCGH to investigate the copy number of genes in other species, such as mouse (12, 38), rat (40), and macaque (41), and between human and other primate species including chimpanzee, bonobo, gorilla, orangutan, and macaque in an attempt to define lineage-specific genes that may aid in understanding genome evolution (42–46).

As the sequence quality of nonhuman genomes increases, a number of species-specific and interspecies databases are emerging including:

- AtSFP (47), the Salk Institute Genome Analysis Laboratory (SIGnAL) Arabidopsis Single Feature Polymorphism database and genome browser (48).
- ChickVD (49), the Beijing Genomics Institute Chicken Variation Database, which so far contains ~2.8 million nonredundant SNPs and 0.3 million indels. These data are available for download and can be viewed in a genome browser (50).
- CNVVdb (51), the Taipei Genomics Research Center at Academia Sinica Copy Number Variations across Vertebrate genomes database that identifies potential interspecies CNVs by finding duplicated regions within a genome (paralogues) and between different genomes (orthologues) from pairwise sequence alignments between 16 vertebrate species (52).

2.1.7. Database of Genomic Variants

The Database of Genomic Variants (DGV) (53) is to date the most comprehensive database for the deposition, retrieval, and visualization of common human structural variation. The database is continuously updated and curated with new data from peer-reviewed research studies. Generally, regions >3 Mb are excluded and those 100 bp–1 kb are displayed in an indel track. Currently, original SV data from 42 publications are available to download. In addition, these variants are mapped to NCBI35, NCBI36 and GRCh37 using the UCSC liftover tool (54) where necessary. These data can be downloaded and also viewed in a genome browser.

2.2. Clinically Significant Structural Variation

Although only a small fraction of structural variants have been experimentally proven to be causative of a disease, there are many variants that have been identified in individuals with a disease phenotype. As many of these variants are rare, only by collating these data and comparing with variants in healthy individuals can we begin to elucidate the significance of these variants and their relationship to disease. Several resources provide access, although often controlled access, to clinically significant structural variation data online and are described below.

2.2.1. AnEUploidy Project

The goal of the AnEUploidy Project (55) is to understand the molecular mechanisms of gene dosage imbalance (aneuploidy) in human health and includes the identification and characterization of novel microaneuploidy syndromes and the establishment of a catalog of CNVs and segmental duplications (SDs) in Europeans. Access to clinical data is under controlled access.

2.2.2. Chromosome Abnormality Database

Chromosome Abnormality Database (CAD) (56) is an online collection of both constitutional and acquired abnormal karyotypes reported by UK Regional cytogenetics centers and holds over 150,000 records collected from all UK NHS laboratories. Access to clinical data is under controlled access.

2.2.3. ECARUCA

ECARUCA (57), the European Cytogeneticists Association Register of Unbalanced Chromosome Aberrations, is a database which collects and provides cytogenetic and clinical information on rare chromosomal disorders, including microdeletions. The Register contains over 4,500 cases with more than 5,500 aberrations and links are provided to view all cases smaller than 30 Mb on NCBI35 in the UCSC or Ensembl genome browsers. Access to clinical data is under controlled access.

2.2.4. DECIPHER

DECIPHER (58), the DatabasE of Chromosomal Imbalance and Phenotype in Humans using Ensembl Resources, is a database of submicroscopic chromosomal imbalance that includes clinical information about chromosomal microdeletions/duplications/insertions, translocations, and inversions. Coordinates are mapped to GRCh37, using NCBI's remapping service, and access to non-consented clinical data is under controlled access.

2.2.5. Mental Retardation

The Database for Mental Retardation Associated CNVs (59) is a publicly accessible test database from the University of Tartu, Estonia, that gathers information about CNVs and related diseases.

2.2.6. Autism

There are several structural variation databases for autism spectrum disorder:

- The Autism Chromosome Rearrangement Database (60) is a collection of hand curated breakpoints and other genomic features, related to autism, taken from publicly available literature, databases, and unpublished data. These data can be viewed in NCB36 coordinates in a genome browser.
- The Autism CNV Database (61) provides CNV data that was obtained using the Affymetrix GeneChip Mapping 10K 2.0 (62) and Affymetrix Whole Genome mapping 10 and 500K microarrays (63). These data can be viewed in NCB35 coordinates in a genome browser.
- The Autism Genetic Database (64) is a comprehensive database for autism susceptibility gene-CNVs integrated with known noncoding RNAs and fragile sites and viewable in a genome browser.

2.3. Cancer

It has been known for many years that somatic changes in cancer cells include gross copy number changes and structural alterations. As such several projects and databases provide access to structural variation data specifically found in cancer and are described below.

2.3.1. The Cancer Genome Project

The Wellcome Trust Sanger Institute (WTSI) Cancer Genome Project (65) is using the human genome sequence and high-throughput mutation detection techniques to identify somatically acquired sequence variants/mutations and hence identify genes critical in the development of human cancers. The results from this work are collated and stored in the Catalogue of Somatic Mutations in Cancer, COSMIC (66), which also contains somatic mutation data published in the scientific literature.

2.3.2. The Cancer Genome Atlas

The National Cancer Institute (NCI) and the National Human Genome Research Institute (NHGRI) launched The Cancer Genome Atlas (TCGA) (67) program to create a comprehensive atlas of the genomic changes involved in more than 20 common types of cancer. The TCGA Data Portal (68) provides a platform for researchers to search, download, and analyze data sets generated by TCGA.

2.3.3. Progenetix

The University of Zurich Progenetix (69) database collects genomic CNV information from comparative genomic hybridization (CGH) experiments of individual cancer and leukemia cases, published in peer-reviewed journals (70).

2.3.4. Cancer Chromosomes

Three databases, the NCI/NCBI SKY/M-FISH & CGH Database, the NCI Mitelman Database of Chromosome Aberrations in Cancer, and the NCI Recurrent Aberrations in Cancer, are now integrated into the NCBI Entrez system as Cancer Chromosomes (71, 72). [Note - this resource is scheduled to be retired; data will be available for search and retrieval via dbVar/DGVa].

3. Limitations

As described above, there are numerous resources for accessing and viewing structural variation. However, when interpreting data, or collating data from different databases, there are several limitations that should be taken into consideration and are described below.

3.1. Choice of Technology

Currently, no single technology can accurately identify all structural variants in a sample. When comparing SV data derived from different methodologies, even if used on the same sample or samples, it is likely that the reported SVs will differ due to the resolution, genome coverage, and variant type detectable by the chosen method. While the decreasing costs of sequencing will eventually make it feasible for laboratories to routinely sequence and genotype whole genomes, the cost is still prohibitive enough to make aCGH and PEM the preferred methods of choice for SV detection in many laboratories.

aCGH generally uses two samples to determine relative gains and losses between the two. Using BAC aCGH, with a low resolution of ~1 Mb, means the identified region may contain several smaller CNVs and the extent of the CNV is often overestimated. Oligonucleotide arrays, with 45–85 bp probes, offer higher resolution than BAC arrays but the exact breakpoints of the structural variants can still not be accurately determined. Resolution of oligo arrays is in the order of one probe every 1–5 kb for whole genome arrays and greater than one probe every 50 bp for custom arrays. Oligo SNP arrays have an average resolution of one probe every 1–5 kb and can also be used to discover segregating deletion variants evident from genotyping patterns of null genotypes, Mendelian inconsistencies, and Hardy–Weinberg disequilibrium (10, 73). However, the detection is limited to areas of the genome containing SNPs, while many regions of known structural variation are sparsely covered by SNPs, and many of the oligo arrays only contain probes to single copy regions of the genome. This is being improved by the new microarrays from, for example, NimbleGen (74) and Illumina (75) that contain >4 million probes and include coverage of novel CNV regions such as segmental duplications, megasatellites, and other unstable regions of the genome.

Paired-end mapping (PEM) uses the end sequences of BACs, fosmids, and more recently DNA fragments from next-generation sequencing technologies, to compare to a reference genome. The advantage of this methodology over aCGH is that it not only allows identification of insertions and deletions, but also allows detection of balanced translocations and inversions, small indels from the end sequence alignments, and novel insert sequences not present in the reference genome, all of which are not amenable to aCGH. However, like aCGH, PEM does not provide exact breakpoint resolution of the structural variant but if the clones are available from a genomic library the inserts can be fully sequenced and the nature of the structural variant is determined.

3.2. Choice of SV Detection Algorithm

In addition to the different technologies available for identifying structural variants, many of which are used in parallel for validation, interpretation of the results and detection of a region as structurally variant is open to analysis, and reanalysis, using a whole suite of different software programs and algorithms. Continually being developed, these programs include BreakPtr (77), CNAG (78), CNVfinder (79), dCHIP (80), GEMCA (81), PennCNV (82), and VariationHunter (83). Although often optimized for a particular methodology or technique, many of these algorithms can be used on the same datasets to help achieve the most accurate and reproducible consensus set of SVs. Many of the array datasets are available via GEO (84) or ArrayExpress (85) and many sequences generated through paired-end sequencing or whole genome sequencing are available via Trace (86) or the Short Read Archive (SRA) (87). These resources provide a great opportunity to allow the data available from online databases to be reanalyzed and reinterpreted using different parameters of the original algorithm or using an additional or novel algorithm.

3.3. Choice of Reference Genome

Structural variants are generally defined as a region on a given reference assembly, e.g., NCBI35 or GRCh37. However, the reference genome against which the SVs were identified may be different. aCGH identifies regions CNV between two different samples but the locations are placed on a reference assembly by virtue of the coordinates of the probes used on the arrays, whereas PEM aligns sequences directly to the reference assembly. Hence, a "Loss" displayed on a genome browser from an aCGH study is not necessarily the same as a "Deletion" displayed on the same reference assembly in the genome browser from a PEM study. Indeed, an aCGH "Loss" may even be the same as a PEM "Duplication." Another major limitation of PEM, and other sequence analysis reliant on the reference assembly for comparison, is that a common loss or gain in the sample could simply reflect the presence of a minor allele in the reference. Therefore, the choice of reference genome should be taken into consideration when collating data across studies, even if reported on the same reference assembly.

3.4. Coordinate Remapping

A major limitation with SVs being reported on a particular genome assembly, e.g., NCBI35, is that they are often not transposable to a new or different assembly, e.g., GRCh37 due to their complicated genomic structure (21, 31, 88). UCSC provides a liftover tool (52) to allow carryover of coordinates to a new assembly but essentially 10–15% of mappings are lost. In order to compare variants from different studies they all need to be mapped reliably to the same genome assembly and effort is underway at NCBI to develop a robust method for remapping http://www.ncbi.nlm.nih.gov/genome/tools/remap.

3.5. Nomenclature

While the field of structural variation is still relatively new, and the methods and analyses are continuously changing, there is still a need for controlled vocabularies to facilitate searching and access of data. For example, depending on the technology, detection algorithm, or the resource the data were submitted to, the same CNV could be described as a gain, duplication, or amplification. In order to compare variants from different studies, it would be of great benefit to the field if, for example, the methodologies and variant types could be defined as a controlled vocabulary and used throughout all the SV online resources and peer-reviewed publications.

3.6. Patient Consent

Many clinical studies are under controlled access due to patient confidentiality. Unfortunately, this means there is a wealth of structural variant data stored in controlled access databases such as dbGaP (89) and EGA (90) that cannot be incorporated into the public databases. As the significance of many of these SVs cannot be determined until compared with other studies, there is a need to deposit de-identified and/or summary information from these controlled access database into the public domain.

4. dbVar and DGVa

As described in the previous sections, there are numerous structural variation databases but also numerous limitations. To address these issues, NCBI and EBI (in collaboration with the DGV) have launched the databases of genomic structural variation dbVar (91) and DGVa (92), respectively, with the aims to:

- Accession and track individual objects by providing study and variant accessions.
- Provide access to raw datasets for reanalysis (via links to, e.g., GEO (82), ArrayExpress (83), Trace Archive (84), and SRA (85)).
- Represent both common and clinically relevant data.
- Use controlled vocabularies where possible to facilitate searching.
- Represent data not on a sequenced genome assembly.
- Provide robust remapping.
- Provide resolution/confidence values to access quality of the data.
- Provide validation data.
- Store genotyping information to distinguish homozygous versus heterozygous variants.
- Store sample information to distinguish germline versus somatic variants.
- Provide summary data for controlled access data in dbGaP and EGA.
- Display data for species other than human.
- Create reference variants.

dbVar and DGVa are both accepting submissions of structural variant data and will provide study and variant accessions for all studies that have been published, submitted to a peer-review journal, or represent unpublished data from large-scale community based projects; currently this includes several large-scale projects such as the HGSV Project (30), 1000 Genomes (33), and the International Standards for Cytogenomic Arrays (ISCA) Consortium database (93). Data are exchanged between dbVar and DGVa at regular intervals and are available at both sites. dbVar and DGVa will continue to develop to meet the needs of the community as the technology and analysis methods for detecting structural variation evolve. Ultimately, these two SV resources promise to greatly enable studies of genetic variation.

Acknowledgments

This research was supported (in part) by the Intramural Research Program of the NIH, National Library of Medicine.

References

1. Lupski, J. R., de Oca-Luna, R. M., Slaugenhaupt, S., Pentao, L., Guzzetta, V., Trask, B. J., Saucedo-Cardenas, O., Barker, D. F., Killian, J. M., Garcia, C. A., Chakravarti, A., and Patel, P. I. (1991) DNA duplication associated with Charcot-Marie-Tooth disease type 1A, *Cell* **66**, 219–232.
2. Koolen, D. A., Vissers, L. E., Pfundt, R., de Leeuw, N., Knight, S. J., Regan, R., Kooy, R. F., Reyniers, E., Romano, C., Fichera, M., Schinzel, A., Baumer, A., Anderlid, B. M., Schoumans, J., Knoers, N. V., van Kessel, A. G., Sistermans, E. A., Veltman, J. A., Brunner, H. G., and de Vries, B. B. (2006) A new chromosome 17q21.31 microdeletion syndrome associated with a common inversion polymorphism, *Nat Genet* **38**, 999–1001.
3. Shaw-Smith, C., Pittman, A. M., Willatt, L., Martin, H., Rickman, L., Gribble, S., Curley, R., Cumming, S., Dunn, C., Kalaitzopoulos, D., Porter, K., Prigmore, E., Krepischi-Santos, A. C., Varela, M. C., Koiffmann, C. P., Lees, A. J., Rosenberg, C., Firth, H. V., de Silva, R., and Carter, N. P. (2006) Microdeletion encompassing MAPT at chromosome 17q21.3 is associated with developmental delay and learning disability, *Nat Genet* **38**, 1032–1037.
4. Weiss, L. A., Shen, Y., Korn, J. M., Arking, D. E., Miller, D. T., Fossdal, R., Saemundsen, E., Stefansson, H., Ferreira, M. A., Green, T., Platt, O. S., Ruderfer, D. M., Walsh, C. A., Altshuler, D., Chakravarti, A., Tanzi, R. E., Stefansson, K., Santangelo, S. L., Gusella, J. F., Sklar, P., Wu, B. L., and Daly, M. J. (2008) Association between Microdeletion and Microduplication at 16p11.2 and Autism, *N Engl J Med.* 358(7), 667–675
5. Gonzalez, E., Kulkarni, H., Bolivar, H., Mangano, A., Sanchez, R., Catano, G., Nibbs, R. J., Freedman, B. I., Quinones, M. P., Bamshad, M. J., Murthy, K. K., Rovin, B. H., Bradley, W., Clark, R. A., Anderson, S. A., O'Connell R, J., Agan, B. K., Ahuja, S. S., Bologna, R., Sen, L., Dolan, M. J., and Ahuja, S. K. (2005) The influence of CCL3L1 gene-containing segmental duplications on HIV-1/AIDS susceptibility, *Science* **307**, 1434–1440.
6. Fanciulli, M., Norsworthy, P. J., Petretto, E., Dong, R., Harper, L., Kamesh, L., Heward, J. M., Gough, S. C., de Smith, A., Blakemore, A. I., Froguel, P., Owen, C. J., Pearce, S. H., Teixeira, L., Guillevin, L., Graham, D. S., Pusey, C. D., Cook, H. T., Vyse, T. J., and Aitman, T. J. (2007) FCGR3B copy number variation is associated with susceptibility to systemic, but not organ-specific, autoimmunity, *Nat Genet* **39**, 721–723.
7. Fellermann, K., Stange, D. E., Schaeffeler, E., Schmalzl, H., Wehkamp, J., Bevins, C. L., Reinisch, W., Teml, A., Schwab, M., Lichter, P., Radlwimmer, B., and Stange, E. F. (2006) A chromosome 8 gene-cluster polymorphism with low human beta-defensin 2 gene copy number predisposes to Crohn disease of the colon, *American journal of human genetics* **79**, 439–448.
8. Hollox, E. J., Huffmeier, U., Zeeuwen, P. L., Palla, R., Lascorz, J., Rodijk-Olthuis, D., van de Kerkhof, P. C., Traupe, H., de Jongh, G., den Heijer, M., Reis, A., Armour, J. A., and Schalkwijk, J. (2008) Psoriasis is associated with increased beta-defensin genomic copy number, *Nat Genet* **40**, 23–25.
9. Huse, K., Taudien, S., Groth, M., Rosenstiel, P., Szafranski, K., Hiller, M., Hampe, J., Junker, K., Schubert, J., Schreiber, S., Birkenmeier, G., Krawczak, M., and Platzer, M. (2008) Genetic Variants of the Copy Number Polymorphic beta-Defensin Locus Are Associated with Sporadic Prostate Cancer, *Tumour Biol* **29**, 83–92.
10. Conrad, D. F., Andrews, T. D., Carter, N. P., Hurles, M. E., and Pritchard, J. K. (2006) A high-resolution survey of deletion polymorphism in the human genome, *Nat Genet* **38**, 75–81.
11. Cooper, G. M., Nickerson, D. A., and Eichler, E. E. (2007) Mutational and selective effects on copy-number variants in the human genome, *Nat Genet* **39**, S22–29.
12. Cutler, G., Marshall, L. A., Chin, N., Baribault, H., and Kassner, P. D. (2007) Significant gene content variation characterizes the genomes of inbred mouse strains, *Genome research* **17**, 1743–1754.

13. Graubert, T. A., Cahan, P., Edwin, D., Selzer, R. R., Richmond, T. A., Eis, P. S., Shannon, W. D., Li, X., McLeod, H. L., Cheverud, J. M., and Ley, T. J. (2007) A high-resolution map of segmental DNA copy number variation in the mouse genome, *PLoS Genet* **3**, e3.
14. Nguyen, D. Q., Webber, C., and Ponting, C. P. (2006) Bias of selection on human copy-number variants, *PLoS Genet* **2**, e20.
15. Wong, K. K., deLeeuw, R. J., Dosanjh, N. S., Kimm, L. R., Cheng, Z., Horsman, D. E., MacAulay, C., Ng, R. T., Brown, C. J., Eichler, E. E., and Lam, W. L. (2007) A comprehensive analysis of common copy-number variations in the human genome, *American journal of human genetics* **80**, 91–104.
16. (2003) The International HapMap Project, *Nature* **426**, 789–796.
17. Sachidanandam, R., Weissman, D., Schmidt, S. C., Kakol, J. M., Stein, L. D., Marth, G., Sherry, S., Mullikin, J. C., Mortimore, B. J., Willey, D. L., Hunt, S. E., Cole, C. G., Coggill, P. C., Rice, C. M., Ning, Z., Rogers, J., Bentley, D. R., Kwok, P. Y., Mardis, E. R., Yeh, R. T., Schultz, B., Cook, L., Davenport, R., Dante, M., Fulton, L., Hillier, L., Waterston, R. H., McPherson, J. D., Gilman, B., Schaffner, S., Van Etten, W. J., Reich, D., Higgins, J., Daly, M. J., Blumenstiel, B., Baldwin, J., Stange-Thomann, N., Zody, M. C., Linton, L., Lander, E. S., and Altshuler, D. (2001) A map of human genome sequence variation containing 1.42 million single nucleotide polymorphisms, *Nature* **409**, 928–933.
18. NCBI – http://www.ncbi.nlm.nih.gov/.
19. HuRef Project – http://huref.jcvi.org/.
20. MapViewer – http://www.ncbi.nlm.nih.gov/mapview/.
21. Wheeler, D. A., Srinivasan, M., Egholm, M., Shen, Y., Chen, L., McGuire, A., He, W., Chen, Y. J., Makhijani, V., Roth, G. T., Gomes, X., Tartaro, K., Niazi, F., Turcotte, C. L., Irzyk, G. P., Lupski, J. R., Chinault, C., Song, X. Z., Liu, Y., Yuan, Y., Nazareth, L., Qin, X., Muzny, D. M., Margulies, M., Weinstock, G. M., Gibbs, R. A., and Rothberg, J. M. (2008) The complete genome of an individual by massively parallel DNA sequencing, *Nature* **452**, 872–876.
22. Wang, J., Wang, W., Li, R., Li, Y., Tian, G., Goodman, L., Fan, W., Zhang, J., Li, J., Zhang, J., Guo, Y., Feng, B., Li, H., Lu, Y., Fang, X., Liang, H., Du, Z., Li, D., Zhao, Y., Hu, Y., Yang, Z., Zheng, H., Hellmann, I., Inouye, M., Pool, J., Yi, X., Zhao, J., Duan, J., Zhou, Y., Qin, J., Ma, L., Li, G., Yang, Z., Zhang, G., Yang, B., Yu, C., Liang, F., Li, W., Li, S., Li, D., Ni, P., Ruan, J., Li, Q., Zhu, H., Liu, D., Lu, Z., Li, N., Guo, G., Zhang, J., Ye, J., Fang, L., Hao, Q., Chen, Q., Liang, Y., Su, Y., San, A., Ping, C., Yang, S., Chen, F., Li, L., Zhou, K., Zheng, H., Ren, Y., Yang, L., Gao, Y., Yang, G., Li, Z., Feng, X., Kristiansen, K., Wong, G. K., Nielsen, R., Durbin, R., Bolund, L., Zhang, X., Li, S., Yang, H., and Wang, J. (2008) The diploid genome sequence of an Asian individual, *Nature* **456**, 60–65.
23. Ahn, S. M., Kim, T. H., Lee, S., Kim, D., Ghang, H., Kim, D. S., Kim, B. C., Kim, S. Y., Kim, W. Y., Kim, C., Park, D., Lee, Y. S., Kim, S., Reja, R., Jho, S., Kim, C. G., Cha, J. Y., Kim, K. H., Lee, B., Bhak, J., and Kim, S. J. (2009) The first Korean genome sequence and analysis: full genome sequencing for a socio-ethnic group, *Genome research* **19**, 1622–1629.
24. Bentley, D. R., Balasubramanian, S., Swerdlow, H. P., Smith, G. P., Milton, J., Brown, C. G., Hall, K. P., Evers, D. J., Barnes, C. L., Bignell, H. R., Boutell, J. M., Bryant, J., Carter, R. J., Keira Cheetham, R., Cox, A. J., Ellis, D. J., Flatbush, M. R., Gormley, N. A., Humphray, S. J., Irving, L. J., Karbelashvili, M. S., Kirk, S. M., Li, H., Liu, X., Maisinger, K. S., Murray, L. J., Obradovic, B., Ost, T., Parkinson, M. L., Pratt, M. R., Rasolonjatovo, I. M., Reed, M. T., Rigatti, R., Rodighiero, C., Ross, M. T., Sabot, A., Sankar, S. V., Scally, A., Schroth, G. P., Smith, M. E., Smith, V. P., Spiridou, A., Torrance, P. E., Tzonev, S. S., Vermaas, E. H., Walter, K., Wu, X., Zhang, L., Alam, M. D., Anastasi, C., Aniebo, I. C., Bailey, D. M., Bancarz, I. R., Banerjee, S., Barbour, S. G., Baybayan, P. A., Benoit, V. A., Benson, K. F., Bevis, C., Black, P. J., Boodhun, A., Brennan, J. S., Bridgham, J. A., Brown, R. C., Brown, A. A., Buermann, D. H., Bundu, A. A., Burrows, J. C., Carter, N. P., Castillo, N., Chiara, E. C. M., Chang, S., Neil Cooley, R., Crake, N. R., Dada, O. O., Diakoumakos, K. D., Dominguez-Fernandez, B., Earnshaw, D. J., Egbujor, U. C., Elmore, D. W., Etchin, S. S., Ewan, M. R., Fedurco, M., Fraser, L. J., Fuentes Fajardo, K. V., Scott Furey, W., George, D., Gietzen, K. J., Goddard, C. P., Golda, G. S., Granieri, P. A., Green, D. E., Gustafson, D. L., Hansen, N. F., Harnish, K., Haudenschild, C. D., Heyer, N. I., Hims, M. M., Ho, J. T., Horgan, A. M., Hoschler, K., Hurwitz, S., Ivanov, D. V., Johnson, M. Q., James, T., Huw Jones, T. A., Kang, G. D., Kerelska, T. H., Kersey, A. D., Khrebtukova, I., Kindwall, A. P., Kingsbury, Z., Kokko-Gonzales, P. I., Kumar, A., Laurent, M. A., Lawley, C. T., Lee, S. E., Lee, X., Liao, A. K., Loch, J. A., Lok, M., Luo, S., Mammen, R. M., Martin, J. W., McCauley, P. G., McNitt, P., Mehta, P., Moon, K. W., Mullens,

J. W., Newington, T., Ning, Z., Ling Ng, B., Novo, S. M., O'Neill, M. J., Osborne, M. A., Osnowski, A., Ostadan, O., Paraschos, L. L., Pickering, L., Pike, A. C., Pike, A. C., Chris Pinkard, D., Pliskin, D. P., Podhasky, J., Quijano, V. J., Raczy, C., Rae, V. H., Rawlings, S. R., Chiva Rodriguez, A., Roe, P. M., Rogers, J., Rogert Bacigalupo, M. C., Romanov, N., Romieu, A., Roth, R. K., Rourke, N. J., Ruediger, S. T., Rusman, E., Sanches-Kuiper, R. M., Schenker, M. R., Seoane, J. M., Shaw, R. J., Shiver, M. K., Short, S. W., Sizto, N. L., Sluis, J. P., Smith, M. A., Ernest Sohna Sohna, J., Spence, E. J., Stevens, K., Sutton, N., Szajkowski, L., Tregidgo, C. L., Turcatti, G., Vandevondele, S., Verhovsky, Y., Virk, S. M., Wakelin, S., Walcott, G. C., Wang, J., Worsley, G. J., Yan, J., Yau, L., Zuerlein, M., Rogers, J., Mullikin, J. C., Hurles, M. E., McCooke, N. J., West, J. S., Oaks, F. L., Lundberg, P. L., Klenerman, D., Durbin, R., and Smith, A. J. (2008) Accurate whole human genome sequencing using reversible terminator chemistry, *Nature* **456**, 53–59.

25. Kim, J. I., Ju, Y. S., Park, H., Kim, S., Lee, S., Yi, J. H., Mudge, J., Miller, N. A., Hong, D., Bell, C. J., Kim, H. S., Chung, I. S., Lee, W. C., Lee, J. S., Seo, S. H., Yun, J. Y., Woo, H. N., Lee, H., Suh, D., Lee, S., Kim, H. J., Yavartanoo, M., Kwak, M., Zheng, Y., Lee, M. K., Park, H., Kim, J. Y., Gokcumen, O., Mills, R. E., Zaranek, A. W., Thakuria, J., Wu, X., Kim, R. W., Huntley, J. J., Luo, S., Schroth, G. P., Wu, T. D., Kim, H., Yang, K. S., Park, W. Y., Kim, H., Church, G. M., Lee, C., Kingsmore, S. F., and Seo, J. S. (2009) A highly annotated whole-genome sequence of a Korean individual, *Nature* **460**, 1011–1015.

26. The Copy Number Variation (CNV) Project – http://www.sanger.ac.uk/humgen/cnv/.

27. Redon, R., Ishikawa, S., Fitch, K. R., Feuk, L., Perry, G. H., Andrews, T. D., Fiegler, H., Shapero, M. H., Carson, A. R., Chen, W., Cho, E. K., Dallaire, S., Freeman, J. L., Gonzalez, J. R., Gratacos, M., Huang, J., Kalaitzopoulos, D., Komura, D., MacDonald, J. R., Marshall, C. R., Mei, R., Montgomery, L., Nishimura, K., Okamura, K., Shen, F., Somerville, M. J., Tchinda, J., Valsesia, A., Woodwark, C., Yang, F., Zhang, J., Zerjal, T., Zhang, J., Armengol, L., Conrad, D. F., Estivill, X., Tyler-Smith, C., Carter, N. P., Aburatani, H., Lee, C., Jones, K. W., Scherer, S. W., and Hurles, M. E. (2006) Global variation in copy number in the human genome, *Nature* **444**, 444–454.

28. UCSC Genome Browser – http://genome.ucsc.edu/.

29. Ensembl Genome Browser – http://www.ensembl.org/.

30. Human Genome Structural Variation Project – http://hgsv.washington.edu/.

31. Kidd, J. M., Cooper, G. M., Donahue, W. F., Hayden, H. S., Sampas, N., Graves, T., Hansen, N., Teague, B., Alkan, C., Antonacci, F., Haugen, E., Zerr, T., Yamada, N. A., Tsang, P., Newman, T. L., Tuzun, E., Cheng, Z., Ebling, H. M., Tusneem, N., David, R., Gillett, W., Phelps, K. A., Weaver, M., Saranga, D., Brand, A., Tao, W., Gustafson, E., McKernan, K., Chen, L., Malig, M., Smith, J. D., Korn, J. M., McCarroll, S. A., Altshuler, D. A., Peiffer, D. A., Dorschner, M., Stamatoyannopoulos, J., Schwartz, D., Nickerson, D. A., Mullikin, J. C., Wilson, R. K., Bruhn, L., Olson, M. V., Kaul, R., Smith, D. R., and Eichler, E. E. (2008) Mapping and sequencing of structural variation from eight human genomes, *Nature* **453**, 56–64.

32. Tuzun, E., Sharp, A. J., Bailey, J. A., Kaul, R., Morrison, V. A., Pertz, L. M., Haugen, E., Hayden, H., Albertson, D., Pinkel, D., Olson, M. V., and Eichler, E. E. (2005) Fine-scale structural variation of the human genome, *Nat Genet* **37**, 727–732.

33. 1000 Genomes Project – http://www.1000genomes.org.

34. The 1000 Genomes Project Consortium (2010) A map of human genome variation from population-scale sequencing, Nature 467, 1061–1073.

35. Mills, R.E., Walter, K., Stewart, C.et. al. (2011) Mapping copy number variation by population-scale genome sequencing, Nature 470, 59–65.

36. The Copy Number Variation project at the Children's Hospital of Philadelphia (CHOP) – http://cnv.chop.edu/.

37. Shaikh, T. H., Gai, X., Perin, J. C., Glessner, J. T., Xie, H., Murphy, K., O'Hara, R., Casalunovo, T., Conlin, L. K., D'Arcy, M., Frackelton, E. C., Geiger, E. A., Haldeman-Englert, C., Imielinski, M., Kim, C. E., Medne, L., Annaiah, K., Bradfield, J. P., Dabaghyan, E., Eckert, A., Onyiah, C. C., Ostapenko, S., Otieno, F. G., Santa, E., Shaner, J. L., Skraban, R., Smith, R. M., Elia, J., Goldmuntz, E., Spinner, N. B., Zackai, E. H., Chiavacci, R. M., Grundmeier, R., Rappaport, E. F., Grant, S. F., White, P. S., and Hakonarson, H. (2009) High-resolution mapping and analysis of copy number variations in the human genome: a data resource for clinical and research applications, *Genome research* **19**, 1682–1690.

38. Chromosome Anomaly Collection – http://www.ngrl.org.uk/Wessex/collection/index.htm.

39. Snijders, A. M., Nowak, N. J., Huey, B., Fridlyand, J., Law, S., Conroy, J., Tokuyasu, T., Demir, K., Chiu, R., Mao, J. H., Jain, A. N., Jones, S. J., Balmain, A., Pinkel, D., and Albertson, D. G. (2005) Mapping segmental and sequence variations among laboratory mice using BAC array CGH, *Genome research* **15**, 302–311.
40. Guryev, V., Saar, K., Adamovic, T., Verheul, M., van Heesch, S. A., Cook, S., Pravenec, M., Aitman, T., Jacob, H., Shull, J. D., Hubner, N., and Cuppen, E. (2008) Distribution and functional impact of DNA copy number variation in the rat, *Nat Genet* **40**, 538–545.
41. Lee, A. S., Gutierrez-Arcelus, M., Perry, G. H., Vallender, E. J., Johnson, W. E., Miller, G. M., Korbel, J. O., and Lee, C. (2008) Analysis of copy number variation in the rhesus macaque genome identifies candidate loci for evolutionary and human disease studies, *Hum Mol Genet* 17(8), 1127–1136.
42. Fortna, A., Kim, Y., MacLaren, E., Marshall, K., Hahn, G., Meltesen, L., Brenton, M., Hink, R., Burgers, S., Hernandez-Boussard, T., Karimpour-Fard, A., Glueck, D., McGavran, L., Berry, R., Pollack, J., and Sikela, J. M. (2004) Lineage-specific gene duplication and loss in human and great ape evolution, *PLoS Biol* **2**, E207.
43. Goidts, V., Armengol, L., Schempp, W., Conroy, J., Nowak, N., Muller, S., Cooper, D. N., Estivill, X., Enard, W., Szamalek, J. M., Hameister, H., and Kehrer-Sawatzki, H. (2006) Identification of large-scale human-specific copy number differences by inter-species array comparative genomic hybridization, *Hum Genet* **119**, 185–198.
44. Locke, D. P., Segraves, R., Carbone, L., Archidiacono, N., Albertson, D. G., Pinkel, D., and Eichler, E. E. (2003) Large-scale variation among human and great ape genomes determined by array comparative genomic hybridization, *Genome research* **13**, 347–357.
45. Perry, G. H., Tchinda, J., McGrath, S. D., Zhang, J., Picker, S. R., Caceres, A. M., Iafrate, A. J., Tyler-Smith, C., Scherer, S. W., Eichler, E. E., Stone, A. C., and Lee, C. (2006) Hotspots for copy number variation in chimpanzees and humans, *Proceedings of the National Academy of Sciences of the United States of America* **103**, 8006–8011.
46. Wilson, G. M., Flibotte, S., Missirlis, P. I., Marra, M. A., Jones, S., Thornton, K., Clark, A. G., and Holt, R. A. (2006) Identification by full-coverage array CGH of human DNA copy number increases relative to chimpanzee and gorilla, *Genome research* **16**, 173–181.
47. AtSFP – The SIGnAL Arabidopsis SNP, Deletion and SFP Database – http://signal.salk.edu/cgi-bin/AtSFP.
48. Borevitz, J. O., Hazen, S. P., Michael, T. P., Morris, G. P., Baxter, I. R., Hu, T. T., Chen, H., Werner, J. D., Nordborg, M., Salt, D. E., Kay, S. A., Chory, J., Weigel, D., Jones, J. D., and Ecker, J. R. (2007) Genome-wide patterns of single-feature polymorphism in Arabidopsis thaliana, *Proceedings of the National Academy of Sciences of the United States of America* **104**, 12057–12062.
49. Wang, J., He, X., Ruan, J., Dai, M., Chen, J., Zhang, Y., Hu, Y., Ye, C., Li, S., Cong, L., Fang, L., Liu, B., Li, S., Wang, J., Burt, D. W., Wong, G. K., Yu, J., Yang, H., and Wang, J. (2005) ChickVD: a sequence variation database for the chicken genome, *Nucleic acids research* **33**, D438–441.
50. Chicken Variation Database (ChickVD) – http://chicken.genomics.org.cn/.
51. Copy Number Variations across Vertebrate genomes (CNVVdb) – http://cnvvdb.genomics.sinica.edu.tw/.
52. Chen, F. C., Chen, Y. Z., and Chuang, T. J. (2009) CNVVdb: a database of copy number variations across vertebrate genomes, *Bioinformatics (Oxford, England)* **25**, 1419–1421.
53. Database of Genomic Variants – http://projects.tcag.ca/variation/.
54. UCSC liftover tool – http://genome.ucsc.edu/cgi-bin/hgLiftOver.
55. AnEUploidy Project – http://www.aneuploidy.eu/.
56. CAD (Chromosome Abnormality Database) – http://www.ukcad.org.uk./cocoon/ukcad/
57. ECARUCA (The European Cytogeneticists Association Register of Unbalanced Chromosome Aberrations) – http://www.ecaruca.net/.
58. DECIPHER (DatabasE of Chromosomal Imbalance and Phenotype in Humans using Ensembl Resources) – https://decipher.sanger.ac.uk/.
59. Database of Mental Retardation Associated CNVs – http://bioinfo.ut.ee/dbcard/.
60. The Autism Chromosome Rearrangement Database – http://projects.tcag.ca/autism/.
61. Autism CNV Database – http://projects.tcag.ca/autism_500k/.
62. Szatmari, P., Paterson, A. D., Zwaigenbaum, L., Roberts, W., Brian, J., Liu, X. Q., Vincent, J. B., Skaug, J. L., Thompson, A. P., Senman, L., Feuk, L., Qian, C., Bryson, S. E., Jones, M. B., Marshall, C. R., Scherer, S. W., Vieland, V. J., Bartlett, C., Mangin, L. V., Goedken, R., Segre, A., Pericak-Vance, M. A., Cuccaro, M. L., Gilbert, J. R., Wright, H. H., Abramson, R.

K., Betancur, C., Bourgeron, T., Gillberg, C., Leboyer, M., Buxbaum, J. D., Davis, K. L., Hollander, E., Silverman, J. M., Hallmayer, J., Lotspeich, L., Sutcliffe, J. S., Haines, J. L., Folstein, S. E., Piven, J., Wassink, T. H., Sheffield, V., Geschwind, D. H., Bucan, M., Brown, W. T., Cantor, R. M., Constantino, J. N., Gilliam, T. C., Herbert, M., Lajonchere, C., Ledbetter, D. H., Lese-Martin, C., Miller, J., Nelson, S., Samango-Sprouse, C. A., Spence, S., State, M., Tanzi, R. E., Coon, H., Dawson, G., Devlin, B., Estes, A., Flodman, P., Klei, L., McMahon, W. M., Minshew, N., Munson, J., Korvatska, E., Rodier, P. M., Schellenberg, G. D., Smith, M., Spence, M. A., Stodgell, C., Tepper, P. G., Wijsman, E. M., Yu, C. E., Roge, B., Mantoulan, C., Wittemeyer, K., Poustka, A., Felder, B., Klauck, S. M., Schuster, C., Poustka, F., Bolte, S., Feineis-Matthews, S., Herbrecht, E., Schmotzer, G., Tsiantis, J., Papanikolaou, K., Maestrini, E., Bacchelli, E., Blasi, F., Carone, S., Toma, C., Van Engeland, H., de Jonge, M., Kemner, C., Koop, F., Langemeijer, M., Hijmans, C., Staal, W. G., Baird, G., Bolton, P. F., Rutter, M. L., Weisblatt, E., Green, J., Aldred, C., Wilkinson, J. A., Pickles, A., Le Couteur, A., Berney, T., McConachie, H., Bailey, A. J., Francis, K., Honeyman, G., Hutchinson, A., Parr, J. R., Wallace, S., Monaco, A. P., Barnby, G., Kobayashi, K., Lamb, J. A., Sousa, I., Sykes, N., Cook, E. H., Guter, S. J., Leventhal, B. L., Salt, J., Lord, C., Corsello, C., Hus, V., Weeks, D. E., Volkmar, F., Tauber, M., Fombonne, E., Shih, A., and Meyer, K. J. (2007) Mapping autism risk loci using genetic linkage and chromosomal rearrangements, *Nat Genet* **39**, 319–328.

63. Marshall, C. R., Noor, A., Vincent, J. B., Lionel, A. C., Feuk, L., Skaug, J., Shago, M., Moessner, R., Pinto, D., Ren, Y., Thiruvahindrapduram, B., Fiebig, A., Schreiber, S., Friedman, J., Ketelaars, C. E., Vos, Y. J., Ficicioglu, C., Kirkpatrick, S., Nicolson, R., Sloman, L., Summers, A., Gibbons, C. A., Teebi, A., Chitayat, D., Weksberg, R., Thompson, A., Vardy, C., Crosbie, V., Luscombe, S., Baatjes, R., Zwaigenbaum, L., Roberts, W., Fernandez, B., Szatmari, P., and Scherer, S. W. (2008) Structural variation of chromosomes in autism spectrum disorder, *American journal of human genetics* **82**, 477–488.
64. Autism Genetic Database – http://wren.bcf.ku.edu/.
65. The Cancer Genome Project – http://www.sanger.ac.uk/genetics/CGP/.
66. COSMIC – http://www.sanger.ac.uk/genetics/CGP/cosmic/.
67. The Cancer Genome Atlas – http://cancergenome.nih.gov/.
68. TCGA Data Portal – http://tcga-data.nci.nih.gov/tcga/homepage.htm.
69. Progenetix – www.progenetix.net/.
70. Baudis, M., and Cleary, M. L. (2001) Progenetix.net: an online repository for molecular cytogenetic aberration data, *Bioinformatics (Oxford, England)* **17**, 1228–1229.
71. Cancer Chromosomes – http://www.ncbi.nlm.nih.gov/cancerchromosomes.
72. Knutsen, T., Gobu, V., Knaus, R., Padilla-Nash, H., Augustus, M., Strausberg, R. L., Kirsch, I. R., Sirotkin, K., and Ried, T. (2005) The interactive online SKY/M-FISH & CGH database and the Entrez cancer chromosomes search database: linkage of chromosomal aberrations with the genome sequence, *Genes, chromosomes & cancer* **44**, 52–64.
73. McCarroll, S. A., Hadnott, T. N., Perry, G. H., Sabeti, P. C., Zody, M. C., Barrett, J. C., Dallaire, S., Gabriel, S. B., Lee, C., Daly, M. J., and Altshuler, D. M. (2006) Common deletion polymorphisms in the human genome, *Nat Genet* **38**, 86–92.
74. NimbleGen – http://www.nimblegen.com.
75. Illumina – http://www.illumina.com.
76. Korbel, J. O., Urban, A. E., Affourtit, J. P., Godwin, B., Grubert, F., Simons, J. F., Kim, P. M., Palejev, D., Carriero, N. J., Du, L., Taillon, B. E., Chen, Z., Tanzer, A., Saunders, A. C., Chi, J., Yang, F., Carter, N. P., Hurles, M. E., Weissman, S. M., Harkins, T. T., Gerstein, M. B., Egholm, M., and Snyder, M. (2007) Paired-end mapping reveals extensive structural variation in the human genome, *Science* **318**, 420–426.
77. BreakPtr – http://breakptr.gersteinlab.org.
78. CNAG – http://www.genome.umin.jp/CNAGtop2.html.
79. http://www.sanger.ac.uk/resources/software/cnvfinder/.
80. dChip – http://biosun1.harvard.edu/complab/dchip/.
81. http://www.genome.rcast.u-tokyo.ac.jp/CNV/gemca_details.html.
82. PennCNV – http://www.neurogenome.org/cnv/penncnv/.
83. VariationHunter – http://compbio.cs.sfu.ca/strvar.htm.
84. GEO (Gene Expression Omnibus) – http://www.ncbi.nlm.nih.gov/geo/.
85. ArrayExpress – http://www.ebi.ac.uk/arrayexpress/.

86. Trace Archive – http://www.ncbi.nlm.nih.gov/Traces/trace.cgi.
87. Short Read Archive (SRA) – http://trace.ncbi.nlm.nih.gov/Traces/sra/sra.cgi.
88. Levy, S., Sutton, G., Ng, P. C., Feuk, L., Halpern, A. L., Walenz, B. P., Axelrod, N., Huang, J., Kirkness, E. F., Denisov, G., Lin, Y., MacDonald, J. R., Pang, A. W., Shago, M., Stockwell, T. B., Tsiamouri, A., Bafna, V., Bansal, V., Kravitz, S. A., Busam, D. A., Beeson, K. Y., McIntosh, T. C., Remington, K. A., Abril, J. F., Gill, J., Borman, J., Rogers, Y. H., Frazier, M. E., Scherer, S. W., Strausberg, R. L., and Venter, J. C. (2007) The diploid genome sequence of an individual human, *PLoS Biol* **5**, e254.
89. dbGaP – http://www.ncbi.nlm.nih.gov/gap.
90. European Genome-phenome Archive (EGA) – http://www.ebi.ac.uk/ega/.
91. dbVar – http://www.ncbi.nlm.nih.gov/dbvar/.
92. DGVa – http://www.ebi.ac.uk/dgva/
93. Miller, D. T., Adam, M. P., Aradhya, S., Biesecker, L. G., Brothman, A. R., Carter, N. P., Church, D. M., Crolla, J. A., Eichler, E. E., Epstein, C. J., Faucett, W. A., Feuk, L., Friedman, J. M., Hamosh, A., Jackson, L., Kaminsky, E. B., Kok, K., Krantz, I. D., Kuhn, R. M., Lee, C., Ostell, J. M., Rosenberg, C., Scherer, S. W., Spinner, N. B., Stavropoulos, D. J., Tepperberg, J. H., Thorland, E. C., Vermeesch, J. R., Waggoner, D. J., Watson, M. S., Martin, C. L., and Ledbetter, D. H. Consensus statement: chromosomal microarray is a first-tier clinical diagnostic test for individuals with developmental disabilities or congenital anomalies, *American journal of human genetics* **86**, 749–764.

Chapter 14

Algorithm Implementation for CNV Discovery Using Affymetrix and Illumina SNP Array Data

Laura Winchester and Jiannis Ragoussis

Abstract

SNP array data can be analysed for the purpose of calling SNP alleles but also for determining the absolute copy number of a certain genomic segment. Here, the method for detecting copy number (CN) change using intensity data from SNP arrays is focused on. Methods incorporating data from the two main genotyping platforms, Affymetrix and Illumina, are described and possible options and problems that may be faced are examined. We discuss the importance of the quality control when using this analysis method and present some guidelines for implementation, both prior and post to algorithm use. A discussion of algorithms available for CN detection is included as well as ideas for further analysis protocols.

Key words: SNP array, Copy number, Detection algorithm, Copy number variant, Illumina, Affymetrix

1. Introduction

1.1. Application of SNP Arrays in CNV Detection

SNP arrays have been designed for allowing the simultaneous genotyping of a very high number of polymorphisms, typically from thousands to millions for a given sample and used successfully for whole genome association studies (1). However, recent interest in structural variation of the genome has led to the investigation of a second use for the data. Intensity data from SNPs can be used to detect copy number (CN) changes and a number of software and methods have been developed and are now available to assist in this procedure. Initially assays were optimised for genotyping SNPs, but recent array designs include specially designed probes to allow us to investigate sites of known variation. These copy number polymorphisms (CNPs) can be assayed in parallel with other SNPs to produce a clearer picture of genome variation. At the time of

Lars Feuk (ed.), *Genomic Structural Variants: Methods and Protocols*, Methods in Molecular Biology, vol. 838, DOI 10.1007/978-1-61779-507-7_14, © Springer Science+Business Media, LLC 2012

writing, the Affymetrix Chip 6.0 and Illumina 1M-Duo have 1,852,600 and 1,119,187 assays, respectively. Increased overall genomic coverage and assay density allows the detection of rare CN changes; copy number variants (CNVs) are events found in less than 1% of the population (2). Detection of these CNVs and their link with disease phenotypes is the focus of the following method.

In this chapter, the use of signal intensity data from two major SNP array platforms, Affymetrix and Illumina, is discussed. Using examples of high coverage chips, Affymetrix Chip6.0 and Illumina 1M Duo, we describe the analysis process from raw intensity files to filtering and investigation of the events detected from this data.

1.2. Introduction of Methods

The protocol described in this chapter has been broken down into several sections for ease of explanation. Since this work is computation based, where appropriate, examples of commands have been included (italic font). The process is outlined by the flow chart in Fig. 1. The first parts of the method have been divided in half to address the differences between the Illumina and Affymetrix output and data handling. The method begins with a description of the raw data and initial quality control of samples. A method using the proprietary software is described but also mentioned are possibilities for a user who does not have access to the original files or the commercial software.

The second section of the methods discusses the available analysis software and presents the authors preferred method using two algorithms as an example. GADA (3) (Genome Alteration Detection Analysis) uses a two stage analysis, Sparse Bayesian Learning is employed to detect breakpoints and CN changes, then Backward Elimination is used to remove false positives from the data. GADA is a flexible tool and when used in the R environment, allows the user to manipulate input data and display results graphically. QuantiSNP (4) is based on a Hidden Markov Model (HMM) and it uses the different copy number as it hidden states to detect changes. It is a command line run tool, which was designed for use on Illumina data and includes a score with each event to allow the user to rank CN changes by likelihood.

The other focus of this chapter is to discuss potential analysis options for the events detected by the algorithms. The "Analysing Output" section describes the output of the software and lists a selection of the techniques and options for analysis of the events found from the algorithm. This includes an important quality control section to eliminate false positives and filter the results.

The last section of the chapter outlines a few of the potential methods for looking at CNVs in more detail. It suggests several software possibilities with academic and commercial options.

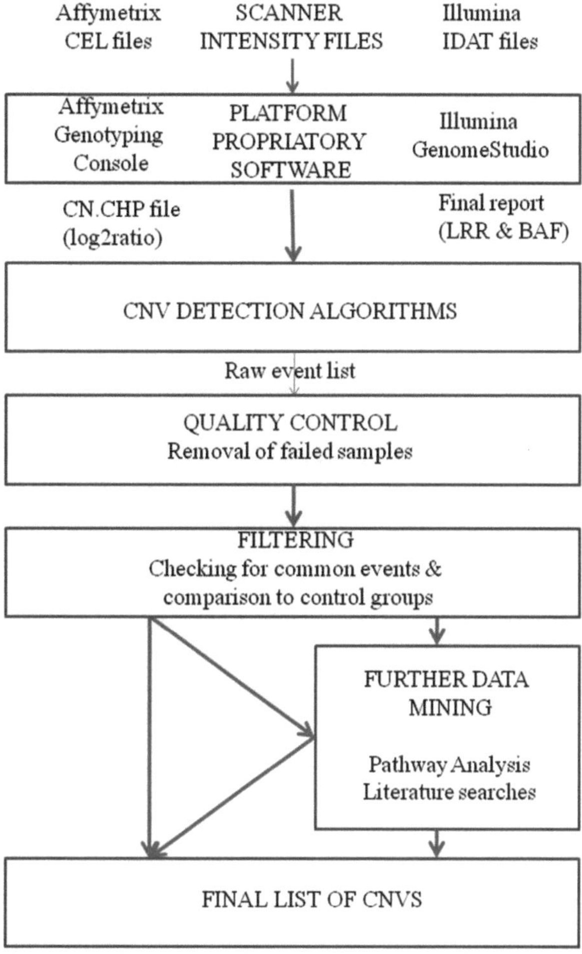

Fig. 1. Flow chart showing the protocol for CNV detection and analysis.

2. Materials

2.1. Hardware

Only computation tools are required for the protocol described in this chapter. A high performance PC with quad core and 64-bit processing is necessary for handling the large data sets produced by high density SNP arrays. As with most computationally complex processes, the more CPU and RAM resources available the quicker and more reliably the algorithms will run. Commercial tools require Microsoft Windows operating system to run. Access to 64 bit LINUX based systems would be an advantage for running the academically developed detection algorithms. Adequate storage is an important factor in preparing for analysis, a single Illumina intensity scan file can be up to 8.6 GB in size (Affymetrix CEL files are 68 MB each).

Raw datasets used are generated by Affymetrix and Illumina genotyping platforms. Use of the proprietary software from Affymetrix (Genotyping Console v4) and Illumina (GenomeStudio v2009.1) is discussed; however, it is not vital to the protocol, as raw data can be modified using freely available techniques for the detection algorithms.

2.2. CNV Detection Algorithms

Two detection algorithms are described in detail for this chapter. QuantiSNP (4) can be downloaded for free from https://sites.google.com/site/quantisnp/. Included within the download is a runtime version of MATLAB required for data processing and installation instructions. The GADA algorithm (3) is available from a similar group site (http://groups.google.com/group/gadaproject) or using a direct site, http://biron.usc.edu/~piquereg/GADA/. The R version will require the free statistical software R (http://www.r-project.org/) and an additional package: plotrix.

2.3. Further Analysis Tools

This chapter describes a number of options for further analysis. Listed in Table 1 are the programs and databases used along with details of location and download options.

3. Methods

3.1. Platform and Sample Selection

1. Picking a suitable sample collection:
   - Selection of patient samples is generally guided by project and disease factors. However, it is important to highlight that each patient set will require a strong control group to eliminate common events from the final results. Access to a suitable collection is vital for full analysis of the dataset.
   - See Note 1 for a detailed discussion of factors effecting CNV detection experimental design.
2. Choosing a SNP array platform:
   - The choice of platform for the study will have the biggest effect on the algorithm choice. Academically developed algorithms in particular tend to be targeted to a single platform. Illumina and Affymetrix data are described in detail for this chapter, but both have been utilised in highly successful studies for CNVs associated with disease (5–8).
   - When choosing a SNP array, it is important to consider the users own needs such as accessibility and prior knowledge of the platform and cost.

Table 1
Further analysis tools

Software or database	Description	Location for download and further information	Reference
Database of Genomic Variants (DGV)	Database resource containing known CNVs	http://projects.tcag.ca/variation/	(17)
Ensembl	Genome Database for several species	http://www.ensembl.org/index.html	(18)
GRAIL (Gene Relationships Across Implicated Loci)	Functional analysis using text based references	http://www.broadinstitute.org/mpg/grail/	(14)
GOStats	Functional analysis using Gene Ontology terms (R package)	http://www.bioconductor.org/packages/release/bioc/html/GOstats.html	(19)
Ingenuity Pathway Analysis	Commercial functional analysis	Available for purchase from http://www.ingenuity.com/	Technical notes
NCBI	US National Library of Medicine site including literature database	http://www.ncbi.nlm.nih.gov/guide/ and http://www.ncbi.nlm.nih.gov/pubmed/	N/A
Nexus	Commercially available analysis suite developed for Copy Number analysis. Includes data viewer and further analysis tools	Available for purchase from http://www.partek.com/	Technical notes
Partek	Commercially available analysis suite including Copy Number analysis. Includes data viewer and support tools	Available for purchase from http://www.biodiscovery.com/index/nexus	Technical notes
PennCNV	Academic detection algorithm and downstream analysis tools	http://www.openbioinformatics.org/penncnv/	(12)
R and Bioconductor	Statistical and graphical modelling environment with applicable packages for multiple functions	http://www.r-project.org/ and http://www.bioconductor.org/	N/A
UCSC Genome Browser	Genome Browser with incorporated database	http://genome.ucsc.edu/	(20)

- Coverage is also a factor in choice as each company has a range of products with different probe sets and locations. Arrays such as Affymetrix Chip 6.0 and Illumina's 1M Duo include CNP specific probes. Other arrays contain probe sets to target specific areas of the genome.

3.2. Using Raw Signal Data

Signal intensity data are used as an input in all detection algorithms. However, each platform outputs the data in a slightly different format. We describe the method to extract and check intensity from raw signal files using proprietary tools.

3.2.1. Affymetrix Data and Genotyping Console

Affymetrix provide two types of software for SNP array analysis; the Genotyping Console (GTC), which is the main proprietary software and the Affymetrix Power Tools (APT) that are a freely available resource for further processing of data. All the APT programs are run from command line and include cross plate normalisation for large sample sets and genotyping tools to be used outside the GTC software. The GTC itself incorporates a number of different algorithms and analysis methods dependant on chip type. For analysis of the Chip 6.0 data, algorithms from the BirdSuite (9) have been incorporated into the software, which are highly recommended for reliable analysis. Basic steps for all GTC data processing remain the same and are described in the following protocol.

1. Import sample intensity data contained in individual CEL files. Samples are then grouped into "In Bounds" and "Out of Bounds" for initial quality control. Proceed with the passed "In Bounds" samples.
   - Related library files and annotation will be required for array type and imported samples can be grouped by batch within a workspace/project.
2. Run Copy Number/LOH tool on all passed samples. Output is shown in batch format but individual samples file should end with .CNCHP.
3. Export data from Copy Number/LOH results sample batch making sure to include log2ratio column in the file for analysis outside the software. See Fig. 2a for example file.
4. Run the segment reporting tool on the Copy Number/LOH results to give an initial overview of the major CN changes.
   - The CNAT tool can be used to detect CN events in a sample; however, its sensitivity is lower than the academically developed detection methods, so it is not recommended for full analysis.

3.2.2. Illumina Data and Genome Studio

The GenomeStudio standard protocol and algorithms are identical for all chip formats. The same clustering method is applied to all probes and these values are used to calculate the Log R Ratio and B Allele Frequency data required for copy number detection.

a

Name	Chr	Position	Log R Ratio	B Allele Freq
200047	2	219793146	0.3014887	0.4718856
200050	2	219797929	0.156415	0.5165015
200052	2	219783037	0.1347295	0.462815
200053	2	219783289	0.2033534	0.505632
200124	7	150564216	0.1792383	0.005483711
200199	22	40847933	-0.0998625	1

b

ProbeSet	Chromosome	Position	CNState	Log2Ratio	SmoothSignal	LOH	Allele Difference
SNP_A-8575125	1	554484	2	-0.141646	1.81754	0	0.929529
SNP_A-8575115	1	554636	2	-0.031719	1.81719	0	-0.950615
SNP_A-8575371	1	557616	2	0.0142129	1.81007	0	-1.23373
CN_524192	1	615321	2	-0.424541	1.74862	null	null
CN_496034	1	696950	2	0.27859	2.06544	null	null

Fig. 2. Input data required for CN detection algorithms. Example files of required input for CN detection algorithms are shown. Image (**a**) is from part of an Affymetrix CN.CHP file. Image (**b**) shows the data format required for Illumina samples.

1. Load intensity data into the GenomeStudio software using a sample sheet to give as many details of name, sex and family relationship and affected status.
   - The corresponding manifest file for array format containing probe information is required to create initial project.
   - Run calling algorithm across all autosomal SNPs; exclude male samples temporarily before running caller on the X chromosome probes.
2. Calculate initial statistics for samples and produce a DNA report to check call rates. All samples should have a call rate greater than 0.99.
3. Flag or eliminate samples that fall outside cut-offs. Family relationships and replicates can be used at this stage to investigate any anomalies.
4. Use the final output report wizard to produce a file containing your passed samples, with other columns containing SNP Name, Chromosome, Position, *B* Allele Frequency, and Log *R* Ratio.
   - Do not remove zeroed SNPs while using the report function, as this will also exclude specialised CN probes. If SNP filtering is carried out, select and highlight the passed assays before proceeding with the report function.
5. Files can then be split or modified according to detection algorithms requirement.
   - Batch analysis for large number of samples can done using LINUX commands or user produced scripts. For simple LINUX modification, exporting results from the Genome Studio Full Data table is recommended by selecting all the

rows and exporting the displayed data to file. The cut function can then be used within a LINUX environment to divide into individual samples. For example, "cut -f 1-6 full_data_table.txt>patient1.txt" will add columns 1–6 to patient1.txt file. See Fig. 2b for example file.

6. For a quick overview of the major events in a sample the CNVPartition plug-in can be run on the data from within GenomeStudio and displayed on the Chromosome Viewer.
   - Results from the CNVPartition are not as sensitive as some of the other analysis options; however, the false positive rate in the events detected is considerably lower.

3.3. CNV Detection Algorithms

There are number of algorithms available for each platform. The use of multiple algorithms for a single data set is recommended (10) to allow the user to confirm events detected by the software. Described here are two examples of academically developed detection tools; however, there are multiple alternative options (see Note 2).

3.3.1. QuantiSNP: Illumina

There are several formats available including a LINUX based program and a plug-in for GenomeStudio, the method described here uses the PC version, QuantiSNPv2.0, run by MS-DOS commands.

INPUT REQUIREMENT: Sample sex and Single text file per sample containing the following columns: SNP Name, Chromosome, Position, Log *R* Ratio, and *B* Allele Frequency.

1. Open the Command Prompt tool.
2. Change location to folder containing the QuantiSNP.exe file (e.g. *cdC:\Program Files\QuantiSNP_v2-0*). Move sample input files in the same folder using windows copy and paste function.
   - Check and record pathway location for GC correction information. This adjusts LRR for local genomic GC content which can cause false calls.
   - Check input file for Build version of SNPs and match this to GC correction information.
   - Create input and output folders for data and make a note of this location.
3. Run launch command (e.g. *quantisnp2.exe --chr [1:23] --outdir "C:\Program Files\\QuantiSNP_v2-0\win32\03072009\binaries\win32\output_test" --sampleid patient1 --gender female --emiters 10 --lsetting 200000 --gcdir "C:\Program Files\QuantiSNP_v2-0\win32\03072009\binaries\win32\b36" --plot --genotype --params params.dat --level levels.dat --input-files "C:\Program Files\ QuantiSNP_v2-0\win32\03072009\binaries\win32\patient1.txt" --chrX 23 –doXcorrect*).

- Details of all available parameters are included in the downloadable manual. The following parameters should contain locations appropriate for the users own computer: *--outdir*, *--input-files*, and *–gcdir*. Other parameters can be adjusted according to run requirements: *--chr* (specifies chromosomes analysed), *--emiters* (number of iterations used in learning model), and *–lsetting* (length used to calculate a transition probability). Sample ID and gender should be filled in as appropriate (*––sampleid*, *--gender*).
- The flexibility to change the Emiters and length setting gives the user a wider detection range. By increasing and decreasing the length parameter, smaller or larger events can be detected, standard size is 200,000. Increasing the emitters or iteration steps will make the processing take longer but also allow the algorithm to fit more accurate model parameters to its data, recommended number for QuantiSNPv2.0 is 10.

4. Check files indicating a successful run. There are four output files for each sample (e.g. patient1.cnv, patient1.qc, patient1.ps, patient1.gn). The CNV calls are contained in the file with .cnv extension and the quality control values in the .qc file and can be opened in text editors. The remaining files contain plots which require a GhostScript viewer and genotypes which can be read and unzipped using LINUX.
   - Using quality control outputs from the software, it is possible to plot results and pick out outliers. Figure 3 shows a single chromosome plot of HapMap sample. Possible outliers have been labelled.
   - High background noise will cause false positive calls; using the standard deviation of the LRR as a guide, outliers can be eliminated. Expected values are between 0.1 and 0.25 (suggested cut-off >0.35). Standard deviation of the BAF is usually between 0.025 and 0.04.

3.3.2. GADA: Affymetrix

GADA is available in two formats; here, we describe the R package and its use. Detailed instructions of additional features are available in a downloadable vignette.

INPUT REQUIREMENT: Data can be imported in different formats as long as the correct column for log2ratio is specified and total number of columns are given. This allows the use of data directly from GTC or data processed using the APT set.

1. Load the GADA tool within the R environment using the command: *library(gada)*.
2. Load the intensity data, e.g. *dataAffy<– setupGADAaffy (file = "patient1_GW6_C.MyTest.CN5.CNCHP.txt", NumCols=8, log2ratioCol=5)*.
   - In this command the *setupGADAaffy* function is used to load the sample file into R and called dataAffy.

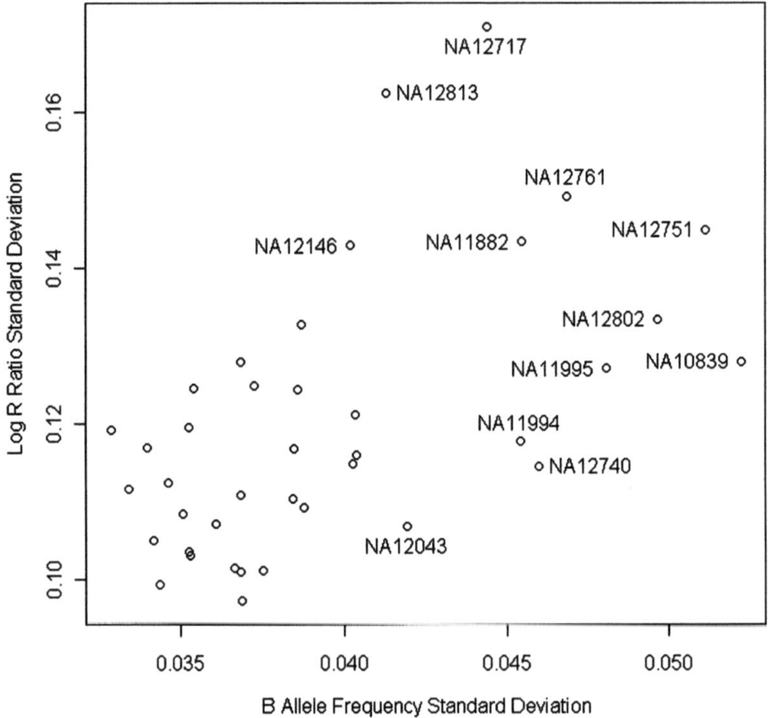

Fig. 3. QuantiSNP quality control measures plot. Plot of Chromosome 3 data for 269 HapMap samples. Possible outliers are labelled. Good quality data have LRR standard deviation values between 0.1 and 0.25 and BAF standard deviations from 0.025 to 0.04.

- Success of data load can be checked using the given name (e.g. *dataAffy*); this will show a summary of the uploaded file including the number of probes for each chromosome.
- Individual chromosomes can also be plotted to check for anomalies such as background noise, e.g. *plotRatio (dataAffy,chr=16)*.

3. STAGE 1: Applying a Sparse Bayesian learning (SBL) model to the data, e.g. *step1<-SBL(dataAffy, estim.sigma2=TRUE)*.
   - In this command we have run SBL model on dataAffy and called the output step1. Results from this command can be displayed by using *step1* as a command.
   - The command is used to find the most likely breakpoints and copy number states for our sample.
4. STAGE2: Eliminate false positives from the data using the Backwards Elimination command, e.g. *step2<-BackwardElimination(step1,T=4.5,MinSegLen=3)*.
   - Parameter T and sigma (Stage1) can be adjusted according to array type and sensitivity required. Increasing the values will give a lower false discovery rate and will reduce detection sensitivity.

5. Output final results using "*summary (step2)*" command. This will produce a list of the events found for the sample.
    - To display the results graphically the *plotRatio(step2)* command can be used. Figure 4 shows an example of several events on Chromosome 16.

3.4. Analysing Output

Outputs from all the algorithm software contain the same basic data: the related sample; the event size and location and SNPs involved. After quality control, the user can prioritise following filtering stages as fits their experimental plan. Certain stages will not be beneficial to every experiment; for example, stringent filtering will also remove a certain amount of valid results.

- The most important stage in analysis is to check the data and filter out any problem samples. Samples containing a high number of events are often composed of false positives intermingled with the true events. At this stage, it is recommended to remove any such samples from further analysis.
    - Summary statistics for sample events such as mean and median are useful to highlight outliers. Frequency of events for a sample will vary according the number of SNPs on the array used therefore is up to the individual to set cut-offs for each dataset.
- Filtering events to remove false positives using detection software guidelines.
    - Remove events containing less than three SNPs.

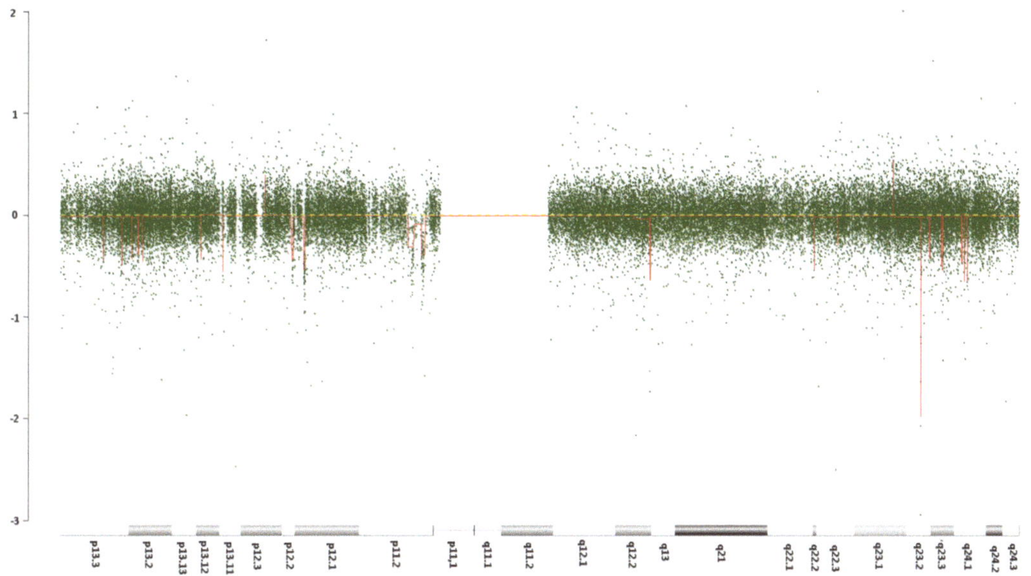

Fig. 4. GADA results for Chromosome 16 HapMap sample. The plot shows assay log2ratio values (*green*) and CN segments (*red*) from Chromosome 16 in a HapMap sample.

- Note 3 consists of additional filtering guidelines for eliminating unreliable event predictions.

- Viewing the results in the UCSC genome browser using a .bed file.
  - UCSC browser resource contains reference sequence and related structural information for full genomes. Sequence information includes genes, SNPs, and data on variation and repeats such as DGV events (see Fig. 5).
  - Files containing CN changes from an experiment can be uploaded using the "Add custom tracks" function found below the genome location entry; these have a standard format that can be produced from detected CNV events using start, finish, and chromosome location. File can be produced manually (see Note 4 for format details) or using a generation script; for example, the PennCNV package contains a script to produce the required output file from a list of events.

- Using the Database of Genomic Variants to remove CNPs from the data.
  - Events recorded in the DGV are common in the population and therefore unlikely to be linked to disorders where a rare event is thought to be the case. CNPs in a dataset can be eliminated by using the DGV as a reference resource.

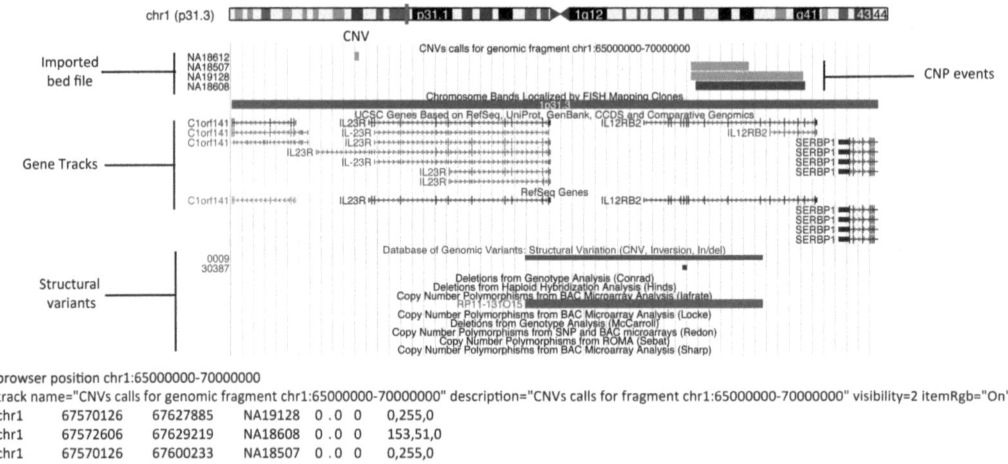


```
browser position chr1:65000000-70000000
track name="CNVs calls for genomic fragment chr1:65000000-70000000" description="CNVs calls for fragment chr1:65000000-70000000" visibility=2 itemRgb="On"
chr1    67570126    67627885    NA19128  0 . 0  0    0,255,0
chr1    67572606    67629219    NA18608  0 . 0  0    153,51,0
chr1    67570126    67600233    NA18507  0 . 0  0    0,255,0
chr1    67397300    67399848    NA18612  0 . 0  0    0,255,0
```

Fig. 5. UCSC Browser Image showing result CNV track. The figure shows an example of the UCSC Browser with a range of default tracks and an example custom track. Gene tracks shown are UCSC genes (*pack*) and RefSeq genes (*pack*) from the Genes and Prediction section. These allow the user to identify the genes in the region selected. Variation and repeat tracks included are Struct Var (*full*) and DGV Struct Var. The CN events shown are from four HapMap samples, three events are a similar size to the DGV event suggesting these events are CNPs. The single event shown is not found in the DGV and could be a CNV.

- The DGV results can be found as a built in track on the UCSC for ease of reference (as shown in Fig. 5).
- Care should be taken when using the DGV for part of the filtering criteria as some diseases may have genes in sites of known variation, for example the 15q13 region, although a CNP site is also of interest in Autism studies (11).

* Identification of genes within listed events.
  - The PennCNV resource (12) contains a dedicated and freely available Perl script which allows the user to process all their events and samples to produce a gene list (options include genes inside or neighbouring each event).
  - For a small number of events it should be possible to use the UCSC browser to identify the relevant genes. Figure 5 shows an example of searching using the location of the CNVs to find genes in the region.
* Comparison between CN detection algorithms.
  - Running more than one algorithm on a dataset allows the user to confirm the presence and breakpoints for events of interest in another analysis. Events confirmed by two detection algorithms are less likely to be false positives.
  - Comparisons can be carried out event by event using a viewing tool (e.g. UCSC Browser) or datasets can be merged.
* Simple conclusions for association with the disease using comparison to a control population.
  - It is critical to compare rare events to a control population to confirm their unique status. Comparison of the frequency of events between the case and control set will show whether the changes at the CNV locus are linked to the disease. Events can be further sub-divided by deletion and duplication to assess CNV type between case and controls.
  - Fisher's Exact test can be used to compare the frequency of a CNV between case and control populations. See Note 6 for an example using R.
  - In large sample sets, it is possible to apply standard genotype association study methodology to analyse (13).
* Non-synonymous SNPs (nsSNPs) analysis.
  - nsSNPs cause a change in amino acid as well as sequence. Therefore, those located within a deletion or CNV will have an increased chance of affecting the overall phenotype by changing protein function. Information about sites where the nsSNP and the CNV coincide should be compared and then checked for the presence of the pathogenic allele.

There are a number of options to streamline the process for researchers who would prefer a less computer intensive analysis method (see Note 5). Illumina and Affymetrix have included their own event detection algorithms within their genotyping software. There are also well-supported commercial products available, which provide algorithm choices and downstream analysis tools.

3.5. Further Data Mining

Final event lists are not always conclusive; often, after cherry picking favourite genes much data is left without exploring the content properly. The following tools are available to look at lists of CNVs together and the genes located within these.

3.5.1. Pathway Analysis

Pathway analysis uses gene ontology annotation to analyse gene function and interaction. Depending on experiment setup you can compare the pathways for parents and children or cases and controls. Genes found within events are imported into the software and the software uses functional information to look for pathways containing greater than expected number of genes. Ingenuity Pathway Analysis is a commercial software which allows the user to compare the function of genes using a number of analysis. The Bioconductor/R environment contains a package, GOstats, which allows similar analysis.

3.5.2. Literature Searching

Manual checking can be done on genes and regions known to the user using searches on the National Centre for Biotechnology Information (NCBI) database site, for example PubMed and NCBI Gene. Ensembl contains detail on gene sequence for discriminating between introns and exons boundaries when whole genes are not affected by the event of interest.

Gene Relationships Across Implicated Loci (GRAIL) statistical analysis (14) looks at biological relationships between genes and regions using PubMed abstracts. It will group genes according to publication links giving the user an alternative approach to standard functional analysis using GO terms. There is a simple online interface for uploading lists and retrieving data at http://www.broadinstitute.org/mpg/grail/.

4. Notes

1. Designing an Experiment: There are a number of factors to consider when designing an experiment. Here, we simply mention those that will impact directly on our copy number detection algorithm use and analysis. Factors such as sample size will impact on the likelihood of finding a causative variant but not necessarily the method of finding. The content of the sample set will have a large impact on the analysis choices. Including parents in a study allows the user to remove inherited events and concentrate only on novel events in a patient.

However, even with the use of parents in the sample it is important to use as large a control set as possible to check the CNV is not common in the general population. As in a situation with any planned case–control study control samples should be as closely matched as possible. Specifically for analysis purposes it helps to match the platform and array used to allow for the same detection algorithms and coverage and therefore a direct comparison of data. A full study using case and control sample set can be very informative and similar rules to a standard association study should be applied when choosing samples and numbers (15). Details of patient's phenotype for any type of study are extremely important and will play a vital role in guiding the final analysis stages in picking relevant genes within events. This chapter focuses on data derived from genomic DNA; results from whole genome amplified samples will require special attention and filtering due to the variation in genotype clustering patterns and signal background noise.

2. Choosing an Algorithm: There are multiple options available for detection of CN changes in a dataset. Choices are often dependent on the platform the data was generated; however, there are a number of other factors to consider. There are two types of statistical method used in copy number detection (16): parametric and non-parametric. Parametric methods include Hidden Markov Model (HMM), which assumes that each hidden state is a different copy number; QuantiSNP is an example of this method. Also included in the parametric methods are algorithms such as GADA, which uses step functions within its model. A common example of the non-parametric method is Circular Binary Segmentation (CBS), which was originally developed for use on array CGH data and uses a technique to continually segment regions until there are no differences in CN remaining. Application of these methods varies; basic modelling can be carried out in statistical environments such as R or using the guided tools available. Tools such as QuantiSNP and GADA incorporate calculations within the run package for ease of use but still require command line manipulation; options for simple directed analysis are discussed in Note 5. Detection tools can also be divided by ease of use and application and Table 2 describes a selection of those available. For a more detailed description of algorithms and their methods see Winchester et al. (10).

   When choosing your algorithms, it is important to take into account the factors in the dataset and analysis plans. Some algorithms provide more flexibility for adjusting parameters, while others have support for further analysis. By analysing with more than one algorithm, the user can take advantage of different features and strengths of individual tools, additionally results can be compared to confirm events of interest. False

Table 2
Copy number change detection algorithms

Software	Platform	Related download site	Description
Birdsuite (9)	Affymetrix only	http://www.broadinstitute.org/science/programs/medical-and-population-genetics/birdsuite/birdsuite-0	Combined tool set to genotype SNPs and CNPs and to detect CNVs. Also incorporated into GTC for easy application on Chip 6.0 format
CNAT	Affymetrix only	http://www.affymetrix.com/support/technical/software_downloads.affx	Affymetrix Proprietary algorithm. Run in GTC. Easy to use and produces general overview of data. Low sensitivity
CNVPartition	Illumina only	http://www.illumina.com/software/illumina_connect.ilmn	Illumina Proprietary algorithm. Version 2.4.4 run in GenomeStudio. Easy to use and produces general overview of data. Low sensitivity
COKGEN (21)	Affymetrix	http://mendel.gene.cwru.edu/laframboiselab/software.php	Analysis framework including normalisation using R. Two stage CNV detection based on optimisation
Dchip SNP (22)	Affymetrix or Illumina	http://biosun1.harvard.edu/complab/dchip/snp.htm	Independent analysis freeware with viewer options
GADA (3)	Affymetrix or Illumina	http://biron.usc.edu/~piquereg/GADA/	Two stage analysis using Sparse Bayesian Learning and backwards elimination to filter results (R and Java formats)
HMMSeg (23)	Multiple	http://noble.gs.washington.edu/proj/hmmseg/	HMM application tool to any genomic data including CNV detection. Needs statistical modelling experience for accurate use
ITALICS (24)	Affymetrix	http://www.bioconductor.org/packages/release/bioc/html/ITALICS.html	Normalisation based algorithm designed to support 100K and 500K arrays
Nexus Biodiscovery	Multiple	Available for purchase from http://www.biodiscovery.com/index/nexus	Commercially available analysis suite developed for Copy Number analysis. Includes data viewer and further analysis tools

(continued)

Table 2 (continued)

Software	Platform	Related download site	Description
Partek	Multiple	Available for purchase from http://www.partek.com/	Commercially available analysis suite including Copy Number analysis. Includes data viewer and support tools
PennCNV (12)	Illumina or Affymetrix	http://www.openbioinformatics.org/penncnv/	HMM detection tool based on Perl script. Includes downstream processing tools and well-supported instructions
QuantiSNP (4)	Illumina or Affymetrix	http://groups.google.co.uk/group/quantisnp http://www.well.ox.ac.uk/quantisnp	HMM detection tool using MATLAB software. Freely available in LINUX or PC formats. Easily modified parameters with posterior measure of confidence for each event detected
SCIMMkit (25)	Illumina	http://droog.gs.washington.edu/scimmkit/	Includes SCIMM applications and SCOUT for rare variants. Modelling algorithm requiring R and Perl to run. Needs statistical modelling experience for accurate use
TriTyper (26)	Illumina	http://www.ludesign.nl/trityper/	Identify deletions and genotype single SNPs with null allele

positives will be present in results produced by a more sensitive algorithm and a user can compare data from two algorithms to eliminate false predictions from the final dataset.

3. Event Filtering and Data Quality Control: After using the main filtering criteria there are a number of other options to be considered in CNV event analysis. Events can be filtered by location, such as proximity to the centromere and telomere or if found in regions of known chromosomal rearrangement. By applying an extreme filter, you can remove all events with a frequency higher than 1%, which will leave a dataset fitting the excepted definition of CNV (2). Event predictions can also be merged in instances where a CNV has been artificially split by an algorithm. Data can be split into groups before checks; events can be divided into deletions and duplications or samples by sex to check for trends in the data (Fig. 6).
4. Designing Bed Files for the UCSC Browser Each bed file must contain certain core columns, and for more elaborate display

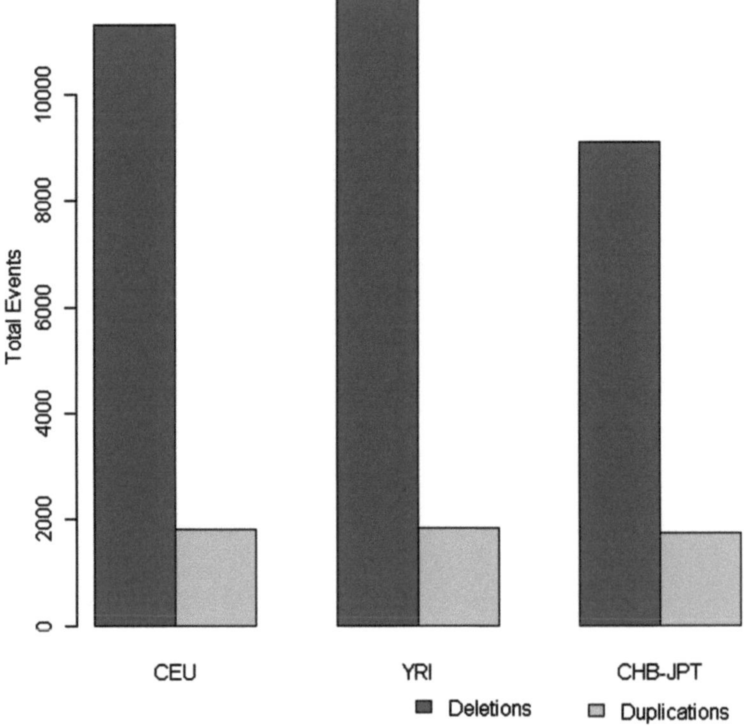

Fig. 6. Graph showing the difference in HapMap populations. Bar chart shows the total number of deletions and duplications for each HapMap population.

functions, refer to the following site: http://genome.ucsc.edu/goldenPath/help/customTrack.html.

Files must start with the following information: track name = "Track:'Patient1'"description = "'Patient1_QuantiSNP'" visibility = 2 itemRgb = "On".

The following columns should contain, without headers: Chromosome (e.g. chr1), Start position, End position (position values should not contain commas), event name (e.g. "Patient 1"), series of numbers explaining track details such as colour (e.g. 130|.|0|0|255,0,0). Columns should be separated by tabs, as should values be separated by "|" symbols.

The file must be saved with a .bed extension; an example of the file is included in Fig. 4.

5. Analysis Options Suited to Researchers Skills: The range of detection algorithms also includes some easily accessible and simple software which allow the user a more standardised approach (see Table 2). Commercially available software gives the user access to well-supported and maintained analysis options. Often as well as including detection algorithms and simple batch processing, there are viewer options and guided analysis stages.

The Biodiscovery Nexus Copy Number suite Version 5 allows the user to apply four different algorithms and compare the output using a chromosome browser viewer. The software implements both types of detection method, the Rank Segmentation method uses CBS and FASST uses HMM modelling (Note 2 discusses these models in more detail). The viewer makes it easy to compare events between samples and groups and to get information about genes and regions without having to use multiple online sites and databases. It also includes filtering and statistical tools alongside gene function analysis based on GO terms.

6. Fisher's Exact Test: Fisher's exact test can be used to assess whether the frequency of an event is different between the case and control populations. The test takes categorical variables and calculates the probability of a difference in proportions between two classifications, where the null hypothesis is that there is no difference between the classifications. The following example uses results from a single CNV; this can be applied to a region or single SNP depending on requirements. This calculation uses the R environment, and no additional library or package is required.
   - Load the data using the following command: *Cnv1<-matrix(c(8,269,4,113), nr=2)*
     - Data can be checked using the data input name e.g., cnv1

```
      [,1] [,2]
[1,]    8    4
[2,]  269  113
```

   - Run the test using the command: *fisher.test(cnv1)*. The result will be displayed as shown.

```
Fisher's Exact Test for Count Data
data:  cnv1
p-value = 0.7553
alternative hypothesis: true odds ratio is not equal to 1
95 percent confidence interval:
 0.219824    3.892075
sample estimates:    odds ratio    0.84055
```

     - In this case the frequency of the event in the case and control population is not significantly different (p-value of 0.7553 is larger than 0.05).

References

1. Ragoussis J. Genotyping technologies for genetic research. Annu Rev Genomics Hum Genet 2009;10:117–33.
2. Feuk L, Carson AR, Scherer SW. Structural variation in the human genome. Nat Rev Genet 2006;7:85–97.
3. Pique-Regi R, Monso-Varona J, Ortega A, Seeger RC, Triche TJ, Asgharzadeh S. Sparse representation and Bayesian detection of genome copy number alterations from microarray data. Bioinformatics 2008;24:309–18.
4. Colella S, Yau C, Taylor JM, et al. QuantiSNP: an Objective Bayes Hidden-Markov Model to detect and accurately map copy number variation using SNP genotyping data. Nucleic Acids Res 2007;35:2013–25.
5. International Schizophrenia Consortium. Rare chromosomal deletions and duplications increase risk of schizophrenia. Nature 2008; 455:237–41.
6. Glessner JT, Wang K, Cai G, et al. Autism genome-wide copy number variation reveals ubiquitin and neuronal genes. Nature 2009;459:569–73.
7. Wain LV, Pedroso I, Landers JE, et al. The role of copy number variation in susceptibility to amyotrophic lateral sclerosis: genome-wide association study and comparison with published loci. PLoS One 2009;4:e8175.
8. Zhang D, Cheng L, Qian Y, et al. Singleton deletions throughout the genome increase risk of bipolar disorder. Mol Psychiatry 2009; 14:376–80.
9. Korn JM, Kuruvilla FG, McCarroll SA, et al. Integrated genotype calling and association analysis of SNPs, common copy number polymorphisms and rare CNVs. Nat Genet 2008;40:1253–60.
10. Winchester L, Yau C, Ragoussis J. Comparing CNV detection methods for SNP arrays. Brief Funct Genomic Proteomic 2009;8: 353–66.
11. Ben-Shachar S, Lanpher B, German JR, et al. Microdeletion 15q13.3: a locus with incomplete penetrance for autism, mental retardation, and psychiatric disorders. J Med Genet 2009;46:382–8.
12. Wang K, Li M, Hadley D, et al. PennCNV: an integrated hidden Markov model designed for high-resolution copy number variation detection in whole-genome SNP genotyping data. Genome Res 2007;17:1665–74.
13. McCarroll SA, Altshuler DM. Copy-number variation and association studies of human disease. Nat Genet 2007;39:S37–42.
14. Raychaudhuri S, Plenge RM, Rossin EJ, et al. Identifying relationships among genomic disease regions: predicting genes at pathogenic SNP associations and rare deletions. PLoS Genet 2009;5:e1000534.
15. Cardon LR, Bell JI. Association study designs for complex diseases. Nat Rev Genet 2001; 2:91–9.
16. Yau C, Holmes CC. CNV discovery using SNP genotyping arrays. Cytogenet Genome Res 2008;123:307–12.
17. Iafrate AJ, Feuk L, Rivera MN, et al. Detection of large-scale variation in the human genome. Nat Genet 2004;36:949–51.
18. Hubbard TJ, Aken BL, Ayling S, et al. Ensembl 2009. Nucleic Acids Res 2009;37:D690–7.
19. Falcon S, Gentleman R. Using GOstats to test gene lists for GO term association. Bioinformatics 2007;23:257–8.
20. Karolchik D, Baertsch R, Diekhans M, et al. The UCSC Genome Browser Database. Nucleic Acids Res 2003;31:51–4.
21. Yavas G, Koyuturk M, Ozsoyoglu M, Gould MP, LaFramboise T. An optimization framework for unsupervised identification of rare copy number variation from SNP array data. Genome Biol 2009;10:R119.
22. Lin M, Wei LJ, Sellers WR, Lieberfarb M, Wong WH, Li C. dChipSNP: significance curve and clustering of SNP-array-based loss-of-heterozygosity data. Bioinformatics 2004;20:1233–40.
23. Day N, Hemmaplardh A, Thurman RE, Stamatoyannopoulos JA, Noble WS. Unsupervised segmentation of continuous genomic data. Bioinformatics 2007;23:1424–6.
24. Rigaill G, Hupe P, Almeida A, et al. ITALICS: an algorithm for normalization and DNA copy number calling for Affymetrix SNP arrays. Bioinformatics 2008;24:768–74.
25. Zerr T, Cooper GM, Eichler EE, Nickerson DA. Targeted interrogation of copy number variation using SCIMMkit. Bioinformatics 2010;26:120–2.
26. Franke L, de Kovel CG, Aulchenko YS, et al. Detection, imputation, and association analysis of small deletions and null alleles on oligonucleotide arrays. Am J Hum Genet 2008;82:1316–33.

Chapter 15

Targeted Screening and Validation of Copy Number Variations

Shana Ceulemans, Karlijn van der Ven, and Jurgen Del-Favero

Abstract

The accessibility of genome-wide screening technologies considerably facilitated the identification and characterization of copy number variations (CNVs). The increasing amount of available data describing these variants, clearly demonstrates their abundance in the human genome. This observation shows that not only SNPs, but also CNVs and other structural variants strongly contribute to genetic variation. Even though not all structural variants have an obvious phenotypic effect, there is evidence that CNVs influence gene dosage and hence can have profound effects on human disease susceptibility, disease manifestation, and disease severity. Therefore, CNV screening and analysis methodologies, specifically focusing on disease-related CNVs are actively progressing.

This chapter specifically describes different techniques currently available for the targeted screening and validation of CNVs. We not only provide an overview of all these CNV analysis methods, but also address their strong and weak points. Methods covered include fluorescence in situ hybridization (FISH), quantitative real-time PCR (qPCR), paralogue ratio test (PRT), molecular copy-number counting (MCC), and multiplex PCR-based approaches, such as multiplex amplifiable probe hybridization (MAPH), multiplex ligation-dependent probe amplification (MLPA), multiplex PCR-based real-time invader assay (mPCR-RETINA), quantitative multiplex PCR of short fluorescent fragments (QMPSF), and multiplex amplicon quantification (MAQ). We end with some general remarks and conclusions, furthermore briefly addressing the future perspectives.

Key words: Copy number variation, CNV analysis, CNV validation, Targeted screening

1. Introduction

A fundamental problem in current biomedical research is providing a link between the observed phenotypic variability and the underlying genetic variation regarding both neutral polymorphic and pathological variation (1). Genetic variations include single nucleotide polymorphisms (SNPs) and structural variations (SVs) of which the latter can be subdivided into microscopic (>3 Mb) and submicroscopic (1 kb to 3 Mb) variations (2). In the broadest

Lars Feuk (ed.), *Genomic Structural Variants: Methods and Protocols*, Methods in Molecular Biology, vol. 838, DOI 10.1007/978-1-61779-507-7_15, © Springer Science+Business Media, LLC 2012

sense, SVs are defined as DNA sequence alterations comprising deletions, duplications (grouped together as CNVs), indels, inversions, and translocations (see Table 1 for definitions). Although SNPs are by far the most frequent source of genetic variation in the human genome, CNVs—although far less numerous than SNPs—affect from 1 kb to several mega bases of DNA per event, which adds up to a significant fraction of the genome several magnitudes larger compared to SNPs (2, 3).

Currently, over 29,000 CNVs have been catalogued, affecting a large fraction of annotated genes (see Database of Genomic Variants: http://projects.tcag.ca/variation/) many of which are potentially disease-causing (4, 5). Gene Ontology (GO) analysis from various studies have revealed that a substantial amount of the affected genes are involved in, e.g., immune responses, drug metabolism, responses to external biotic stimuli, neurophysiological

Table 1
Terminology for structural variants

Type of variant	Description
Structural variation (SV)	Group of genomic alterations involving segments of DNA typically larger than 1 kb. A structural variant can be quantitative (copy number variation), positional (translocation), or orientational (inversion). The term is used as a neutral descriptor with nothing implied about frequency, association with disease or phenotype
Balanced rearrangement	Represents a change in the position (translocation) or orientation (inversion) of a DNA segment, and is not thought to be associated with either gain or loss of genetic material
Unbalanced rearrangement	Rearrangement which leads to a net gain (duplication) or loss (deletion) of DNA in a genome
Copy number variation (CNV)	A DNA segment of at least 1 kb in size, that is present at a variable copy number compared to a reference genome. These quantitative variations can be either copy number gains (insertions and duplications) or losses (deletions)
Insertion/deletion variant (InDel)	Refers to the addition (insertion) and/or removal (deletion) of DNA base pairs ranging in size from 1 to 1,000. InDels > 1 kb in size are often also called CNVs
Translocation	A chromosome abnormality caused by the rearrangement (transfer) of a DNA segment between chromosomes, not leading to a change in total DNA content. Translocations can be either intra- or interchromosomal
Inversion	A chromosomal rearrangement in which a DNA segment is reversed in orientation compared to a reference genome, and which is mostly not associated with alterations in copy number

processes, and brain development (2, 5). This new appreciation of CNVs as a frequent and inherent feature of the human genome has the potential to revolutionize our understanding of genetics, population genetics, and genetic epidemiology.

CNVs can affect gene expression in a variety of ways, which may be reflected in their phenotypic effects. In the simplest case, CNVs spanning one or more genes lead to variations in gene expression through dosage effects, where gain or losses in copy number increases or decreases expression levels, respectively (5). Other causes are disruption of the gene coding sequences (removal of exons or by affecting regulatory elements), deletions, or duplications outside the coding sequences which may affect gene expression through position effects (6). The effect of CNVs on gene expression and their potential disruptive effect on gene structure and function suggest that they are likely to make a considerable contribution to human phenotypic variability and human diseases (4). Over the years, numerous examples of the involvement of CNVs in the etiology of monogenetic and complex disorders have accumulated, including important common disorders, such as autism, Alzheimer's disease, cancer, epilepsy, and schizophrenia (for an extensive review, see refs. 3, 4, 7).

Discovery of the abundance of submicroscopic CNVs has mainly depended upon comparative genomic hybridization (CGH) platforms and microarray-based SNP arrays. Array-CGH was first developed by Pinkel et al. (8), covering the genome at 1 Mb resolution. Although not originally designed for this purpose, SNP-arrays were used early on for CNV detection (9). The latest generation SNP arrays contain up to one million SNPs and additional probes yielding the best genome coverage possible for genome-wide CNV detection (10). However, SNP arrays are still relatively expensive and not ideal when thousands of samples need to be screened. Moreover, the power of SNP arrays is limited by their resolution by a lower detection limit of ~10–25 kb (11). Since the initial reports on the presence of unexpected high numbers of CNVs per human genome about 5 years ago (12, 13), hundreds of new CNVs have been detected in diagnostic and research labs worldwide. Validation and replication of these CNVs is of high priority. Validation can be interpreted as the observation of the same CNV among multiple individuals or the confirmation of the CNV in the same individual with one or more independent, most preferably different platform-based technologies. However (array-based), whole genome screening techniques lack the flexibility of analyzing individual genes or sequences of interest, nor are they able to detect copy number neutral variants, such as balanced rearrangements (14). To overcome these disadvantages and fully explore the impact of CNVs on human diseases, high-resolution, high-throughput, and low cost methods are required to validate and replicate the increasing number of

CNVs, including CNVs <1 kb in size. These are of particular importance to genetic medicine as they are far more numerous than the larger CNVs (3).

In this review, we specifically focus on the currently available methods for routine and targeted CNV detection, validation, and replication covering the basic principles, workflow, data analysis, and their respective strong and weak points.

2. Targeted Screening of CNVs

Over the years, different methods have been introduced to detect genomic imbalances. Conventional chromosomal analysis (karyotyping) allows detection of chromosomal aberrations larger than 3 Mb. The introduction of fluorescence in situ hybridization (FISH) has led to the identification of subtle cryptic chromosomal abnormalities (15). FISH, a molecular cytogenetic technique to visualize specific sequences in the nucleus and/or on metaphase chromosomes, is an extremely useful method for the detection of chromosomal abnormalities, such as aneuploidy and chromosome rearrangements. This is done by hybridizing fluorescently labeled DNA probes to metaphase chromosome spreads (resolution 1–3 Mb) or interphase nuclei (50 kb to 2 Mb). The highest resolution is obtained by fiber-FISH (5–500 kb), where probes are visualized on mechanically stretched chromosome fibers and which is currently the preferred method to precisely determine the genomic structure of complex CNVs (16). Nowadays, FISH combined with multiple probes labeled in different colors (multicolor FISH), is widely used in clinical diagnostics as a screening tool to confirm the presence of CNVs and other SVs, such as balanced translocations, in either patients or nonsymptomatic carriers. However, FISH has several limitations. Locus-specific probes are expensive and the procedure is time-consuming and labor intensive. Furthermore, only a limited number of chromosomal loci can be screened in a single experiment and the identification of an abnormality is highly dependent on the DNA probe used, specifically on its size and hybridization localization. Indeed, some patients with uniquely localized (small) microdeletions or duplications may yield normal clinical FISH findings and will go undiagnosed due to the mapping of the abnormality outside the area covered by the (relatively large) FISH probe. To conclude, FISH is an accurate and powerful technique, but its use as a high-throughput (diagnostic) analysis method is hampered by its high consumable cost, hands on time, and the requirement of experienced personnel. Therefore, there is a need for fast, simple, and flexible methods for targeted CNV analysis that ideally can be applied in a high-throughput setting. There are several techniques and platforms available, although PCR-based approaches have been the most widely applied, because of their ease of use.

2.1. Real-Time Quantitative PCR

The first PCR-based technique used for targeted CNV analysis was real-time quantitative PCR (qPCR). qPCR combines "traditional" endpoint PCR with fluorescent detection technologies to record the accumulation of amplicons in "real time" during PCR cycling. Fluorescence monitoring systems for qPCR consist of hydrolysis probes (e.g., TaqMan (17), Scorpions (18), and FRET probes (19)), hybridization probes (20), or DNA-intercalating agents (e.g., SYBR green (21)). Irrespective of the detection technique used, with the accumulation of target sequences during PCR, more probes and agents are hydrolyzed, hybridized, or intercalated and the fluorescence signal increases. Amplicon quantification relies on the observation that the fractional cycle number (Ct), at which the amount of an amplified target amplicon reaches a fixed threshold, is directly related to the amount of starting target. So a higher or lower starting copy number of a genomic DNA target will respectively result in a significant earlier or later increase in fluorescence, and thus, in a decreased or increased Ct (22). Briefly, there are primarily two types of qPCR analysis: "relative quantitation" (comparing the target and control by calculating the difference in cycle number, ΔCt) and "standard-curve quantitation" (a standard curve of "knowns" is used to quantify the "unknowns"). The qPCR technique offers great flexibility and adaptability, and can be carried out in a closed system, thus eliminating the risk of PCR and sample contamination and does not require postprocessing of PCR products. Therefore, it has become one of the most popular methods for CNV analysis (23). Although real-time qPCR has added true quantitative ability to the power of PCR, there are some important limitations to the application of qPCR for CNV analysis. A first restrictive concern is that the precision of CNV determination greatly varies with the number of replicates (23). Optimized PCRs achieve a doubling of template with each cycle. Because of this doubling per cycle, distinguishing differences smaller than twofold requires reliable assessment of fractional cycle differences. Given the respective system standard deviations (SD), in most reported studies running each sample in triplicate sufficed to accurately distinguish these twofold differences typical for deletions (24–26). In their paper however, Weaver et al. (23) show a steep increase in the number of replications when distinguishing two from three copies (1.5-fold, eight replicates), three from four copies (1.33-fold, 12 replicates), and four from five copies (1.25-fold, 18 replicates). In this respect, the development of high-throughput qPCR microfluidic platforms enables better quantitative resolution due to the possibility of including more replicates per sample. Although most research papers investigate CNVs in a range of 0–4 copies in a diploid genome, several recent studies have demonstrated that some genes or groups of genes can show a CNV between 0 and 10 copies, e.g., for human complement C4 (27) and α- and β-defensin (28). When a higher number of repeats are present, the quantitative differences not only become gradually smaller (increasing the need for extensive replication as mentioned above), but also an intrinsic underestimation of the actual genomic copy number is reported (22, 27). This second limitation of qPCR can be overcome through the use of "standard-curve quantitation" by using controls that sufficiently span the whole range of diploid copy numbers to yield reliable standard curves. A third and last disadvantage of qPCR we wish to mention is that the number of loci that can be analyzed in a single tube is currently restricted to four or less by fluorescence profiles and optical properties of hardware (29).

Given these limitations, we value the power of qPCR for CNV analysis in studies measuring a small number of CNVs within the range of 0–4 copies in large sample sets. Among patients however, disease-related CNVs are rarely identical, with variable breakpoints. Therefore, to acquire a complete CNV profile for every patient, several qPCRs are required resulting in an expensive and time-consuming analysis. To overcome this problem, quantitative PCR approaches with a higher degree of multiplexing (mPCR), yielding for each CNV region several fragments with either a unique length and/or color, have been designed. In the following sections, recently developed mPCR-based approaches for the targeted detection and validation of CNVs are further discussed.

2.2. Paralogue Ratio Test

To obtain reproducible results with multiplex PCR (mPCR)-based CNV analysis methods, great care needs to be taken in matching the amplification properties of target and reference amplicons. Many of the problems of accuracy and reproducibility associated with mPCR can be avoided if the target and reference amplicons are similar. On this basis, the paralogue ratio test (PRT) method was recently developed. This comparative mPCR method is based on the use of a pair of precisely designed primers located in a dispersed repeat sequence. This primer pair simultaneously amplifies products from both target and reference loci that are subsequently separated and quantified by internal sequence differences (30). Separation of the amplicons can be performed by several methods, such as restriction digestion, pyrosequencing, or capillary electrophoresis.

In summary, the PRT method appears to be comparable in accuracy to other PCR-based approaches, and provides a rapid and inexpensive method for copy number analysis, only requiring small amounts (10–20 ng) of genomic DNA. PRT analysis is thus suitable for CNV typing in large patient-control studies. The major disadvantage of this method is, however, its dependence of the presence of homologous repeat sequences in and outside copy variable regions, hampering the use of PRT as a generic method for large scale, high-throughput CNV analysis.

2.3. Molecular Copy-Number Counting

Single molecule PCR (smPCR) allows DNA analysis starting from single DNA molecules, obtained by diluting template DNA to an average of <1 target molecule per aliquot (31). In this setting, individual aliquots can only be positive or negative for the target sequence, hence the terms "digital-PCR" or molecular copy-number counting (MCC) (32). The MCC approach is based on smPCR analysis of genomes and starts with diluting the template DNA in aliquots containing less than one haploid genome in microtiter plates followed by a first round of limited cycle mPCR on each aliquot to amplify the target sequences of interest. Next, the acquired amplification products are split into replica plates followed by a second round of PCR with a single semi-nested primerset specific for each target sequence. The resulting PCR products are separated by gel electrophoresis and scored for the presence or absence of the specific target sequence (33). Through the simultaneous MCC testing of multiple target loci and reference loci expected to be at normal copy, the relative copy number of target loci can be inferred (34). It is shown that MCC cannot only distinguish a 1:2 variation, but can also accurately assess copy number variation (CNV) extending over an eightfold range (0.5–4 copies/haploid genome). However, it should be noted that MCC assumes that all the copies of the loci tested are amplified with approximately the same efficiency. This is clearly not the case when degraded DNA is used. To overcome this limitation, microdissection MCC (μMCC), a modification of MCC, was developed (35). By ensuring that all amplicons are of short, uniform length (100–120 bp), μMCC allows a precise assessment of CNVs even with small amounts of severely fragmented DNA extracted from, for example, formalin-fixed paraffin-embedded clinical biopsies (35).

In summary, MCC is a fast and flexible method for characterizing a wide range of chromosomal aberrations with the capacity to simultaneously analyze multiple target loci, and is highly tolerant to poor-quality and limited amounts of template DNA (μMCC).

Recently, the Fluidigm Corporation (www.fluidigm.com) developed a unique integrated nanofluidic system, the digital array, allowing CNV analysis based on the MCC principle. This new system accurately quantitates CNVs of interest in DNA samples based on the fact that single DNA molecules are randomly distributed in the 9,216 reaction chambers of the digital array (36). The concentration of any sequence in a DNA sample (copies/μl) can thus be calculated using the numbers of positive chambers that contain at least one copy of that sequence. To make sure that the difference in copy number in different samples are true observations, the "relative copy number" of a sequence, i.e., number of copies of that sequence per haploid genome, is used. This can be interpreted as the ratio of the copy number of a target sequence to the copy number of a single copy reference sequence (two copies per cell) in a DNA sample, which is always one per haploid genome. Since two fluorescent dyes are used on the same digital array, it is possible to simultaneously quantitate both sequences in the same nanowell. Furthermore, the accuracy of the results is only subject to the random distribution of the molecules and can be improved by using multiple replicates per sample.

To summarize, CNV analysis using the nanofluidic biochip is fast, easy to perform, and the data obtained are easy to interpret. So, the digital array provides a new, reliable, and robust platform to study gene- and sequence-specific CNVs, providing a much greater discrimination power than qPCR and at the same time allowing cost- and time-effective high-throughput analysis of large sample sets.

2.4. Multiplex Amplifiable Probe Hybridization

Multiplex amplifiable probe hybridization (MAPH) relies on sequence-specific probe hybridization to genomic DNA immobilized on a nylon filter, followed by washing off and amplification of the hybridized probes, and subsequent quantitative analysis of the resulting PCR products (37). Thus, MAPH is not only a hybridization based, but also an mPCR-based CNV detection method. MAPH can be used as a targeted approach to analyze DNA losses or gains at known genomic locations (e.g., disease-related genes *DMD*, *PMP22*, and *SNRPN*) or for de novo CNV detection anywhere in the genome (e.g., subtelomeric or subcentromeric regions) (38, 39). In more detail, genomic DNA (1 μg) is fixed to a nylon membrane and hybridized with a set of amplifiable probes corresponding to the target sequences to be detected. These amplifiable probes are generated by cloning the target sequences into a plasmid vector, followed by amplification of the cloned sequences using primers directed to the vector, resulting in probes of different sizes that are all flanked by the same primer sequence. MAPH probes can be specifically designed for any gene (exons), locus, telomere, chromosomal segment, whole chromosome, or the total human genome at extremely high resolution, enabling sensitive CNV detection of genomic DNA sequences as small as 100–150 bp. However, the true effective resolution of MAPH is given by the size of the probes (40). Probe selection is a key step toward the successful application of MAPH and is based on a number of important criteria: probe sequences should be unique in the human genome, nonrepetitive, nonpolymorphic, well localized, small-sized (preferably around 100–600 bp) with similar GC% (around 50%) (39). One additional restriction on probe design is that probes that are intended to be multiplexed must differ sufficiently in size to be resolved by electrophoresis (either gel or capillary). Furthermore, caution is warranted since MAPH probes are inherently amplifiable, and thus poses a significant risk of contamination compared to multiplex ligation-dependent probe amplification (MLPA) probes, which only become amplifiable after the ligation step (see Subheading 2.5) (41).

After hybridization, the membrane is rigorously washed to remove unbound probes followed by stripping the bound amplifiable probes from the membrane. The amount of recovered probe is proportional to the copy number of the target sequence in the genomic DNA. Eluted probes are then amplified using a universal primer pair and size-separated by capillary gel electrophoresis. Changes in peak heights, relative to controls, can be detected to indicate CNVs. Although different probes may be amplified with slightly different efficiencies, the proportional contribution from each locus reflects its copy number in the sample (37).

MAPH also possesses several limitations, besides being constrained by the core elements of the assay, such as probe sequence specificity, PCR efficiency, and hybridization kinetics. The washing steps in the MAPH technique, necessary to remove unbound probes introduce a contamination risk (41).

A recent improvement in MAPH, addressing these major limitations, is the introduction of microarrays, resulting in the development of microarray-based MAPH (array-MAPH). This method combines the flexibility, high specificity, and sensitivity of MAPH with the potential of highly accurate and high-resolution genomic analysis (multiplex ability) provided by a microarray format (40, 42, 43). After hybridization and washing, the recovered probes are labeled and hybridized to an oligonucleotide microarray, therefore allowing the simultaneous measurement of copy numbers at a larger scale. This array-based detection allows fragments to be identified and quantified based on sequence rather than on size, therefore enabling much higher order multiplexes. Patsalis et al. (43) showed the possibility to simultaneously analyze a large number of probes (~700). Nevertheless, array-MAPH involves two hybridization steps, the first one on filters and the second one on arrays, in total requiring 2 µg of genomic DNA, whereas only 0.5–1 µg is needed for gel-based MAPH (43).

The latest MAPH adaptation, quadruplex MAPH (QuadMAPH), offers a fourfold increase in the number of loci that can be tested simultaneously (44). This format has been particularly designed to allow thorough CNV screening of 100–200 kb regions at high resolution (1–2 kb) in a large number of genomic DNA samples. The alternative MAPH probe generation is established by preparing probe sets by shotgun cloning of random short fragments from a BAC clone covering the region of interest. Therefore, using four different BAC clones allows the generation of four independent MAPH probe sets and subsequently a substantial increase in the number of targets that can be investigated in a single four-color assay. Tyson et al. could also demonstrate that multiplexing four probe sets does not result in a loss of accuracy (44).

To summarize, MAPH is accurate, reliable, and cost-effective in detecting small genomic changes, with up to 40 probes that can be multiplexed and resolved by gel electrophoresis simultaneously (41). However, large amounts of genomic DNA (>1 µg) are needed, which have to be fixed to a solid carrier. Also, amplifiable probe preparation is cumbersome since it requires cloning steps.

Taken together, these steps have a high contamination risk, are labor-intensive, time-consuming and require skilled personnel making MAPH less amendable as a high-throughput, low cost CNV analysis method.

2.5. Multiplex Ligation-Dependent Probe Amplification

MLPA is by far the most used mPCR-based method for CNV analysis. The principle of MLPA is similar to MAPH with the main difference that hybridization is performed in solution and that an extra ligation step is incorporated to further increase specificity (45). In detail, each MLPA probe consists of two oligonucleotide hemiprobes, one synthetic and one derived from the single-stranded M13 bacteriophage. These oligonucleotides hybridize to adjacent sites of the target sequence. Each hemiprobe is flanked by universal PCR primer sites and one of the hemiprobes also has a "stuffer" sequence allowing each probe set to have different fragment lengths. Following solution-based hybridization of the MLPA probes to genomic DNA, the two hybridized hemiprobes are ligated resulting in a proportional relation between the number of joined primers and the target copy number. After denaturation, PCR amplification is carried out using a single universal dye-labeled primerset. Finally, the resulting PCR products, each with a unique size (130–480 bp) are separated by capillary gel electrophoresis followed by data analysis to identify CNVs. Since the relative quantity of each of the PCR products is proportional to the number of copies of the target sequence, results are given as allele copy numbers as compared to normal controls.

MLPA is specifically developed to screen up to 50 (on average 20–40) independent loci simultaneously, with results typically available after 1–3 days. MLPA is also scalable for the study of large populations (29). In addition, many disease-specific kits—e.g., targeting the disease-related genes *APP*, *BRCA1–2*, *SCN1A*—are commercially available from MRC Holland (www.mlpa.com). Since MLPA is liquid based, it is well suited for automation using multiwell formats, which makes it more flexible for higher throughput analysis than MAPH. SNPs at the binding site of the target-specific sequence in the MLPA probes (20–30 nucleotides) can prevent efficient hybridization and ligation, which can result in inconclusive or false detection of deletions. For this reason, all single probe based deletions found by MLPA (or any other detection method for that manner) should be confirmed by an independent method (41).

To circumvent the laborious cloning-based preparation of stuffer containing hemiprobes, alternative approaches using synthetic generated 5′ and 3′ hemiprobes have been developed. Although chemical synthesis has limits, oligonucleotide synthesis up to 100 nucleotides in length is now routine. The strategy of direct oligonucleotide synthesis has already been successfully applied for mutation detection and refinement of deletions (46, 47).

The 5′ and 3′ hemiprobes are designed to be of similar length and the PCR product size is than modulated either by increasing the length of the target-specific sequences (47) or by adding a stuffer sequence to both half-probes (48). The inherent limitations of this approach are caused by the short synthesis lengths of commercially available oligonucleotides, hereby limiting the number of probes that can be combined in one assay. Since synthetic MLPA assays contain, on average only about ten probes, a *two-color MLPA assay* can increase the multiplexing capacity of synthetic MLPA assays (46). This approach thus increases the multiplexing capacity by a factor 2, though each probe set must include its own control probes, and results from the two different MLPA probe sets must be analyzed separately.

A modified MLPA assay, termed methylation-specific MLPA (MS-MLPA), has been designed to simultaneously detect CNVs and epigenetic aberrations (CpG methylation) (49, 50). The MS-MLPA probe design is slightly different, given that the sequences targeted by MS-MLPA probes contain a restriction site for the HhaI endonuclease that specifically recognizes unmethylated GCGC sequences. After the hybridization step, the MLPA reaction is split into two tubes of which one tube is processed as a standard MLPA reaction, providing information on CNVs while the other tube is incubated with the ligase and the HhaI endonuclease. Digested probes (unmethylated) cannot be amplified exponentially during PCR and hence will not produce a signal on capillary electrophoresis. The methylated DNA sample on the other hand, is protected against HhaI digestion and the ligated probes will generate a detectable PCR product.

Recently, a third MLPA approach emerged called MLGA which is based on the selector technology (51). Genomic DNA is first digested with a restriction enzyme, generating fragments of different sizes needed for their discrimination. These restriction fragments are denatured and circularized by MLGA probes (i.e., a target-specific selector probe and a universal vector oligonucleotide) together with the DNA ligase. Next, PCR is performed using universal primers that hybridize to a sequence in the vector oligonucleotide. Thus, MLGA decreases probe amplification background, and the use of shorter probes allows for a cost-efficient design resulting in increased reaction kinetics, significantly lowering the total assay time (51).

Even more recently, array technology was applied to further increase the performance of MLPA. This *array-MLPA* format uses amplification products of essentially uniform size (100–120 bp) and distinguishes them by the incorporation of tag sequences, increasing multiplexity up to 124 probes (52). Thus, array-MLPA significantly increases the potential for high multiplexed assays, and eliminates the need to have amplification products of different sizes for detection and quantification by capillary electrophoresis (49).

An additional advantage is the automated array processing, measurement and data analysis, making array-MLPA a highly reproducible method.

To conclude, MLPA is a reliable, reproducible and low cost method that can be used for large-scale, high-throughput CNV analysis. However, the MLPA procedure requires several steps which render the method vulnerable to contamination. The main disadvantage is the design of highly multiplexed assays (>40 probes) which requires a cloning-based construction of hemiprobes that can take up to 4–6 weeks.

2.6. Single Reaction Closed Tube Multiplex PCR: QMPSF and MAQ

Charbonnier et al. (53) described a simple semiquantitative procedure based on the multiplex PCR of short fluorescent fragments (QMPSF). This method is a multiplex PCR amplification method based on trial and error multiplexing of PCR primers. QMPSF is therefore limited in the number of amplicons that can be amplified per mPCR reaction (approximately 12 amplicons per reaction). Recently, a similar mPCR method was developed based on a specifically designed PCR multiplexing algorithm capable of designing mPCR reactions containing up to 50 amplicons. This method has been termed multiplex amplicon quantification (MAQ). MAQ is highly comparable to the previous described MLPA method, with the important difference that the MAQ experimental setup is a closed tube assay that consists of a single PCR reaction resulting in a considerable reduction in labor and time as compared to MLPA. The strictly controlled MAQ primer design results in mPCR amplification of all amplicons under the same PCR conditions with the absence of spurious amplification products. The optimization process of the mPCR reactions is therefore reduced to merely adapting the primer concentration to obtain uniform amplification signals (54–56).

Together with the target amplicons, reference amplicons (regions of known copy number which are not expected to vary) are amplified in the same mPCR reaction followed by capillary gel electrophoresis-based fragment separation. Calculation of the dosage quotient (DQ) is done by comparing normalized peak areas between the test individual and control individuals. To facilitate data analysis of MAQ assays, specific analysis software (MAQ-S) has been developed. MAQ-S allows calculation and visualization of DQs starting directly from the raw data files (www.multiplicon.com/tools/maq-s).

Several commercial MAQ assays—e.g., for the detection of CNVs in the CMT1A/HNPP region (34 amplicon mPCR), in the amyloid precursor protein region (37 amplicons), and for aneuploidy detection of chromosomes (45 amplicons)—are currently available (www.multiplicom.com).

In conclusion, MAQ is a single, closed tube mPCR method allowing simultaneous amplification of up to 50 amplicons per

PCR reaction. MAQ assays provide the same assay multiplexity as MLPA assays, but reduce the handling complexity to the level of a standard PCR reaction. Therefore, MAQ is a powerful asset in high-throughput CNV analysis because of its ease of use, accuracy, and low cost.

3. Concluding Remarks

Recent demonstration of the considerable plasticity of the human genome holds the potential to revolutionize our understanding of clinical genetics and genetic epidemiology. Currently, our knowledge of CNVs is still incomplete and more research is needed for a better understanding of the impact of CNVs on phenotypic variation and human diseases. As reviewed in this chapter, to date, a wide range of molecular techniques for the detection and analysis of CNVs are available: some methods have the power to screen the whole genome, others are designed to analyze one or a few loci, some may detect balanced rearrangements, while others only unbalanced rearrangements. An overview of the characteristics comparing all techniques mentioned above is provided in Table 2. The method of choice depends on project specific factors and the questions to answer (57). Usually, compromises have to be made regarding the number of samples, number of CNVs to analyze, resolution, cost, and throughput. PCR-based methods are usually least demanding, and the degree of multiplexing often plays a decisive role, knowing that in most cases throughput is inversely related to the multiplexing grade.

A major problem in the PCR-based measurement of CNVs at the locus level is that the higher the copy number in a genome, the more difficult it is to resolve them. This bias is mainly due to differences in the kinetics at the early and late stages of PCR amplification (22). In addition, none of the PCR-based CNV analysis techniques described above is suitable to detect balanced rearrangements. Advanced technologies using highly sensitive reporter molecules and detection tools to exclude PCR amplification are being developed, such as automated fiber-FISH methods (58, 59) and NanoString's nCounter technology (60), but it is still too early to say whether they will be successful.

The most complete way to identify structural variants is an in silico approach by comparing DNA sequences of different sources. One advantage of this method is that all types of variants can be detected, including balanced rearrangements, such as inversions and translocations. However, the success of this computational method depends on the availability of the whole-genome sequence of the individual under study. With the development of next-generation sequencing (NGS) technologies, it will be possible

Table 2
Detailed comparison of the different methods available for the targeted detection and validation of CNVs

Method		FISH	Real-time qPCR	PRT	MCC	MAPH	MLPA	QMPSF	MAQ
Required amount of DNA		N/A[a]	10–20 ng	10–20 ng	<<10 ng[b]	≥1 μg	20–100 ng	10–50 ng	10–50 ng
Multiplex capabilities		No	Max 4	Max 4	>100[b]	Up to 40	Up to 50	Up to 12	Up to 50
Assay design and optimization		1–2 weeks	1–3 days	1–3 days	1–3 days	2–6 weeks	2–6 weeks	2–4 weeks	1 week
Protocol duration (days)		2–3	<1	<1	1–2	1–3	1–3	<1	<1
Minimal resolution (bp)		[c]	<100	<100	~100	~150	~100	~100	~100
Applicability[d]	Scenario 1	×××	×××	×××	××	×	×	×	×
	Scenario 2	×	×	×	××	×××	×××	××	×××
	Scenario 3	×	×	×	××	××	××	××	×××

FISH fluorescence in situ hybridization, *qPCR* quantitative PCR, *PRT* paralogue ratio test, *MCC* molecular copy-number counting, *MAPH* multiplex amplifiable probe hybridization, *MLPA* multiplex ligation-dependent probe amplification, *QMPSF* quantitative multiplex PCR of short fluorescent fragments, *MAQ* multiplex amplicon quantification; N/A not applicable;

[a]N/A: not applicable; FISH can be performed on different specimen types: peripheral blood, amniotic fluid, bone marrow, chorionic villi, lymph node, products of conception, tumor, skin biopsy, urine sample, and FFPE tissue sections

[b]See refs. 34, 35

[c]FISH on metaphase chromosome spreads: 1–3 Mb; on interphase nuclei: 50 kb–2 Mb; fiber-FISH: 5–500 kb (or 1–400 kb on fully condensed chromatin (66))

[d]A higher number of crosses "×", indicates a more appropriate method. Applicability scenario 1: analysis of one CNV in five individuals; applicability scenario 2: analysis of >20 CNVs in 100 individuals; and applicability scenario 3: analysis of >20 CNVs in >1,000 individuals (high-throughput)

in the near future, to cost-effectively obtain a personal genome sequence allowing for an "in silico" identification of interindividual (structural) variants. The different NGS platforms to date, Roche/454, Illumina/Solexa, ABI/SOLid, Complete Genomics, and Helicos are all high-throughput and timesaving sequencing technologies which have significantly accelerated the detailed analysis of whole genomes. Furthermore, NGS technologies make it possible to determine both the CNV structure as well as the breakpoint sequences, the latter clearly facilitating direct screening by breakpoint PCR. Several new computational approaches, using the increasing load of sequencing information from massive parallel sequencing projects, for the detection and finemapping of structural variants have been and are still being developed, such as BreakPtr (61) and event-wise testing using read depth of coverage (11).

The operational cost of whole genome sequencing is currently still too high for routine sequencing of complete human genomes. Therefore, implementing techniques for resequencing of targeted (disease) genes requires high-throughput front-end template enrichment methods. The ease of use, speed, and high multiplex potential of the described MAQ technology renders it as an ideal front-end method for parallelizable sequencing projects of a moderate number of amplicons (50–500) (62). To further explore the enormous potential of NGS, the development of new technologies for target-enrichment is of high priority. Recent developments include microdroplet PCR enrichment which enables 1.5 million amplifications in parallel (63), DNA sequence capture, and enrichment using microarray- or liquid-based hybridization approaches (64, 65). We further predict that with ongoing efforts to further reduce sequencing costs with the set goal of the 1,000$ genome, that soon CNV research, NGS and computational approaches (re) using the great load of sequencing information of different sequencing projects will facilitate and provide a new, very powerful tool for genome-wide CNV analysis.

References

1. Henrichsen,C.N., Chaignat,E. and Reymond,A. (2009) Copy number variants, diseases and gene expression. *Hum.Mol.Genet.*, **18**, R1–R8.
2. Feuk,L., Carson,A.R. and Scherer,S.W. (2006) Structural variation in the human genome. *Nat.Rev.Genet.*, 7, 85–97.
3. Beckmann,J.S., Sharp,A.J. and Antonarakis,S.E. (2008) CNVs and genetic medicine (excitement and consequences of a rediscovery). *Cytogenet.Genome Res.*, **123**, 7–16.
4. Lachman,H.M. (2008) Copy variations in schizophrenia and bipolar disorder. *Cytogenet. Genome Res.*, **123**, 27–35.
5. de Smith,A.J., Walters,R.G., Froguel,P. and Blakemore,A.I. (2008) Human genes involved in copy number variation: mechanisms of origin, functional effects and implications for disease. *Cytogenet.Genome Res.*, **123**, 17–26.
6. Stranger,B.E., Forrest,M.S., Dunning,M., Ingle,C.E., Beazley,C., Thorne,N., Redon,R., Bird,C.P., de Grassi,A., Lee,C. *et al.* (2007) Relative impact of nucleotide and copy number variation on gene expression phenotypes. *Science*, **315**, 848–853.
7. Gu,W. and Lupski,J.R. (2008) CNV and nervous system diseases – what's new? *Cytogenet. Genome Res.*, **123**, 54–64.

8. Pinkel,D., Segraves,R., Sudar,D., Clark,S., Poole,I., Kowbel,D., Collins,C., Kuo,W.L., Chen,C., Zhai,Y. *et al.* (1998) High resolution analysis of DNA copy number variation using comparative genomic hybridization to microarrays. *Nat.Genet.*, **20**, 207–211.
9. Komura,D., Shen,F., Ishikawa,S., Fitch,K.R., Chen,W., Zhang,J., Liu,G., Ihara,S., Nakamura,H., Hurles,M.E. *et al.* (2006) Genome-wide detection of human copy number variations using high-density DNA oligonucleotide arrays. *Genome Res.*, **16**, 1575–1584.
10. Yau,C. and Holmes,C.C. (2008) CNV discovery using SNP genotyping arrays. *Cytogenet. Genome Res.*, **123**, 307–312.
11. Yoon,S., Xuan,Z., Makarov,V., Ye,K. and Sebat,J. (2009) Sensitive and accurate detection of copy number variants using read depth of coverage. *Genome Res.*, **19**, 1586–1592.
12. Iafrate,A.J., Feuk,L., Rivera,M.N., Listewnik,M.L., Donahoe,P.K., Qi,Y., Scherer,S.W. and Lee,C. (2004) Detection of large-scale variation in the human genome. *Nat.Genet.*, **36**, 949–951.
13. Sebat,J., Lakshmi,B., Troge,J., Alexander,J., Young,J., Lundin,P., Maner,S., Massa,H., Walker,M., Chi,M. *et al.* (2004) Large-scale copy number polymorphism in the human genome. *Science*, **305**, 525–528.
14. Wu,X. and Xiao,H. (2009) Progress in the detection of human genome structural variations. *Sci.China C.Life Sci.*, **52**, 560–567.
15. Patsalis,P.C., Evangelidou,P., Charalambous,S. and Sismani,C. (2004) Fluorescence in situ hybridization characterization of apparently balanced translocation reveals cryptic complex chromosomal rearrangements with unexpected level of complexity. *Eur.J.Hum.Genet.*, **12**, 647–653.
16. Florijn,R.J., Bonden,L.A., Vrolijk,H., Wiegant,J., Vaandrager,J.W., Baas,F., Den Dunnen,J.T., Tanke,H.J., van Ommen,G.J. and Raap,A.K. (1995) High-resolution DNA Fiber-FISH for genomic DNA mapping and colour bar-coding of large genes. *Hum.Mol. Genet.*, **4**, 831–836.
17. Wilke,K., Duman,B. and Horst,J. (2000) Diagnosis of haploidy and triploidy based on measurement of gene copy number by real-time PCR. *Hum.Mutat.*, **16**, 431–436.
18. Solinas,A., Brown,L.J., McKeen,C., Mellor,J.M., Nicol,J., Thelwell,N. and Brown,T. (2001) Duplex Scorpion primers in SNP analysis and FRET applications. *Nucleic Acids Res.*, **29**, E96.
19. Shengqi,W., Xiaohong,W., Suhong,C. and Wei,G. (2002) A new fluorescent quantitative polymerase chain reaction technique. *Anal. Biochem.*, **309**, 206–211.
20. Klein,D. (2002) Quantification using real-time PCR technology: applications and limitations. *Trends Mol.Med.*, **8**, 257–260.
21. Simpson,D.A., Feeney,S., Boyle,C. and Stitt,A.W. (2000) Retinal VEGF mRNA measured by SYBR green I fluorescence: A versatile approach to quantitative PCR. *Mol.Vis.*, **6**, 178–183.
22. Lee,J.H. and Jeon,J.T. (2008) Methods to detect and analyze copy number variations at the genome-wide and locus-specific levels. *Cytogenet.Genome Res.*, **123**, 333–342.
23. Weaver,S., Dube,S., Mir,A., Qin,J., Sun,G., Ramakrishnan,R., Jones,R.C. and Livak,K.J. (2010) Taking qPCR to a higher level: Analysis of CNV reveals the power of high throughput qPCR to enhance quantitative resolution. *Methods*, **50**, 271–276.
24. Norskov,M.S., Frikke-Schmidt,R., Loft,S. and Tybjaerg-Hansen,A. (2009) High-throughput genotyping of copy number variation in glutathione S-transferases M1 and T1 using real-time PCR in 20,687 individuals. *Clin.Biochem.*, **42**, 201–209.
25. Rose-Zerilli,M.J., Barton,S.J., Henderson,A.J., Shaheen,S.O. and Holloway,J.W. (2009) Copy-number variation genotyping of GSTT1 and GSTM1 gene deletions by real-time PCR. *Clin.Chem.*, **55**, 1680–1685.
26. Weksberg,R., Hughes,S., Moldovan,L., Bassett,A.S., Chow,E.W. and Squire,J.A. (2005) A method for accurate detection of genomic microdeletions using real-time quantitative PCR. *BMC.Genomics*, **6**, 180.
27. Wu,Y.L., Savelli,S.L., Yang,Y., Zhou,B., Rovin,B.H., Birmingham,D.J., Nagaraja,H.N., Hebert,L.A. and Yu,C.Y. (2007) Sensitive and specific real-time polymerase chain reaction assays to accurately determine copy number variations (CNVs) of human complement C4A, C4B, C4-long, C4-short, and RCCX modules: elucidation of C4 CNVs in 50 consanguineous subjects with defined HLA genotypes. *J. Immunol.*, **179**, 3012–3025.
28. Nuytten,H., Wlodarska,I., Nackaerts,K., Vermeire,S., Vermeesch,J., Cassiman,J.J. and Cuppens,H. (2009) Accurate determination of copy number variations (CNVs): application to the alpha- and beta-defensin CNVs. *J.Immunol.Methods*, **344**, 35–44.
29. Gouas,L., Goumy,C., Veronese,L., Tchirkov,A. and Vago,P. (2008) Gene dosage methods as diagnostic tools for the identification of chromosome abnormalities. *Pathol.Biol.(Paris)*, **56**, 345–353.

30. Armour,J.A., Palla,R., Zeeuwen,P.L., den Heijer,M., Schalkwijk,J. and Hollox,E.J. (2007) Accurate, high-throughput typing of copy number variation using paralogue ratios from dispersed repeats. *Nucleic Acids Res.*, **35**, e19.

31. Dear,P.H. and Cook,P.R. (1993) Happy mapping: linkage mapping using a physical analogue of meiosis. *Nucleic Acids Res.*, **21**, 13–20.

32. McCaughan,F. and Dear,P.H. (2010) Single-molecule genomics. *J.Pathol.*, **220**, 297–306.

33. Daser,A., Thangavelu,M., Pannell,R., Forster,A., Sparrow,L., Chung,G., Dear,P.H. and Rabbitts,T.H. (2006) Interrogation of genomes by molecular copy-number counting (MCC). *Nat.Methods*, **3**, 447–453.

34. McCaughan,F. (2009) Molecular copy-number counting: potential of single-molecule diagnostics. *Expert.Rev.Mol.Diagn.*, **9**, 309–312.

35. McCaughan,F., Darai-Ramqvist,E., Bankier,A.T., Konfortov,B.A., Foster,N., George,P.J., Rabbitts,T.H., Kost-Alimova,M., Rabbitts,P.H. and Dear,P.H. (2008) Microdissection molecular copy-number counting (microMCC)–unlocking cancer archives with digital PCR. *J.Pathol.*, **216**, 307–316.

36. Qin,J., Jones,R.C. and Ramakrishnan,R. (2008) Studying copy number variations using a nanofluidic platform. *Nucleic Acids Res.*, **36**, e116.

37. Armour,J.A., Sismani,C., Patsalis,P.C. and Cross,G. (2000) Measurement of locus copy number by hybridisation with amplifiable probes. *Nucleic Acids Res.*, **28**, 605–609.

38. Hollox,E.J., Atia,T., Cross,G., Parkin,T. and Armour,J.A. (2002) High throughput screening of human subtelomeric DNA for copy number changes using multiplex amplifiable probe hybridisation (MAPH). *J.Med.Genet.*, **39**, 790–795.

39. Patsalis,P.C., Kousoulidou,L., Sismani,C., Mannik,K. and Kurg,A. (2005) MAPH: from gels to microarrays. *Eur.J.Med.Genet.*, **48**, 241–249.

40. Gibbons,B., Datta,P., Wu,Y., Chan,A. and Al Armour,J. (2006) Microarray MAPH: accurate array-based detection of relative copy number in genomic DNA. *BMC.Genomics*, **7**, 163.

41. Sellner,L.N. and Taylor,G.R. (2004) MLPA and MAPH: new techniques for detection of gene deletions. *Hum.Mutat.*, **23**, 413–419.

42. Kousoulidou,L., Mannik,K., Sismani,C., Zilina,O., Parkel,S., Puusepp,H., Tonisson,N., Palta,P., Remm,M., Kurg,A. *et al.* (2008) Array-MAPH: a methodology for the detection of locus copy-number changes in complex genomes. *Nat.Protoc.*, **3**, 849–865.

43. Patsalis,P.C., Kousoulidou,L., Mannik,K., Sismani,C., Zilina,O., Parkel,S., Puusepp,H., Tonisson,N., Palta,P., Remm,M. *et al.* (2007) Detection of small genomic imbalances using microarray-based multiplex amplifiable probe hybridization. *Eur.J.Hum.Genet.*, **15**, 162–172.

44. Tyson,J., Majerus,T.M., Walker,S. and Armour,J.A. (2009) Quadruplex MAPH: improvement of throughput in high-resolution copy number screening. *BMC.Genomics*, **10**, 453.

45. Schouten,J.P., McElgunn,C.J., Waaijer,R., Zwijnenburg,D., Diepvens,F. and Pals,G. (2002) Relative quantification of 40 nucleic acid sequences by multiplex ligation-dependent probe amplification. *Nucleic Acids Res.*, **30**, e57.

46. White,S.J., Vink,G.R., Kriek,M., Wuyts,W., Schouten,J., Bakker,B., Breuning,M.H. and Den Dunnen,J.T. (2004) Two-color multiplex ligation-dependent probe amplification: detecting genomic rearrangements in hereditary multiple exostoses. *Hum.Mutat.*, **24**, 86–92.

47. Stern,R.F., Roberts,R.G., Mann,K., Yau,S.C., Berg,J. and Ogilvie,C.M. (2004) Multiplex ligation-dependent probe amplification using a completely synthetic probe set. *Biotechniques*, **37**, 399–405.

48. Kozlowski,P., Roberts,P., Dabora,S., Franz,D., Bissler,J., Northrup,H., Au,K.S., Lazarus,R., Domanska-Pakiela,D., Kotulska,K. *et al.* (2007) Identification of 54 large deletions/duplications in TSC1 and TSC2 using MLPA, and genotype–phenotype correlations. *Hum. Genet.*, **121**, 389–400.

49. Kozlowski,P., Jasinska,A.J. and Kwiatkowski,D.J. (2008) New applications and developments in the use of multiplex ligation-dependent probe amplification. *Electrophoresis*, **29**, 4627–4636.

50. Nygren,A.O., Ameziane,N., Duarte,H.M., Vijzelaar,R.N., Waisfisz,Q., Hess,C.J., Schouten,J.P. and Errami,A. (2005) Methylation-specific MLPA (MS-MLPA): simultaneous detection of CpG methylation and copy number changes of up to 40 sequences. *Nucleic Acids Res.*, **33**, e128.

51. Isaksson,M., Stenberg,J., Dahl,F., Thuresson,A.C., Bondeson,M.L. and Nilsson,M. (2007) MLGA – a rapid and cost-efficient assay for gene copy-number analysis. *Nucleic Acids Res.*, **35**, e115.

52. Zeng,F., Ren,Z.R., Huang,S.Z., Kalf,M., Mommersteeg,M., Smit,M., White,S., Jin,C.L., Xu,M., Zhou,D.W. *et al.* (2008) Array-MLPA: comprehensive detection of deletions and duplications and its application to DMD patients. *Hum.Mutat.*, **29**, 190–197.
53. Charbonnier,F., Raux,G., Wang,Q., Drouot,N., Cordier,F., Limacher,J.M., Saurin,J.C., Puisieux,A., Olschwang,S. and Frebourg,T. (2000) Detection of exon deletions and duplications of the mismatch repair genes in hereditary nonpolyposis colorectal cancer families using multiplex polymerase chain reaction of short fluorescent fragments. *Cancer Res.*, **60**, 2760–2763.
54. Sleegers,K., Brouwers,N., Gijselinck,I., Theuns,J., Goossens,D., Wauters,J., Del Favero,J., Cruts,M., Van Duijn,C.M. and Van Broeckhoven,C. (2006) APP duplication is sufficient to cause early onset Alzheimer's dementia with cerebral amyloid angiopathy. *Brain*, **129**, 2977–2983.
55. Suls,A., Claeys,K.G., Goossens,D., Harding,B., Van Luijk,R., Scheers,S., Deprez,L., Audenaert,D., Van Dyck,T., Beeckmans,S. *et al.* (2006) Microdeletions involving the SCN1A gene may be common in SCN1A-mutation-negative SMEI patients. *Hum. Mutat.*, **27**, 914–920.
56. Sutrala,S.R., Goossens,D., Williams,N.M., Heyrman,L., Adolfsson,R., Norton,N., Buckland,P.R. and Del-Favero,J. (2007) Gene copy number variation in schizophrenia. *Schizophr.Res.*, **96**, 93–99.
57. Aten,E., White,S.J., Kalf,M.E., Vossen,R.H., Thygesen,H.H., Ruivenkamp,C.A., Kriek,M., Breuning,M.H. and Den Dunnen,J.T. (2008) Methods to detect CNVs in the human genome. *Cytogenet.Genome Res.*, **123**, 313–321.
58. Sieben,V.J., Debes Marun,C.S., Pilarski,P.M., Kaigala,G.V., Pilarski,L.M. and Backhouse,C.J. (2007) FISH and chips: chromosomal analysis on microfluidic platforms. *IET.Nanobiotechnol.*, **1**, 27–35.
59. Sieben,V.J., Debes-Marun,C.S., Pilarski,L.M. and Backhouse,C.J. (2008) An integrated microfluidic chip for chromosome enumeration using fluorescence in situ hybridization. *Lab Chip.*, **8**, 2151–2156.
60. Geiss,G.K., Bumgarner,R.E., Birditt,B., Dahl,T., Dowidar,N., Dunaway,D.L., Fell,H.P., Ferree,S., George,R.D., Grogan,T. *et al.* (2008) Direct multiplexed measurement of gene expression with color-coded probe pairs. *Nat.Biotechnol.*, **26**, 317–325.
61. Korbel,J.O., Urban,A.E., Affourtit,J.P., Godwin,B., Grubert,F., Simons,J.F., Kim,P.M., Palejev,D., Carriero,N.J., Du,L. *et al.* (2007) Paired-end mapping reveals extensive structural variation in the human genome. *Science*, **318**, 420–426.
62. Goossens,D., Moens,L.N., Nelis,E., Lenaerts,A.S., Glassee,W., Kalbe,A., Frey,B., Kopal,G., De Jonghe,P., De Rijk,P. *et al.* (2009) Simultaneous mutation and copy number variation (CNV) detection by multiplex PCR-based GS-FLX sequencing. *Hum.Mutat.*, **30**, 472–476.
63. Tewhey,R., Warner,J.B., Nakano,M., Libby,B., Medkova,M., David,P.H., Kotsopoulos,S.K., Samuels,M.L., Hutchison,J.B., Larson,J.W. *et al.* (2009) Microdroplet-based PCR enrichment for large-scale targeted sequencing. *Nat. Biotechnol.*, **27**, 1025–1031.
64. Chou,L.S., Liu,C.S., Boese,B., Zhang,X. and Mao,R. (2010) DNA sequence capture and enrichment by microarray followed by next-generation sequencing for targeted resequencing: neurofibromatosis type 1 gene as a model. *Clin.Chem.*, **56**, 62–72.
65. Ng,S.B., Buckingham,K.J., Lee,C., Bigham,A.W., Tabor,H.K., Dent,K.M., Huff,C.D., Shannon,P.T., Jabs,E.W., Nickerson,D.A. *et al.* (2010) Exome sequencing identifies the cause of a Mendelian disorder. *Nat.Genet.*, **42**, 30–35.
66. Raap,A.K., Florijn,R.J., Blonden,L.A.J., Wiegant,J., Vaandrager,J.W., Vrolijk,H., den Dunnen,J., Tanke,H.J. and van Ommen,G.J. (1996) Fiber FISH as a DNA Mapping Tool. *Methods*, **9**, 67–73.

Chapter 16

High-Resolution Copy Number Profiling by Array CGH Using DNA Isolated from Formalin-Fixed, Paraffin-Embedded Tissues

Hendrik F. van Essen and Bauke Ylstra

Abstract

We describe protocols to acquire high-quality DNA from formalin-fixed, paraffin-embedded (FFPE) tissues for the use in array comparative genome hybridization (CGH). Formalin fixation combined with paraffin embedding is routine procedure for solid malignancies in the diagnostic practice of the pathologist. As a consequence, large archives of FFPE tissues are available in pathology institutes across the globe. This archival material is for many research questions an invaluable resource, with long-term clinical follow-up and survival data available. FFPE is, thus, highly attractive for large genomics studies, including experiments requiring samples for test/learning and validation. Most larger array CGH studies have, therefore, made use of FFPE material and show that CNAs have tumor- and tissue-specific traits (Chin et al. Cancer Cell 10: 529–541, 2006; Fridlyand et al. BMC Cancer 6: 96, 2006; Weiss et al. Oncogene 22: 1872–1879, 2003; Jong et al. Oncogene 26: 1499–1506, 2007). The protocols described are tailored to array CGH of FFPE solid malignancies: from sectioning FFPE blocks to specific cynosures for pathological revisions of sections, DNA isolation, quality testing, and amplification. The protocols are technical in character and elaborate up to the labeling of isolated DNA while further processes and interpretation and data analysis are beyond the scope.

Key words: Formalin fixed, paraffin embedded, Chromosomal DNA, Microarray, Copy number aberrations, Archival tissue, Sodium thiocyanate, Array comparative genome hybridization

1. Introduction

Pathological assessments have relied primarily on morphological and histological indicators which permitted general classifications into morphologic subtypes. Thereby, pathologists have only taken few molecular markers into account, such as expression or chromosomal copy number. Array techniques allow for the simultaneous analysis of the expression of thousands of genes or chromosomal positions in parallel. Thereby, arrays offer unprecedented

Lars Feuk (ed.), *Genomic Structural Variants: Methods and Protocols*, Methods in Molecular Biology, vol. 838,
DOI 10.1007/978-1-61779-507-7_16, © Springer Science+Business Media, LLC 2012

opportunities to obtain comprehensive molecular signatures of malignancies in one single experiment.

Profiling at the expression level has the limitation that the levels of RNA are rapidly altered by environmental influences, such as temperature, cell cycle, or circadian rhythms. Additionally, acquisition of high-quality RNA and optimal transport are problematic because of its inherent instability. In contrast, tumor DNA is relatively stable, easy to transport, and can be obtained from formalin-fixed, paraffin-embedded (FFPE) tissue blocks. Together with the fact that chromosomal aberrations are a hallmark of cancer makes chromosomal copy number profiling the ideal technique for pathological assessments. Array comparative genome hybridization (CGH) offers the opportunity to globally profile the presence of chromosomal copy number aberrations in tumors at high resolution in constitutional or tumor DNA samples (see Fig. 1). The array CGH technique has already allowed a deeper insight into the biology of a variety of tumor types and in the near future will undoubtedly prove to be a key technology leading to a better understanding of cancer (5).

Next to the application in ongoing research projects, we have also implemented chromosomal copy number profiling using the array CGH technique on DNA isolated from FFPE tissues in our

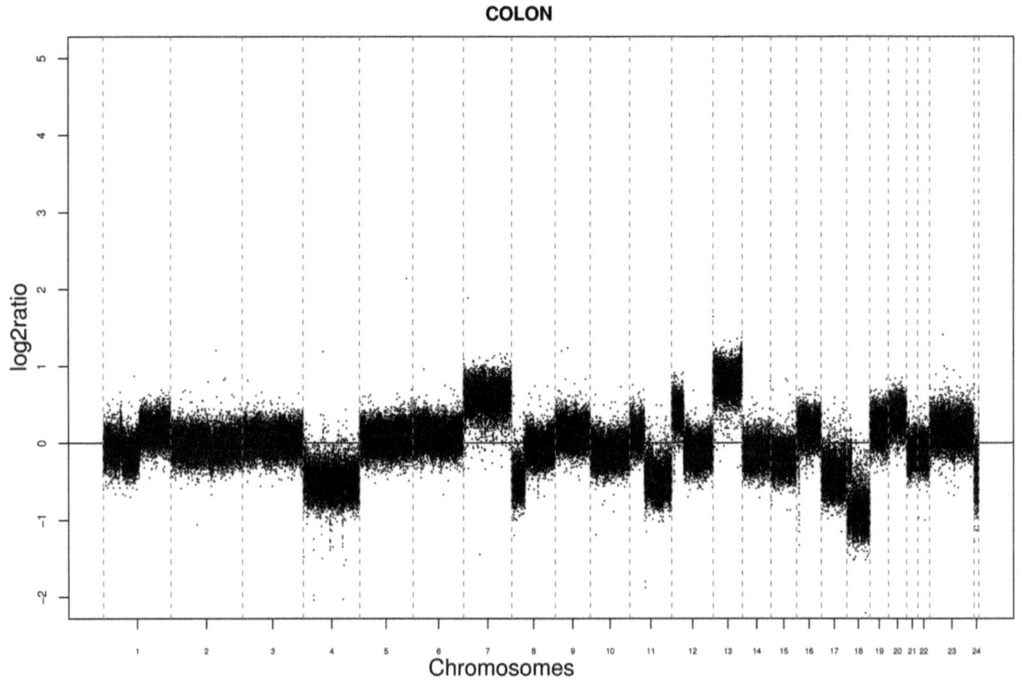

Fig. 1. An example of an array CGH profile of a direct hybridization of colon tumor versus patient's normal tissue labeled with Cy3 and Cy5 fluorescent dyes, respectively. *Y-axis*: Log2 tumor to normal ratios of all chromosomes. *X-axis*: Array elements plotted according to chromosomal position (1 – *Y*). *Vertical dashed lines*: Transition between chromosomes. Typical characteristics of colon tumors are visible, such as gains of chromosomes 13 and 20 (4).

diagnostic practice. Profiles of double tumors are compared to aid the pathologist in distinguishing second primary tumors from recurrence (6). Pivotal in all these processes is the quality of the DNA isolates. The described protocol, therefore, has an emphasis on the isolation procedures and quality assessment of the DNA. Often enough, DNA isolated from FFPE, which was fully functional for PCR, failed in array CGH experiments. For array CGH experiments, the integrity of DNA is crucial in terms of genome-wide representation. The actual fragment size is incidentally less important, although fragments less than 300 bp seem to perform less (7). Furthermore, the enzymatic accessibility of the isolates needs to be even throughout the genome. Fundamentally, an important step in the procedure is the removal of DNA cross-links introduced by formalin during the fixation of the tissue. Therefore, an overnight incubation was introduced with the chaotropic agent sodium thiocyanate (NaSCN) in the DNA isolation procedure (8). Furthermore, we have chosen to avoid phenol/chloroform in our procedures. Although phenol/chloroform treatment of DNA is highly effective for the removal of (DNA bound) proteins, it has several disadvantages. First, the toxicity of both phenol and chloroform requires specific safety measures for handling and disposal. Furthermore, the phase separation step is difficult to standardize and as a consequence phenol/chloroform is harder to implement in the diagnostic routine. Finally, traces of phenol/chloroform can severely inhibit enzymatic reactions in labeling and/or amplification procedures. The described protocols make use of column-based purification methods which avoid both these obstacles, thus making it easier to implement in routine (diagnostic) processes and standard operating procedures. At the same time, we have not observed any deterioration in the quality of the final results, the array CGH profiles, since we have moved from phenol/chloroform to column-based procedures. We realize that the isolation procedures we recommend are long and elaborate, but they have given the best and most reproducible results in our hands and those of our collaborators, where alternative procedures have failed. Despite the intricate procedures, not all samples succeed. This variability may be dependent on several factors. First, the DNA in the FFPE blocks may already be of a less-than-optimal quality. Dropout varies between tissues; we and others have experienced much dropout with gastric samples (up to 30%), whereas the dropout with lung and glioma was intermediary in independent projects and esophageal and colorectal tumors tend to perform much better (roughly 10% dropout). Second, fixation procedures seem to play an important role, where quality seems to be dependent on fixation time. With more recent cases, we have found that smaller tissue blocks, such as biopsies, tend to yield good-quality DNA. It seems that tissue samples of a smaller size, in which the fixative infiltrates faster through the tissue, seem to preserve the DNA better, thus yielding a better quality. Third, the age of the FFPE blocks is another important item. Frequently, we have been able to get useful profiles with archival tissue of up to 20 years old. The use of buffer in the fixative seems to play an important role in the preservation of the DNA during fixation. A large number of pathology laboratories introduced buffered formalin in the late eighties, which may explain that older samples often do not perform well. Fourth, DNA degradation due to late fixation or necrosis is another important factor that can explain variable results. The fifth and final factor is the pathological re-review of the H&E sections—as described in our protocol. In our experience and those of others (9), a minimum of 40–70% of tumor cells are required. Another advantage is that the pathologist can easily demarcate a smaller area of the tumor to enrich the tumor cells in the sections for the DNA isolation. This also ensures that necrotic areas in the sections are avoided.

Some protocols recommend fragmentation (shearing) of the DNA prior to labeling. We have not experienced any advantage using fragmented or full-length DNA in this or other protocols, for either fresh or FFPE material. To simplify the protocols, we therefore avoid fragmentation. Other manufacturers recommend nonenzymatic labeling of DNA isolated from paraffin blocks. Although this has theoretical advantages, in our hands this promise has not come true; enzymatic labeling versus nonenzymatic labeling has repeatedly yielded very high-quality profiles (10). Nonenzymatic labeling has the disadvantage that it generally requires more input DNA, whereas sample quantity is often a restriction when processing clinical samples. In conclusion, the labeling and hybridization procedures of array CGH are not the issue, but the quality of the DNA input is.

Copy number variation (CNV) is the last item that requires serious attention and consideration before the onset of any project involving solid malignancies. The issue is independent whether one works with FFPE or fresh material, but pertinent to high-resolution arrays (i.e., >30K chromosomal positions). Copy number-variable regions are parts of the genome that can vary between healthy individuals—for definitions, see Feuk et al. (11). Since CNVs are germ-line variations, they are not DNA copy number aberrations (CNAs) that occur somatically and are thus specific to the tumor. Although either CNVs or CNAs may be important to the development to the tumor, ideally germ-line and somatic copy number variations should be detected separately. Technically, this is simple to achieve; if tumor DNA is hybridized against its own matched normal, only somatic variations (CNAs) will be detected. Obviously, CNVs are the same in tumor and normal DNA and are cancelled out in the hybridization. Although technically simple, in our experience, matched normal DNA is not always easy to obtain. When possible, use normal tissue from a different FFPE block, although DNA extracted from the resection margins can be used. The danger here is that chromosomal aberrations can be present in tissue close to the tumor cells which may look normal to the pathologist, but is not normal in terms of chromosomal aberrations. When in doubt, the normal channel may be computationally compared to a pooled reference to check for aberrations as described in detail by Buffart et al. (12). If no matched normal DNA is available, a pooled reference can be used, although it is important to realize that this profile is "CNV contaminated."

We present validated protocols for the recovery of genomic DNA from archival FFPE tissue specimens and its application in array CGH for the detection of chromosomal CNAs on a high-resolution end genome-wide scale.

2. Materials

2.1. Genomic DNA Isolation from FFPE Material

1. Microtome.
2. Bovine serum albumin (BSA).
3. Superfrost uncoated glass slides.
4. Xylene.
5. Ethanol absolute, 96 and 70%.
6. Hematoxylin.
7. Eosin.
8. Microscope coverslips.
9. Depex mounting medium.
10. Methanol.
11. Scalpel or needle.
12. Pipettes.
13. Centrifuge.
14. Vortex.
15. 1 M NaSCN (i.e., Sigma) (see Notes 1–3).
16. 1.5-ml safe-lock Eppendorf tube.
17. Spectrophotometer.
18. Qiagen, QIAamp DNA Micro kit containing:
    (a) ATL buffer.
    (b) Proteinase K: 20 mg/ml in water.
    (c) AL buffer.
    (d) AW1 buffer.
    (e) AW2 buffer.
    (f) AE buffer.

2.2. Isothermal Amplification

1. BioScore™ Screening and Amplification Kit (Enzo Life Sciences), containing:
   (a) Primers.
   (b) Enzyme (Klenow fragment of DNA Polymerase); keep on ice at all times.
   (c) Stop buffer.
   (d) Deoxynucleotide mix.
   (e) Nuclease-free water.
2. MinElute purification columns (QIAgen).
3. Thermocycler strip (i.e., BIOplatics).
4. PCR machine with heated lid.
5. Cold block/ice bath.

2.3. Removal of Uncoupled Deoxynucleotides

1. 1.5-ml Eppendorf tubes.
2. QIAgen, QIAquick kit (QIAgen), containing:
   (a) Binding buffer PB.
   (b) Wash buffer PE.
   (c) Elution buffer EB.

3. Methods

3.1. Genomic DNA Isolation from FFPE Tissue

3.1.1. Cut and Mount Sections from FFPE Blocks

A visual representation of the DNA isolation procedures is shown in Fig. 2. First, the paraffin-embedded tissue block is secured to the microtome. Align the block with the cutting blade such that the least amount of tissue is lost. Cut the first section at a thickness of 3–5 μm for H&E staining. This is then followed by subsequent sections of 10 μm. For each tumor sample, a minimum of 4–6 sections is recommended, each containing 1 cm^2 of tissue, but isolation from as little as one 5-μm section has given high-quality results in our hands. If available, DNA from an FFPE tissue sample which contains normal cells (from the same patient) should be isolated (see Subheading 1). Again, cut a 3–5-μm section for H&E staining followed by 2–3 sections at 10 μm.

1. Cut 4- and 10-μm sections.
2. Mount all the sections with BSA 0.1% on uncoated glass slides.
3. Dry the slides overnight at 37–40°C.

3.1.2. H&E Staining

H&E staining is performed according to standard procedures in the pathological laboratory.

1. Place the slides in xylene, three baths, 5 min each.
2. Place the slides in ethanol absolute, two baths, 5 min each.

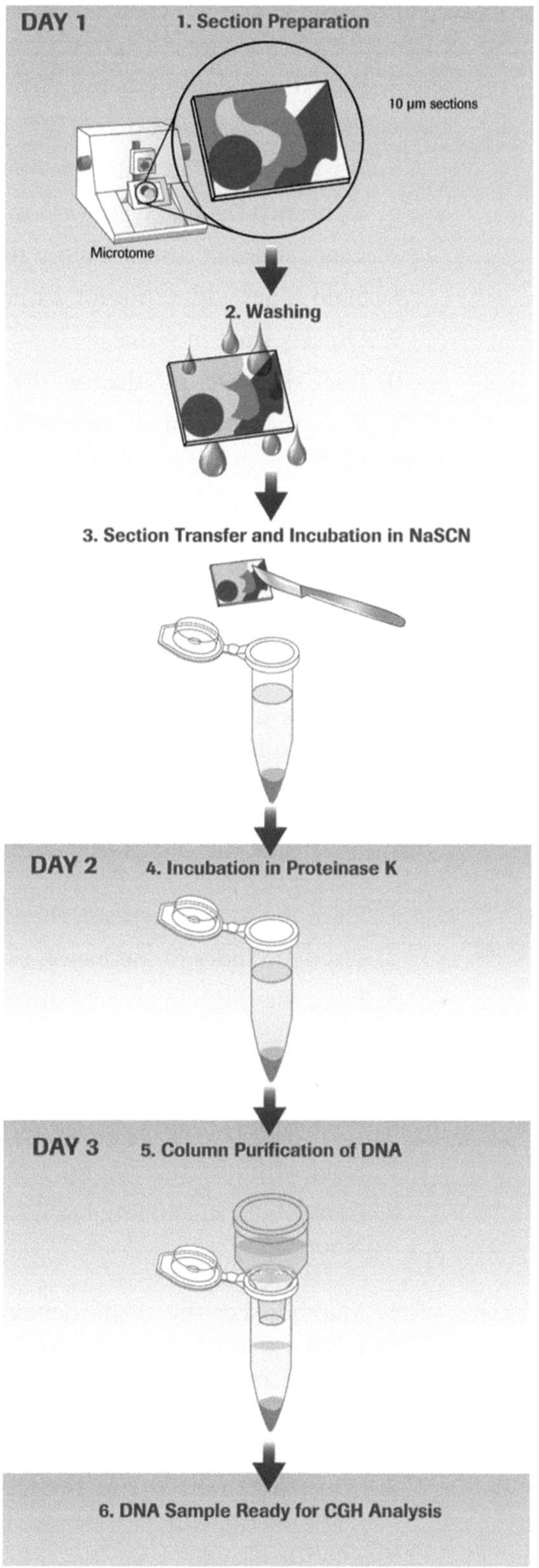

Fig. 2. Visual representation of the 3-day process of genomic DNA isolation from FFPE tissue as described methods in Subheading 3.1 (with permission from the "Technical Note: Array CGH using DNA from Formalin-Fixed Paraffin-Embedded (FFPE) Tissue Samples" by Roche NimbleGen, Inc. ©).

3. Place the slides in ethanol 96% for 1 min.
4. Place the slides in ethanol 70% for 1 min.
5. Wash slides in tap water for 30 s.
6. Stain with hematoxylin for 5 min.
7. Leave slides in running tap water to blue up for 10 min.
8. Stain slides with eosin for 2 min.
9. Quickly rinse in water.
10. Place the slides in ethanol 70% for 30 s.
11. Place the slides in ethanol 96% for 30 s.
12. Place the slides in ethanol absolute, two baths, 3 min each.
13. Place the slides in xylene, two baths, 3 min each.
14. Add Depex to stained sections followed by a cover glass and leave to dry.

3.1.3. Review of H&E Slides by Pathologist

After staining, the H&E slides are ready for review by a pathologist, who demarcates the tumor area and determines the percentage of tumor cells in that area. The latter is a visual estimation of the amount of tumor cells compared to surrounding tissue architecture. A minimum of 70% of tumor cells is recommended, although useful results have been obtained with as little as 30% estimated tumor cells, in our hands (see Note 4).

3.1.4. Hematoxylin Staining of 10-μm Slides Before DNA Isolation

1. Place the slides in xylene, three baths, 7 min each.
2. Place the slides in methanol, two baths, 7 min each.
3. Place the slides in ethanol absolute, two baths, 2 min each.
4. Place the slides in ethanol 96% for 1 min.
5. Place the slides in ethanol 70% for 1 min.
6. Wash slides in tap water for 30 s.
7. Stain with hematoxylin for 1–2 min.
8. Leave slides in running tap water to blue up for 5–10 min (see Note 5).

3.1.5. DNA Isolation from Hematoxylin-Stained Tissue Sections

1. Macrodissect the tissue demarcated in the H&E section for DNA isolation from the glass slide. Macrodissection is performed by scraping the area of interest from the glass slide using a needle or scalpel. Transfer the cells to a clean 1.5-ml safe-lock Eppendorf tube (see Note 6).
2. Add 500–1,000 μl 1 M NaSCN to the collected sample in the safe-lock Eppendorf tube (see Note 7).
3. Mix by inverting (do not vortex).
4. Place the Eppendorf tube in a heat block and incubate overnight at 38–40°C (see Note 8).

5. Spin down at full speed (20,000 × *g*; 14,000 rpm) for 60 min or longer (tissue might not always form a pellet).
6. Carefully pipette off the NaSCN and discard.
7. Spin down at full speed (20,000 × *g*; 14,000 rpm) for 1 min.
8. Remove the remaining NaSCN and discard.
9. Add 160 µl ATL Buffer (QIAmp DNA micro-kit, Qiagen) to the sample in the safe-lock Eppendorf tube.
10. Add 40 µl of 20 mg/ml Proteinase K (QIAmp DNA micro-kit, Qiagen) to the sample in the safe-lock Eppendorf tube (see Note 10).
11. Vortex the Eppendorf tubes containing the tissue sample for 10–15 s.
12. Place the Eppendorf tube in a heat block and incubate overnight at 55°C (see Note 9).
13. Vortex the Eppendorf tube containing the lysate for 10–15 s (see Note 11).
14. Spin down briefly (20,000 × *g*; 14,000 rpm).
15. Incubate at 98°C for 10 min.
16. Spin down the Eppendorf tube at full speed (20,000 × *g*; 14,000 rpm) for 1 min.
17. Add 200 µl AL Buffer (QIAmp DNA microkit, Qiagen) and vortex for 15 s.
18. Add 200 µl ethanol and vortex for 15 s.
19. Incubate at room temperature for 5 min.
20. Spin down the Eppendorf tube at full speed (20,000 × *g*; 14,000 rpm) for 1 min.
21. Transfer up to 600 µl lysate to the QIAamp MinElute column (QIAmp DNA micro-kit, Qiagen).
22. Spin down the column (6,000 × *g*; 8,000 rpm) for 1 min. Place the QIAamp MinElute column in a clean 2-ml collection tube (see Note 12); discard flow-through.
23. Add 500 µl AW1 buffer (QIAmp DNA micro-kit, Qiagen) to the column (see Note 13).
24. Spin down the column for 1 min (6,000 × *g*; 8,000 rpm). Place the column in a new tube (see Note 12), and discard the tube containing the flow-through.
25. Add 500 µl AW2 buffer (QIAmp DNA micro-kit, Qiagen) to the column (see Note 14).
26. Spin down the column for 1 min (6,000 × *g*; 8,000 rpm) (see Note 12).
27. Place the column in a new tube, and discard the tube containing the flow-through.

28. Spin down the column for 3 min at full speed to dry the membrane (see Note 12).

    Place the QIAamp MinElute column in a properly marked Eppendorf tube, discard the collection tube.
29. Add 20 to 30 μl (depending on sample size) of AE buffer (QIAmp DNA micro-kit, Qiagen) to the membrane of the column and incubate at room temperature for a minimum 5 min.
30. Spin down the column at full speed (20,000 × *g*; 14,000 rpm) for 3 min.
31. Discard the column, close the Eppendorf tube and continue with DNA measurement (see Note 15) or store at 4°C for short term or at –20°C for long term.

3.2. Whole Genome Isothermal DNA Amplification

3.2.1. Method

1. Mix together in a nuclease free thermocycler strip:

Nuclease-free water	19 – *X* μl
Primers	20 μl
100 ng genomic DNA	*X* μl
Total	39 μl (see Note 16)

2. Mix gently by flicking the thermocycler strip.
3. Centrifuge briefly.
4. Place thermocycler strip in a PCR machine and heat at 95–99°C for 10 min.
5. Place thermocycler strip on ice for 5 min.
6. Centrifuge briefly, return to ice and keep on ice in steps 7 and 8.
7. Add 10 μl deoxynucleotide mix.
8. Add 1 μl Klenow.
9. Mix by gently flicking the thermocycler strip.
10. Centrifuge briefly.
11. Incubate in a PCR machine at 37°C for 1 h.
12. Centrifuge briefly.
13. Add 5 μl of stop buffer.
14. Mix by gently flicking the thermocycler strip.
15. Centrifuge briefly.
16. Store sample at –20°C or continue with removal of uncoupled deoxynucleotides.

3.2.2. Removal of Uncoupled Deoxynucleotides (Qiaquick PCR Purification Kit, QIAgen)

1. Transfer sample to a clean 1.5-ml Eppendorf tube.
2. Add 275 μl of binding buffer to the sample.
3. Transfer the buffered sample to the column.

4. Spin down the column for 1 min (16,000 × *g*; 13,000 rpm) and discard flow-through. Place the column back in the collection tube.
5. Add 500 µl of wash buffer (PE).
6. Spin down the column for 1 min (16,000 × *g*; 13,000 rpm) and discard flow-through. Place the column back in the collection tube (see Note 12).
7. Add another 500 µl of wash buffer (PE).
8. Spin down the column for 1 min (16,000 × *g*; 13,000 rpm) and place the column in a clean 1.5-ml—properly marked—Eppendorf tube; discard the collection tube containing the flow-through.
9. Add 25 µl of elution buffer (EB) to the membrane of the column.
10. Incubate at RT for 1 min.
11. Spin down the column for 1 min (16,000 × *g*; 13,000 rpm).
12. Add another 25 µl of elution buffer to the membrane of the column.
13. Incubate at RT for 1 min.
14. Spin down the column for 1 min (16,000 × *g*; 13,000 rpm).
15. Discard the column and close the Eppendorf tube.
16. Continue with measurement of DNA (see Note 15) or store at 4°C for short term or at −20°C for long term.

4. Notes

1. Any time a supplier is preceded by "i.e.," we suggest a possible supplier for that product as currently used in our laboratory. Whenever a specific supplier is not preceded by "i.e.," we believe that the item should be obtained from mentioned supplier to ensure the best results.
2. Unless stated otherwise, all chemicals used are of "high grade" or "molecular biology grade." The latter one is preferred above the "high grade."
3. Unless stated otherwise, all chemicals should be prepared in water that has resistivity of 18.2 MΩ cm and a total organic content of less than five parts per billion. This standard is referred to as "water" in this text.
4. Although re-review seems arbitrary, it forms a major bottleneck in our research projects, especially when large sets of samples are presented to the pathologist.

5. Hematoxylin-stained 10-μm sections must be dissected from the slide as soon as possible.
6. Macrodissection of hematoxylin stained 10-μm sections works best when sections are still slightly wet.
7. A volume of 500 μl NaSCN is sufficient for 1 cm^2 of tissue. If the tissue size exceeds 1 cm^2, the volume of NaSCN can be increased to a maximum of 1,000 μl.
8. Make sure that all the tissue is in the liquid at all times.
9. Vortex regularly during the incubation step at 55°C.
10. If the isolation procedure is started in the morning, then add another volume of 20 μl proteinase K at the end of the day. If the isolation procedure is started in the *afternoon*, then add another volume of 20 μl proteinase K in the morning of day 3 and continue in the afternoon.
11. Check if all tissue is digested; this means that the lysate should be clear. If tissue is not digested, add another volume of 20 μl proteinase K, vortex for 15 s, and leave for a minimum of 4 h.
12. Make sure that the column tip is free of droplets after spin down. If the column tip is not free of droplets, then spin down again.
13. Wash buffer 1 (AW1) must be prepared according to the manufacturer's protocol.
14. Wash buffer 2 (AW2) must be prepared according to the manufacturer's protocol.
15. DNA measurement is done with the NanoDrop-1000 and 260/280 and 260/230 values are used to determine the quality of the sample.
16. Add nuclease-free water, primers, and DNA in the same sequence as listed in the protocol.

References

1. Chin, K., DeVries, S., Fridlyand, J., Spellman, P.T., Roydasgupta, R., Kuo, W.L., et al. (2006) Genomic and transcriptional aberrations linked to breast cancer pathophysiologies. *Cancer Cell* **10**, 529–541.
2. Fridlyand, J., Snijders, A.M., Ylstra, B., Li, H., Olshen, A., Segraves, R., et al. (2006) Breast tumor copy number aberration phenotypes and genomic instability. *BMC Cancer* **6**, 96.
3. Weiss, M.M., Kuipers, E.J., Postma, C., Snijders, A.M., Siccama, I., Pinkel, D., et al. (2003) Genomic profiling of gastric cancer predicts lymph node status and survival. *Oncogene* **22**, 1872–1879.
4. Jong, K., Marchiori, E., van der Vaart, A., Chin, S.F., Carvalho, B., Tijssen, M., et al. (2007) Cross-platform array comparative genomic hybridization meta-analysis separates hematopoietic and mesenchymal from epithelial tumors. *Oncogene* **26**, 1499–1506.
5. Costa, J.L., Meijer, G., Ylstra, B., Caldas, C. (2008) Array comparative genomic hybridization copy number profiling: a new tool for translational research in solid malignancies. *Semin. Radiat. Oncol.* **18**, 98–104.
6. Gallegos Ruiz, M.I., van Cruijsen, H., Smit, E.F., Grünberg, K., Meijer, G.A., Rodriguez, J.A., et al. (2007) Genetic heterogeneity in

patients with multiple neoplastic lung lesions: a report of three cases. *J. Thorac. Oncol.* **2**, 12–21.

7. Buffart, T.E., Tijssen, M., Krugers, T., Carvalho, B., Smeets, S.J., Brakenhoff, R.H., et al. (2007) DNA quality assessment for array CGH by isothermal whole genome amplification. *Cell Oncol.* **29**, 351–359.
8. Hopman, A.H., van Hooren, E., van de Kaa, C.A., Vooijs, P.G., Ramaekers, F.C. (1991) Detection of numerical chromosome aberrations using in situ hybridization in paraffin sections of routinely processed bladder cancers. *Mod. Pathol.* **4**, 503–513.
9. Johnson, N.A., Hamoudi, R.A., Ichimura, K., Liu, L., Pearson, D.M., Collins, V.P., et al. (2006) Application of array CGH on archival formalin-fixed paraffin-embedded tissues including small numbers of microdissected cells. *Lab Invest* **86**, 968–978.
10. van de Wiel, M.A., Brosens, R., Eilers, P.H., Kumps, C., Meijer, G.A., Menten, B., et al. (2009) Smoothing waves in array CGH tumor profiles. *Bioinformatics* **25**, 1099–1104.
11. Feuk, L., Carson, A.R., Scherer, S.W. (2006) Structural variation in the human genome. *Nat. Rev. Genet.* 7, 85–97.
12. Buffart, T.E., Israeli, D., Tijssen, M., Vosse, S.J., Mrsić, A., Meijer, G.A., et al. (2008) Across array comparative genomic hybridization: a strategy to reduce reference channel hybridizations. *Genes Chromosomes. Cancer* **47**, 994–1004.

Chapter 17

Characterizing and Interpreting Genetic Variation from Personal Genome Sequencing

Anna C.V. Johansson and Lars Feuk

Abstract

Since the completion of the human genome project, there has been enormous progress in the development of novel technologies for DNA sequencing. The advent of next-generation sequencing technologies now makes it possible to sequence an entire human genome in one or a few experiments. As a consequence, several individual human genomes have now been fully sequenced, using different experimental strategies. Although the protocols differ between the various sequencing technologies, the challenges of analyzing the data, calling variation, and interpreting the results are similar for all platforms. Here, we give an overview of the human genome sequencing projects completed to date. The strategies for aligning sequence reads and extracting information about different types of genetic variation from the sequence data are discussed. Identification of structural variation, such as copy number variation and insertion-deletion variants, can be complex, and there are a plethora of algorithms and analysis tools available. We also give an overview of the challenge of interpreting the whole-genome sequence data both from a technical and clinical perspective.

Key words: Copy number variation, Indel, Personal genomics, Whole-genome sequencing

1. Introduction

The idea of sequencing the entire human genome was established long before the DNA sequence of any organism was completed. The entire sequence of the diploid human genome consists of six billion base pairs divided into 46 chromosomes; 22 autosome pairs, and two sex chromosomes. With the technologies available at the time when the idea of sequencing the entire human genome was formed, the cost and feasibility of sequencing the human genome seemed insurmountable, and was therefore dismissed by many researchers.

At a meeting in Santa Fe in 1986, researchers discussed a plan for the sequencing of the human genome, which led to the

Lars Feuk (ed.), *Genomic Structural Variants: Methods and Protocols*, Methods in Molecular Biology, vol. 838, DOI 10.1007/978-1-61779-507-7_17, © Springer Science+Business Media, LLC 2012

announcement of the human genome initiative (1). The plans were further formalized when the National Institutes of Health (NIH) joined and additional funding was secured. This led to the launch of the international consortium called the Human Genome Project (HGP) in 1990, with plans to finish the project in 15 years. The US Department of Energy and the US National Institutes of Health founded the $3-billion project and the consortium included geneticists in the USA, United Kingdom, France, Germany, Japan, China, and India. The main objectives of the effort were to determine a haploid human genomic sequence and to discover all the human genes (at that time, it was estimated that there would be approximately 100,000 genes) and make the sequences accessible for further biological studies.

In 1998, researcher Craig Venter and his company Celera launched a similar, privately funded quest. This project was intended to proceed at a faster pace and at a lower cost than the publically funded project. The competition forced the HGP to modify their strategy and timeline and improve the speed of the sequencing. In June 2000, a working draft of the human genome was released and announced jointly by the US president Bill Clinton and British Prime Minister Tony Blair. In February 2001, two parallel publications were released, one in Nature by the HGP (2) and one in Science by Celera (3), describing the methods used to produce the draft sequence and initial analysis of the sequence. An improved draft was announced in 2004, filling in gaps and improving the assembly (4).

Within the HGP, the sequencing was based on a strategy that in retrospect seems laborious and time-consuming, but was the best option considering the technology available at the start of the project. First, the genome was divided into smaller pieces, approximately 150,000 bp in length. These pieces were inserted into vectors known as bacterial artificial chromosomes (BACs) that were in turn inserted into bacteria and amplified by the bacterial DNA replication machinery. Large libraries of BAC clones were created, covering the entire human genome. Each of the BACs was sequenced individually by further dividing them and using Sanger sequencing on each fragment (5). The information from the sequenced fragments is combined to reconstruct first the individual BACs, then overlapping BACs, and eventually entire chromosomes. Celera used an alternative method known as a whole-genome shotgun sequencing, where the entire genome is divided into smaller fragments that are sequenced directly. For this strategy, the assembly process to locate and orient the sequenced fragments within the chromosome relies entirely on sequence information. Assembling the shotgun sequence data represented a challenging informatics problem. In order to complete the Celera draft assembly, the publically available data from HGP was used in the assembly process. Later, the power of the whole-genome shotgun approach to produce a high-quality draft at modest levels of coverage has been demonstrated in a thorough comparison between the different assemblies presented in 2004 (6). Many genomes of other species have since been fully sequenced using the whole-genome shotgun strategy.

1.1. Limitations of the Reference Sequence

The reference assembly has been an excellent tool for the research community and the HGP has led to countless important discoveries. However, the high-quality draft released in 2004 still had a number of sequence gaps and regions difficult to represent as a haploid consensus. The gaps in the sequence can be divided into two categories, euchromatic sequence (24.4 Mb), and highly repetitive heterochromatic regions, including centromeres and telomeres (estimated to be ~200 Mb). The heterochromatic regions were never intended to be targets of the HGP. As the work proceeded, the boundaries of the regions that could be sequenced and those that could not became more difficult to define. At the end of HGP, existing gaps were divided into those that could be closed with existing clones but might be difficult to assemble, and those that could not be traversed by cloned fragments (reviewed by Eichler et al. in 2004 (7)). Many of the remaining gaps in the sequence are due to large structural variants. Clones representing different haplotypes of a polymorphic region can often not be correctly assembled, thereby resulting in a gap (8). For some polymorphic regions of the genome, e.g. the MHC locus on chromosome 6, the reference assembly therefore contains two separate and equally correct assemblies representing the different variant haplotypes. Recently, a working group called The Genome Reference Consortium was formed to correct errors and fill existing gaps in the human reference assembly.

One of the drawbacks of the reference assembly is that it originates from a large number of DNA sources (i.e., BAC libraries were created from multiple individuals), which are anonymous (2). There are no cell lines or original biological material from these individuals, thus there is no way to perform independent validation of the sequences present in the assembly. It also means that the reference sequence is a mosaic, representing a mix of different genomes and may thus contain combinations of alleles that do not exist in the general population.

1.2. Shift to Sequencing of Individual Human Genomes

With the development of next-generation sequencing (NGS) technologies, the time and money needed to produce sequence data have decreased dramatically. We are now entering into the era of personalized genomics and entire genomes, both from humans and other organisms, are now being sequenced at an unprecedented rate. The cloning steps are now circumvented and millions of sequence reads are produced in parallel. Currently, there are four main NGS platforms available (technologies by 454, ABi SOLiD, Solexa, and Illumina) and several others in beta-testing and under development. All the existing methods are based on a combination of chemical and enzymatic reactions to sequence fragments in a highly parallel fashion and use image analysis for the read out. The sequence reads can then either be mapped to a reference sequence or used for de novo assembly. The entire genomic sequence of different individuals theoretically allows for detection of all types of structural variation in all size ranges, from single nucleotide polymorphisms (SNPs) to large copy number variations (CNVs), inversions, and translocations. A summary of the main studies up to date on whole-genome sequencing of individuals is found in Table 1 and these studies are also further described in the following sections.

1.3. The First Individual Human Genomes

The assemblies presented by the HGP (2, 4) and by Celera (3) are both based on several individuals. The first study of a complete assembled diploid human genomic sequence of an individual human (J. Craig Venter) was presented in 2007 by Levy et al. (9). They used traditional Sanger sequencing technology to perform end-sequencing of fragment libraries of known size. The long sequence reads and paired-end information allowed the authors to perform de novo assembly of the Venter genome. The resulting sequence is referred to as the Huref assembly and was based on an average 7.5-fold sequence coverage of the genome. As the Huref assembly was created independently of the National Center for Biotechnology Information (NCBI) reference assembly, variation calling in Venter's genome was based on an assembly comparison strategy. This variation discovery strategy led to the detection of 3.2 million SNPs and >800,000 insertion/deletions (indels).

The following year the first human genome of a single individual (James D. Watson) sequenced by using an NGS technology (454-sequencing) was published by Wheeler et al. (10). The method produced 250 bp long reads in a highly parallel fashion, resulting in 7.4× coverage of the genome. The sequence reads were mapped to the reference assembly and variants were called based on these alignments. Approximately the same number of SNPs (3.32 million) was reported in the Watson genome, as had been detected in the Venter genome (9). However, significantly fewer indels were found, primarily due to the shorter read length (250 bp compared to 750 bp) and the fact that single fragments, rather than sequence end-pairs, were used.

1.4. Re-Sequencing of Individuals Using NGS Methods

In November 2008, two different studies were published presenting whole-genome sequencing results obtained by the new Illumina technology for high-throughput sequencing with reads as short as 35 bp (11, 12). Traditionally, longer reads of 400–800 bp had been the standard, but the availability of high-quality reference sequences makes it possible to use shorter reads for re-sequencing and thereby allows for the development of more efficient sequencing

Table 1
Summary of whole-genome sequencing studies to date

Personal genome	Reference	Platform	Base coverage (fold)	SNPs millions (SNP calling tool)
J. Craig Venter	(9)	Automated Sanger	7.5	3.21
James D. Watson	(10)	454	7.4	3.32 (BLAT)
YRI male (NA18507)	(11)	Illumina	40.6	3.83 (MAQ) 4.14 (ELAND)
Han Chinese male	(12)	Illumina	36	3.07 (SOAP)
Korean male (AK1)	(14)	Illumina	27.8	3.45 (GSNAP)
Korean male (SJK)	(15)	Illumina	29.0	3.44 (MAQ)
Khoisan male	(16)	454; Illumina	10.2; 12.3	4.05 (Newbler, MAQ)
Desmond Tutu	(16)	Solid	30	3.62 (Corona Lite)
YRI male (NA18507)	(17)	Solid	17.9	3.87 (Corona Lite)
CEU male (NA07022)	(19)	Complete Genomics	87	3.07 (SOM)
YRI female (NA19240)	(19)	Complete Genomics	63	4.04 (SOM)
CEU male (NA20431)	(19)	Complete Genomics	45	2.90 (SOM)
Stephen R. Quake	(20)	Helicos	28	2.81 (IndexDP)
European male	(21)	Complete Genomics	88	3.66 (SOM)
European female	(21)	Complete Genomics	51	
European male	(21)	Complete Genomics	52	
European female	(21)	Complete Genomics	54	
AML female (tumor) AML female (normal)	(22)	Illumina Illumina	32.7 13.9	3.81 (MAQ) 2.92 (MAQ)
AML male (tumor) AML male (normal)	(23)	Illumina Illumina	23.3 21.3	3.46 (MAQ) 3.45 (MAQ)
Palaeo-Eskimo Saqqaq individual	(29)	Illumina	20	2.2 (SNPest)

technologies based on massive parallelization and reverse terminator chemistry. Researchers can simply map short read NGS data to known reference genomes, avoiding expensive and laborious methods required for generating long fragments necessary for de novo assembly.

With short read-lengths the computational methods to analyze the data and the accuracy of the alignment method becomes highly important. As part of one of the sequencing studies performed by Illumina themselves, (11) a new aligner called ELAND was developed. The study demonstrates the abilities of the new sequencing method by sequencing human flow-sorted X-chromosomes. The method was subsequently further scaled to determine the sequence of a male Yoruba individual from Ibadan (NA18507), Nigeria. The whole-genome sequencing yielded 40× average genome coverage and the alignments led to detection of a total of 3.83 million SNPs. The number of SNPs discovered was higher than previous individual genomes, which would be expected in an individual of African origin compared to genomes of European ancestry.

Results presented by Wang et al. around the same time (12) also utilized the Illumina sequencing technology to perform whole-genome sequencing of the diploid sequence of a healthy Han Chinese individual. Sequence was generated to a depth of 36× coverage, and 3.07 million SNPs were identified. The publications by Bentley et al. and Wang et al. were the first two studies presenting whole-genome sequencing of non-European individuals, while at the same time illustrating the ability of NGS-methods, producing very short reads, to sequence large eukaryotic genomes in the presence of a reference genome. Notably, the shorter read length makes it more difficult to identify indels, and especially insertions into the reference assembly, as compared to traditional Sanger sequencing (13).

In May 2009, Ahn et al. (14) presented whole-genome re-sequencing data for a Korean individual (AK1). The Illumina technology was used to sequence the genome to 29× coverage and relevant SNPs and indels were subsequently extracted. The results were compared to results obtained from the recently sequenced Chinese individual (12) as well as to HuRef (Venter) (9) and to Watson (9, 11, 12). As expected, the Korean individual had more variants in common with the Chinese genome than with Venter or Watson. However, there are still significant genetic differences in terms of SNPs between the two Asian individuals.

Another Korean whole-genome sequencing study (Korean individual SJK) was presented in August the same year (15). Here, a combination of whole-genome shotgun sequencing (to a coverage of 28×), targeted BAC sequencing and custom-designed high-resolution array comparative genomic hybridization (CGH) was used. The BAC clones were aimed at regions commonly affected by CNVs and sequenced to a very high coverage (151×) also using the Illumina technology. The result of the combined strategies is improved accuracy of SNP, indel and CNV detection and consequently a highly annotated sequence of the single Korean individual.

In February 2010, Schuster et al. (16) published a study presenting the complete genome sequences of a Khoisan indigenous hunter–gatherer from Namibia's Kalahari Desert and of a Bantu individual from South Africa (Desmond Tutu), together with protein-coding regions (exomes) from these two and three additional individuals from other hunter–gatherer groups from the Kalahari. These are believed to represent the oldest lineages of modern humans. Since these genomes were expected to diverge more than the previously sequenced genomes from the reference assembly, a de novo assembly was constructed of one of the Khoisan individual based on 350 bp reads to a coverage of 10.2× produced by the 454 technology. The whole genome of the Bantu individual was sequenced using SOLiD to 30× coverage. The protein coding regions for all the five participants of the study were sequenced to at least 16-fold coverage using SOLiD.

Consistent with the view that southern Africans are among the most divergent human populations, more SNPs were identified (see Table 1) compared to what has been reported in other individual human genomes. The Kalahari populations also seem more genetically different from each other, in terms of nucleotide substitutions, than typical Asians and Europeans.

1.5. Re-Sequencing Projects Focused on a Specific Sequencing Platform

A number of additional examples of whole-genome re-sequencing of single individuals have been published to illustrate the efficiency of new sequencing platforms. In 2009, McKernan et al. presented the first whole-genome sequencing using the SOLiD technology (17). An individual of Yoruba decent was sequenced to 17.9× coverage. The same sample had previously been extensively genotyped as part of the HapMap project (18) (sample NA18507) and the variants detected from the whole-genome re-sequencing could therefore be validated by comparison to earlier results. Interestingly, this was the same sample that had been previously sequenced by Bentley et al. (11) but McKernan et al. performed no comparison to the results of that study. Another recent example is a sequencing platform based on a nanoarray short-read sequencing-by-ligation technology launched by the company Complete Genomics (19). The authors claim that the method allows the sequencing of an entire genome for $4400 for sequencing consumables, as compared to >$100 million of the first human genomes and cost decreasing as low as $48,000 for the first genomes sequenced using NGS methods (9–11, 15, 17). The validity of the method was illustrated by sequencing three genomes from the HapMap (18) and the 1000 Genomes Project to 45×–87× coverage based on 31–35 base long mate-paired reads. High confidence SNPs and indels were identified and compared to previously determined variations for the same samples. A different methodological approach was used by Pushkarev et al. (20) who have performed single-molecule sequencing to characterize the genome of a male of European descent to a total of 28× coverage. The study reported 2.8 million SNPs with an error rate of less than 1% as validated by Sanger sequencing. They used the first commercial release of the technique from Helicos Biosciences, which allows the user to follow ~one billion individual molecules in real time as they are sequenced over the course of a week.

1.6. Whole-Genome Re-Sequencing of a Family

A natural next step after sequencing individuals is to sequence all members of a family, enabling inheritance analysis and more accurate identification of genes responsible for Mendelian disorders, reduction of sequencing errors, direct measurement of mutation rates and exact determination of localization for recombination events. Roach et al. (21) reported on the sequencing of a family, including two parents and their two offspring. The data was used to infer SNPs, indels, CNVs, and other types of variations. Both children carry two recessive disorders and the genes responsible for these were successfully identified. The power of sequencing families was also illustrated by showing that the number of false positives variants and the number of genes with candidate causative mutations, drops exponentially as the number of family members increase. In this study, the authors were also able to make the first direct measurement of the human mutation rate across the full genome and found, in the family that they sequenced, a rate of 1.1×10^{-8} per position per haploid genome.

1.7. Sequencing of Cancer Genomes

In cancer studies, whole-genome sequencing can be used for unbiased discovery of somatic mutations that alter protein-coding genes. If both the tumor cells and their normal counterpart from the same individual are sequenced, acquired differences can be identified. The first example including whole-genome sequencing was presented in 2008 by Ley et al. (22) who sequenced the tumor cells and normal skin cells for a female patient with acute myeloid leukemia (AML). Ten genes were discovered with acquired mutations, two of which were previously described and eight that were novel with as yet unknown function. The study demonstrate the power of whole-genome sequencing to discover novel cancer associated mutations and have since been followed by a number of studies on different forms of cancer. Some examples include a follow up on the same disorder (AML) but with higher coverage in a male patient (23), two individuals with lung cancer (24), a small-cell lung cancer cell line from the type of cancer that is usually associated with smoking to elucidate the mutational effect of this lifestyle (25), malignant melanoma, and lymphoblastoid cell lines (26).

In April 2010, the International Cancer Genome Consortium (ICGC) presented their policies and described their plans to keep track of the data relating to large-scale cancer genome studies of all major cancers in adults and children—a total of 50 different cancer types and/or subtypes (27).

1.8. Sequencing of Ancient DNA

From an evolutionary perspective, modern genomics is to a large extent limited by lack of biological material that would enable direct measures to uncover past human genetic diversity. To directly access such data, sequencing of ancient DNA samples is required. Two projects have been published up to date, mitochondrial (28) and genomic sequence of an individual from an extinct Paleo-Eskimo culture (29) and mitochondrial (30) and nuclear genome sequence of the Neanderthal (31–33).

In the first case, very high-quality DNA with limited damage was extracted from 4,000-year-old permafrost preserved hair found on Greenland. The genome was sequenced to 20-fold coverage and the resulting sequence data was of enough high quality to allow for SNP detection and population analysis. The resulting SNPs were compared to other contemporary populations and gave valuable evidence for migration patterns of populations into Greenland.

The extraction of Neanderthal DNA from bones dating back several tens of thousands of years is significantly more challenging, but the research group lead by Svante Pääbo has developed techniques to make it possible (34) and have presented complete sequence for the mitochondrial DNA (30) and a draft sequence for the nuclear Neanderthal genome (31). The current nuclear genome sequence covers approximately 60% of the entire genome. By selective re-sequencing of 14,000 protein-coding gene segments that differ between humans and chimpanzees for three Neanderthal individuals and five present day humans, valid comparisons could be made. A number of genomic regions were identified that may have been affected by positive selection in ancestral modern humans, including genes involved in metabolism and in cognitive and skeletal development. The Neanderthal genome sequence allows us to identify features unique to present-day humans relative to other, now extinct, hominins. Neanderthals share more genetic variants with present-day humans in Eurasia than with present-day humans in sub-Saharan Africa, suggesting that gene flow from Neanderthals into the ancestors of non-Africans occurred after the migration out of Africa, but before the divergence of the major Eurasian population groups.

1.9. Exome Sequencing

Massively parallel DNA sequencing technology platforms have rendered the whole-genome re-sequencing of individual humans increasingly practical, but the cost remains a key consideration. An alternative approach involves the targeted re-sequencing of a region of interest using NGS, e.g., all protein-coding sequences, which is known as exome sequencing. This strategy has been shown to work very well for disease studies, especially when searching for alleles underlying rare Mendelian disorders, since a clear majority of allelic variants known to underlie such disorders disrupt protein-coding sequences. Protein coding genes constitutes about 1% of the human genome, but harbor 85% of the mutations with large effects on disease-related traits (35).

Before the actual sequencing reaction is performed, the DNA in the target sample corresponding to the exons has to be extracted. The three most commonly used strategies for target enrichment are either PCR-based, array-based hybrid capture or in solution target capture. The latter two are the most commonly used for full exomes while PCR is primarily used for sequencing of a small number of target regions in large sample sets. The first analysis of an individual exome was performed on Craig Venter's DNA (36). However, this was done by simply extracting the exon sequences from the previous whole-genome sequencing study (9). The authors found 12,500 coding variants in the individual's genome out of which the largest fraction are nonsynonymous SNPs (10,400) and of the remaining SNPs and indels, the majority is also believed to be common and neutral to protein function.

The power of exome sequencing to detect rare disease causing variants have been illustrated in two related papers by Ng et al. (35, 37) where the exomes of a number of individuals have been sequenced. The first paper (35) focused on four unrelated individuals with a rare monogenic recessive disorder called Freeman-Sheldon syndrome. By sequencing over 30 MB of coding sequence in each of these individuals as well as in eight individuals from the HapMap project, both rare and common variants could be detected. The variants in the four affected individuals were further filtered against both dbSNP and previously sequenced HapMap exomes to exclude variants not associated with the disease, and the resulting candidate genes could be identified. However, this was only a proof-of-principle study, as the causal gene for the disorder had been identified previously. To follow up on this, the same research group used the same strategy to identify candidate genes for another disorder believed to be recessive, Millers syndrome (37). The exomes of two affected siblings as well as two unrelated affected individuals were sequenced. A candidate gene could be identified and was further validated using Sanger sequencing. This again demonstrated that exome sequencing of a small number of affected individuals is a powerful, efficient and cost-effective strategy to markedly reduce the pool of candidate genes for rare monogenic disorders and may in many cases directly lead to the identification of the disease causing mutations.

Another recent report demonstrates how exome sequencing can be used to make a genetic diagnosis for patients with a previously undiagnosed disease. In this case, the patient was a 5-month-old child with an unidentified genetic illness causing intestinal problems. Instead of a kidney-associated genetic disorder as the physician originally suspected, the identification of the disease causing mutation by exome sequencing showed that the infant suffered from congenital chloride diarrhea (38).

If previous linkage studies or some other methods have been able to narrow the region of interest, targeted re-sequencing of only that candidate region is a powerful method to identify the actual causative gene variants. An example is a large-scale re-sequencing of X-chromosome coding exons in 208 families with X-linked mental retardation (XMLR). The screen identified nine genes implicated in XMLR, but also highlighted potential difficulties in whole-genome sequencing screens, such as the fact that loss of function of 1% or more of X-chromosome genes is found in healthy controls without any obvious effect of phenotype. Other reports of NGS—sequencing of specific regions include two different studies on nonsyndromic deafness, the DFNB79 variant (39) and the DFNB82 variant (40), Fowler syndrome which is a rare prenatal lethal disorder associated with progressive destruction of central nervous system tissue (41) and in the recessive monogenic disorder AR Ataxia (42). This is a rapidly evolving field and new disease genes are reported every month. However, as the cost of sequencing drops, it may eventually be more affordable to sequence the entire genome of an individual than to specifically target the exome.

1.10. Human Pan-Genome

By definition, a pan-genome includes the core genome containing genes present in all individuals of a species, a dispensable genome containing genes present in two or more subpopulations, and finally unique genes specific to single subpopulation. Li et al. (43) have made an attempt to build a sequence map of the human genome by comparing de novo assemblies from one Chinese (12) and one Yoruba individual with the NCBI reference genome. About 5 Mb of novel sequence was found in each of the two individual genomes and a total of 19–40 Mb of sequence is estimated to be missing from the human reference. These results indicate that complete sequencing and assembly of personal genomes may result in a larger number of various types of genetic variation to be identified. Similar data was previously shown in a comparison of the Celera assembly and the reference assembly (44), as well as in the sequencing and assembly of the Venter genome. It has also been shown that these differences are due both to assembly problems and polymorphism, with the latter explaining a large fraction of the sequences differences (45). Hence, it is important to carry out de novo assembly on more human genomes to discover the common polymorphic sequences in the human population and to obtain a complete human pan-genome.

2. Methods

2.1. Identification of Variation in the Human Genome

A large number of different strategies have been developed to identify structural genetic variation from sequence data. This ranges from direct analysis of the sequence of a single human genome, pair-wise comparison between individual genomes and analysis of data from large numbers of samples. Here, we go through some of the major approaches that have been used to date.

2.2. Assembly Comparison

The first whole-genome sequencing studies were conducted with the aim to build a complete assembly. These assemblies could then be compared with previous genome assemblies to identify structural variations in different size ranges. In a comparison of an assembly from Celera with the NCBI Build 35 assembly a large number of variations could be identified, including 1.5 million SNPs, a total of 22,167 bp representing indels<= 10 bp and 23, 859, 805 bp of unmatched sequence representing larger differences between the assemblies (44). Such a comparison was also included in the first individual whole-genome sequencing study, where a de novo assembly of the genomes was build (9, 11, 12).

2.3. Importance of Mapping

The first step in a re-sequencing study, after receiving the results from the sequencer, is to accurately map the many millions of short reads to a reference assembly. Since all subsequent analysis of the sequence data is based on accurate mapping and also accurate statistics from the mapping, this is a very crucial step. A number of different algorithms have been developed aiming to optimize speed and precision. Some mapping tools are specific to a certain NGS platform while others are more generally applicable. There are a few important points to take into account before determining which alignment algorithm to chose. How well does it handle data from the technology that produced the data? Which output formats are available and which format is desired for downstream analysis? If the aim is to identify short indels, it is important to know how well the alignment tool handles gaps in the sequence.

2.4. Identification of SNPs and Indels SNPs and Short Indels

Most mapping strategies allow for limited number of mismatches in the aligned reads. The challenge of an SNP detecting algorithm is to separate which of these mismatches that are the result of sequencing errors and which ones are true SNPs. Most mapping procedures have genotype calling as a more or less integrative part of the mapping procedure. An example is the mapping algorithm MAQ (mapping with qualities) (46) that takes the mapping quality of each individual read into account and outputs both mapped reads, indels, and SNPs. Other methods include the mapping algorithm Mosaik which uses Gigabayes for SNP detection (described and used in (47)), SAMtools that can detect SNPs and indels from different types of mapping data (48) and the efficient mapping algorithm SOAP that also detects SNPs (49, 50).

2.5. Strategies to Identify Structural Variants

From sequence data there are three main approaches to find indels and CNVs, mate-pair mapping, read-depth analysis, and breakpoint mapping. Most platforms have the ability to sequence the two ends of fragments of predetermined size. One benefit of sequencing the ends of fragments of known size is that there is a prior assumption of the distance at which the two end sequences should map to the reference assembly. Any locus where the clusters of paired-end sequences map to the reference genome at distances different from the expected is an indication of an insertion/deletion in the donor genome. Read-depth analysis is based on finding fluctuations in the read coverage, which can be indicative of deletions and duplications. A third option to identify variants is to map sequenced reads to known breakpoints in the reference. An overview of these three strategies is shown in Fig. 1, and described in more detail below.

2.5.1. Mate-Pair Mapping

In a mate-pair run, the ends of DNA fragments of known size are sequenced while the information about which pairs that belongs together is retained. The sequenced ends are then mapped to the reference assembly, which gives ample possibilities to locate variations. Depending on the library preparation method, the read-pair can be the sequenced ends of a rather short segment, 200–600 bp, often referred to as paired-end reads, or the reads can be located

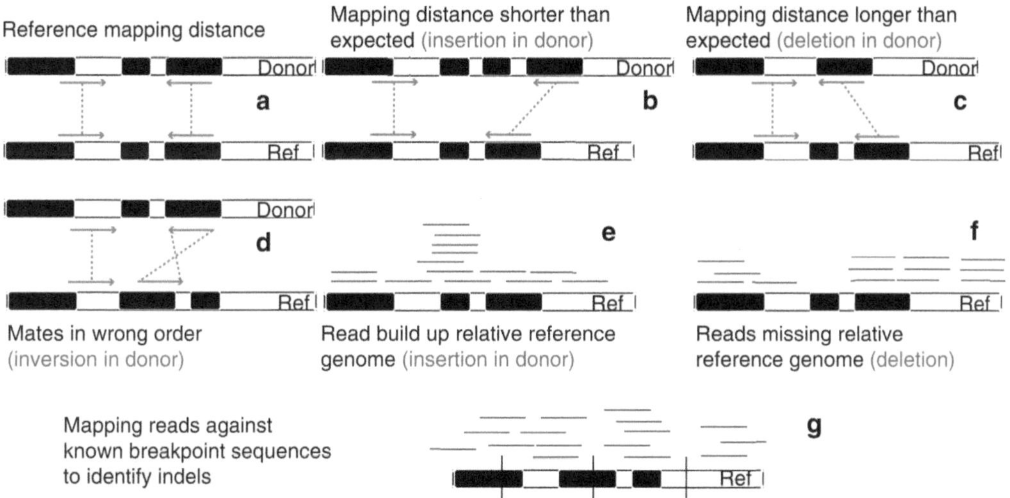

Fig. 1. Different methods to detect copy number variations in NGS-data. (**a**–**d**) insertions, deletions, and inversions from inconsistent mapping of mate-pairs. (**e**, **f**) insertions and deletions detected by larger/smaller read depth than expected and (**g**) variants found by using read mapping against breakpoint-libraries.

further apart, 2–10 kb, separated by a size selected insert. The latter are commonly referred to as mate-pairs, but the terms are often used interchangeable. The same general approach can be used for detecting variants independent of the distance between the two reads.

If the two reads in a pair map at a distance closer together than expected or further apart than expected, this could be indicative of an insertion and deletion, respectively, in the donor genome (Fig. 1b, c). Inversions can be efficiently identified since one of the ends in that case is found to map in the wrong orientation relative the reference genome (Fig. 1d). The resolution of the variations that can be found depends on the expected distance between the sequenced mates, the width of the size distribution of insert lengths and to some extent also the coverage of the re-sequenced genomes. An insertion in the donor genome that is larger than the expected insert size can never be found by using paired mates, but a deletion can, in theory, be very large and still be detected. For all sequencing platforms, there is always a distribution of the insert lengths between the sequenced mates. As a limit of what is considered as a normal insert length, a rule of thumb is to use two standard deviations from the average. If there is a wide distribution, two standard deviations will correspond to a rather large span in terms of number of nucleotides and thus limit the resolution of the analysis. The coverage of reads in the region also needs to be sufficiently high to be able to call variations with high confidence.

The first sequence-based study that used mate-pair data to find CNVs in the human genome was published in 2005 by Tuzun et al. (51). They used paired-end clones from a fosmid library mapped to the reference sequence to identify variations. By detecting sites where the mapping was inconsistent from the expected, i.e., the ends where closer than 32 kb or more distant than 48 kb or in the wrong order, a total of 297 inserts, deletions and inversions >8 kb were detected. The same method was used by Kidd et al. in 2008 (52), but covering eight individuals and refining the locations of a total of 1,695 structural variation >8 kb.

Statistically robust methods have been developed to perform the analysis on mate-pair data with different expected insert lengths. The number of methods with similar but slightly different applications and prerequisites is growing fast and the task of producing a comprehensive list is almost impossible. Some examples of algorithms include a method called Modil that can detect indels with high precision (53), PEMer (Paired-end Mapper) for mapping variations at high confidence and then analyze them with a built in database (54), BreakDancer that uses mate-pair data to detect variations in all size ranges (55) and NovelSeq that can discover the content and location of long sequence insertions using paired-end sequencing data (56).

2.5.2. Read-Depth

Individual reads can either be fragments where each read is independent, or they can come from mate-pair runs. When mapped to a reference assembly, both types can be used to detect structural variations from the position-dependent read depth. A build up of reads in a region is an indication of extra copies of that region and no reads in a region indicate a deletion in the donor. The read depth of mapped reads varies over the genome due to gaps in the assembly, repetitive sequences, uneven base composition, as well as technical difficulties to produce a totally unbiased set of reads from the donor. This is in some respects analogous to having absolute microarray intensity data from a single sample, and proper interpretation of the data requires a reference sample. When the data is normalized against a reference, nonrandom biases from both the experiment and the sequence read mapping are minimized. The approach can be used to detect very large events, but works less well for detecting breakpoints than mate-pair-based methods. The strength of an event is determined by the coverage and the size of the event, i.e. the larger the event, the stronger the support. Campbell et al. (24) was the first to use this strategy events to detect CNVs between tumor and healthy samples of the same individuals.

Different methods have been developed that compares two samples in statistically robust manner to minimize the number of false positives and to assign confidence scores to the candidate regions. A fast, conceptually very simple and accurate method is called CNV-seq (57), where the input is the number of mapped reads as function of coordinate in the reference. This makes CNV-seq easy to use independent of the mapping algorithm used for the sequencing project.

2.5.3. Break-Point Detection

A helpful step toward understanding of the mechanisms underlying the formation of new structural variants is to characterize the exact breakpoints and the regions surrounding the variant (58). Libraries of breakpoints for common variants can also be used to identify variations from NGS data, either as a validation step to check if the candidate region is already known or by mapping short reads against the breakpoint sequence directly (59, 60). The strategy is based on taking all reads that did not map onto the reference assembly, and instead map them directly against the library of breakpoint sequences. If the reads align across the breakpoint, the presence of that variant in the donor genome can be inferred.

An alternative approach utilizing the breakpoint concept is based on the fact that a short read overlapping the breakpoint of an insertion or deletion will not be possible to map to the reference sequence. By splitting the sequence read, the part that is common between the donor and the reference can be mapped and the breakpoint of the indel identified. The approach has been implemented in two methods. The Pindel algorithm (61) uses mate-pair information to identify the expected mapping position for a read. The SplitSeek tool (62) was originally developed as a method to find splice junctions in RNA-Seq, but it works equally well for defining indel junctions in genomic data.

3. Discussion

3.1. Sequencing Compared to Array-Based Strategies

Traditionally, CNVs and large indels have been mapped using various array-based strategies, either CGH or SNP arrays (63). Compared to NGS approaches, mapping variants using arrays is significantly faster and more affordable. However, arrays also have limitations that can be overcome by using sequence-based strategies. The main benefit of using sequence-based detection of variations is that it is mainly an unbiased approach that can find variations in all size ranges as well as balanced rearrangements, such as translocations and inversions. The sequence data is also less biased in terms of genome coverage while arrays only provide information about the sequences corresponding to the probes present on the array. The array designs are also based on the reference genome while sequencing equally well covers sequences not present in the reference assembly. However, usually the NGS reads are mapped against a reference sequence, which in some respect limits the possibility to find variations in segments not covered by the reference. It is expected that 19–40 Mb of sequence is missing for the human reference (43), and the number is most likely much larger for less complete assemblies. Another benefit of sequencing is that it works better for differentiating exact copy numbers when there are many copies. Arrays signal intensities are not linear with increasing copy number and can get saturated while NGS data has a more direct dose–response (64). The main limitation of sequencing studies is still the cost and the time aspect, which is also where array studies are still the better choice—especially for large-scale studies, including a large number of individuals. The drawback with arrays is that you can only find what you are looking for, i.e., only unbalanced variations (gains and losses) within the regions covered by the probes on the array can be identified. Even with the new arrays with many millions of probes, there is still a significant limitation in the resolution compared to a standard NGS run.

3.2. Clinical Interpretation of Whole-Genome Sequencing Data

Whole-genome sequences of identified human individuals have now been described in a number of studies. The main aim of these studies has been proof-of-principle for the approach as such, rather than attempting to identify specific genotype–phenotype correlations. However, going forward whole-genome sequencing may become just another test that the doctor can order, and direct-to-consumer (DTC) genetic tests are already widely available. The question is what information we can extract from a whole-genome sequence, and what we have learned from the complete genomes sequenced to date.

One strategy when sequencing patients is to focus the analysis only on variants that may be relevant for the disease in question. Professor James Lupski, who is one of the most prominent figures in clinical genetics today, had his own genome sequenced to find the cause of his Charcot-Marie-Tooth Neuropathy (65). The analysis uncovered 9,069 nonsynonymous variants. A comparison of these variants with known disease mutations showed that 21 variants were previously described as causative of Mendelian disease. However, 16 of these were heterozygous, thereby corresponding to the expected load of recessive disease causing variants. The remaining five must be either erroneously defined as disease causing, or have reduced penetrance. By specifically examining genes linked to neuropathy, coding variants were detected, and the patient was found to be compound heterozygous for one of these (*SH3TC2*). Genotyping of these variants in additional family members clearly showed that this was the causative gene.

Another question is what information that can be extracted from sequencing a healthy individual, with no specific disease or phenotype in mind. As mentioned above, it is possible to extract information about carrier status for recessive and incompletely penetrant dominant Mendelian disorders. In addition, there are more than 1,000 DNA variants identified that have been associated with diseases or traits (66), and more than 650 drug response phenotypes (67). The majority of these have a very limited impact on phenotype, but can be used to calculate absolute risk for developing a disease or carrying a specific trait. The majority of these variants are targeted in genotyping services offered by DTC companies, such as 23 and Me and Navigenics (68). Although these variants can be used to get a general picture of the susceptibility risk for certain disorders, the odds ratio of most of these variants is very low and therefore has poor predictive value at the level of the individual. Better predictions can be made if the genomic data is incorporated with other data, such as clinical characteristics, diet, and exercise, and it was recently shown that a thorough analysis does provide limited, but still very useful, clinical information for the individual patient (69). Clearly, we are only at the beginning of a rapidly emerging field, and as more genomes are sequenced and more rare variants are detected, the interpretation of the data only gets better.

3.3. Future Perspectives

The first individual human genome was published in 2007. In the few years since then there has been an enormous increase in sequencing capacity. Any reasonably funded research group now has the ability to submit a sample for whole-genome sequencing. The major challenge going forward is therefore not the technology and sequencing per se, but rather the analysis and interpretation of the data. The methods for variation discovery are still being perfected, and there is a lack of databases that gather information on rare variants to facilitate analysis and interpretation. One major project aiming to provide the appropriate control data for whole-genome sequencing is the 1000 Genomes Project (http://www.1000Genomes.org). The current goal of this project is to sequence 2,500 genomes from 27 populations. The data is made available to the research community and provides an important resource toward understanding whole-genome sequence data. If the genomics field continues to develop at the pace of the last few years, we should have data for tens of thousands of genomes over the next 2–3 years.

The major challenge of personal genome sequencing going forward is to interpret the data in a way that is clinically useful to the individual. Currently, we can provide answers to individuals suffering from Mendelian disorders and that is a great start. However, we can say very little about complex traits. In every genome sequenced, there are hundreds of nonsense mutations and missense mutations predicted to be damaging and we have no idea how to interpret that data as the knowledge of the function of most genes is still very limited. As technology gets better and whole-genome sequencing gets even more affordable, the increase in data and collections of rare variants facilitates our interpretation. There is no doubt that whole-genome sequencing will be offered to consumers who are willing to pay in the near future, but the step to make genome sequencing another routine analysis in the doctor's office is still far away. That requires developments not only in sequence data production and interpretation, but also in ethics and counseling.

References

1. DeLisi, C. (2008) Meetings that changed the world: Santa Fe 1986: Human genome baby-steps, *Nature 455*, 876–877.
2. Lander, E. S., Linton, L. M., Birren, B., Nusbaum, C., Zody, M. C., Baldwin, J., Devon, K., Dewar, K., Doyle, M., FitzHugh, W., Funke, R., Gage, D., Harris, K., Heaford, A., Howland, J., Kann, L., Lehoczky, J., LeVine, R., McEwan, P., McKernan, K., Meldrim, J., Mesirov, J. P., Miranda, C., Morris, W., Naylor, J., Raymond, C., Rosetti, M., Santos, R., Sheridan, A., Sougnez, C., Stange-Thomann, N., Stojanovic, N., Subramanian, A., Wyman, D., Rogers, J., Sulston, J., Ainscough, R., Beck, S., Bentley, D., Burton, J., Clee, C., Carter, N., Coulson, A., Deadman, R., Deloukas, P., Dunham, A., Dunham, I., Durbin, R., French, L., Grafham, D., Gregory, S., Hubbard, T., Humphray, S., Hunt, A., Jones, M., Lloyd, C., McMurray, A., Matthews, L., Mercer, S., Milne, S., Mullikin, J. C., Mungall, A., Plumb, R., Ross, M., Shownkeen, R., Sims, S., Waterston, R. H., Wilson, R. K., Hillier, L. W., McPherson, J. D., Marra, M. A., Mardis, E. R., Fulton, L. A., Chinwalla, A. T., Pepin, K. H., Gish, W. R., Chissoe, S. L., Wendl, M. C., Delehaunty, K. D., Miner, T. L., Delehaunty, A., Kramer, J. B., Cook, L. L., Fulton, R. S.,

Johnson, D. L., Minx, P. J., Clifton, S. W., Hawkins, T., Branscomb, E., Predki, P., Richardson, P., Wenning, S., Slezak, T., Doggett, N., Cheng, J. F., Olsen, A., Lucas, S., Elkin, C., Uberbacher, E., Frazier, M., Gibbs, R. A., Muzny, D. M., Scherer, S. E., Bouck, J. B., Sodergren, E. J., Worley, K. C., Rives, C. M., Gorrell, J. H., Metzker, M. L., Naylor, S. L., Kucherlapati, R. S., Nelson, D. L., Weinstock, G. M., Sakaki, Y., Fujiyama, A., Hattori, M., Yada, T., Toyoda, A., Itoh, T., Kawagoe, C., Watanabe, H., Totoki, Y., Taylor, T., Weissenbach, J., Heilig, R., Saurin, W., Artiguenave, F., Brottier, P., Bruls, T., Pelletier, E., Robert, C., Wincker, P., Smith, D. R., Doucette-Stamm, L., Rubenfield, M., Weinstock, K., Lee, H. M., Dubois, J., Rosenthal, A., Platzer, M., Nyakatura, G., Taudien, S., Rump, A., Yang, H., Yu, J., Wang, J., Huang, G., Gu, J., Hood, L., Rowen, L., Madan, A., Qin, S., Davis, R. W., Federspiel, N. A., Abola, A. P., Proctor, M. J., Myers, R. M., Schmutz, J., Dickson, M., Grimwood, J., Cox, D. R., Olson, M. V., Kaul, R., Shimizu, N., Kawasaki, K., Minoshima, S., Evans, G. A., Athanasiou, M., Schultz, R., Roe, B. A., Chen, F., Pan, H., Ramser, J., Lehrach, H., Reinhardt, R., McCombie, W. R., de la Bastide, M., Dedhia, N., Blocker, H., Hornischer, K., Nordsiek, G., Agarwala, R., Aravind, L., Bailey, J. A., Bateman, A., Batzoglou, S., Birney, E., Bork, P., Brown, D. G., Burge, C. B., Cerutti, L., Chen, H. C., Church, D., Clamp, M., Copley, R. R., Doerks, T., Eddy, S. R., Eichler, E. E., Furey, T. S., Galagan, J., Gilbert, J. G., Harmon, C., Hayashizaki, Y., Haussler, D., Hermjakob, H., Hokamp, K., Jang, W., Johnson, L. S., Jones, T. A., Kasif, S., Kaspryzk, A., Kennedy, S., Kent, W. J., Kitts, P., Koonin, E. V., Korf, I., Kulp, D., Lancet, D., Lowe, T. M., McLysaght, A., Mikkelsen, T., Moran, J. V., Mulder, N., Pollara, V. J., Ponting, C. P., Schuler, G., Schultz, J., Slater, G., Smit, A. F., Stupka, E., Szustakowski, J., Thierry-Mieg, D., Thierry-Mieg, J., Wagner, L., Wallis, J., Wheeler, R., Williams, A., Wolf, Y. I., Wolfe, K. H., Yang, S. P., Yeh, R. F., Collins, F., Guyer, M. S., Peterson, J., Felsenfeld, A., Wetterstrand, K. A., Patrinos, A., Morgan, M. J., de Jong, P., Catanese, J. J., Osoegawa, K., Shizuya, H., Choi, S., and Chen, Y. J. (2001) Initial sequencing and analysis of the human genome, *Nature 409*, 860–921.

3. Venter, J. C., Adams, M. D., Myers, E. W., Li, P. W., Mural, R. J., Sutton, G. G., Smith, H. O., Yandell, M., Evans, C. A., Holt, R. A., Gocayne, J. D., Amanatides, P., Ballew, R. M., Huson, D. H., Wortman, J. R., Zhang, Q., Kodira, C. D., Zheng, X. H., Chen, L., Skupski, M., Subramanian, G., Thomas, P. D., Zhang, J., Gabor Miklos, G. L., Nelson, C., Broder, S., Clark, A. G., Nadeau, J., McKusick, V. A., Zinder, N., Levine, A. J., Roberts, R. J., Simon, M., Slayman, C., Hunkapiller, M., Bolanos, R., Delcher, A., Dew, I., Fasulo, D., Flanigan, M., Florea, L., Halpern, A., Hannenhalli, S., Kravitz, S., Levy, S., Mobarry, C., Reinert, K., Remington, K., Abu-Threideh, J., Beasley, E., Biddick, K., Bonazzi, V., Brandon, R., Cargill, M., Chandramouliswaran, I., Charlab, R., Chaturvedi, K., Deng, Z., Di Francesco, V., Dunn, P., Eilbeck, K., Evangelista, C., Gabrielian, A. E., Gan, W., Ge, W., Gong, F., Gu, Z., Guan, P., Heiman, T. J., Higgins, M. E., Ji, R. R., Ke, Z., Ketchum, K. A., Lai, Z., Lei, Y., Li, Z., Li, J., Liang, Y., Lin, X., Lu, F., Merkulov, G. V., Milshina, N., Moore, H. M., Naik, A. K., Narayan, V. A., Neelam, B., Nusskern, D., Rusch, D. B., Salzberg, S., Shao, W., Shue, B., Sun, J., Wang, Z., Wang, A., Wang, X., Wang, J., Wei, M., Wides, R., Xiao, C., Yan, C., Yao, A., Ye, J., Zhan, M., Zhang, W., Zhang, H., Zhao, Q., Zheng, L., Zhong, F., Zhong, W., Zhu, S., Zhao, S., Gilbert, D., Baumhueter, S., Spier, G., Carter, C., Cravchik, A., Woodage, T., Ali, F., An, H., Awe, A., Baldwin, D., Baden, H., Barnstead, M., Barrow, I., Beeson, K., Busam, D., Carver, A., Center, A., Cheng, M. L., Curry, L., Danaher, S., Davenport, L., Desilets, R., Dietz, S., Dodson, K., Doup, L., Ferriera, S., Garg, N., Gluecksmann, A., Hart, B., Haynes, J., Haynes, C., Heiner, C., Hladun, S., Hostin, D., Houck, J., Howland, T., Ibegwam, C., Johnson, J., Kalush, F., Kline, L., Koduru, S., Love, A., Mann, F., May, D., McCawley, S., McIntosh, T., McMullen, I., Moy, M., Moy, L., Murphy, B., Nelson, K., Pfannkoch, C., Pratts, E., Puri, V., Qureshi, H., Reardon, M., Rodriguez, R., Rogers, Y. H., Romblad, D., Ruhfel, B., Scott, R., Sitter, C., Smallwood, M., Stewart, E., Strong, R., Suh, E., Thomas, R., Tint, N. N., Tse, S., Vech, C., Wang, G., Wetter, J., Williams, S., Williams, M., Windsor, S., Winn-Deen, E., Wolfe, K., Zaveri, J., Zaveri, K., Abril, J. F., Guigo, R., Campbell, M. J., Sjolander, K. V., Karlak, B., Kejariwal, A., Mi, H., Lazareva, B., Hatton, T., Narechania, A., Diemer, K., Muruganujan, A., Guo, N., Sato, S., Bafna, V., Istrail, S., Lippert, R., Schwartz, R., Walenz, B., Yooseph, S., Allen, D., Basu, A., Baxendale, J., Blick, L., Caminha, M., Carnes-Stine, J., Caulk, P., Chiang, Y. H., Coyne, M., Dahlke, C., Mays, A., Dombroski, M., Donnelly, M., Ely, D., Esparham, S., Fosler, C., Gire, H., Glanowski, S., Glasser, K., Glodek, A., Gorokhov, M., Graham, K., Gropman, B., Harris, M., Heil, J., Henderson, S., Hoover, J., Jennings, D., Jordan, C., Jordan, J., Kasha, J., Kagan, L., Kraft, C., Levitsky, A., Lewis, M.,

Liu, X., Lopez, J., Ma, D., Majoros, W., McDaniel, J., Murphy, S., Newman, M., Nguyen, T., Nguyen, N., Nodell, M., Pan, S., Peck, J., Peterson, M., Rowe, W., Sanders, R., Scott, J., Simpson, M., Smith, T., Sprague, A., Stockwell, T., Turner, R., Venter, E., Wang, M., Wen, M., Wu, D., Wu, M., Xia, A., Zandieh, A., and Zhu, X. (2001) The sequence of the human genome, *Science 291*, 1304–1351.

4. Consortium, I. H. G. S. (2004) Finishing the euchromatic sequence of the human genome, *Nature 431*, 931–945.
5. Sanger, F., Nicklen, S., and Coulson, A. R. (1977) DNA sequencing with chain-terminating inhibitors, *Proc Natl Acad Sci USA 74*, 5463–5467.
6. Istrail, S., Sutton, G. G., Florea, L., Halpern, A. L., Mobarry, C. M., Lippert, R., Walenz, B., Shatkay, H., Dew, I., Miller, J. R., Flanigan, M. J., Edwards, N. J., Bolanos, R., Fasulo, D., Halldorsson, B. V., Hannenhalli, S., Turner, R., Yooseph, S., Lu, F., Nusskern, D. R., Shue, B. C., Zheng, X. H., Zhong, F., Delcher, A. L., Huson, D. H., Kravitz, S. A., Mouchard, L., Reinert, K., Remington, K. A., Clark, A. G., Waterman, M. S., Eichler, E. E., Adams, M. D., Hunkapiller, M. W., Myers, E. W., and Venter, J. C. (2004) Whole-genome shotgun assembly and comparison of human genome assemblies, *Proc Natl Acad Sci USA 101*, 1916–1921.
7. Eichler, E. E., Clark, R. A., and She, X. (2004) An assessment of the sequence gaps: unfinished business in a finished human genome, *Nat Rev Genet 5*, 345–354.
8. Bovee, D., Zhou, Y., Haugen, E., Wu, Z., Hayden, H. S., Gillett, W., Tuzun, E., Cooper, G. M., Sampas, N., Phelps, K., Levy, R., Morrison, V. A., Sprague, J., Jewett, D., Buckley, D., Subramaniam, S., Chang, J., Smith, D. R., Olson, M. V., Eichler, E. E., and Kaul, R. (2008) Closing gaps in the human genome with fosmid resources generated from multiple individuals, *Nat Genet 40*, 96–101.
9. Levy, S., Sutton, G., Ng, P. C., Feuk, L., Halpern, A. L., Walenz, B. P., Axelrod, N., Huang, J., Kirkness, E. F., Denisov, G., Lin, Y., MacDonald, J. R., Pang, A. W., Shago, M., Stockwell, T. B., Tsiamouri, A., Bafna, V., Bansal, V., Kravitz, S. A., Busam, D. A., Beeson, K. Y., McIntosh, T. C., Remington, K. A., Abril, J. F., Gill, J., Borman, J., Rogers, Y. H., Frazier, M. E., Scherer, S. W., Strausberg, R. L., and Venter, J. C. (2007) The diploid genome sequence of an individual human, *PLoS Biol 5*, e254.
10. Wheeler, D. A., Srinivasan, M., Egholm, M., Shen, Y., Chen, L., McGuire, A., He, W., Chen, Y. J., Makhijani, V., Roth, G. T., Gomes, X., Tartaro, K., Niazi, F., Turcotte, C. L., Irzyk, G. P., Lupski, J. R., Chinault, C., Song, X. Z., Liu, Y., Yuan, Y., Nazareth, L., Qin, X., Muzny, D. M., Margulies, M., Weinstock, G. M., Gibbs, R. A., and Rothberg, J. M. (2008) The complete genome of an individual by massively parallel DNA sequencing, *Nature 452*, 872–876.
11. Bentley, D. R., Balasubramanian, S., Swerdlow, H. P., Smith, G. P., Milton, J., Brown, C. G., Hall, K. P., Evers, D. J., Barnes, C. L., Bignell, H. R., Boutell, J. M., Bryant, J., Carter, R. J., Keira Cheetham, R., Cox, A. J., Ellis, D. J., Flatbush, M. R., Gormley, N. A., Humphray, S. J., Irving, L. J., Karbelashvili, M. S., Kirk, S. M., Li, H., Liu, X., Maisinger, K. S., Murray, L. J., Obradovic, B., Ost, T., Parkinson, M. L., Pratt, M. R., Rasolonjatovo, I. M., Reed, M. T., Rigatti, R., Rodighiero, C., Ross, M. T., Sabot, A., Sankar, S. V., Scally, A., Schroth, G. P., Smith, M. E., Smith, V. P., Spiridou, A., Torrance, P. E., Tzonev, S. S., Vermaas, E. H., Walter, K., Wu, X., Zhang, L., Alam, M. D., Anastasi, C., Aniebo, I. C., Bailey, D. M., Bancarz, I. R., Banerjee, S., Barbour, S. G., Baybayan, P. A., Benoit, V. A., Benson, K. F., Bevis, C., Black, P. J., Boodhun, A., Brennan, J. S., Bridgham, J. A., Brown, R. C., Brown, A. A., Buermann, D. H., Bundu, A. A., Burrows, J. C., Carter, N. P., Castillo, N., Chiara, E. C. M., Chang, S., Neil Cooley, R., Crake, N. R., Dada, O. O., Diakoumakos, K. D., Dominguez-Fernandez, B., Earnshaw, D. J., Egbujor, U. C., Elmore, D. W., Etchin, S. S., Ewan, M. R., Fedurco, M., Fraser, L. J., Fuentes Fajardo, K. V., Scott Furey, W., George, D., Gietzen, K. J., Goddard, C. P., Golda, G. S., Granieri, P. A., Green, D. E., Gustafson, D. L., Hansen, N. F., Harnish, K., Haudenschild, C. D., Heyer, N. I., Hims, M. M., Ho, J. T., Horgan, A. M., Hoschler, K., Hurwitz, S., Ivanov, D. V., Johnson, M. Q., James, T., Huw Jones, T. A., Kang, G. D., Kerelska, T. H., Kersey, A. D., Khrebtukova, I., Kindwall, A. P., Kingsbury, Z., Kokko-Gonzales, P. I., Kumar, A., Laurent, M. A., Lawley, C. T., Lee, S. E., Lee, X., Liao, A. K., Loch, J. A., Lok, M., Luo, S., Mammen, R. M., Martin, J. W., McCauley, P. G., McNitt, P., Mehta, P., Moon, K. W., Mullens, J. W., Newington, T., Ning, Z., Ling Ng, B., Novo, S. M., O'Neill, M. J., Osborne, M. A., Osnowski, A., Ostadan, O., Paraschos, L. L., Pickering, L., Pike, A. C., Chris Pinkard, D., Pliskin, D. P., Podhasky, J., Quijano, V. J., Raczy, C., Rae, V. H., Rawlings, S. R., Chiva Rodriguez, A., Roe, P. M., Rogers, J., Rogert Bacigalupo, M. C., Romanov, N., Romieu, A., Roth, R. K., Rourke, N. J., Ruediger, S. T., Rusman, E., Sanches-Kuiper, R. M., Schenker,

M. R., Seoane, J. M., Shaw, R. J., Shiver, M. K., Short, S. W., Sizto, N. L., Sluis, J. P., Smith, M. A., Ernest Sohna Sohna, J., Spence, E. J., Stevens, K., Sutton, N., Szajkowski, L., Tregidgo, C. L., Turcatti, G., Vandevondele, S., Verhovsky, Y., Virk, S. M., Wakelin, S., Walcott, G. C., Wang, J., Worsley, G. J., Yan, J., Yau, L., Zuerlein, M., Mullikin, J. C., Hurles, M. E., McCooke, N. J., West, J. S., Oaks, F. L., Lundberg, P. L., Klenerman, D., Durbin, R., and Smith, A. J. (2008) Accurate whole human genome sequencing using reversible terminator chemistry, *Nature 456*, 53–59.

12. Wang, J., Wang, W., Li, R., Li, Y., Tian, G., Goodman, L., Fan, W., Zhang, J., Li, J., Guo, Y., Feng, B., Li, H., Lu, Y., Fang, X., Liang, H., Du, Z., Li, D., Zhao, Y., Hu, Y., Yang, Z., Zheng, H., Hellmann, I., Inouye, M., Pool, J., Yi, X., Zhao, J., Duan, J., Zhou, Y., Qin, J., Ma, L., Li, G., Zhang, G., Yang, B., Yu, C., Liang, F., Li, W., Li, S., Ni, P., Ruan, J., Li, Q., Zhu, H., Liu, D., Lu, Z., Li, N., Guo, G., Ye, J., Fang, L., Hao, Q., Chen, Q., Liang, Y., Su, Y., San, A., Ping, C., Yang, S., Chen, F., Li, L., Zhou, K., Ren, Y., Yang, L., Gao, Y., Yang, G., Li, Z., Feng, X., Kristiansen, K., Wong, G. K., Nielsen, R., Durbin, R., Bolund, L., Zhang, X., and Yang, H. (2008) The diploid genome sequence of an Asian individual, *Nature 456*, 60–65.
13. Pang, A. W., MacDonald, J. R., Pinto, D., Wei, J., Rafiq, M. A., Conrad, D. F., Park, H., Hurles, M. E., Lee, C., Venter, J. C., Kirkness, E. F., Levy, S., Feuk, L., and Scherer, S. W. (2010) Towards a comprehensive structural variation map of an individual human genome, *Genome Biol 11*, R52.
14. Ahn, S. M., Kim, T. H., Lee, S., Kim, D., Ghang, H., Kim, D. S., Kim, B. C., Kim, S. Y., Kim, W. Y., Kim, C., Park, D., Lee, Y. S., Kim, S., Reja, R., Jho, S., Kim, C. G., Cha, J. Y., Kim, K. H., Lee, B., Bhak, J., and Kim, S. J. (2009) The first Korean genome sequence and analysis: full genome sequencing for a socio-ethnic group, *Genome Res 19*, 1622–1629.
15. Kim, J. I., Ju, Y. S., Park, H., Kim, S., Lee, S., Yi, J. H., Mudge, J., Miller, N. A., Hong, D., Bell, C. J., Kim, H. S., Chung, I. S., Lee, W. C., Lee, J. S., Seo, S. H., Yun, J. Y., Woo, H. N., Lee, H., Suh, D., Kim, H. J., Yavartanoo, M., Kwak, M., Zheng, Y., Lee, M. K., Kim, J. Y., Gokcumen, O., Mills, R. E., Zaranek, A. W., Thakuria, J., Wu, X., Kim, R. W., Huntley, J. J., Luo, S., Schroth, G. P., Wu, T. D., Kim, H., Yang, K. S., Park, W. Y., Church, G. M., Lee, C., Kingsmore, S. F., and Seo, J. S. (2009) A highly annotated whole-genome sequence of a Korean individual, *Nature 460*, 1011–1015.
16. Schuster, S. C., Miller, W., Ratan, A., Tomsho, L. P., Giardine, B., Kasson, L. R., Harris, R. S., Petersen, D. C., Zhao, F., Qi, J., Alkan, C., Kidd, J. M., Sun, Y., Drautz, D. I., Bouffard, P., Muzny, D. M., Reid, J. G., Nazareth, L. V., Wang, Q., Burhans, R., Riemer, C., Wittekindt, N. E., Moorjani, P., Tindall, E. A., Danko, C. G., Teo, W. S., Buboltz, A. M., Zhang, Z., Ma, Q., Oosthuysen, A., Steenkamp, A. W., Oostuisen, H., Venter, P., Gajewski, J., Zhang, Y., Pugh, B. F., Makova, K. D., Nekrutenko, A., Mardis, E. R., Patterson, N., Pringle, T. H., Chiaromonte, F., Mullikin, J. C., Eichler, E. E., Hardison, R. C., Gibbs, R. A., Harkins, T. T., and Hayes, V. M. (2010) Complete Khoisan and Bantu genomes from southern Africa, *Nature 463*, 943–947.
17. McKernan, K. J., Peckham, H. E., Costa, G. L., McLaughlin, S. F., Fu, Y., Tsung, E. F., Clouser, C. R., Duncan, C., Ichikawa, J. K., Lee, C. C., Zhang, Z., Ranade, S. S., Dimalanta, E. T., Hyland, F. C., Sokolsky, T. D., Zhang, L., Sheridan, A., Fu, H., Hendrickson, C. L., Li, B., Kotler, L., Stuart, J. R., Malek, J. A., Manning, J. M., Antipova, A. A., Perez, D. S., Moore, M. P., Hayashibara, K. C., Lyons, M. R., Beaudoin, R. E., Coleman, B. E., Laptewicz, M. W., Sannicandro, A. E., Rhodes, M. D., Gottimukkala, R. K., Yang, S., Bafna, V., Bashir, A., MacBride, A., Alkan, C., Kidd, J. M., Eichler, E. E., Reese, M. G., De La Vega, F. M., and Blanchard, A. P. (2009) Sequence and structural variation in a human genome uncovered by short-read, massively parallel ligation sequencing using two-base encoding, *Genome Res 19*, 1527–1541.
18. Consortium, T. I. H. (2003) The International HapMap Project, *Nature 426*, 789–796.
19. Drmanac, R., Sparks, A. B., Callow, M. J., Halpern, A. L., Burns, N. L., Kermani, B. G., Carnevali, P., Nazarenko, I., Nilsen, G. B., Yeung, G., Dahl, F., Fernandez, A., Staker, B., Pant, K. P., Baccash, J., Borcherding, A. P., Brownley, A., Cedeno, R., Chen, L., Chernikoff, D., Cheung, A., Chirita, R., Curson, B., Ebert, J. C., Hacker, C. R., Hartlage, R., Hauser, B., Huang, S., Jiang, Y., Karpinchyk, V., Koenig, M., Kong, C., Landers, T., Le, C., Liu, J., McBride, C. E., Morenzoni, M., Morey, R. E., Mutch, K., Perazich, H., Perry, K., Peters, B. A., Peterson, J., Pethiyagoda, C. L., Pothuraju, K., Richter, C., Rosenbaum, A. M., Roy, S., Shafto, J., Sharanhovich, U., Shannon, K. W., Sheppy, C. G., Sun, M., Thakuria, J. V., Tran, A., Vu, D., Zaranek, A. W., Wu, X., Drmanac, S., Oliphant, A. R., Banyai, W. C., Martin, B., Ballinger, D. G., Church, G. M., and Reid, C. A. (2010) Human genome sequencing using

unchained base reads on self-assembling DNA nanoarrays, *Science 327*, 78–81.

20. Pushkarev, D., Neff, N. F., and Quake, S. R. (2009) Single-molecule sequencing of an individual human genome, *Nat Biotech 27*, 847–850.
21. Roach, J. C., Glusman, G., Smit, A. F., Huff, C. D., Hubley, R., Shannon, P. T., Rowen, L., Pant, K. P., Goodman, N., Bamshad, M., Shendure, J., Drmanac, R., Jorde, L. B., Hood, L., and Galas, D. J. (2010) Analysis of genetic inheritance in a family quartet by whole-genome sequencing, *Science 328*, 636–639.
22. Ley, T. J., Mardis, E. R., Ding, L., Fulton, B., McLellan, M. D., Chen, K., Dooling, D., Dunford-Shore, B. H., McGrath, S., Hickenbotham, M., Cook, L., Abbott, R., Larson, D. E., Koboldt, D. C., Pohl, C., Smith, S., Hawkins, A., Abbott, S., Locke, D., Hillier, L. W., Miner, T., Fulton, L., Magrini, V., Wylie, T., Glasscock, J., Conyers, J., Sander, N., Shi, X., Osborne, J. R., Minx, P., Gordon, D., Chinwalla, A., Zhao, Y., Ries, R. E., Payton, J. E., Westervelt, P., Tomasson, M. H., Watson, M., Baty, J., Ivanovich, J., Heath, S., Shannon, W. D., Nagarajan, R., Walter, M. J., Link, D. C., Graubert, T. A., DiPersio, J. F., and Wilson, R. K. (2008) DNA sequencing of a cytogenetically normal acute myeloid leukaemia genome, *Nature 456*, 66–72.
23. Mardis, E. R., Ding, L., Dooling, D. J., Larson, D. E., McLellan, M. D., Chen, K., Koboldt, D. C., Fulton, R. S., Delehaunty, K. D., McGrath, S. D., Fulton, L. A., Locke, D. P., Magrini, V. J., Abbott, R. M., Vickery, T. L., Reed, J. S., Robinson, J. S., Wylie, T., Smith, S. M., Carmichael, L., Eldred, J. M., Harris, C. C., Walker, J., Peck, J. B., Du, F., Dukes, A. F., Sanderson, G. E., Brummett, A. M., Clark, E., McMichael, J. F., Meyer, R. J., Schindler, J. K., Pohl, C. S., Wallis, J. W., Shi, X., Lin, L., Schmidt, H., Tang, Y., Haipek, C., Wiechert, M. E., Ivy, J. V., Kalicki, J., Elliott, G., Ries, R. E., Payton, J. E., Westervelt, P., Tomasson, M. H., Watson, M. A., Baty, J., Heath, S., Shannon, W. D., Nagarajan, R., Link, D. C., Walter, M. J., Graubert, T. A., DiPersio, J. F., Wilson, R. K., and Ley, T. J. (2009) Recurring Mutations Found by Sequencing an Acute Myeloid Leukemia Genome, *N Engl J Med 361*, 1058–1066.
24. Campbell, P. J., Stephens, P. J., Pleasance, E. D., O'Meara, S., Li, H., Santarius, T., Stebbings, L. A., Leroy, C., Edkins, S., Hardy, C., Teague, J. W., Menzies, A., Goodhead, I., Turner, D. J., Clee, C. M., Quail, M. A., Cox, A., Brown, C., Durbin, R., Hurles, M. E., Edwards, P. A. W., Bignell, G. R., Stratton, M. R., and Futreal, P. A. (2008) Identification of somatically acquired rearrangements in cancer using genome-wide massively parallel paired-end sequencing, *Nat Genet 40*, 722–729.
25. Pleasance, E. D., Stephens, P. J., O‚ÄôMeara, S., McBride, D. J., Meynert, A., Jones, D., Lin, M.-L., Beare, D., Lau, K. W., Greenman, C., Varela, I., Nik-Zainal, S., Davies, H. R., Ordonez, G. R., Mudie, L. J., Latimer, C., Edkins, S., Stebbings, L., Chen, L., Jia, M., Leroy, C., Marshall, J., Menzies, A., Butler, A., Teague, J. W., Mangion, J., Sun, Y. A., McLaughlin, S. F., Peckham, H. E., Tsung, E. F., Costa, G. L., Lee, C. C., Minna, J. D., Gazdar, A., Birney, E., Rhodes, M. D., McKernan, K. J., Stratton, M. R., Futreal, P. A., and Campbell, P. J. (2010) A small-cell lung cancer genome with complex signatures of tobacco exposure, *Nature 463*, 184–190.
26. Pleasance, E. D., Cheetham, R. K., Stephens, P. J., McBride, D. J., Humphray, S. J., Greenman, C. D., Varela, I., Lin, M.-L., Ordonez, G. R., Bignell, G. R., Ye, K., Alipaz, J., Bauer, M. J., Beare, D., Butler, A., Carter, R. J., Chen, L., Cox, A. J., Edkins, S., Kokko-Gonzales, P. I., Gormley, N. A., Grocock, R. J., Haudenschild, C. D., Hims, M. M., James, T., Jia, M., Kingsbury, Z., Leroy, C., Marshall, J., Menzies, A., Mudie, L. J., Ning, Z., Royce, T., Schulz-Trieglaff, O. B., Spiridou, A., Stebbings, L. A., Szajkowski, L., Teague, J., Williamson, D., Chin, L., Ross, M. T., Campbell, P. J., Bentley, D. R., Futreal, P. A., and Stratton, M. R. (2010) A comprehensive catalogue of somatic mutations from a human cancer genome, *Nature 463*, 191–196.
27. Consortium, T. I. C. G. (2010) International network of cancer genome projects, *Nature 464*, 993–998.
28. Gilbert, M. T., Kivisild, T., Gronnow, B., Andersen, P. K., Metspalu, E., Reidla, M., Tamm, E., Axelsson, E., Gotherstrom, A., Campos, P. F., Rasmussen, M., Metspalu, M., Higham, T. F., Schwenninger, J. L., Nathan, R., De Hoog, C. J., Koch, A., Moller, L. N., Andreasen, C., Meldgaard, M., Villems, R., Bendixen, C., and Willerslev, E. (2008) Paleo-Eskimo mtDNA genome reveals matrilineal discontinuity in Greenland, *Science 320*, 1787–1789.
29. Rasmussen, M., Li, Y., Lindgreen, S., Pedersen, J. S., Albrechtsen, A., Moltke, I., Metspalu, M., Metspalu, E., Kivisild, T., Gupta, R., Bertalan, M., Nielsen, K., Gilbert, M. T., Wang, Y., Raghavan, M., Campos, P. F., Kamp, H. M., Wilson, A. S., Gledhill, A., Tridico, S., Bunce, M., Lorenzen, E. D., Binladen, J., Guo, X., Zhao, J., Zhang, X., Zhang, H., Li, Z., Chen, M., Orlando, L., Kristiansen, K., Bak, M., Tommerup, N.,

Bendixen, C., Pierre, T. L., Gronnow, B., Meldgaard, M., Andreasen, C., Fedorova, S. A., Osipova, L. P., Higham, T. F., Ramsey, C. B., Hansen, T. V., Nielsen, F. C., Crawford, M. H., Brunak, S., Sicheritz-Ponten, T., Villems, R., Nielsen, R., Krogh, A., Wang, J., and Willerslev, E. (2010) Ancient human genome sequence of an extinct Palaeo-Eskimo, *Nature 463*, 757–762.

30. Green, R. E., Malaspinas, A. S., Krause, J., Briggs, A. W., Johnson, P. L., Uhler, C., Meyer, M., Good, J. M., Maricic, T., Stenzel, U., Prufer, K., Siebauer, M., Burbano, H. A., Ronan, M., Rothberg, J. M., Egholm, M., Rudan, P., Brajkovic, D., Kucan, Z., Gusic, I., Wikstrom, M., Laakkonen, L., Kelso, J., Slatkin, M., and Paabo, S. (2008) A complete Neandertal mitochondrial genome sequence determined by high-throughput sequencing, *Cell 134*, 416–426.
31. Green, R. E., Krause, J., Briggs, A. W., Maricic, T., Stenzel, U., Kircher, M., Patterson, N., Li, H., Zhai, W., Fritz, M. H., Hansen, N. F., Durand, E. Y., Malaspinas, A. S., Jensen, J. D., Marques-Bonet, T., Alkan, C., Prufer, K., Meyer, M., Burbano, H. A., Good, J. M., Schultz, R., Aximu-Petri, A., Butthof, A., Hober, B., Hoffner, B., Siegemund, M., Weihmann, A., Nusbaum, C., Lander, E. S., Russ, C., Novod, N., Affourtit, J., Egholm, M., Verna, C., Rudan, P., Brajkovic, D., Kucan, Z., Gusic, I., Doronichev, V. B., Golovanova, L. V., Lalueza-Fox, C., de la Rasilla, M., Fortea, J., Rosas, A., Schmitz, R. W., Johnson, P. L., Eichler, E. E., Falush, D., Birney, E., Mullikin, J. C., Slatkin, M., Nielsen, R., Kelso, J., Lachmann, M., Reich, D., and Paabo, S. (2010) A draft sequence of the Neandertal genome, *Science 328*, 710–722.
32. Green, R. E., Krause, J., Ptak, S. E., Briggs, A. W., Ronan, M. T., Simons, J. F., Du, L., Egholm, M., Rothberg, J. M., Paunovic, M., and Paabo, S. (2006) Analysis of one million base pairs of Neanderthal DNA, *Nature 444*, 330–336.
33. Noonan, J. P., Coop, G., Kudaravalli, S., Smith, D., Krause, J., Alessi, J., Chen, F., Platt, D., Paabo, S., Pritchard, J. K., and Rubin, E. M. (2006) Sequencing and analysis of Neanderthal genomic DNA, *Science 314*, 1113–1118.
34. Burbano, H. A., Hodges, E., Green, R. E., Briggs, A. W., Krause, J., Meyer, M., Good, J. M., Maricic, T., Johnson, P. L. F., Xuan, Z., Rooks, M., Bhattacharjee, A., Brizuela, L., Albert, F. W., de la Rasilla, M., Fortea, J., Rosas, A., Lachmann, M., Hannon, G. J., and Paabo, S. (2010) Targeted Investigation of the Neandertal Genome by Array-Based Sequence Capture, *Science 328*, 723–725.
35. Ng, S. B., Turner, E. H., Robertson, P. D., Flygare, S. D., Bigham, A. W., Lee, C., Shaffer, T., Wong, M., Bhattacharjee, A., Eichler, E. E., Bamshad, M., Nickerson, D. A., and Shendure, J. (2009) Targeted capture and massively parallel sequencing of 12 human exomes, *Nature 461*, 272–276.
36. Ng, P. C., Levy, S., Huang, J., Stockwell, T. B., Walenz, B. P., Li, K., Axelrod, N., Busam, D. A., Strausberg, R. L., and Venter, J. C. (2008) Genetic variation in an individual human exome, *PLoS Genet 4*, e1000160.
37. Ng, S. B., Buckingham, K. J., Lee, C., Bigham, A. W., Tabor, H. K., Dent, K. M., Huff, C. D., Shannon, P. T., Jabs, E. W., Nickerson, D. A., Shendure, J., and Bamshad, M. J. (2010) Exome sequencing identifies the cause of a mendelian disorder, *Nat Genet 42*, 30–35.
38. Choi, M., Scholl, U. I., Ji, W., Liu, T., Tikhonova, I. R., Zumbo, P., Nayir, A., Bakkaloglu, A., Ozen, S., Sanjad, S., Nelson-Williams, C., Farhi, A., Mane, S., and Lifton, R. P. (2009) Genetic diagnosis by whole exome capture and massively parallel DNA sequencing, *Proc Natl Acad Sci USA 106*, 19096–19101.
39. Rehman, A. U., Morell, R. J., Belyantseva, I. A., Khan, S. Y., Boger, E. T., Shahzad, M., Ahmed, Z. M., Riazuddin, S., Khan, S. N., and Friedman, T. B. (2010) Targeted capture and next-generation sequencing identifies C9orf75, encoding taperin, as the mutated gene in nonsyndromic deafness DFNB79, *Am J Hum Genet 86*, 378–388.
40. Walsh, T., Shahin, H., Elkan-Miller, T., Lee, M. K., Thornton, A. M., Roeb, W., Abu Rayyan, A., Loulus, S., Avraham, K. B., King, M.-C., and Kanaan, M. Whole Exome Sequencing and Homozygosity Mapping Identify Mutation in the Cell Polarity Protein GPSM2 as the Cause of Nonsyndromic Hearing Loss DFNB82, *The American Journal of Human Genetics In Press, Corrected Proof.*
41. Lalonde, E., Albrecht, S., Ha, K. C. H., Jacob, K., Bolduc, N., Polychronakos, C., Dechelotte, P., Majewski, J., and Jabado, N. (2010) Unexpected allelic heterogeneity and spectrum of mutations in Fowler syndrome revealed by next-generation exome sequencing, *Human Mutation 9999*, n/a.
42. Hoischen, A., Gilissen, C., Arts, P., Wieskamp, N., Vliet, W. v. d., Vermeer, S., Steehouwer, M., Vries, P. d., Meijer, R., Seiqueros, J., Knoers, N. V. A. M., Buckley, M. F., Scheffer, H., and Veltman, J. A. (2010) Massively parallel sequencing of ataxia genes after array-based enrichment, *Human Mutation 31*, 494–499.

43. Li, R., Li, Y., Zheng, H., Luo, R., Zhu, H., Li, Q., Qian, W., Ren, Y., Tian, G., Li, J., Zhou, G., Zhu, X., Wu, H., Qin, J., Jin, X., Li, D., Cao, H., Hu, X., Blanche, H., Cann, H., Zhang, X., Li, S., Bolund, L., Kristiansen, K., Yang, H., Wang, J., and Wang, J. (2010) Building the sequence map of the human pan-genome, *Nat Biotech 28*, 57–63.

44. Khaja, R., Zhang, J., MacDonald, J. R., He, Y., Joseph-George, A. M., Wei, J., Rafiq, M. A., Qian, C., Shago, M., Pantano, L., Aburatani, H., Jones, K., Redon, R., Hurles, M., Armengol, L., Estivill, X., Mural, R. J., Lee, C., Scherer, S. W., and Feuk, L. (2006) Genome assembly comparison identifies structural variants in the human genome, *Nat Genet 38*, 1413–1418.

45. Kidd, J. M., Sampas, N., Antonacci, F., Graves, T., Fulton, R., Hayden, H. S., Alkan, C., Malig, M., Ventura, M., Giannuzzi, G., Kallicki, J., Anderson, P., Tsalenko, A., Yamada, N. A., Tsang, P., Kaul, R., Wilson, R. K., Bruhn, L., and Eichler, E. E. (2010) Characterization of missing human genome sequences and copy-number polymorphic insertions, *Nat Meth 7*, 365–371.

46. Li, H., Ruan, J., and Durbin, R. (2008) Mapping short DNA sequencing reads and calling variants using mapping quality scores, *Genome Res 18*, 1851–1858.

47. Hillier, L. W., Marth, G. T., Quinlan, A. R., Dooling, D., Fewell, G., Barnett, D., Fox, P., Glasscock, J. I., Hickenbotham, M., Huang, W., Magrini, V. J., Richt, R. J., Sander, S. N., Stewart, D. A., Stromberg, M., Tsung, E. F., Wylie, T., Schedl, T., Wilson, R. K., and Mardis, E. R. (2008) Whole-genome sequencing and variant discovery in C. elegans, *Nat Meth 5*, 183–188.

48. Li, H., Handsaker, B., Wysoker, A., Fennell, T., Ruan, J., Homer, N., Marth, G., Abecasis, G., Durbin, R., and Genome Project Data Processing Subgroup. (2009) The Sequence Alignment/Map format and SAMtools, *Bioinformatics 25*, 2078–2079.

49. Li, R., Yu, C., Li, Y., Lam, T.-W., Yiu, S.-M., Kristiansen, K., and Wang, J. (2009) SOAP2: an improved ultrafast tool for short read alignment, *Bioinformatics 25*, 1966–1967.

50. Li, R., Li, Y., Kristiansen, K., and Wang, J. (2008) SOAP: short oligonucleotide alignment program, *Bioinformatics 24*, 713–714.

51. Tuzun, E., Sharp, A. J., Bailey, J. A., Kaul, R., Morrison, V. A., Pertz, L. M., Haugen, E., Hayden, H., Albertson, D., Pinkel, D., Olson, M. V., and Eichler, E. E. (2005) Fine-scale structural variation of the human genome, *Nat Genet 37*, 727–732.

52. Kidd, J. M., Cooper, G. M., Donahue, W. F., Hayden, H. S., Sampas, N., Graves, T., Hansen, N., Teague, B., Alkan, C., Antonacci, F., Haugen, E., Zerr, T., Yamada, N. A., Tsang, P., Newman, T. L., Tuzun, E., Cheng, Z., Ebling, H. M., Tusneem, N., David, R., Gillett, W., Phelps, K. A., Weaver, M., Saranga, D., Brand, A., Tao, W., Gustafson, E., McKernan, K., Chen, L., Malig, M., Smith, J. D., Korn, J. M., McCarroll, S. A., Altshuler, D. A., Peiffer, D. A., Dorschner, M., Stamatoyannopoulos, J., Schwartz, D., Nickerson, D. A., Mullikin, J. C., Wilson, R. K., Bruhn, L., Olson, M. V., Kaul, R., Smith, D. R., and Eichler, E. E. (2008) Mapping and sequencing of structural variation from eight human genomes, *Nature 453*, 56–64.

53. Lee, S., Hormozdiari, F., Alkan, C., and Brudno, M. (2009) MoDIL: detecting small indels from clone-end sequencing with mixtures of distributions, *Nat Meth 6*, 473–474.

54. Korbel, J., Abyzov, A., Mu, X., Carriero, N., Cayting, P., Zhang, Z., Snyder, M., and Gerstein, M. (2009) PEMer: a computational framework with simulation-based error models for inferring genomic structural variants from massive paired-end sequencing data, *Genome Biology 10*, R23.

55. Chen, K., Wallis, J. W., McLellan, M. D., Larson, D. E., Kalicki, J. M., Pohl, C. S., McGrath, S. D., Wendl, M. C., Zhang, Q., Locke, D. P., Shi, X., Fulton, R. S., Ley, T. J., Wilson, R. K., Ding, L., and Mardis, E. R. (2009) BreakDancer: an algorithm for high-resolution mapping of genomic structural variation, *Nat Meth 6*, 677–681.

56. Hajirasouliha, I., Hormozdiari, F., Alkan, C., Kidd, J. M., Birol, I., Eichler, E. E., and Sahinalp, S. C. (2010) Detection and characterization of novel sequence insertions using paired-end next-generation sequencing, *Bioinformatics 26*, 1277–1283.

57. Xie, C., and Tammi, M. T. (2009) CNV-seq, a new method to detect copy number variation using high-throughput sequencing, *BMC Bioinformatics 10*, 80.

58. Conrad, D. F., Bird, C., Blackburne, B., Lindsay, S., Mamanova, L., Lee, C., Turner, D. J., and Hurles, M. E. (2010) Mutation spectrum revealed by breakpoint sequencing of human germline CNVs, *Nat Genet 42*, 385–391.

59. Lam, H. Y., Mu, X. J., Stutz, A. M., Tanzer, A., Cayting, P. D., Snyder, M., Kim, P. M., Korbel, J. O., and Gerstein, M. B. (2010) Nucleotide-resolution analysis of structural variants using BreakSeq and a breakpoint library, *Nat Biotechnol 28*, 47–55.

60. Korbel, J. O., Urban, A. E., Affourtit, J. P., Godwin, B., Grubert, F., Simons, J. F., Kim, P. M., Palejev, D., Carriero, N. J., Du, L., Taillon, B. E., Chen, Z., Tanzer, A., Saunders, A. C. E., Chi, J., Yang, F., Carter, N. P., Hurles, M. E., Weissman, S. M., Harkins, T. T., Gerstein, M. B., Egholm, M., and Snyder, M. (2007) Paired-End Mapping Reveals Extensive Structural Variation in the Human Genome, *Science 318*, 420–426.

61. Ye, K., Schulz, M. H., Long, Q., Apweiler, R., and Ning, Z. (2009) Pindel: a pattern growth approach to detect break points of large deletions and medium sized insertions from paired-end short reads, *Bioinformatics 25*, 2865–2871.

62. Ameur, A., Wetterbom, A., Feuk, L., and Gyllensten, U. (2010) Global and unbiased detection of splice junctions from RNA-seq data, *Genome Biol 11*, R34.

63. Feuk, L., Carson, A. R., and Scherer, S. W. (2006) Structural variation in the human genome, *Nat Rev Genet 7*, 85–97.

64. Chiang, D. Y., Getz, G., Jaffe, D. B., O'Kelly, M. J., Zhao, X., Carter, S. L., Russ, C., Nusbaum, C., Meyerson, M., and Lander, E. S. (2009) High-resolution mapping of copy-number alterations with massively parallel sequencing, *Nat Methods 6*, 99–103.

65. Lupski, J. R., Reid, J. G., Gonzaga-Jauregui, C., Rio Deiros, D., Chen, D. C., Nazareth, L., Bainbridge, M., Dinh, H., Jing, C., Wheeler, D. A., McGuire, A. L., Zhang, F., Stankiewicz, P., Halperin, J. J., Yang, C., Gehman, C., Guo, D., Irikat, R. K., Tom, W., Fantin, N. J., Muzny, D. M., and Gibbs, R. A. Whole-genome sequencing in a patient with Charcot-Marie-Tooth neuropathy, *N Engl J Med 362*, 1181–1191.

66. Frazer, K. A., Murray, S. S., Schork, N. J., and Topol, E. J. (2009) Human genetic variation and its contribution to complex traits, *Nat Rev Genet 10*, 241–251.

67. Klein, T. E., Chang, J. T., Cho, M. K., Easton, K. L., Fergerson, R., Hewett, M., Lin, Z., Liu, Y., Liu, S., Oliver, D. E., Rubin, D. L., Shafa, F., Stuart, J. M., and Altman, R. B. (2001) Integrating genotype and phenotype information: an overview of the PharmGKB project. Pharmacogenetics Research Network and Knowledge Base, *Pharmacogenomics J 1*, 167–170.

68. Ng, P. C., Murray, S. S., Levy, S., and Venter, J. C. (2009) An agenda for personalized medicine, *Nature 461*, 724–726.

69. Ashley, E. A., Butte, A. J., Wheeler, M. T., Chen, R., Klein, T. E., Dewey, F. E., Dudley, J. T., Ormond, K. E., Pavlovic, A., Morgan, A. A., Pushkarev, D., Neff, N. F., Hudgins, L., Gong, L., Hodges, L. M., Berlin, D. S., Thorn, C. F., Sangkuhl, K., Hebert, J. M., Woon, M., Sagreiya, H., Whaley, R., Knowles, J. W., Chou, M. F., Thakuria, J. V., Rosenbaum, A. M., Zaranek, A. W., Church, G. M., Greely, H. T., Quake, S. R., and Altman, R. B. (2010) Clinical assessment incorporating a personal genome, *Lancet 375*, 1525–1535.

Chapter 18

Massively Parallel Sequencing Approaches for Characterization of Structural Variation

Daniel C. Koboldt, David E. Larson, Ken Chen, Li Ding, and Richard K. Wilson

Abstract

The emergence of next-generation sequencing (NGS) technologies offers an incredible opportunity to comprehensively study DNA sequence variation in human genomes. Commercially available platforms from Roche (454), Illumina (Genome Analyzer and Hiseq 2000), and Applied Biosystems (SOLiD) have the capability to completely sequence individual genomes to high levels of coverage. NGS data is particularly advantageous for the study of structural variation (SV) because it offers the sensitivity to detect variants of various sizes and types, as well as the precision to characterize their breakpoints at base pair resolution. In this chapter, we present methods and software algorithms that have been developed to detect SVs and copy number changes using massively parallel sequencing data. We describe visualization and *de novo* assembly strategies for characterizing SV breakpoints and removing false positives.

Key words: Next-generation sequencing, Paired-end sequencing, 454, Illumina, Solexa, Abi solid, Insertions, Deletions, Duplications, Inversions, Translocations, Indels, Copy number variants

1. Introduction

Massively parallel sequencing technologies have fundamentally changed the study of genetics and genomics. New instruments from Roche (454), Illumina (Genome Analyzer and Hiseq 2000), and Applied Biosystems (SOLiD) generate millions of DNA sequence reads in a single run, enabling researchers to address questions with unprecedented speed (1). These next-generation sequencing (NGS) technologies make it feasible to sequence entire genomes to high levels of coverage in a matter of weeks. Indeed, the complete genomes of several individuals have been sequenced on new platforms (2–9), and ambitious efforts like the 1,000 Genomes Project (http://www.1000genomes.org) aim to add thousands more, offering an unprecedented survey of DNA sequence variation in humans.

Lars Feuk (ed.), *Genomic Structural Variants: Methods and Protocols*, Methods in Molecular Biology, vol. 838, DOI 10.1007/978-1-61779-507-7_18, © Springer Science+Business Media, LLC 2012

NGS has enabled powerful new approaches for the detection of copy number variation (CNV) and structural variation (SV) in the human genome. Compared to array-based methods, NGS has demonstrated higher sensitivity, in terms of the types and sizes of variants that can be detected. Furthermore, sequencing enabled the precise definition of SV breakpoints, information that is critical for assessing functional impact and inferring likely mutational mechanisms of origin.

Most current approaches to sequence-based SV detection extend seminal work by Volik et al. (10) and Raphael et al. (11). Their method, first presented in 2003, applied end-sequence profiling (ESP) of bacterial artificial chromosomes to map structural rearrangements in cancer cell lines. The ESP method requires sequencing both ends of a genomic fragment of known size (e.g., a 200-kb BAC insert) and then mapping the end-sequence pair to a reference sequence. Fragments overlapping SV events result in paired sequences that map to different parts of the reference genome, possibly another chromosome entirely. In 2005, Tuzun et al. (12) used this approach to systematically discover SVs in a human genome, reporting hundreds of intermediate-sized variants, including insertions, deletions, and inversions.

Paired-end sequencing on NGS platforms has enabled detection of CNV and SV in the human genome at unprecedented scale and throughput, and at a substantially reduced cost. Korbel et al. (13) developed a paired-end mapping (PEM) approach for the Roche/454 platform and used it to fine-map more than 1,000 SVs in two human genomes. Campbell et al. (14) used Illumina paired-end sequencing to characterize genomic rearrangements in cancer cell lines. Massively parallel sequencing has since been employed to systematically characterize large-scale variation in individual (2, 6, 8, 13) and cancer (14–17) genomes.

Although NGS platforms are well-suited to CNV and SV detection, analysis of NGS data presents substantial bioinformatics challenges due to the relatively short read lengths (36–250 bp) and the unprecedented volume of data. In this chapter, we discuss the tools and methods that have been developed for NGS data analysis – including alignment, assembly, variant detection, and visualization – for the characterization of CNV and structural variation.

2. Materials

2.1. Massively Parallel Sequencing Data

Three commercially available NGS platforms have been successfully applied to the discovery of CNV/SV in humans, and will be the focus of this review: the 454 platform (Roche), the GenomeAnalyzer/HiSeq2000 platform (Illumina), and the SOLiD platform (Life Technologies).

2.1.1. Roche (454) Data

The Roche (454) Genome Sequencer FLX (18) utilizes massively parallel pyrosequencing of DNA fragments that are amplified *en masse* by emulsion PCR (1). The current FLX Titanium chemistry produces up to 1.25 million reads per run, with read lengths of ~400 bp. A single run yields almost half a *billion* base pairs (0.5 gb) of high-quality sequence. While the 454 platform is prone to indel errors near runs of multiple nucleotides (homopolymers), the substitution error rate is very low, and the relatively long reads are well-suited to alignment or *de novo* assembly.

2.1.2. Illumina (Solexa) Data

The Illumina (Solexa) Genome Analyzer IIx utilizes sequencing-by-synthesis of surface-amplified DNA fragments (3). While the reads produced on the Illumina platform were initially quite short (~32–40 bp), the current instrument yields ~500 million 100-bp reads (50 gb) per run. The astonishing sequence throughput of Illumina machines present substantial informatics challenges, particularly the alignment of short sequences of imperfect quality to large reference sequences. Indeed, an entire generation of novel algorithms (Maq, BWA, Novoalign, Bowtie, and others) has been developed to address the analysis challenges of Illumina sequencing.

2.1.3. ABI SOLiD Data

The Applied Biosystems SOLiD sequencer uses a unique process catalyzed by DNA ligase, in which oligo adapter-linked DNA fragments are coupled to magnetic beads and amplified by emulsion PCR (6). The current instrument (SOLiD 3) produces up to 50 gb of high-quality sequence per slide in the form of 35- or 50-bp reads. A unique advantage of the ABI SOLiD platform is its di-base encoding scheme, in which each base is effectively called twice in a sequencing read. The availability of two calls per base makes it possible to distinguish between sequencing errors and true variation, thereby improving the overall accuracy of reads from this platform.

2.1.4. Paired-End Libraries

The optimal dataset for SV detection is paired-end sequence data, produced by sequencing both ends of randomly sheared DNA fragments that have been size-selected by gel electrophoresis or other techniques. Because both reads in a pair come from a linear DNA fragment of known (approximate) size, their relative distance and orientation when mapped to a reference sequence can indicate the presence of underlying structural variation.

2.2. Bioinformatics Resources

Most NGS platforms include basecalling software from the manufacturer. Systematic mapping and/or assembly of NGS data require additional software. A selection of the tools available for short read alignment, assembly, and data handling is provided in Table 1.

2.2.1. Computing Resources

It should be noted that the processing of large massively parallel sequencing datasets, even with advanced algorithms, requires

Table 1
Software tools and algorithms for alignment, SAM/BAM integration, copy number estimation, SV detection, and *de novo* assembly using NGS data

Software	Description	URL
Read alignment/mapping		
Maq	Widely used mapping algorithm for short NGS reads	http://www.maq.sourceforge.net
BWA/BWASW	Burrows-Wheeler Aligner for mapping short (bwa) or long (bwasw) reads	http://www.bio-bwa.sourceforge.net
Bowtie	Ultrafast short read aligner for Illumina data	http://www.bowtie-bio.sourceforge.net
BFAST	Customizable BLAT-like read mapping tool for NGS data	http://www.genome.ucla.edu/bfast
SHRiMP	Efficient Smith-Waterman aligner for short read data	http://www.compbio.cs.toronto.edu/shrimp
SOAP	Short oligo analysis package for alignment and variant calling	http://www.soap.genomics.org.cn/
mrFAST	Maps read to all possible locations for duplication/CNV detection.	http://www.mrfast.sourceforge.net
SSAHA2	Sequence Search and Alignment by Hashing Algorithm	http://www.sanger.ac.uk/resources/software/ssaha2
SAM/BAM file operations		
SAMtools	A suite of tools for manipulating NGS data in SAM/BAM format	http://www.samtools.sourceforge.net
Picard	A suite of Java tools for validating and de-duplicating SAM/BAM files	http://www.picard.sourceforge.net
Copy number estimation		
EWT	CNV calling with event-wise testing (EWT)	http://www.genome.cshlp.org/content/19/9/1586
SegSeq	CNV calling with local changepoint analysis and merging	http://www.broad.mit.edu/cancer/pub/solexa_copy_numbers
CMDS	Recurrent CNA calling in sample populations	https://www.dsgweb.wustl.edu/qunyuan/software/cmds
Structural variation detection		
BreakDancer	SV prediction tool for paired-end Illumina data	http://www.genome.wustl.edu/tools/cancer-genomics
GASV	Geometric method for SV detection	http://www.cs.brown.edu/people/braphael/software.html
Pindel	Indel prediction tool for paired-end NGS data	http://www.ebi.ac.uk/~kye/pindel
De novo assembly		
ABySS	A *de novo*, parallel, paired-end sequence assembler for short reads	http://www.bcgsc.ca/platform/bioinfo/software/abyss
Velvet	A *de novo* sequence assembler for short reads	http://www.ebi.ac.uk/~zerbino/velvet
TIGRA	*De novo* assembly of SV breakpoints	http://www.genome.wustl.edu/tools/cancer-genomics

(continued)

Table 1 (continued)

Software	Description	URL
Visualization		
Circos	Circular visualization of genome and comparative genomics data	http://www.mkweb.bcgsc.ca/circos/
IGV	BAM-driven integrative genomics viewer for NGS data	http://www.broadinstitute.org/igv
LookSeq	Web-based tool for visualization and analysis of sequence alignments	http://www.sanger.ac.uk/resources/software/lookseq
Pairoscope	BAM-driven visualization of predicted structural variants	http://www.pairoscope.sourceforge.net
Savant	Desktop visualization tool that represents paired-end reads for SV identification.	http://www.compbio.cs.toronto.edu/savant

substantial computational resources. At the very least, a dedicated Linux/UNIX/MacOSX server (preferably 64-bit) with at least 2 GB of RAM will be required for data processing. Alignment of just a single lane of Illumina paired-end data (25–30 million read pairs) to the human genome takes anywhere from 4 to 24 h, depending on the hardware, data quality, read length, and mapping parameters.

2.2.2. Short Read Aligners

The critical first step to analysis of NGS data is alignment to a reference sequence. Several short read mapping tools have been developed to address the relatively short read lengths and sheer volume of data produced by NGS platforms. These have been reviewed extensively elsewhere (19); a selection of short read aligners is provided in Table 1. Widely used aligners for Illumina and ABI SOLiD data include Maq (20), BWA (21), Bowtie (22), BFAST (23), and SHRiMP (24). For longer reads generated on the Roche/454 platform, SSAHA2 (25) and BWASW (26) are widely used. Perhaps the most important consideration in selecting an alignment algorithm is its compatibility with the SAM format specification, which is described below.

2.2.3. SAM/BAM Format Specification

The specification of the sequence alignment/map (SAM) format and its compressed binary equivalent (BAM) has simplified NGS data analysis dramatically (27). In essence, SAM format (http://www.samtools.sourceforge.net/SAM1.pdf) enables storing large numbers of sequences (reads), along with their alignments to a reference genome or assembly, in a single file. SAM/BAM files are compact, but flexible enough to accommodate multiple NGS data types and software algorithms. The SAM specification has been widely adopted by the NGS bioinformatics community; many aligners can output directly into SAM/BAM, and a number of freely available tools, including SAMtools ((27), http://www.samtools.

sourceforge.net) and Picard (http://www.picard.sourceforge.net) provide the functions to view, merge, sort, index, filter, assemble, and perform other operations on SAM/BAM files.

2.3. Ancillary CNV/SV Data

To fine-tune and evaluate the performance of SV detection, a set of known structural events can be valuable. Spectral karyotyping data and even cytogenetic analysis can reveal chromosomal deletions, duplications, or translocations that should manifest in the sequence data. Copy number changes and structural variants inferred from high-resolution array data (SNP or CGH), if available, serve as true positive controls; studies have shown that sequence-based approaches are highly sensitive for identifying SVs detected by array-based methods (28).

3. Methods

Massively parallel sequencing technologies have enabled genome-wide characterization of CNV and structural variation in a single experiment (Fig. 1). It begins with the alignment of sequence reads

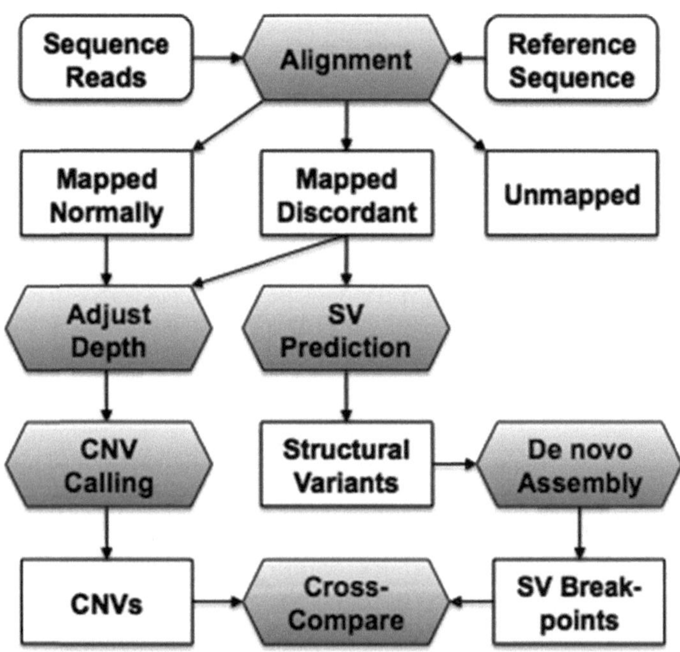

Fig. 1. Detection of copy number variation (CNV) and structural variation (SV) by massively parallel sequencing. When aligned to a reference sequence, read pairs are classified as mapped normally, mapped discordantly, or unmapped. Both classes of mapped reads are used for inference of read depth and then CNV calling. Discordant pairs are utilized for prediction of SVs, which are further resolved by *de novo* assembly using both discordant and unmapped reads.

to a reference sequence. In paired-end sequencing, the alignment process yields three smaller datasets: (1) reads mapped with "correct" pairing, with expected orientation and distance between mates, (2) reads mapped in pairs deemed "discordant" in terms of distance and/or orientation, or where only one read is mapped, and (3) reads that are not mapped at all. Once all possible reads have been mapped, the resulting read depth (after some normalization) serves as a quantitative measure of genome-wide copy number. Segmentation algorithms and other methods have been developed to call copy number variants (CNVs) from read depth. For SV detection, most approaches first examine reads in set (2) above whose discordant mapping suggests the presence of underlying variation.

3.1. Copy Number Analysis

Massively parallel sequencing data can be used to infer DNA CNVs throughout the genome. Most sequencing-based approaches for CNV detection partition the genome into nonoverlapping bins and use the read depth (RD) to look for regions that differ in copy number. In contrast to methods for SV detection, estimates of copy number typically utilize read pairs that map uniquely to the genome with correct spacing and orientation. Fragment-end (unpaired) reads may also be used.

3.1.1. Addressing Mapping Bias

To address issues related to read mapping bias, it is important to correct for varying levels of uniqueness across the genome. Campbell et al. (14) performed this correction after in silico simulations of Illumina 2×35 bp paired-end reads, which were mapped to the genome using Maq (20). The genome was divided into nonoverlapping windows of varying widths such that each window contained a 425 uniquely mapped reads, equivalent to ~15 kb of mapped sequence.

3.1.2. Addressing GC Content Bias

Sequencing coverage on NGS platforms is influenced by G+C content. On the Illumina platform, average sequence coverage is significantly reduced for regions with particularly low (<20%) or high (>60%) G+C content (3). Yoon et al. (29) addressed G+C bias by segmenting the genome into 100-bp windows, and adjusting each window's read counts based on the observed deviation in coverage for a given G+C percentage.

3.1.3. Circular Binary Segmentation

The GC-adjusted read depth within defined windows serves as a quantitative measurement of genome copy number. Thus, copy number changes can be detected using the same types of segmentation algorithms that were developed for SNP or CGH microarray data. Campbell et al. adapted a circular binary segmentation algorithm for SNP array data to detect statistically significant copy number changes. Their adapted algorithm, implemented in R as the "DNAcopy" library of the Bioconductor project (http://www.bioconductor.org), takes the normalized read count for each window as input and estimates both the copy number in each region and the boundaries (change points) defining the copy number change. To evaluate the accuracy of their method, Campbell et al. compared CNV calls based on uniquely mapped, correctly paired reads to predictions of structural variants from aberrantly mapped read pairs. Their CNV algorithm identified all nine tandem duplications in a cancer cell line (ranging in size from <1 kb to 2.7 Mb) and correctly predicted the breakpoints to within 30 kb (14).

3.1.4. Event-Wise Testing

Yoon et al. developed a novel three-step algorithm, called event-wise testing (EWT), to identify CNVs from read depth data (29). First, EWT rapidly searches the genome for specific classes of small events that meet predefined criteria for statistical significance. Then, clusters of small events with a copy number changed in the same direction are merged together into a single CNV call. Events with a low absolute difference in read depth (between 0.75 and 1.25 times the mean) are removed. Finally, the significance of each merged event is assessed with a one-sided *Z*-test. Application of EWT to simulated data, as well as real data from the 1,000 Genomes Project, suggested a good sensitivity for events larger than 1 kb, and high specificity (75–89%) for CNV calls that were polymorphic in multiple individuals (29).

3.1.5. SegSeq

Chiang et al. (30) applied massively parallel sequencing to map copy number alterations in tumor cell lines. Their method, called SegSeq, combines local change-point analysis with a merging procedure that joins adjacent segments. Tumor and normal read counts for the full segments are used to determine statistical significance. When applied to three tumor cell lines, their algorithm identified 194, 126, and 15 copy number alterations, compared to 153, 93, and 18 alterations identified from SNP array data. The results from both methods were highly concordant, but the sequencing approach yielded better breakpoint resolution and a higher dynamic range for copy number estimation (30).

3.2. Mapping Segmental Duplications

Duplicated regions play important roles in genetic and phenotypic variation, but have been refractory to characterization due to their repetitive structure and high copy number in the human genome. Alkan et al. (31) developed an approach to comprehensively map these regions using NGS. Their algorithm, mrFAST, places NGS reads to all possible locations in the reference genome, which is critical to accurately determine the copy number of duplicated sequences. When applied to three human genomes, this approach revealed that individuals differ in copy number for 73–87 genes on average, differences that largely correspond to segmental duplications (31).

3.3. Detection of Structural Variation

Numerous algorithms have been developed in recent years to identify and characterize SVs from massively parallel sequencing data. Many of these leverage paired-end sequencing, in which a fragment of known size is sequenced at both ends. Read pairing information not only improves the accuracy of read mapping (32), but can be used to infer the presence of underlying structural variation (33) (see Fig. 2). Several diverse yet robust methods have been successfully applied to SV detection using paired-end sequencing datasets.

3.3.1. Geometric Analysis

Raphael et al. (34) have introduced a geometric approach to the detection, classification, and comparison of SVs from sequencing data. Their method, called Geometric Analysis of Structural Variants (GASV), represents putative SVs as polygons in a plane, and employs a computational geometry algorithm to merge read pairs supporting the same event. When applied to the genomes of nine individuals already characterized for SVs, GASV provided better localization of SV breakpoints. The authors also applied GASV to data from several cancer cell lines, and compared these results to those from normal individuals to distinguish germline and somatic variants.

3.3.2. Pindel

Pindel (35) employs a pattern growth approach to detect the breakpoints of large insertions and deletions using NGS data. First, all reads are mapped to the reference genome using SSAHA2 (25). The mapping results are examined to identify read pairs for which only one end was placed in the reference. Using the mapped location as an anchor point, Pindel splits the unmapped read into

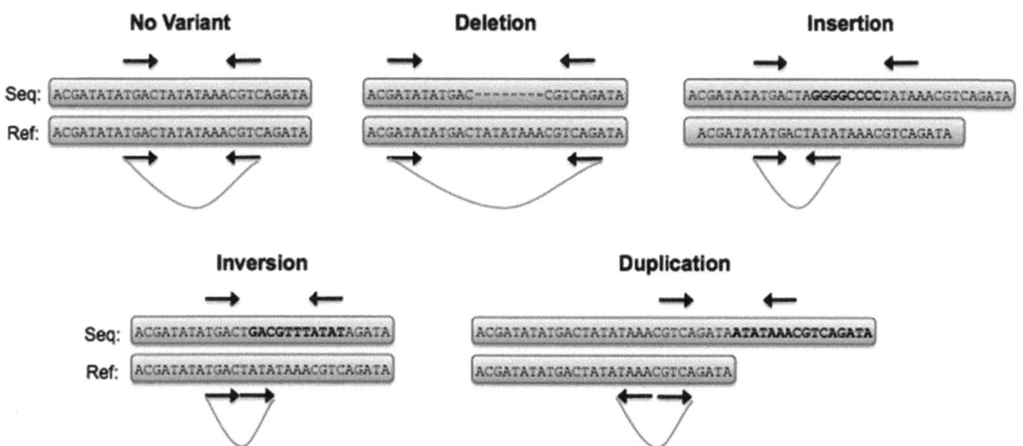

Fig. 2. Detecting structural variants by paired-end mapping. The distance between and relative orientation of associated read pairs suggests specific classes of SVs. Deletions produce reads that map more distant from one another than expected while insertions have the opposite effect. Read pairs spanning breakpoints of inversions and duplications have altered distance and orientation while read pairs spanning translocations (not shown) will map to different chromosomes.

two (for deletions) or three (for insertions) fragments and attempts to map them individually to the anchored subregion of the reference genome. The maximum deletion size is a user-provided parameter that decides the size of the region that will be searched (typically 1–10,000 bp), while the maximum insertion size corresponds to read length (16 bp for 36-bp reads).

3.3.3. BreakDancer

Our group has developed a discovery pipeline for SVs that conducts *de novo* prediction and in silico confirmation using Illumina paired-end data. The *de novo* prediction program, BreakDancer (36), consists of two complementary algorithms.

The first algorithm (BreakDancerMax) identifies anomalously mapped read pairs (ARPs) whose ends are mapped in unexpected distances or orientations. It searches for genomic regions that anchor significantly more ARPs than expected by chance, and derives putative SVs from one or more regions that are interconnected by at least two ARPs. The confidence score for each SV prediction is estimated using a Poisson model that takes into consideration the number of supporting ARPs, the size of the anchoring regions, and the coverage of the genome. BreakDancerMax outputs five types of SVs: insertions, large deletions (>100 bp), inversions, intrachromosomal rearrangements, and interchromosomal translocations.

The second algorithm (BreakDancerMini) predicts small indels 10–100 bp by examining the normally mapped read pairs (NRPs) that are ignored by BreakDancerMax. It employs a Kolmogorov-Smirnov test to identify indel-containing regions and makes SV predictions using procedures similar to BreakDancerMax. Both BreakDancerMax and BreakDancerMini can be applied to a pool of DNA samples to identify common and novel variants. To further resolve breakpoints and reduce false positives, we perform *de novo* assembly on all predicted SVs (see Subheading 5).

3.4. Detection of Fusion Transcripts by RNA-Seq

Gene fusion events that result from chromosomal rearrangements are a common form of somatic mutation in human cancers (37). Massively parallel sequencing of cDNA libraries, or RNA-Seq, enables the identification of gene fusions despite the high background of abundant housekeeping genes. Maher et al. (38) developed pipeline for gene fusion discovery that utilizes data from two NGS platforms: long fragment-end reads from the Roche/454 platform and short paired-end reads from the Illumina GAII platform. This hybrid approach proved a powerful system for gene fusion discovery by transcriptomes sequencing, as exemplified by the detection of multiple gene fusions in cancer cell lines and tissues. Levin et al. (39) applied targeted RNA-Seq to 467 cancer-related genes in K-562, a well-characterized chronic myeloid leukemia (CML) cell line. Using paired-end Illumina sequencing, they identified the known BCR-ABL fusion transcript as well as several novel gene fusions.

3.5. Visualization

Although many methods for SV detection have been published for NGS data, there remains a paucity of tools for visualizing predicted SVs in the context of supporting data and relevant genome annotations.

3.5.1. Integrative Genomics Viewer and Savant

The Integrated Genomics Viewer (IGV, http://www.broadinstitute.org/igv) is a BAM-driven visualization tool for NGS data that displays read depth at each position, and color-codes reads according to the chromosome of their mate pairs; as such, it can be used to infer translocations and SNP/indel variants. The Savant Genome Browser ((40), http://www.compbio.cs.toronto.edu/savant) offers similar features, but also provides a novel representation of paired end reads to assist the identification of structural variation.

3.5.2. LookSeq and Circos

The Web-based application LookSeq (41) offers visualization of NGS data by sorting paired reads according to insert size; apparent valleys and peaks in the visualization indicate insertions and deletions, respectively. Currently, LookSeq does not support the visualization of other SVs, such as translocations and inversions. These variants (along with insertions and deletions) can be visualized using the circular graphing tool Circos (42), but require preformatting by the user.

3.5.3. Pairoscope

We recently developed a visualization tool (Pairoscope, http://www.pairoscope.sourceforge.net) to display inversions, duplications, and translocations detected in NGS data. Pairoscope accepts standard sequence BAM format files as input, as well as an "annotation BAM" file containing gene/transcript information. Using mate-pair information embedded in the sequence BAM file, Pairoscope generates three plots for each region that is requested (Fig. 3). First, it displays an auto-scaled graph of read depth per base to show the sequencing coverage across the region. SVs associated with copy number changes (insertions, deletions, and duplications) are indicated by changes in read depth. Second, Pairoscope plots a "bar-code" graph indicating the aberrantly mapped read pairs in the region. Each aberrant read is represented as a vertical bar, color-coded by aberration class. If both reads in a pair are present in the region, they are linked by a line arc. When visualized in Pairoscope, translocations are apparent as a series of mate-pair arcs linking the regions of two separate chromosomes. Inversions are apparent as two separate clusters of arcs at the breakpoints of the inverted sequence whose colors indicate aberrant orientation of read pairs. Because mapping quality can play a large part in determining the veracity of an SV prediction, Pairoscope allows the user to filter displayed read pairs according to mapping quality. Third, Pairoscope displays exon information for a region of interest. The required input is an annotation BAM file with custom tags indicating transcript-specific information. By displaying predicted SVs in the

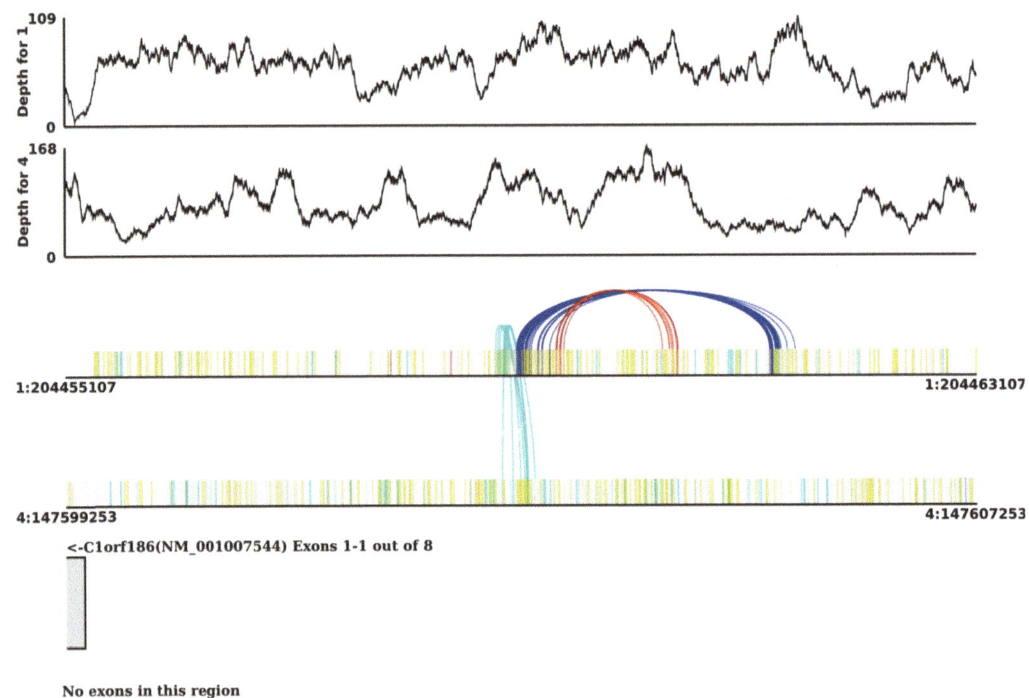

Fig. 3. Pairoscope graph of predicted inversion and translocation. The top two tracks show the read depth for these regions. The second two show color-coded read pairs indicating the orientation of abnormal read pairs. Single bars in yellow represent reads whose mate did not map. The cyan arcs support a translocation between chromosomes 1 and 4. The red and blue arcs indicate reads oriented in a forward–forward and reverse–reverse orientation, respectively. The gene model shown in the last track indicate this translocation occurs upstream of the translocation start site of an open reading frame.

context of protein-coding genes, Pairoscope allows for the analysis and interpretation of how variants may affect gene structure. This feature is particularly valuable for identifying translocations that create fusion genes, which are prevalent in many cancers (43).

4. Conclusions

In summary, NGS technologies have become powerful tools for the characterization of copy number and structural variation in human genomes. Sequence-based detection of CNV and SV is particularly appealing for tumor genomes. Somatic rearrangements resulting in fusion genes (e.g., BCR-ABL) are both common to many tumor types, and promising as candidates for targeted therapy. Ambitious efforts like the Cancer Genome Atlas project (TCGA) will sequence hundreds of tumor genomes in the coming years; detection of somatic structural and copy number alterations in such datasets will be critical for characterizing the full set of acquired genetic changes underlying tumor development and growth.

As the throughput and number of NGS platforms continues to grow, so too does the set of tools for sequence alignment, variant detection, *de novo* assembly, visualization of high-throughput sequencing data. These and other analysis tools, in conjunction with the availability of hundreds or thousands of individual genome sequences, will undoubtedly shed light on the full spectrum of genetic variation in humans, from single base changes (SNPs) to structural events spanning millions of base pairs.

5. Notes

Mapping reads to a reference genome represents the critical first step in analyzing NGS data. Yet when aligning short sequences of imperfect quality to large reference genomes, it is important to realize that a considerable fraction of reads maps incorrectly or not at all. While longer read lengths (~75–100 bp) and read pairing information will help address this issue, the improvements will be marginal. Currently, as many as 5–15% of reads from a high quality paired-end sequencing run cannot be uniquely mapped to the human genome. An obvious approach to improving read alignment rates is to use a more sensitive aligner, such as Novoalign for Illumina data or SHRiMP (http://www.compbio.cs.toronto.edu/shrimp) for ABI SOLiD data. Unfortunately, more sensitive aligners also increase computational load and execution time. A better solution might be a two-step mapping strategy, in which a fast aligner (Bowtie or BWA) quickly places 70–80% of reads in a dataset, and then a more sensitive aligner is applied to the remaining unplaced reads. The output of most aligners can be converted to SAM/BAM format and merged together using SAMtools (http://www.samtools.sourceforge.net).

Reads that map incorrectly to the reference sequence are a more difficult and potentially more worrisome problem to address. Many aligners now provide a "mapping quality" score, a log-scaled numerical representation of the confidence that a read is correctly mapped. Filtering reads by mapping quality may remove alignment-related artifacts. In our experience with Illumina data, 45 and 70 make good cutoffs for BWA and Maq, respectively. Another approach to remove misaligned reads prior to SV detection is to identify and remove troublesome regions of the reference sequence, such as centromeres, telomeres, and regions enriched for tandem duplications. The disadvantage of this approach, of course, is that such regions are often enriched for structural variation (28, 44).

CNV detection using read depth (RD) and SV detection with PEM each have different advantages and limitations. RD-based methods can utilize both fragment-end and paired-end data to infer CNVs, and often detect certain classes of SV – segmental duplications, for example – that are refractory to PEM. In contrast, PEM approa-ches can identify copy-number-neutral events (e.g., inversions) and novel insertions to the reference genome that would be missed by RD approaches alone. Both approaches are limited in their ability to characterize variation in highly repetitive regions, where short reads cannot be uniquely mapped with high accuracy. Ultimately, a combined approach of RD and PEM methods yields the most comprehensive information about structural variation in a sequenced genome.

While traditional ESP approaches have the advantage of sequencing an entire clone to resolve complex structural events, they are also costly and labor-intensive compared to massively parallel sequencing. As the range of available insert sizes for paired-end sequencing (250 bp–3 kbp) continues to grow, so too does the sensitivity for SVs. Currently, as many as 90% of large deletions detected by ESP of BACs on capillary sequencers are detectable with long insert libraries on next-generation platforms. Of course, long insert libraries typically require more input DNA, which is disadvantageous when samples are in limited supply. This practical reality, combined with statistical analyses of the contributions of various insert sizes to SV detection, suggest that a combined library approach of small and large insert sizes offers the greatest probability of resolving SVs on NGS platforms (6).

Some classes of SVs, such as inversions and translocations, are more difficult to detect and validate using current technologies. Furthermore, studies of structural variation in humans have shown that SV events are enriched near duplicated or repetitive regions of the human genome (28, 44) that are refractory to accurate mapping of short NGS reads. Detecting these variants with high specificity require further analysis and filtering.

To remove false positives from alignment artifacts, and to precisely define the breakpoints of predicted SVs, we perform *de novo* assembly of all read pairs that have at least one end mapped to the predicted intervals. Our internally developed short read assembler TIGRA (Chen et al. unpublished) returns the exact locations of SVs and the nucleotide sequences that span each SV's breakpoint(s). TIGRA has achieved confirmation rates as high as 93%.

Validating SVs predicted by NGS is a necessary step, but can be difficult because these variants are often flanked by repetitive sequence and encompass hundreds or thousands of base pairs. The first phase of the Human Genome Structural Variation (HGSV) Project (28), while reliant on traditional ESP approaches, employed multiple validation strategies that are suitable for SVs predicted from NGS data.

First, discordant fosmids whose apparent insert sizes suggested the presence of underlying SV were subjected to four complete restriction digests and resequenced via ESP. In NGS, this is analogous to sequencing multiple paired-end libraries with varying insert sizes, an approach which statistical theory suggests is most likely to resolve SV breakpoints (6).

A second validation strategy is to design custom high-density oligonucleotide arrays targeting specific insertions and deletions. Such custom arrays not only offer tools for validation, but also offer the opportunity to screen for validated SVs in other samples or populations. Orthogonal datasets, such as those described in Subheading 2 offer a third avenue of SV validation. High-density SNP arrays, spectral karyotyping, and even cytogenetic screens can provide evidence of large structural variants to corroborate predictions from NGS data.

Acknowledgments

We thank John Wallis for insightful discussions on structural variant analysis. We are also grateful for the support of the medical genomics, analysis pipeline, and technology development groups of the Genome Institute at Washington University in St. Louis.

References

1. Mardis, E.R. (2008). The impact of next-generation sequencing technology on genetics. *Trends Genet.* 24(3): p. 133–41.
2. Ahn, S.M., T.H. Kim, S. Lee, et al. (2009). The first Korean genome sequence and analysis: full genome sequencing for a socio-ethnic group. *Genome Res.* 19(9): p. 1622–9.
3. Bentley, D.R., S. Balasubramanian, H.P. Swerdlow, et al. (2008). Accurate whole human genome sequencing using reversible terminator chemistry. *Nature.* 456(7218): p. 53–9.
4. Drmanac, R., A.B. Sparks, M.J. Callow, et al. Human genome sequencing using unchained base reads on self-assembling DNA nanoarrays. *Science.* 327(5961): p. 78–81.
5. Kim, J.I., Y.S. Ju, H. Park, et al. (2009). A highly annotated whole-genome sequence of a Korean individual. *Nature.* 460(7258): p. 1011–5.
6. McKernan, K.J., H.E. Peckham, G.L. Costa, et al. (2009). Sequence and structural variation in a human genome uncovered by short-read, massively parallel ligation sequencing using two-base encoding. *Genome Res.* 19(9): p. 1527–41.
7. Pushkarev, D., N.F. Neff, and S.R. Quake (2009). Single-molecule sequencing of an individual human genome. *Nat Biotechnol.* 27(9): p. 847–52.
8. Wang, J., W. Wang, R. Li, et al. (2008). The diploid genome sequence of an Asian individual. *Nature.* 456(7218): p. 60–5.
9. Wheeler, D.A., M. Srinivasan, M. Egholm, et al. (2008). The complete genome of an individual by massively parallel DNA sequencing. *Nature.* 452(7189): p. 872–6.
10. Volik, S., S. Zhao, K. Chin, et al. (2003). End-sequence profiling: sequence-based analysis of aberrant genomes. *Proc Natl Acad Sci U S A.* 100(13): p. 7696–701.
11. Raphael, B.J., S. Volik, C. Collins, et al. (2003). Reconstructing tumor genome architectures. *Bioinformatics.* 19 Suppl 2: p. ii162–71.
12. Tuzun, E., A.J. Sharp, J.A. Bailey, et al. (2005). Fine-scale structural variation of the human genome. *Nat Genet.* 37(7): p. 727–32.
13. Korbel, J.O., A.E. Urban, J.P. Affourtit, et al. (2007). Paired-end mapping reveals extensive structural variation in the human genome. *Science.* 318(5849): p. 420–6.
14. Campbell, P.J., P.J. Stephens, E.D. Pleasance, et al. (2008). Identification of somatically acquired rearrangements in cancer using genome-wide massively parallel paired-end sequencing. *Nat Genet.* 40(6): p. 722–9.
15. Stephens, P.J., D.J. McBride, M.L. Lin, et al. (2009). Complex landscapes of somatic rearrangement in human breast cancer genomes. *Nature.* 462(7276): p. 1005–10.
16. Pleasance, E.D., P.J. Stephens, S. O'Meara, et al. A small-cell lung cancer genome with complex

signatures of tobacco exposure. *Nature*. 463(7278): p. 184–90.

17. Pleasance, E.D., R.K. Cheetham, P.J. Stephens, et al. A comprehensive catalogue of somatic mutations from a human cancer genome. *Nature*. 463(7278): p. 191–6.
18. Margulies, M., M. Egholm, W.E. Altman, et al. (2005). Genome sequencing in microfabricated high-density picolitre reactors. *Nature*. 437(7057): p. 376–80.
19. Li, H. and N. Homer A survey of sequence alignment algorithms for next-generation sequencing. *Brief Bioinform*.
20. Li, H., J. Ruan, and R. Durbin (2008). Mapping short DNA sequencing reads and calling variants using mapping quality scores. *Genome Res*. 18(11): p. 1851–8.
21. Li, H. and R. Durbin (2009). Fast and accurate short read alignment with Burrows-Wheeler transform. *Bioinformatics*. 25(14): p. 1754–60.
22. Langmead, B., C. Trapnell, M. Pop, et al. (2009). Ultrafast and memory-efficient alignment of short DNA sequences to the human genome. *Genome Biol*. 10(3): p. R25.
23. Homer, N., B. Merriman, and S.F. Nelson (2009). BFAST: an alignment tool for large scale genome resequencing. *PLoS One*. 4(11): p. e7767.
24. Rumble, S.M., P. Lacroute, A.V. Dalca, et al. (2009). SHRiMP: accurate mapping of short color-space reads. *PLoS Comput Biol*. 5(5): p. e1000386.
25. Ning, Z., A.J. Cox, and J.C. Mullikin (2001). SSAHA: a fast search method for large DNA databases. *Genome Res*. 11(10): p. 1725–9.
26. Li, H. and R. Durbin Fast and accurate long-read alignment with Burrows-Wheeler transform. *Bioinformatics*. 26(5): p. 589–95.
27. Li, H., B. Handsaker, A. Wysoker, et al. (2009). The Sequence Alignment/Map format and SAMtools. *Bioinformatics*. 25(16): p. 2078–9.
28. Kidd, J.M., G.M. Cooper, W.F. Donahue, et al. (2008). Mapping and sequencing of structural variation from eight human genomes. *Nature*. 453(7191): p. 56–64.
29. Yoon, S., Z. Xuan, V. Makarov, et al. (2009). Sensitive and accurate detection of copy number variants using read depth of coverage. *Genome Res*. 19(9): p. 1586–92.
30. Chiang, D.Y., G. Getz, D.B. Jaffe, et al. (2009). High-resolution mapping of copy-number alterations with massively parallel sequencing. *Nat Methods*. 6(1): p. 99–103.
31. Alkan, C., J.M. Kidd, T. Marques-Bonet, et al. (2009). Personalized copy number and segmental duplication maps using next-generation sequencing. *Nat Genet*. 41(10): p. 1061–7.
32. Koboldt, D.C. (2009). Short Read Aligners. *MassGenomics*. http://www.massgenomics.org/short-read-aligners.
33. Hormozdiari, F., C. Alkan, E.E. Eichler, et al. (2009). Combinatorial algorithms for structural variation detection in high-throughput sequenced genomes. *Genome Res*. 19(7): p. 1270–8.
34. Sindi, S., E. Helman, A. Bashir, et al. (2009). A geometric approach for classification and comparison of structural variants. *Bioinformatics*. 25(12): p. i222–30.
35. Ye, K., M.H. Schulz, Q. Long, et al. (2009). Pindel: a pattern growth approach to detect break points of large deletions and medium sized insertions from paired-end short reads. *Bioinformatics*. 25(21): p. 2865–71.
36. Chen, K., J.W. Wallis, M.D. McLellan, et al. (2009). BreakDancer: an algorithm for high-resolution mapping of genomic structural variation. *Nat Methods*. 6(9): p. 677–81.
37. Futreal, P.A., L. Coin, M. Marshall, et al. (2004). A census of human cancer genes. *Nat Rev Cancer*. 4(3): p. 177–83.
38. Maher, C.A., C. Kumar-Sinha, X. Cao, et al. (2009). Transcriptome sequencing to detect gene fusions in cancer. *Nature*. 458(7234): p. 97–101.
39. Levin, J.Z., M.F. Berger, X. Adiconis, et al. (2009). Targeted next-generation sequencing of a cancer transcriptome enhances detection of sequence variants and novel fusion transcripts. *Genome Biol*. 10(10): p. R115.
40. Fiume, M., V. Williams, A. Brook, et al. Savant: genome browser for high-throughput sequencing data. *Bioinformatics*. 26(16): p. 1938–44.
41. Manske, H.M. and D.P. Kwiatkowski (2009). LookSeq: a browser-based viewer for deep sequencing data. *Genome Res*. 19(11): p. 2125–32.
42. Krzywinski, M., J. Schein, I. Birol, et al. (2009). Circos: an information aesthetic for comparative genomics. *Genome Res*. 19(9): p. 1639–45.
43. Bashir, A., S. Volik, C. Collins, et al. (2008). Evaluation of paired-end sequencing strategies for detection of genome rearrangements in cancer. *PLoS Comput Biol*. 4(4): p. e1000051.
44. Eichler, E.E., D.A. Nickerson, D. Altshuler, et al. (2007). Completing the map of human genetic variation. *Nature*. 447(7141): p. 161–5.

Index

A

B

C

D

E

F

G

H

I

L

M

Lars Feuk (ed.), *Genomic Structural Variants: Methods and Protocols*, Methods in Molecular Biology, vol. 838,
DOI 10.1007/978-1-61779-507-7, © Springer Science+Business Media, LLC 2012

MIX
Papier aus verantwortungsvollen Quellen
Paper from responsible sources
FSC® C105338

If you have any concerns about our products,
you can contact us on
ProductSafety@springernature.com

In case Publisher is established outside the EU,
the EU authorized representative is:
Springer Nature Customer Service Center GmbH
Europaplatz 3, 69115 Heidelberg, Germany

Printed by Libri Plureos GmbH
in Hamburg, Germany